New Challenges in the Monitoring, Risk Assessment and Management of Pesticides and Biocides in the "One Health Era"

New Challenges in the Monitoring, Risk Assessment and Management of Pesticides and Biocides in the "One Health Era"

Guest Editor

Teresa D'Amore

MDPI

Basel • Beijing • Wuhan • Barcelona • Belgrade • Novi Sad • Cluj • Manchester

Guest Editor
Teresa D'Amore
Department of Food Safety,
Nutrition and Veterinary
Public Health
Istituto Superiore di
Sanità—National Institute
of Health
Rome
Italy

Editorial Office
MDPI AG
Grosspeteranlage 5
4052 Basel, Switzerland

This is a reprint of the Special Issue, published open access by the journal *Journal of Xenobiotics* (ISSN 2039-4713), freely accessible at: https://www.mdpi.com/journal/jox/special_issues/75WX3S4UF0.

For citation purposes, cite each article independently as indicated on the article page online and as indicated below:

Lastname, A.A.; Lastname, B.B. Article Title. *Journal Name* **Year**, *Volume Number*, Page Range.

ISBN 978-3-7258-7809-3 (Hbk)
ISBN 978-3-7258-7810-9 (PDF)
https://doi.org/10.3390/books978-3-7258-7810-9

Cover image courtesy of Teresa D'Amore

Contents

About the Editor

Teresa D'Amore

Teresa D'Amore is a Research Scientist at the Italian National Institute of Health (Istituto Superiore di Sanità, ISS), Department of Food Safety, Nutrition and Public Veterinary Health. As a specialist in risk assessment and management of chemicals, she has consolidated expertise in food safety and nutraceuticals, bioanalytical and biomedical applications of mass spectrometry, green analytical method development and validation, toxicology, and public health.

Her research activity spans several interconnected fields, including chemical food safety, contaminants and cancer, nutraceuticals and oncology, as well as the development, application, and implementation of new approach methodologies as tools for next-generation risk assessment. She has contributed to numerous national and international research programs, serving as Principal Investigator, co-investigator, and collaborator in ministerial and international research projects, European partnerships, multidisciplinary scientific networks, and as a member of expert working groups.

As the author of more than 80 peer-reviewed publications, book chapters, and scientific contributions, she also serves as an editor and associate editor for international journals. She has received multiple recognitions and awards for her scientific excellence and outstanding peer-review activity.

Preface

Pesticides and biocides hold a complex and controversial position at the intersection of agricultural productivity, environmental protection, and public health. While their technological function is firmly established, their safety assessment increasingly challenges traditional toxicological paradigms and demands continuous re-evaluation to account for evolving use patterns, multiple exposure routes, and scientific and technological advances that improve the identification and interpretation of long-term effects.

In this context, chemical risk assessment is a continuously adapting process, shaped by scientific innovation, regulatory evolution and societal expectations, rather than a static exercise.

The One Health framework offers a compelling approach to addressing these challenges. It recognizes the connections between human, animal, and environmental health, and encourages risk assessment strategies that transcend disciplinary boundaries and more accurately reflect the complexity of the real world. For pesticides and biocides in particular, the One Health perspective highlights the importance of integrating analytical innovation, environmental monitoring, food safety evaluation, and mechanistic toxicology within coherent and transparent assessment frameworks.

This Special Issue was conceived to showcase current scientific approaches that support this transition. Rather than promoting a single methodology or regulatory vision, the collection reflects the diversity of tools now contributing to modern chemical safety assessment, from advanced analytical techniques and probabilistic exposure modeling to non-animal approaches and emerging biological mitigation strategies. Together, these contributions demonstrate how the generation of evidence is evolving beyond traditional endpoints towards data-rich, mechanism-informed and sustainability-oriented approaches.

However, this body of work also raises awareness of ongoing challenges. There are still knowledge gaps in our understanding of combined exposures, the chronic effects of low doses, and how emerging hazard classes behave across different environmental compartments. Closing these gaps will require technological innovation, as well as closer dialogue between scientists, regulators, and stakeholders to ensure that new methods are scientifically credible and relevant to regulation.

I hope that this Special Issue will encourage reflection on how pesticide and biocide assessment can evolve alongside societal priorities, scientific capabilities, and the broader goals of environmental and public health protection. The contributions in this collection are designed to promote a risk assessment and management approaches that are more forward-looking, comprehensive, and in line with the principles of One Health.

I firmly believe that this can be achieved by facilitating interdisciplinary collaboration, adopting integrated methodological approaches, and promoting continuous dialogue between research and policy. One Health is definitely the way forward.

I would like to extend my most sincere gratitude to all contributing authors for the dedication exhibited in their work, as well as to the reviewers for their meticulous and constructive evaluations. I would also like to thank the editorial and administrative staff for their consistent support throughout the development of this Special Issue, particularly Ms. Julie Shi, Managing Editor, whose professionalism and dedication were instrumental in its completion.

To my beloved dad. All good things are due to you.

Teresa D'Amore
Guest Editor

Journal of *Xenobiotics*

MDPI

Editorial

New Challenges in the Monitoring, Risk Assessment, and Management of Pesticides and Biocides in the "One Health Era"

Teresa D'Amore [1,2]

[1] Department of Food Safety, Nutrition and Veterinary Public Health, Istituto Superiore di Sanità (ISS)—Italian National Institute of Health, Viale Regina Elena 299, 00161 Rome, Italy; teresa.damore@iss.it
[2] Department of Pharmacy, University of Naples, Federico II, 80131 Naples, Italy

1. Introduction

Pesticides and biocides remain indispensable chemicals for agriculture, food safety, public health, and industrial applications, as they safeguard crop yields, control disease vectors, and maintain high hygiene standards. However, their presence in food, feed, animals, and humans raises significant concerns for health and the environment, making their monitoring, assessment, and management a continuing public health priority [1,2].

The regulation and management of pesticides and biocides represents one of the most complex and innovation-driven fields of chemical governance. It is particularly relevant in the context of the One Health approach, which acknowledges the inherent interconnectedness of human, animal, and environmental health. Globally, a variety of frameworks exist that impose stringent requirements to ensure safety and their controlled use under specific conditions. In Europe (EU), the Regulations (EC) No. 1107/2009 (Plant Protection Products Regulation—PPPR) and No. 283/2013 concern the general conditions and requirements for the placing on the market of plant protection products; the Regulation (EU) No. 528/2012 (Biocidal Products Regulation—BPR) relates to biocidal products; and the Regulation (EC) No. 1272/2008 is pertinent to the classification, packaging, and labeling (CLP) of substances and mixtures. In the United States (US), the Federal Insecticide, Fungicide, and Rodenticide Act (FIFRA) and the Toxic Substances Control Act (TSCA), as well as the Federal Food, Drug, and Cosmetic Act (FFDCA), empower the Environmental Protection Agency (EPA) to establish tolerances—defined as maximum legally permissible levels—for pesticide residues in food. The EPA is also responsible for conducting risk assessment for the registration of pesticides. Conversely, in the EU, the responsibility of risk assessment of active substances falls under the mandates of the European Food Safety Authority (EFSA), while the European Commission and Member States take risk management decisions, including the approval of active substances and setting of maximum residue levels (MRLs) for pesticide residues in food and feed, listed in the Regulation (EC) No. 396/2005 [3,4].

Recent regulatory developments have introduced new hazard classes, including endocrine disruptors (ED), PBT/vPvB, and PMT/vPvM, which have significantly increased the complexity of data requirements for pesticides and biocides. These changes necessitate the implementation of advanced analytical strategies, mechanistic studies, and integrated risk assessments that extend beyond the scope of traditional toxicology. The identification of endocrine-disrupting properties, for example, necessitates detailed mode-of-action evidence and adverse outcome characterization, as reflected in EU criteria and US screening programs [5–7].

Evidently, this topic is multidisciplinary, spanning food chemistry, agricultural and biological science, toxicology, ecotoxicology, exposure science, risk assessment, and public health. It involves a globally interconnected network of actors, including researchers, academics, policy makers, authorities (e.g., EFSA, ECHA, ECDC, EPA, FDA, OECD), and others, working toward harmonized standards and coordinated assessments.

Initiatives such as "One Substance—One Assessment" in the EU and international collaborations under OECD exemplify efforts to streamline these processes. Similarly, in an effort to strength the challenges to ensure the best protection of human and environmental health, global strategies have been implemented, including the United Nations 2030 Agenda for Sustainable Development, the Farm to Fork Strategy, and the Chemicals Strategy for Sustainability [8,9].

Furthermore, the framework governing pesticides and biocides has been identified as a catalyst for scientific and technological innovation. The implementation of New Approach Methodologies (NAMs), Adverse Outcome Pathways (AOPs), in silico modeling, and integrated testing strategies is leading to a paradigm shift in risk assessment methodologies, thereby enabling the transition toward Next Generation Risk Assessment (NGRA). Programs such as the EU Partnership for the Assessment of Risks from Chemicals (PARC) and analogous global initiatives aspire to expedite the incorporation of these instruments into regulatory practice [10–12]. Despite the advancements, there are still obstacles that must be surmounted if the objective of attaining complete regulatory acceptance and harmonization of these approaches is to be realized.

In the context of the One Health, the monitoring, risk assessment, and management of pesticides and biocides necessitates not only adherence to evolving regulatory frameworks but also the development of innovative, science-based solutions that address emerging hazards and complex exposure scenarios. This Special Issue is devoted to the exploration of the aforementioned challenges and opportunities. The compendium comprises 15 peer-reviewed papers: ten research articles, three review articles, one perspective article, and one systematic review article. The contributions listed are noteworthy for their high quality and broad scope, covering a wide range of subjects, including analytical innovations, regulatory perspectives, food safety, toxicology, ecotoxicology, and emerging hazard classifications. Accordingly, the Special Issue offers a comprehensive overview of current progress and future directions in this field.

1.1. Analytical Innovations

Accurate detection of pesticide/biocide residues is the cornerstone of risk assessment and regulatory compliance. The advances in analytical chemistry have significantly improved sensitivity, specificity, and throughput for residue monitoring. High-resolution mass spectrometry (HRMS), liquid chromatography–tandem mass spectrometry (LC–MS/MS), and ion chromatography are very powerful tools for quantifying highly polar pesticides in complex matrices such as food and pollinator tissues. These methods not only meet stringent validation criteria but also enable multicenter harmonization, ensuring reproducibility across laboratories.

Beyond conventional residue analysis, innovative approaches such as biomimetic chromatography offer high-throughput screening for bioaccumulation potential and aquatic toxicity. The correlation between chromatographic hydrophobicity indices with bioconcentration factors and lethal concentration thresholds provides rapid proxies for ecotoxicological endpoints, reducing reliance on animal testing. Similarly, computational models, including Monte Carlo simulations and QSAR-based descriptors, are gaining traction as predictive tools for aquatic toxicity, offering ethical and cost-effective alternatives to traditional assays.

These analytical innovations support a paradigm shift toward integrated monitoring systems that combine chemical analytics with biological indicators. Honeybees, for instance, are increasingly recognized as bioindicators of environmental contamination, and validated

 https://doi.org/10.3390/jox16010035

methods for pesticide detection in bee matrices strengthen pollinator protection strategies. Such advancements comply with One Health principles by linking residue monitoring to environmental and human health.

1.2. Regulatory Perspectives and New Requirements

The European Union has established one of the most comprehensive regulatory frameworks for pesticide and biocide risk assessment. Yet, significant challenges persist. Current methodologies often fall short in addressing cumulative exposure, mixture toxicity, and low-dose effects—issues that are increasingly relevant given the complexity of real-world exposures. Recent policy developments show the need for integrated approaches that incorporate biomonitoring data, mechanistic toxicology, and environmental modeling into regulatory decision-making.

Endocrine-disrupting chemicals (EDCs) and reprotoxic substances remain at the forefront of regulatory concern. While progress has been made in hazard identification, gaps persist in implementing standardized criteria and testing protocols. The lack of formal recognition of emerging hazard classes, such as immunotoxicity, has been identified as a key area for improvement, with the necessity for expanded classification systems that reflect evolving scientific evidence being emphasized.

Another critical regulatory trend is the transition toward animal-free toxicology. Advances in in vitro systems, omics technologies, and computational modeling offer promising alternatives, but their integration into regulatory frameworks requires harmonized validation and acceptance criteria.

Furthermore, the renewal and approval processes for bioactive substances must adapt to incorporate sustainability considerations.

1.3. Toxicology and Hazard Classification

The toxicological research in this Special Issue evidences the complexity of pesticide-induced health effects. Systematic reviews of widely used fungicides reveal consistent cytotoxicity across mammalian cell lines, with oxidative stress and mitochondrial dysfunction emerging as central mechanisms. Apoptosis, endoplasmic reticulum stress, and genotoxicity further underscore the multifaceted nature of pesticide toxicity. Despite these findings, toxicokinetic data remain sparse, particularly with the regard to chronic exposure and cumulative effects—a gap that hampers accurate risk characterization.

Human biomonitoring studies provide compelling evidence of widespread pesticide and biocide exposure, from occupational settings to urban populations. Urinary analyses reveal stark exposure gradients, with farmworkers exhibiting the highest contamination levels, followed by para-occupational and general populations. Sociodemographic factors such as education, water sources, and protective practices significantly influence exposure risk, highlighting the interplay between behavioral and environmental determinants.

These data reinforce the urgency of refining hazard classifications to capture emerging endpoints. Immunotoxicity, neurotoxicity, and endocrine disruption demand systematic incorporation into regulatory schemes. Moreover, cumulative risk assessment frameworks must evolve to account for synergistic effects among chemical mixtures, moving beyond single-substance evaluations toward holistic exposure models.

1.4. Food Safety and Dietary Exposure Assessment

Food safety remains a central pillar of the One Health approach, requiring robust surveillance of active substance residues in high consumption commodities and transparent risk characterization across diverse populations. The contributions in this Special Issue

 https://doi.org/10.3390/jox16010035

demonstrate the efficacy of longitudinal monitoring, market surveillance, and probabilistic exposure modelling in elucidating chemical contaminants occurrence data, cumulative risks, and vulnerable subgroups.

Across multiple staple crops, multi residue contamination has emerged as a recurrent feature rather than an exception. Long term surveillance of tomatoes over four years—using a fully validated LC MS/MS MRM method in line with SANTE guidance—consistently detected mixtures dominated by fungicides, followed by insecticides and acaricides. The occurrence of up to five residues in a single sample shows the importance of moving beyond single substance assessments to evaluate mixture toxicity and aggregate exposure. Notably, the environmental mixture analysis (via Concentration Addition and Independent Action) returned MCR values ≥ 1, indicating that the mixture contribution to overall toxicity can exceed that of the single most toxic component, albeit only slightly in the scenarios examined. Consumer dietary risk modeling for adults and toddlers did not flag immediate health concerns, but the evidence nonetheless reinforces the need to systematically include cumulative exposure assessments when evaluating products prone to multiresidue findings.

A large-scale market surveillance of onions ($n \approx 5700$ across 2021–2024) complements these insights with temporal and compositional dynamics. Residue free rates and EU MRL noncompliance fluctuated year to year, while fungicides consistently dominated the residue profile, with insecticides increasing over time—a shift likely reflecting evolving agronomic practices and pest pressures. Multiple residues peaked above 30% in certain years, and several active substances exceeded MRLs. Still, chronic dietary exposure modeled via EFSA PRIMo 3.1 remained well below ADI thresholds across population groups, with the highest signal (e.g., chlorpyrifos) staying firmly under critical levels even in high consumption regions. These outcomes demonstrate the value of commodity-specific surveillance over extended windows, revealing both compliance trends and the real-world relationship between occurrence, regulatory limits, and modeled health impacts.

A complementary matrix assessment of root and tuber vegetables (potatoes, carrots, celery, radishes, horseradish, ginger, onions, leeks) adds critical population-level nuance. Using validated HPLC MS/MS and GC MS/MS methods, investigators detected 33 distinct pesticides, with multiple residues present in a subset and MRL exceedances recorded in others. Acute and chronic dietary risks were quantified with hazard quotients (HQ) and hazard indices (HI), supported by Monte Carlo simulations to capture variability and uncertainty. While individual HQs remained below 1, cumulative acute exposure approached ~63% of ARfD in children (and ~51% in adults) for specific commodities (e.g., ginger, celery), whereas cumulative chronic exposure remained within acceptable bounds (max ~11% ADI in children, ~6% in adults). Vulnerable groups of population and consumption rates as major determinants of chronic risk variability.

Methodologically, these studies exemplify best practice agreement with EU guidance (method validation, PRIMo modeling) while demonstrating how cumulative risk, mixture contribution (MCR), and probabilistic methods (Monte Carlo) can be operationalized for routine surveillance.

1.5. Ecotoxicology and Environmental Monitoring

The environmental issues associated with the utilization of pesticides and biocides are of equal significance to those related to human health. Studies on aquatic ecosystems reveal nuanced interactions between chemical stressors and environmental variables. Glyphosate-based herbicides, for example, alter fungal sporulation patterns in streams, with species-specific responses that affect decomposition processes and nutrient cycling. Lithium, an emerging contaminant, exacerbates these effects, illustrating the complexity of multi-stressor scenarios.

https://doi.org/10.3390/jox16010035

Pollinators, particularly honeybees, remain sentinel species for agrochemical impacts. Advanced analytical methods now enable precise quantification of highly polar pesticides in bee matrices, supporting risk assessments that extend beyond human health to ecosystem integrity.

Similarly, marine monitoring strategies employing biofilm mesocosms, passive samplers, and grab water analyses provide complementary insights into contaminant dynamics in coastal environments. These approaches reveal the prevalence of contaminants of emerging concern, such as UV filters and pharmaceuticals, underscoring the need for integrated surveillance systems.

Finally, bioremediation continues to be a promising strategy for mitigating environmental contamination. Microbial degradation of fluorinated pyrethroids demonstrates the potential of harnessing natural processes to detoxify persistent pollutants. The identification of cytochrome P450 enzymes in bacterial strains opens new frontiers for enzymatic remediation, offering sustainable solutions to pesticide persistence.

2. Conclusions and Future Perspectives

The research presented in this Special Issue advances the science and policy of pesticide and biocide assessment and management under the One Health paradigm.

Indeed, it is imperative to acknowledge that, when integrated approaches prove effective, several notable success stories emerge and illustrate how improved assessment and monitoring of both regulated products and contaminants can translate into tangible mitigation measures and more sustainable agricultural practices. For instance, the phase-out of highly hazardous organophosphates in the EU and US, following strengthened toxicological and exposure assessments, has led to measurable reductions in dietary and occupational risk. In a similar manner, strengthened ecotoxicological weight of evidence approaches, incorporated chronic toxicity endpoints, residue quantification in pollinators, and field-realistic exposure scenarios, were decisive in the EU restrictions of neonicotinoids, actions that subsequently led to the adoption of integrated pest management (IPM) strategies and stimulated the development of low-risk plant protection products. In parallel, repeated MRL exceedances detected in national residue monitoring plans have led growers to adjust pre-harvest intervals, optimize application timing, and adopt drift-reduction technologies, demonstrating how analytical evidence can directly guide agricultural practice.

Powerful take-home messages emerge:

- Integration is imperative. Analytical innovations, toxicological evaluation, and regulatory frameworks must converge to address the complexity of real-world exposures. This requires harmonized methodologies, interoperable data systems, and cross-sector collaboration.
- Knowledge gaps must be prioritized. Chronic exposure, mixture toxicity, and emerging hazard classes remain underexplored. Long-term and multigenerational studies are essential to capture cumulative effects and inform precautionary policies/mitigation measures.
- Innovation must accelerate. Predictive models, high-throughput screening tools, and omics-based assays offer pathways to reduce animal testing and enhance mechanistic understanding. Their regulatory acceptance is contingent upon the implementation of rigorous validation processes and effective stakeholder engagement strategies.
- Sustainability is non-negotiable. The transition to low-risk plant protection products, along with bioremediation strategies, links chemical management with biodiversity and climate objectives. Policy frameworks must incentivize these innovations.

https://doi.org/10.3390/jox16010035

- One Health is the way. In terms of future projections, the One Health approach offers a unifying lens through which to navigate these challenges.

Finally, it is evident that the future of pesticide and biocide management will depend on the seamless convergence of analytical chemistry, exposure science, mechanistic toxicology, NAM-based hazard evaluation, and responsive regulatory actions. Only when research, regulation and policy evolve jointly can we achieve the highest level of protection for entire world health under the One Health framework. The trajectory of the past decade shows that significant progress is achievable, and the most impactful advances are still to come.

Acknowledgments: The author acknowledges Paolo Stacchini (Department of Food Safety, Nutrition and Veterinary Public Health, Istituto Superiore di Sanità—ISS, Rome) for his advice and support provided during the preparation of this Editorial.

Conflicts of Interest: The author declares no conflicts of interest.

Abbreviations

The following abbreviations are used in this manuscript:

ADI	Acceptable Daily Intake
AOP	Adverse Outcome Pathway
ARfD	Acute Reference Dose
BPR	Biocidal Products Regulation
CECs	Contaminants of Emerging Concern
CLP	Classification, Labeling and Packaging
ED/EDCs	Endocrine Disruptor/Endocrine-Disrupting Chemicals
ECHA	European Chemicals Agency
ECDC	European Centre for Disease Prevention and Control
EFSA	European Food Safety Authority
EPA	Environmental Protection Agency
EU	European Union
FFDCA	Federal Food, Drug, and Cosmetic Act
FIFRA	Federal Insecticide, Fungicide, and Rodenticide Act
GC-MS/MS	Gas Chromatography Tandem Mass Spectrometry
HI	Hazard Index
HPLC-MS/MS	High-Performance Liquid Chromatography Tandem Mass Spectrometry
HQ	Hazard Quotient
HRMS	High-Resolution Mass Spectrometry
IC-HRMS	Ion Chromatography-High-Resolution Mass Spectrometry
LC-MS/MS	Liquid Chromatography Tandem Mass Spectrometry
MCR	Mixture Contribution Ratio
MRL	Maximum Residue Level
NAMs	New Approach Methodologies
NGRA	Next Generation Risk Assessment
OECD	Organization for Economic Co-operation and Development
PARC	Partnership for the Assessment of Risks from Chemicals
PBT	Persistent, Bioaccumulative and Toxic
PMT	Persistent, Mobile and Toxic
PPPR	Plant Protection Products Regulation
PRIMo	Pesticide Residue Intake Model
QSAR	Quantitative Structure

 https://doi.org/10.3390/jox16010035

SANTE	Directorate-General for Health and Food Safety (European Commission)
TSCA	Toxic Substances Control Act
vPvB	very Persistent and very Bioaccumulative
vPvM	very Persistent and very Mobile

List of Contributions

1. Pacini, T.; Verdini, E.; Orsini, S.; Russo, K.; Mauti, T.; Gasparini, M.; Borgia, M.; Angelone, B.; D'Amore, T.; Pecorelli, I. Development and Validation of LC–MS/MS and IC–HRMS Methods for Highly Polar Pesticide Detection in Honeybees: A Multicenter Study for the Determination of Pesticides in Honeybees to Support Pollinators and Environmental Protection. *J. Xenobiotics* **2025**, *15*, 95. https://doi.org/10.3390/jox15040095.
2. Botnaru, A.A.; Lupu, A.; Morariu, P.C.; Nedelcu, A.H.; Morariu, B.A.; Di Gioia, M.L.; Lupu, V.V.; Dragostin, O.M.; Caba, I.-C.; Anton, E.; et al. Innovative Analytical Approaches for Food Pesticide Residue Detection: Towards One Health-Oriented Risk Monitoring. *J. Xenobiotics* **2025**, *15*, 151. https://doi.org/10.3390/jox15050151.
3. Ciura, K. Biomimetic Chromatography as a High-Throughput Tool for Screening Bioaccumulation and Acute Aquatic Toxicity of Pesticides. *J. Xenobiotics* **2026**, *16*, 4. https://doi.org/10.3390/jox16010004.
4. Toropova, A.P.; Toropov, A.A.; Benfenati, E. Monte Carlo Simulation of Pesticide Toxicity for Rainbow Trout (Oncorhynchus Mykiss) Using New Criteria of Predictive Potential. *J. Xenobiotics* **2025**, *15*, 82. https://doi.org/10.3390/jox15030082.
5. Calliera, M.; Capri, E.; Suciu, N.A.; Trevisan, M. Advancing the One Health Framework in EU Plant Protection Product Regulation: Challenges and Opportunities. *J. Xenobiotics* **2025**, *15*, 200. https://doi.org/10.3390/jox15060200.
6. Buonsenso, F. Scientific and Regulatory Perspectives on Chemical Risk Assessment of Pesticides in the European Union. *J. Xenobiotics* **2025**, *15*, 173. https://doi.org/10.3390/jox15050173.
7. Fagundes, T.R.; Coradi, C.; Vacario, B.G.L.; de Morais Valentim, J.M.B.; Panis, C. Global Evidence on Monitoring Human Pesticide Exposure. *J. Xenobiotics* **2025**, *15*, 187. https://doi.org/10.3390/jox15060187.
8. Vulpe, C.-B.; Iachimov-Datcu, A.-D.; Pujicic, A.; Agachi, B.-V. A Systematic Review on the Toxicology of European Union-Approved Triazole Fungicides in Cell Lines and Mammalian Models. *J. Xenobiotics* **2025**, *15*, 208. https://doi.org/10.3390/jox15060208.
9. Ben Khadda, Z.; Radu, A.-F.; El Balkhi, S.; Mustapha, F.; El Karmoudi, Y.; Bungau, G.; Marquet, P.; Sqalli Houssaini, T.; Achour, S. From Farmworkers to Urban Residents: Mapping Multi-Class Pesticide Exposure Gradients in Morocco via Urinary Biomonitoring. *J. Xenobiotics* **2025**, *15*, 120. https://doi.org/10.3390/jox15040120.
10. Atzei, A.; Bouakline, H.; Corrias, F.; Angioni, A. Four-Year Monitoring Survey of Pesticide Residues in Tomato Samples: Human Health and Environmental Risk Assessment. *J. Xenobiotics* **2025**, *15*, 171. https://doi.org/10.3390/jox15050171.
11. Malhat, F.; Shokr, S.; Heikal, S.; Zidan, N.E.-H. Temporal Trends, Multiple Residue Incidence, and Chronic Health Risk of Pesticides in Egyptian Onions: A Four-Year Market Surveillance. *J. Xenobiotics* **2025**, *15*, 192. https://doi.org/10.3390/jox15060192.
12. Lučić, M.; Onjia, A. Prioritization and Sensitivity of Pesticide Risks from Root and Tuber Vegetables. *J. Xenobiotics* **2025**, *15*, 125. https://doi.org/10.3390/jox15040125.
13. Rodrigues, J.; Gerós, H.; Côrte-Real, M.; Cássio, F. Do Isopropylammonium Glyphosate and LiCl Impact the Spore Diversity and Functions of Aquatic Fungi Involved in Plant Litter Decomposition in Streams? *J. Xenobiotics* **2025**, *15*, 65. https://doi.org/10.3390/jox15030065.
14. Díaz-Montaña, E.J.; Domínguez-Gil, S. Multisampling Strategies for Determining Contaminants of Emerging Concern (CECs) in the Marine Environment. *J. Xenobiotics* **2025**, *15*, 149. https://doi.org/10.3390/jox15050149.
15. Khan, M.F.; Liao, J.; Liu, Z.; Chugh, G. Bacterial Cytochrome P450 Involvement in the Biodegradation of Fluorinated Pyrethroids. *J. Xenobiotics* **2025**, *15*, 58. https://doi.org/10.3390/jox15020058.

References

1. Warne, M.S.J.; Reichelt-Brushett, A. Pesticides and Biocides. In *Marine Pollution—Monitoring, Management and Mitigation*; Reichelt-Brushett, A., Ed.; Springer Nature: Cham, Switzerland, 2023; pp. 155–184.
2. Budzinski, H.; Couderchet, M. Environmental and Human Health Issues Related to Pesticides: From Usage and Environmental Fate to Impact. *Environ. Sci. Pollut. Res.* **2018**, *25*, 14277–14279. [CrossRef] [PubMed]
3. Pelaez, V.; da Silva, L.R.; Araújo, E.B. Regulation of Pesticides: A Comparative Analysis*. *Sci. Public Policy* **2013**, *40*, 644–656. [CrossRef]
4. Gensch, L.; Jantke, K.; Rasche, L.; Schneider, U.A. Pesticide Risk Assessment in European Agriculture: Distribution Patterns, Ban-Substitution Effects and Regulatory Implications. *Environ. Pollut.* **2024**, *348*, 123836. [CrossRef] [PubMed]
5. Gouin, T.; Bitsch, A.; van Duursen, M.; Escher, S.E.; Hamers, T. Informing the Decision-Making Process for Potential PMT/VPvM Chemicals through the Adoption of a Risk-Based Prioritization Framework: The ZeroPM Approach. *Environ. Sci. Eur.* **2024**, *36*, 208. [CrossRef]
6. Touchais, G.; Rey, M.; Journeau, O.A.; Hurtaud-Pessel, D.; Laurentie, M.; Mompelat, S. Risk ranking of biocidal active substances in the food chain: A case study on food of animal origin. *Food Control* **2026**, *179*, 111542. [CrossRef]
7. Andersson, N.; Arena, M.; Auteri, D.; Barmaz, S.; Grignard, E.; Kienzler, A.; Lepper, P.; Lostia, A.M.; Munn, S.; Parra Morte, J.M.; et al. Guidance for the Identification of Endocrine Disruptors in the Context of Regulations (EU) No 528/2012 and (EC) No 1107/2009. *EFSA J.* **2018**, *16*, e05311. [CrossRef] [PubMed]
8. D'Amore, T.; Smaoui, S.; Varzakas, T. Chemical Food Safety in Europe Under the Spotlight: Principles, Regulatory Framework and Roadmap for Future Directions. *Foods* **2025**, *14*, 1628. [CrossRef] [PubMed]
9. Schneider, K.; Barreiro-Hurle, J.; Rodriguez-Cerezo, E. Pesticide Reduction amidst Food and Feed Security Concerns in Europe. *Nat. Food* **2023**, *4*, 746–750. [CrossRef] [PubMed]
10. Miccoli, A.; Marx-Stoelting, P.; Braeuning, A. The Use of NAMs and Omics Data in Risk Assessment. *EFSA J.* **2022**, *20*, e200908. [CrossRef] [PubMed]
11. Niemann, L.; Choi, J.; Kneuer, C. Tewes Tralau Traditional and Novel Approaches to Derive Health-Based Guidance Values for Pesticides. *Curr. Opin. Food Sci.* **2023**, *54*, 101091. [CrossRef]
12. D'Amore, T.; Testai, E.; Di Consiglio, E. Advancing pesticide risk assessment: The role of adverse outcome pathways and new approach methodologies—What's next? *Ann. Ist. Super. Sanità* 2026, *in press*.

https://doi.org/10.3390/jox16010035

Journal of *Xenobiotics*

Article

Development and Validation of LC–MS/MS and IC–HRMS Methods for Highly Polar Pesticide Detection in Honeybees: A Multicenter Study for the Determination of Pesticides in Honeybees to Support Pollinators and Environmental Protection

Tommaso Pacini [1,*], Emanuela Verdini [1], Serenella Orsini [1], Katia Russo [2], Tabita Mauti [2], Mara Gasparini [3], Marialuisa Borgia [3], Barbara Angelone [3], Teresa D'Amore [4,5] and Ivan Pecorelli [1]

1 Chemistry Department, Istituto Zooprofilattico Sperimentale dell'Umbria e delle Marche "Togo Rosati", Via G. Salvemini, 1, 06126 Perugia, Italy; e.verdini@izsum.it (E.V.); s.orsini@izsum.it (S.O.); i.pecorelli@izsum.it (I.P.)
2 Direzione Operativa Chimica, Istituto Zooprofilattico Sperimentale del Lazio e della Toscana "M. Aleandri", Via Appia Nuova 1411, 00178 Rome, Italy; katia.russo@izslt.it (K.R.); tabita.mauti@izslt.it (T.M.)
3 Food and Feed Chemistry Department, Istituto Zooprofilattico Sperimentale della Lombardia e dell'Emilia-Romagna "Bruno Ubertini", Via Bianchi, 9, 25124 Brescia, Italy; mara.gasparini@izsler.it (M.G.); marialuisa.borgia@izsler.it (M.B.); barbara.angelone@izsler.it (B.A.)
4 Laboratory of Preclinical and Translational Research, IRCCS CROB, Basilicata Oncology Referral Center, Rionero in Vulture, 85028 Naples, Italy; teresa.damore@crob.it
5 Department of Pharmacy, University of Naples Federico II, 80131 Naples, Italy
* Correspondence: t.pacini@izsum.it

Abstract: The widespread use of agrochemicals raises concerns about environmental impacts, particularly on pollinators, such as bees, which serve as bioindicators of contamination. Developing methods to assess contamination risks in bioindicators supports regulatory frameworks, including EU regulations on the maximum residue limits (MRLs) for pesticides in food and the environment. This study presents the development and validation of two complementary analytical methods (LC–MS/MS and IC–HRMS) for highly polar pesticide (HPP) detection and quantification in bee matrices. Both methods were validated according to document SANTE/11312/2021 v2. LC–MS/MS was validated with a limit of quantification (LOQ) of 0.005 mg/kg for all the analytes. Repeatability at 0.005, 0.010, 0.020, and 0.100 mg/kg showed RSD_r from 1.6% to 19.7% and recoveries between 70% and 119%. Interlaboratory precision at 0.020 mg/kg across two labs showed RSD_R from 5.5% to 13.6%, with recoveries between 91% and 103%. The IC–HRMS method achieved LOQs of 0.01 mg/kg (glufosinate, N-acetyl glufosinate, MPPA, glyphosate, N-acetyl glyphosate, N-acetyl AMPA) and 0.1 mg/kg (fosetyl, phosphonic acid, AMPA), with mean recoveries in repeatability conditions from 84% to 114% and RSD_r from 2% to 14%. Intralaboratory precision showed mean recoveries from 87% to 119%, with RSD_{wR} values between 10% and 18%. These methods enable accurate monitoring of HPP contamination, supporting risk assessment and sustainable agriculture.

Keywords: polar pesticides; honeybees; LC–MS/MS validation; IC–HRMS validation

1. Introduction

The widespread use of agrochemicals has significantly increased in recent decades, raising concerns about their environmental and ecological impact. Among non-target organisms, pollinators such as honeybees are particularly susceptible to agrochemical exposure, threatening their survival and the vital ecosystem services they provide, particularly pollination. The European Parliament and other regulatory bodies have emphasized the urgent need to mitigate pesticide-related risks by promoting sustainable agricultural practices that protect environmental health, pollinator populations, and biodiversity [1]. Effective monitoring of pesticide contamination in pollinators is crucial for understanding environmental exposure and implementing strategies to safeguard these insects and the ecosystems they support.

Pesticide monitoring in pollinators must account for both the parent compounds applied in fields and their metabolites, which can form through environmental degradation and/or metabolic processes. While several studies have confirmed the presence of pesticide residues in pollinators, highlighting their role as effective bioindicators for environmental contamination, significant gaps still remain in the assessment of highly polar pesticides (HPPs), such as glyphosate, ethephon, fosetyl, and glufosinate, in bee populations [2,3]. Despite their high water solubility and extensive use in modern agriculture, these substances are often overlooked in pesticide monitoring programs aimed at assessing their presence in the environment, which traditionally focus on more lipophilic pesticides [4]. Therefore, the inclusion of polar pesticides in bioindicator studies is essential for comprehensive environmental risk assessment and the protection of pollinators.

Although validated methods for detecting highly polar pesticides in food, foodstuffs, and animal-origin products such as honey exist, there is a lack of established methodologies specifically tailored for their detection in pollinator insects. Previous studies demonstrated the effectiveness of high-performance liquid chromatography coupled with tandem mass spectrometry (HPLC–MS/MS) and ion chromatography coupled with high-resolution mass spectrometry (IC–HRMS) for analyzing polar pesticide residues in honey, fruit, and vegetables [5–7]. These studies have laid the groundwork for the development of more comprehensive pesticide monitoring approaches, but their applicability to unconventional substrates such as bees remains unexplored.

The complex nature of bee matrices presents unique analytical challenges, including the presence of biological substances such as wax and pollen, which can interfere with pesticide detection. Traditional multi-residue methods (e.g., QuEChERS) are unsuitable for the analysis of highly polar compounds due to their low recoveries and high matrix interferences. Consequently, adapting the existing techniques to the use in bee matrices requires substantial modifications to improve sensitivity, selectivity, and extraction efficiency [8]. The validation of novel HPLC–MS/MS and IC–HRMS methods for bee matrices is essential to bridge this analytical gap and enable robust, large-scale monitoring of polar pesticide contamination in pollinators.

For several polar pesticides, the legal definition of the maximum residue levels (MRLs) encompasses their metabolites, as these byproducts have a significant role in the risk assessment of the precursor in terms of toxicity and high yield of metabolization. For instance, regarding glyphosate, in 2019, the European Food Safety Authority (EFSA) conducted a review of its MRLs considering two different definitions for assessing animal-based commodities: one for monitoring purposes, including the sum of glyphosate, aminomethylphosphonic acid (AMPA), and N-acetyl-glyphosate, and another for risk assessment purposes, comprising the sum of glyphosate, AMPA, N-acetyl-glyphosate, and N-acetyl-AMPA, all expressed as glyphosate. However, due to insufficient data on the presence of glyphosate and its metabolites in animal-derived products, these MRLs remain provisional [9].

To address this gap, the EFSA emphasizes the need for confirmatory analytical methods for detecting glyphosate, AMPA, and N-acetyl-glyphosate in fat, kidneys, and liver, as well as methods for verifying the presence of AMPA and N-acetyl-glyphosate across all relevant matrices [9].

In this framework, member states conduct multiannual control programs (MACPs) for pesticide residues to ensure compliance with MRLs and assess consumer exposure to pesticide residues in both plant- and animal-based foods [10]. Regulation (EU) 2024/989, which governs the ongoing multi-annual control plan (2025–2027), mandates analyses of glyphosate and glufosinate ammonium in specific animal-derived products such as swine fat and cow milk (2025), bovine liver and poultry fat (2026), and chicken eggs and bovine fat (2027). Additionally, the plan mandates monitoring for other polar pesticides, such as ethephon and fosetyl, in plant-based foods [10].

In addition to the European legislation focusing on controlling highly polar pesticide residues in food, the EFSA and other regulatory bodies have established guidelines to monitor and manage these substances in environmental matrices such as water and soil. The EFSA has identified data gaps concerning the impact of water treatment processes on pesticide residues, emphasizing the need to assess the nature of these residues in treated water and their potential risks to human health and the environment [11]. Furthermore, the EFSA conducts risk assessments of the MRLs of active substances such as pesticides, considering their potential effects on non-target organisms, including bees [12].

Despite the considerable attention given to defining these limits, specific MRLs for polar pesticides in environmental indicators such as honeybees are not explicitly provided in the current EFSA guidelines. While the EFSA has established a specific protection goal (SPG) for honeybees, setting a 10% maximum acceptable colony size reduction following pesticide exposure, this does not equate to setting the MRLs for individual pesticides in bees, indirectly indicating a regulatory gap in directly linking residue levels to risk assessments for pollinators [13].

The development and validation of reliable analytical methods for polar pesticide monitoring in bees are of paramount importance for conducting wide-scale risk assessments of environmental contamination. The impact of pesticide exposure on bee health extends beyond direct toxicity, influencing critical physiological and behavioral aspects, such as gut microbiome composition, gene expression, foraging behavior, thermoregulation, and flight patterns [14–16]. Through the improvement of methods for the detection and quantification of polar pesticides in pollinators, researchers and regulatory bodies might better assess the risks associated with pesticide co-occurrence and inform policies aimed at reducing environmental contamination.

Implementing validated analytical methods also aligns with the principles of sustainable agriculture by promoting pesticide management strategies that minimize ecological harm. Reliable data on pesticide residue levels in pollinators can guide risk mitigation efforts, such as reducing pesticide application rates, implementing buffer zones, and promoting the use of less harmful alternatives. Additionally, such methods can support the enforcement of regulatory frameworks, including the European Union's regulations on the maximum residue limits (MRLs) for pesticides in food and environmental matrices [17–19].

Recent data highlight the alarming consequences of pollinator decline linked to pesticide exposure and broader environmental contamination. The loss of pollinator biodiversity has far-reaching ecological and economic implications, affecting crop yields, food security, and ecosystem stability. Studies have demonstrated that pesticide-induced mortality and sublethal effects contribute to colony collapse disorder (CCD) and population declines in wild pollinators [20–22]. Additionally, the accumulation of pesticide residues in pollinators has been correlated with increased mortality rates, weakened immune responses, and reproductive impairments.

Given the critical role of pollinators in agricultural and natural ecosystems, it is imperative to implement comprehensive monitoring programs that encompass both traditional and highly polar pesticides.

The importance of developing new analytical methods for detecting various classes of pesticide residues in honeybees is evident from an in-depth review of the literature concerning the analysis of pesticides in honeybees, honey, and other hive products. In 2002, Morzycka developed a simple method based on solid-phase dispersion for quantifying 12 different non-polar pesticides in honeybees, using capillary GC coupled with NPD. The author noted that this method was suitable only for detecting low- and medium-polarity pesticides in honeybees [23].

In 2010, Kamel refined an analytical method for quantifying neonicotinoid pesticides and their metabolites in honeybees and bee products using LC–MS/MS, aiming to better understand their role in pollinator decline [24]. A major advancement came in 2015 with the work of Kiljanek et al., who developed a multi-residue method to detect 200 pesticides, such as neonicotinoids and pyrethroids, and their relative metabolites using GC–MS/MS and LC–MS/MS in combination with the QuEChERS and d-SPE techniques. However, this method also primarily targeted non-polar pesticides [25].

In 2017, Balsebre et al. developed a matrix solid-phase dispersion method combined with GC–FPD, GC–ECD, and GC–MS to detect 12 non-polar pesticides in honeybees from central Chile [26], and in 2018, Tosi et al. validated a method for the detection and quantification of 66 non-polar pesticides in pollen samples collected in Italy through UPLC–MS/MS [27]. Additional methods have since been developed for detecting pesticides, mainly neonicotinoids, in bee products such as bee pollen, beeswax, royal jelly, and honey, as reported by Valverde et al. in 2022 [28].

Starting in 2023, efforts to develop methods for detecting polar pesticides in beehive products emerged. Butovskaya et al. and Jesus et al. contributed to this area with studies on glyphosate and other polar pesticides. In the first study, the occurrence of glyphosate and other polar pesticides was assessed in honey, analyzing samples coming from two different Italian regions, while in the second one, a new method was validated to quantify highly polar anionic pesticides in beehive products (honey and pollen) [29,30].

That same year, Murcia-Morales et al. developed a method for detecting 74 non-polar pesticide residues using API strips, comparing the results with analyses conducted on honeybees [31]. More recently, in 2024, Ozols et al. conducted an occurrence study in Latvia, detecting and quantifying 21 non-polar pesticides using LC–MS/MS [32]. In 2025, Rampazzo et al. introduced a new IC–HRMS method for detecting glyphosate and glufosinate [33].

Despite the numerous studies focused on pesticide detection in honeybees and hive products, there remains a lack of validated methods for the simultaneous determination of multiple polar pesticides in the honeybee matrix.

Polar pesticides are known to be particularly susceptible to significant matrix effects. While several studies have investigated polar pesticides in animal-derived products and others have focused on non-polar pesticides in bees, there is currently a lack of research specifically addressing polar pesticides in insect matrices such as bees. Therefore, it is essential to validate and confirm the performance of the analytical method across a wide range of matrices to ensure its robustness and reliability in such complex and underexplored sample types.

The separation of polar pesticides using HPLC–MS/MS and IC–HRMS has historically presented significant challenges, primarily due to matrix co-extractives commonly found in foods of animal and plant origin [34,35]. This issue is particularly pronounced in the honeybee matrix, which is highly complex owing to its composition of proteins, lipids, enzymes, and chitin. These components can adversely affect analyte recovery and retention, as well as cause ion suppression or enhancement during mass spectrometric analysis. To mitigate these challenges, the separation parameters for both methods were optimized using a comprehensive, score-based approach, as detailed in our previously published work [36].

This study aims to overcome the lack in the literature, developing and validating two novel analytical methods, HPLC–MS/MS and IC–HRMS, for the detection of polar pesticide residues in bees. By addressing the current methodological gap, this research contributes to the advancement of environmental risk assessments and supports the transition toward more sustainable and pollinator-friendly agricultural practices. Ultimately, protecting pollinators from pesticide exposure is essential for maintaining biodiversity, ensuring food security, and preserving ecological balance.

2. Materials and Methods

The validation of the analytical methods was conducted by three different official laboratories [37], utilizing distinct techniques based on their available instrumentation. The Environmental Contaminants Laboratory at Istituto Zooprofilattico della Lombardia ed Emilia Romagna "Bruno Ubertini" (IZSLER), Brescia, Italy, performed analyses using ion chromatography (IC) combined with a high-resolution mass spectrometer (HRMS), while the Pesticides, Mycotoxins, and Plant Toxins Laboratory at Istituto Zooprofilattico dell'Umbria e delle Marche "Togo Rosati" (IZSUM), Perugia, Italy, along with the Environmental Contaminants Laboratory at Istituto Zooprofilattico del Lazio e della Toscana "M. Aleandri" (IZSLT), Rome, Italy, employed ultra-high-performance liquid chromatography (UHPLC) coupled with a triple quadrupole mass spectrometer (QqQ). Significant efforts were dedicated to optimizing chromatographic conditions, selecting suitable solvents for pesticide extraction, and refining sample purification procedures.

2.1. Chemicals and Reagents (LC + IC)

Reference standard solutions of AMPA, N-acetyl-AMPA, ethephon, glufosinate, glyphosate, N-acetyl-glyphosate, ethephon hydroxy (HEPA), 3-methylphosphonicpropionic acid (MPPA), and phosphonic acid, in water/acetonitrile (9:1 v/v) (1000 µg/mL); fosetyl-aluminum and N-acetyl-glufosinate (NAG) in water/acetonitrile (9:1 v/v) (100 µg/mL); and isotopically labelled internal standard (ILIS) solutions of AMPA 13C 15N, ethephon D4, glufosinate D3, N-acetyl-glyphosate D3, HEPA D4, MPPA D3, fosetyl-aluminum D15, NAG D3, and 18O3 phosphonic acid, in water/acetonitrile (9:1 v/v) (1000 g/mL), were purchased from Lab Instruments Srl (Castellana Grotte, Italy), HPC Standards GmbH (Cunnersdorf, Germany), Toronto Research Chemicals (Vaughan, ON, Canada), and Dr Ehrenstorfer (LGC Standards, Teddington, UK). The ILIS solution of glyphosate 2-13C, 15N (1000 µg/mL) in water was purchased from Cambridge Isotope Laboratories, Inc. (Tewksbury, MA, USA). Ethylenediaminetetraacetic acid disodium salt dihydrate (EDTA), 99–102% in purity, was obtained from Merck (Darmstadt, Germany). Methanol (MeOH; 99.9%), n-hexane (99%), formic acid (99%), and acetonitrile (99.9%) were obtained from Carlo Erba Reagents Srl (Milan, Italy). All solvents were of LC–MS or analytical grade. Unless otherwise specified, Milli-Q purified water (Millipore, Merck KgaA, Darmstadt, Germany) was used for sample preparation and analysis.

2.2. Samples (LC + IC)

A total of 314 bee samples, each weighing at least 10 g, were collected by several beekeepers across six Italian regions (Lombardy, Emilia Romagna, Umbria, Tuscany, Marche, and Lazio) from June to December 2023 on days with favorable weather conditions, typically during the morning hours to ensure consistent sampling. Among the samples collected, 22% originated from hives with a high number of dead or dying bees, while the remaining 78% came from healthy hives. For the validation studies, samples of live bees from apparently healthy hives were used. These samples had been previously analyzed to confirm the absence of detectable levels of the target analytes, in accordance with document SANTE/11312/2021 v2 [18]. Figure 1 illustrates the geographical distribution of the hives exhibiting higher mortality rates.

Figure 1. Sampling areas with elevated honeybee hive mortality rates (in red) and average honeybee hive mortality rate (in green).

As shown in the map, regions with lower hive mortality are predominantly located in areas characterized by extensive agricultural practices, which may be associated with a reduced risk of pesticide exposure.

2.3. *Analytical Methods*

Two analytical methods, previously used in the laboratories for the identification and quantitation of polar pesticides in other kinds of matrices, were used as a reference for the development of the new ones:

1. Liquid chromatography coupled with tandem mass spectrometry (LC–MS/MS) for the detection of glyphosate, ethephon, glufosinate, fosetyl, and polar pesticide metabolites such as AMPA, N-acetyl-AMPA, HEPA, MPPA, and NAG in animal-derived products [6].
2. Ion chromatography coupled with high-resolution mass spectrometry (IC–HRMS) for the detection of the abovementioned compounds plus N-acetyl-glyphosate and phosphonic acid in vegetables, fruit, and honey [5].

2.3.1. LC–MS/MS Method

The LC–MS/MS method for the detection and quantification of polar pesticides in animal-derived products was modified to be suitable for the bee matrix. After development, the method was fully validated to detect HPPs and their respective metabolites in honeybees, achieving sensitivity, recoveries (70–117%), repeatability (<20%), within-laboratory reproducibility (<20%), and experimental measurement uncertainty below 50% as required by the document SANTE/11312/2021 v2 criteria [18].

Reference Materials

The method provides the detection and quantification of 9 different target analytes (polar pesticides and their respective metabolites) with a LOQ of 0.005 mg/kg. The target analytes and isotopically labeled internal standards (ILIS) are reported in Table S1.

Three working solutions (WS) were prepared by appropriately diluting the stock and reference solutions to the following concentrations: (1) WS1: 1 μg/mL for AMPA, ethephon, fosetyl, glyphosate, HEPA, glufosinate, MPPA, NAG, and N-acetyl-AMPA; (2) WS2: 0.125 μg/mL for AMPA, ethephon, fosetyl, glyphosate, HEPA, glufosinate, MPPA, NAG, and N-acetyl-AMPA; (3) WS3: 0.01 μg/mL for AMPA, ethephon, fosetyl, glyphosate, HEPA, glufosinate, MPPA, NAG, and N-acetyl-AMPA. A mixed ILIS solution for spiking (WSIS1) was prepared by mixing commercial individual ILIS stock solutions to obtain a final concentration equal to 5 μg/mL for all the internal standards reported in Table S1. WSIS1 was diluted 40 times to prepare WSIS2, which was used for the matrix-matched calibration curve. Six-point matrix-matched calibration solutions, including isotopically labelled internal standards, were prepared by mixing appropriate volumes of WSIS2, WS2, and WS3 solutions with water and blank sample extract. The final volume of each calibrant solution was 500 μL.

Sample Preparation Protocol

The samples were stored at −20 °C until the analysis. Two grams of homogenized samples were weighed using a technical balance into polypropylene (PP) tubes with screw caps, and ILIS (20 μL of WSIS1) were added, along with 10 mL of methanol acidified with formic acid (1% *v*/*v*) and 8 mL of Milli-Q water. The samples were shaken for 5 min and then frozen at −20 °C for 10 min, followed by freezing at −80 °C for an additional 10 min. The frozen samples were centrifuged using a refrigerated ultracentrifuge at 15,000 RCF for 10 min at 0 °C. After centrifugation, a small portion of the organic layer, approximately 1.5 mL, was transferred to a plastic tube and then ultracentrifuged again at 15,000 RCF for an additional 10 min at 0 °C. Finally, the extract was filtered using a PTFE filter (13 mm, 0.22 μm) and diluted 1:1 with 250 μL of Milli-Q water in plastic vials, preparing it for LC–MS/MS analysis. Positive samples for validation purposes were prepared by spiking blank honeybee samples before the extraction procedure with 10 μL of WS1 and 20 μL of WSIS1. In each batch, one blank sample (without any addition of standard) was processed accordingly to prepare the six-point matrix-matched calibration curve.

Chromatographic and Spectrometric Conditions

The chromatographic method, previously developed and validated for different animal origin matrices, such as bovine fat and liver, chicken eggs, and cow milk [6], was modified to be suitable for honeybee matrices. The analysis was performed using a Triple Quadrupole AB Sciex 6500+ mass spectrometer (AB Sciex, Pte Ltd., Nijmegen, The Netherlands), coupled with a Nexera X2 UHPLC (IZSUM) and an Agilent 1290 HPLC (IZSLT). Chromatographic separation was achieved using an Anionic Polar Pesticide Column (2.1 × 10 mm, 5 μm), maintained at 50 °C. The mobile phase consisted of 0.9% formic acid in water (mobile phase A) and acetonitrile acidified with 0.9% formic acid (mobile phase B). The mobile phases' composition started at 10% A, and the separation of HPPs was obtained by gradient as reported in Table S2. The total run time was 20 min, with a flow rate of 0.5 mL/min and an injection volume of 10 μL. The chromatographic protocol and mass spectrometer parameters are reported in Table S2; in Table S3, the electronic settings for each transition are indicated.

2.3.2. IC–HRMS Method

The development and validation of the IC–HRMS method were carried out by modifying and adapting a previously validated method, developed in 2020, for the detection and quantification of polar pesticides in fruit, vegetables, and honey [5]. Honeybees, as an unconventional matrix, were studied and validated according to the criteria established in document SANTE/11312/2021 v2, confirming that the newly developed procedure is selective and robust [18].

Reference Materials

The method provides the detection and quantification of 10 different target analytes (polar pesticides and their respective metabolites) with different LOQs:

- LOQ = 0.01 mg/kg for glyphosate, N-acetyl-glyphosate, N-acetyl-AMPA, glufosinate, MPPA, and NAG.
- LOQ = 0.05 mg/kg for ethephon.
- LOQ = 0.10 mg/kg for fosetyl, phosphonic acid, and AMPA.

The target analytes, along with the respective measurement intervals and relative ILIS, are reported in Table S4.

Two working solutions (WS) in water/acetonitrile (9:1, *v*/*v*) were prepared by appropriately diluting the stock and reference solutions to the following concentrations: WS1: 1 μg/mL for glyphosate, NAG, N-acetyl AMPA, gluphosinate, N-acetyl glyphosinate, MPPA; 5 μg/mL for AMPA and ethephon and 10 μg/mL for fosetyl and phosphonic acid; WS2: 0.1 μg/mL for glyphosate, NAG, N-acetyl AMPA, gluphosinate, N-acetyl glyphosinate, MPPA; 0.5 μg/mL for AMPA and ethephon and 1 μg/mL for fosetyl and phosphonic acid. The ILIS for spiking in water/acetonitrile (9:1, *v*/*v*), ILIS1, was prepared by mixing commercial individual ILIS stock solutions to obtain a final concentration equal to 10 μg/mL for all the internal standards reported in Table S4. ILIS1 was diluted 10 times to prepare ILIS2 in water/acetonitrile (9:1, *v*/*v*), which was used for the matrix-matched calibration curve. Six-point matrix-matched calibration solutions, including isotopically labelled internal standards, were prepared by mixing appropriate volumes of the WS1, WS2, and ILIS2 solutions with the blank sample extract.

Sample Preparation Protocol

All the samples were stored at −60 °C for 24 h before homogenization. Afterward, one gram of the homogenized bee sample was extracted in 50 mL polypropylene tubes with 10 mL of a water:acetonitrile (1:1, *v*/*v*) solution containing 1% (*v*/*v*) formic acid and manually shaken. To ensure analytical accuracy, we monitored extraction efficiency and evaluated matrix effects; 20 μL of isotopically labeled internal standard solution ILIS1 in water/acetonitrile (9:1, *v*/*v*) at 10 μg/mL was added to each sample; the mixture was then mechanically shaken and centrifuged at 4000 RCF for 20 min at 4 °C. Subsequently, 750 μL of the supernatant were transferred to 15 mL polypropylene tubes and diluted with 3 mL of ultrapure water; this dilution was followed by the addition of hexane saturated with acetonitrile. The mixture was manually shaken to enhance phase separation and then centrifuged under the same conditions. Then, 1.5 mL of the lower aqueous phase was carefully collected and passed through a Sep-Pak Plus C18 cartridge and a 0.2 μm syringe PTFE filter in tandem. The resulting filtered solution was transferred to injection vials for final analysis.

Chromatographic and Mass Spectrometric Conditions

The ion chromatography method, previously developed and validated for different matrices, such as fruit, vegetables, and honey [5], has been modified and adapted for honeybee matrices. The analysis was performed using an Ionic Chromatograph IC 5000+ Dionex (Thermo Fisher Scientific, Waltham, Massachusetts, USA) coupled with a high-resolution mass spectrometer Q-Extractive Focus. The ion chromatography conditions are reported in Table S5.

The signals of the analytes in mass spectrometry coupled to the IC were acquired in the full mass mode, with negative ionization HESI. Additional parameters are reported in Table S6. For all the compounds, accurate mass and at least two MS2 fragment ions were detected with a relative mass error < 5 ppm. The inclusion list, in the instrumental method setting, contained retention time windows in which each precursor ion was expected, with their corresponding collision energies.

2.4. Validation Protocol

2.4.1. LC–MS/MS Validation

The validation protocol was carried out following point G7 and of document SANTE/11312/2021 v2 [18]. The following parameters were evaluated for validation purposes: linearity range, LOQ, recovery, precision under repeatability conditions as RSD_r, and inter-laboratory precision in reproducibility conditions as RSD_R. The same validation protocol was followed by both laboratories: IZSUM and IZSLT. The applied interlaboratory approach has wide-ranging use, highlighted in the literature as the preferred way to carry out the validation [38].

Calibration Curves and Linearity Ranges

Matrix-matched calibration curves were built as described in Section 2.3.1. The six calibration points were 0.0002, 0.0005, 0.0010, 0.0025, 0.0050, and 0.010 mg/kg. These ranges were selected considering a dilution factor of 20 applied during the extraction/purification process of all the samples. Calibration curves were obtained by plotting the response factor (RF) versus concentration (mg/kg) for each analyte. The RF was calculated as the ratio between the peak area of the analyte and the corresponding ILIS for all pesticides, except N-acetyl-AMPA, which was quantified without an isotopically labeled internal standard due to its unavailability. For linearity evaluation, calibration curves were injected on three different days spread over two weeks. The ordinary least squares (OLS) method, including the point (0;0), was used to determine the calibration function. For each calibration point, the back-calculated concentration (BCC) deviation, that is, the deviation of the calculated concentration by the calibration function (Cmeasured) from the true concentration (Ctrue), was evaluated using the formula:

$$\text{Deviation} = \frac{(\text{Cmeasured} - \text{Ctrue})}{\text{Ctrue}} \times 100$$

According to document SANTE/11312/2021 v2 [18], the calibration curve was considered linear when, for each point, the BCC deviation did not exceed ±20%. Linearity ranges fulfilled the requirements for both IZSUM and IZSLT.

Recovery

Recovery experiments were independently performed by the two laboratories on four different spiking levels (0.005, 0.010, 0.020, and 0.100 mg/kg). The mass concentration was calculated using the following formula:

$$\text{Cx}\left(\frac{\text{mg}}{\text{kg}}\right) = \frac{(\text{RF} - \text{a})}{\text{b}} \times \text{DF}/1000$$

where:

- RF = response factor (peak area ratio of the analyte/internal standard; peak area for N-acetyl-AMPA only).
- a = slope of the calibration curve (μg^{-1}).
- b = intercept of the calibration curve.
- DF = dilution factor (20).

The analytical parameters were evaluated by analyzing blank samples of bees fortified at four different concentration levels, as reported in Table S7. Level 1 was the estimated LOQ, level 2 was set at 2× LOQ, level 3—at 4 × LOQ, and level 4, the highest, was set at 20 × LOQ for all the target pesticides.

Validation experiments confirmed compliance with the performance criteria (recoveries between 70–120% and precision values lower than or equal to 20%). For inter-laboratory precision (RSD_R), blank samples of homogenized bees were fortified at 0.020 mg/kg and analyzed by 3 different operators over 12 different days in two laboratories: Laboratory of Pesticides, Mycotoxins, and Plant Toxins (IZSUM) and the Environmental Contaminants Laboratory (IZSLT).

2.4.2. IC–HRMS Validation

Instrumental linearity was assessed using six concentration levels in solvent-based solutions and matrix-matched standards. Specificity was evaluated by analyzing reagent blanks and blank samples to confirm no significant interferences near the characteristic retention times of each pesticide. Six fortified samples at two concentrations were analyzed to determine the limit of quantification (LOQ), recovery (trueness), and precision; matrix effects and linearity were assessed by comparing analytical response variations between solvent standard solutions and the matrix-matched calibration curve. Validation for IC–HRMS was conducted by IZSLER.

Calibration Curves and Linearity Ranges

Linearity was evaluated in matrix-matched standards using six calibration levels, obtained by mixing the appropriate amount of reference standards. The calibration points for each analyte are reported in Table S8. The analyses were performed using the response factor (RF) approach, which measures the relative mass response of an analyte compared to its ILIS; N-acetyl-AMPA was compared using ethephon hydroxy D4. AA and CA represent the area and amount of pesticide, respectively, while AIS and CIS correspond to the area and amount of the internal standard. The formula used is as follows:

$$RF = \frac{A_A \times C_{is}}{A_{IS} \times C_A}$$

RF was calculated for each calibration point and each pesticide. The average RF and its standard deviation were determined during the validation study. To assess RF acceptability, an internal criterion based on the standard deviation (% RSD) was established, requiring RF to remain at or below 20%.

During method development and validation, the feasibility of quantifying sample amounts using RF, derived from a matrix-matched calibration curve, was evaluated.

LOQ, Repeatability, and Within-Laboratory Reproducibility

The analytical parameters were evaluated through the analysis of blank bee samples fortified at two different concentration levels, as reported in Table S9. Level 1 was the estimated

LOQ, level 2 was set at 10 × LOQ for all the target pesticides, with the exception of AMPA, for which the higher level was set at 5 × LOQ due to the method's sensitivity for this analyte.

For the concentration levels defined above, validation experiments confirmed compliance with the performance criteria (recoveries within the 70–120% range and precision values lower than or equal to 20%) for all the pesticides. For the calculation of within-laboratory precision (RSD_{wR}), blank samples of homogenized bees were fortified with each analyte at the LOQ level. A total of 18 replicates were considered through an ongoing precision evaluation.

Measurement Uncertainty

According to Appendix C of the SANTE document, for the three laboratories, the relative expanded measurement uncertainty (U') was calculated, as requested by ISO/IEC 17025 [39]. The calculation was based on intra-laboratory validation/QC data as outlined in Appendix C, document SANTE/11312/2021 v2 [18]. The relative expanded measurement uncertainty was calculated through the application of a coverage factor (k = 2) (level of confidence approx. 95%) to the relative combined uncertainty (u'), determined by evaluating the relative standard deviation of within-laboratory reproducibility and the method bias.

The method and laboratory bias were calculated through recovery experiments conducted under an intra-laboratory reproducibility study, according to the following equation:

$$u'(\text{bias}) = \sqrt{\text{RMS}_{\prime}(\text{bias})} = \sqrt{\text{mean}^2_{\text{bias}} + \text{SD.P}^2_{\text{bias}}}$$

where:

$\text{mean}_{\text{bias}}$ = mean of the corresponding bias

$\text{SD.P}_{\text{bias}}$ = population standard deviation of the corresponding bias (calculated using the STDEV.P function in Excel, v.2505 build 16.0.18827.20102).

3. Results and Discussion

3.1. Analytical Method for Determination of Polar Pesticides in Bees Compared to the QuPPe-AO Method

As reported in Section 2.3, the reference analytical methods initially used were previously validated elsewhere for both LC/MS-MS [6], and IC-HRMS [5]. These two methods were developed for conventional food substrates, such as animal origin foods different from insects, as well as fruit, vegetables, and honey, based on the EURL-SRM QuPPe-AO-method v.3.3 [40]. However, some modifications were necessary to improve the recovery rate of polar pesticides extracted from this matrix, which is characterized by a high presence of wax and pollen, both of which can interfere with pesticide detection. To overcome this limitation, the extraction protocol was modified by introducing an additional ultracentrifugation step at low temperature (0 °C) at 15,000 RCF after transferring the extract into a 1.5 mL tube. To evaluate the optimal chromatographic conditions in LC–MS/MS for the separation, detection, and quantification of polar pesticides, three different chromatographic columns were tested: Hypercarb, Raptor Polar X, and Anionic Polar Pesticide. A score-based methodology was applied to determine the most suitable chromatographic column, with the Anionic Polar Pesticide column emerging as the optimal choice [36]. For IC-HRMS application, extraction with acidified water appeared to be the best option to ensure adequate compatibility with both the chromatographic columns and the gradient necessary for analyte elution. After extraction, an additional cleaning step using hexane saturated with acetonitrile was required to remove typical interferents from the material such as bees.

3.2. LC–MS/MS Method Performance Verification

A full in-house validation of the LC–MS/MS method was performed, including an inter-laboratory precision assessment through analysis conducted in two different laboratories: IZSUM and IZSLT. The following analytes were included: HEPA, MPPA, AMPA, ethephon, fosetyl, glufosinate, glyphosate, NAG, N-acetyl-AMPA. The validated parameters, according to SANTE/11312/2021 v2, included linearity ranges, recovery rates (%), repeatability (RSD_r), and inter-laboratory precision (RSD_R) [18].

3.2.1. Linearity Ranges

The linearity ranges were independently confirmed by IZSUM and IZSLT using the BCC method, yielding fully satisfactory results for all the analyzed polar pesticides. Validation was performed at six levels: 0.0, 0.0002, 0.0005, 0.0010, 0.0025, 0.0050, and 0.0100 mg/kg, corresponding to a range of 0.004–0.200 mg/kg in honeybees, considering a dilution factor of 20 in the extraction method. For all spiking levels and in both laboratories, the deviation of BCC remained 20%. This satisfies the linearity criterion according to Table 4, point G of SANTE/11312/2021 v2 [18].

3.2.2. Limit of Quantification (LOQ)

In the absence of specific MRLs for polar pesticides in honeybees, the LOQ, according to the SANTE regulation, was set to 0.005 mg/kg, between the second and the third point of the linearity range. This value was confirmed experimentally as the lowest spiked level meeting the identification and method performance criteria for recovery and precision. The LOQ for each analyte was independently validated by IZSUM and IZSLT [18].

3.2.3. Repeatability (RSD_r)

The evaluation of method precision under repeatability conditions, independently performed by IZSUM and IZSLT, demonstrated that for each analyte, the relative standard deviation (RSD) was below 20%, ranging between 1.6% and 19.7%, meeting the SANTE requirements at four different concentration levels (LOQ, 2× LOQ, 4× LOQ, and 20× LOQ) in six replicates. Additionally, recovery values ranged from 70% to 119%, fully complying with the SANTE acceptability criteria (point G, Table 4) [18]. Repeatability data are reported in Table 1.

Table 1. Repeatability data for each analyte at 4 levels of concentration, 0.005 mg/kg (LOQ), 0.010 mg/kg (2× LOQ), 0.020 mg/kg (4× LOQ), 0.100 mg/kg (20× LOQ).

Analyte	Level (mg/kg) (n = 6)	IZSUM		IZSLT	
		Average Recovery %	RSD_r	Average Recovery %	RSD_r
AMPA	0.005	93	4.0	102	10.6
	0.010	97	2.5	92	3.6
	0.020	101	5.0	88	6.4
	0.100	101	3.5	114	4.1
Ethephon	0.005	91	10.1	99	7.8
	0.010	93	3.4	110	3.0
	0.020	95	5.8	97	1.8
	0.100	95	4.3	115	1.6

Table 1. *Cont.*

Analyte	Level (mg/kg) (n = 6)	IZSUM		IZSLT	
		Average Recovery %	RSD_r	Average Recovery %	RSD_r
Fosetyl	0.005	97	6.9	92	3.5
	0.010	87	5.2	98	6.0
	0.020	105	5.3	106	7.1
	0.100	101	3.3	106	3.8
Glufosinate	0.005	104	6.0	98	10.5
	0.010	96	8.8	92	6.6
	0.020	105	13.8	85	3.3
	0.100	99	4.3	98	6.3
Glyphosate	0.005	114	4.7	102	3.5
	0.010	102	4.6	115	5.7
	0.020	119	11.0	101	10.0
	0.100	98	10.8	108	3.4
HEPA	0.005	95	5.7	100	3.6
	0.010	84	6.8	111	4.2
	0.020	87	7.3	106	6.0
	0.100	102	4.4	117	1.8
MPPA	0.005	95	19.2	103	7.3
	0.010	99	3.6	94	3.1
	0.020	109	12.0	105	3.7
	0.100	104	6.6	111	2.3
N-acetyl-glufosinate	0.005	79	10.4	97	10.3
	0.010	95	6.7	74	14.8
	0.020	104	7.7	81	6.4
	0.100	104	3.4	99	3.0
N-acetyl-AMPA	0.005	79	7.7	93	19.7
	0.010	70	3.7	112	12.7
	0.020	91	3.9	108	4.2
	0.100	102	2.6	118	2.6

3.2.4. Inter-Laboratory Precision (RSD_R)

The inter-laboratory reproducibility study was conducted on the honeybee samples spiked at 0.020 mg/kg in two different laboratories (IZSUM and IZSLT) by three different operators over four different days each (n = 12). The results confirmed the accuracy of the analytical method, with recovery values between 91% and 103% and RSD_R values between 5.5% and 13.6%, meeting the SANTE precision criteria (point G, Table 4) [18]. Inter-laboratory precision data are reported in Table 2.

The experimental relative expanded measurement uncertainties (U′) were below 50% for all the analyzed pesticides, thus remaining within the maximum expanded uncertainty limit set by the SANTE criteria [18].

Table 2. Inter-laboratory precision summary. The spiking level was 0.020 mg/kg for each analyte.

Analyte (0.020 mg/kg)	Day	IZSUM		IZSLT		
		Operator 1	Operator 2	Operator 3	Mean	
		Recovery %	Recovery %	Recovery %	Average Recovery %	RSD_R
AMPA	Day 1	107	89	77	93	7.2
	Day 2	87	100	87		
	Day 3	100	92	90		
	Day 4	102	102	93		
Ethephon	Day 1	84	100	98	91	6.9
	Day 2	88	95	94		
	Day 3	90	96	97		
	Day 4	71	80	98		
Fosetyl	Day 1	109	96	112	101	5.0
	Day 2	92	96	93		
	Day 3	97	100	106		
	Day 4	99	94	114		
Glufosinate	Day 1	101	92	85	93	7.5
	Day 2	93	101	89		
	Day 3	91	107	87		
	Day 4	99	90	80		
Glyphosate	Day 1	89	97	98	98	5.5
	Day 2	91	91	87		
	Day 3	102	103	116		
	Day 4	88	109	107		
HEPA	Day 1	99	107	111	103	6.6
	Day 2	90	80	99		
	Day 3	102	117	114		
	Day 4	96	114	109		
MPPA	Day 1	101	107	104	103	7.9
	Day 2	102	80	110		
	Day 3	102	117	109		
	Day 4	87	114	104		
N-acetyl-glufosinate	Day 1	104	96	82	91	9.8
	Day 2	90	94	76		
	Day 3	98	91	77		
	Day 4	97	105	81		
N-acetyl-AMPA	Day 1	92	95	115	94	13.6
	Day 2	67	86	105		
	Day 3	99	91	109		
	Day 4	78	76	113		

A chromatogram showing the detected and quantified analytes with the LC–MS/MS method is reported in Figure 2.

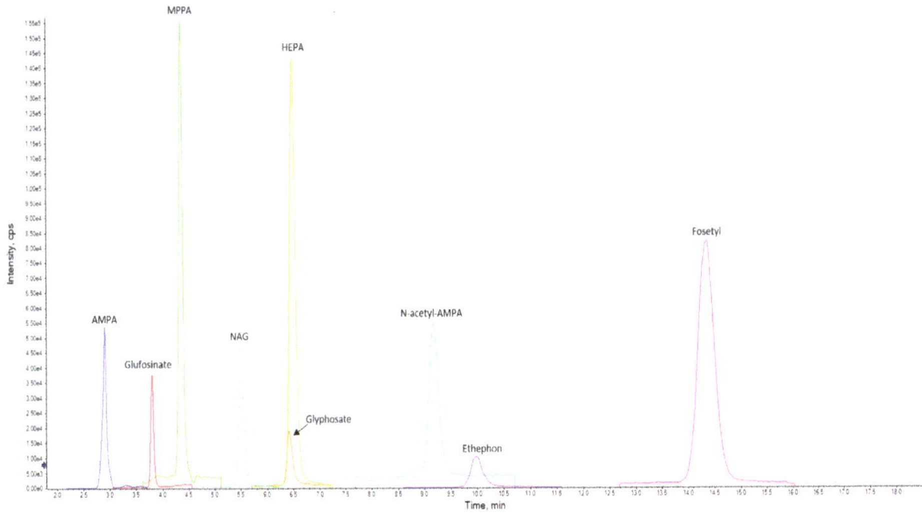

Figure 2. LC–MS/MS chromatogram of the matrix matched standard obtained with the present method. The spiking level was 0.100 mg/kg for each analyte (XIC).

3.3. IC–HRMS Method Performance Verification

A full in-house validation of the IC–HRMS method was performed by IZSLER, confirming compliance with all the required SANTE parameters, for the following analytes: MPPA, AMPA, ethephon, fosetyl, glufosinate, glyphosate, NAG, N-acetyl-glyphosate, N-acetyl-AMPA, and phosphonic acid. The validated parameters included linearity ranges, recovery rates (%), repeatability (RSD_r), and within-laboratory reproducibility (RSD_{WR}).

3.3.1. Linearity Ranges

Linearity was confirmed by IZSLER using the BCC method, yielding fully satisfactory results for all the polar pesticides. The validation was performed at six levels, depending on the defined LOQ, as reported in Section 2.4.2:

- for fosetyl and phosphonic acid, the linearity was confirmed between 0.002 to 0.500 mg/kg;
- for ethephon and AMPA, the linearity was confirmed between 0.001 and 0.250 mg/kg;
- for glufosinate, N-acetyl-glufosinate, MPPA, glyphosate, N-acetyl-glyphosate, and N-acetyl AMPA, the linearity was confirmed between 0.0002 and 0.050 mg/kg.

The deviation of BCC, for all spiking levels, remained within ±20%, fulfilling the linearity criterion according to the SANTE document, as previously mentioned [18].

3.3.2. Limit of Quantification (LOQ)

In the absence of specific MRLs for HPPs in honeybees, the LOQ, according to the SANTE regulation, was set at the lowest concentration satisfying the validation criteria reported in point G, Table 4 of the SANTE document [18]. After the confirmation of the accomplishment of the criteria designed for the highest concentration level of repeatability, recovery range between 70% and 120% and RSD_r less than or equal to 20%, the LOQ was defined by dividing the highest identified level by a factor of 10 and subsequently verifying the compliance with these two criteria at the lowest level experimentally for each analyte but AMPA, for which the LOQ was raised to 0.1 mg/kg.

3.3.3. Repeatability (RSD_r)

The evaluation of method precision in repeatability conditions performed by IZSLER revealed that for each considered analyte, the RSD_r was lower than 20%, in a range between 2.0% and 14.0%, meeting the SANTE requirements for two different concentration levels (LOQ and 10× LOQ), except for AMPA, which confirmed the requirements of the criteria at a concentration level of 5× LOQ in six different replicates. Additionally, fully satisfactory recovery percentage values were obtained, ranging from 84% to 114%, in compliance with the acceptability criteria established by the SANTE document, point G, Table 4 [18]. The repeatability data are reported in Table 3.

Table 3. Repeatability data for each analyte at 2 levels of concentration: LOQ and 5× LOQ for AMPA, LOQ and 10× LOQ for the other polar pesticides.

Analyte	Level (mg/kg) (n = 6)	IZSLER	
		Average Recovery %	RSD_r
AMPA	0.10	93	6.0
	0.50	90	6.0
Ethephon	0.05	106	14.0
	0.50	98	10.0
Fosetyl	0.10	103	4.0
	1.00	108	3.0
Glufosinate	0.01	93	8.0
	0.10	98	3.0
Glyphosate	0.01	84	5.0
	0.10	93	4.0
MPPA	0.01	109	3.0
	0.10	114	4.0
N-acetyl-AMPA	0.01	100	7.0
	0.10	109	3.0
NAG	0.01	96	6.0
	0.10	110	2.0
N-acetyl-glyphosate	0.01	107	14.0
	0.10	85	6.0
Phosphonic acid	0.10	102	3.0
	1.00	100	2.0

3.3.4. Within-Laboratory Precision (RSD_{wR})

Data obtained from the within-laboratory reproducibility study of samples from honeybees, spiked at the LOQ, through 18 replicates from ongoing analyses, confirmed the accuracy of the analytical method. Recovery values ranged from 87% to 119%, with RSD_{wR} values between 10% and 18%, which is lower than 20% in any case, according to the precision criteria reported in point G, Table 4 of the SANTE document [18]. Within-laboratory precision data are reported in Table 4.

Table 4. Within-laboratory precision summary. The spiking level was at the LOQ for each analyte.

Analyte (LOQ) (n = 18)	IZSLER	
	Average Recovery %	$\mathbf{RSD_{wR}}$
AMPA	87	17.0
Etephon	103	18.0
Fosetyl	95	10.0
Glufosinate	94	11.0
Glyphosate	87	12.0
MPPA	97	10.0
N-acetyl-AMPA	119	16.0
NAG	93	10.0
N-acetyl-glyphosate	95	15.0
Phosphonic acid	98	12.0

Experimental relative expanded measurement uncertainties (U′) remained below 50%, in compliance with the SANTE requirements [18].

Chromatograms showing the detected and quantified analytes with the IC–HRMS method are reported in Figure 3.

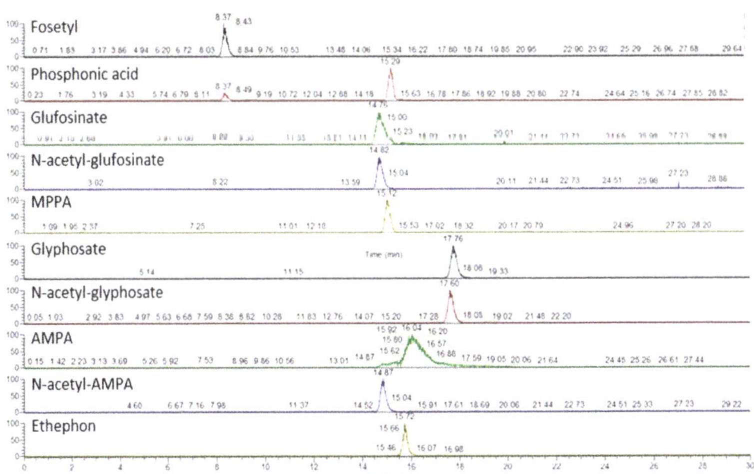

Figure 3. IC–HRMS chromatograms of the matrix-matched standard obtained with the present method. Spiking level 3 (see Table S8).

3.3.5. Measurement Uncertainty

To apply a default uncertainty of 50%, laboratories must demonstrate an experimental expanded MU (U′) value of ≤50%. The experimental U′ values based on intra-laboratory QC data for individual pesticides in honeybees are reported in Table 5.

Table 5. Experimental U′ for each laboratory.

	U′ (Expanded MU) (%)		
	IZSUM	**IZSLT**	**IZSLER**
AMPA	20	22	38
Ethephon	35	15	46
Fosetyl	15	18	49
Glufosinate	20	21	36
Glyphosate	22	9	37
HEPA	17	7	N.D.
MPPA	34	16	42
N-acetyl-AMPA	39	39	48
NAG	19	22	41
N-acetyl-glyphosate	N.D.	N.D.	35
Phosphonic Acid	N.D.	N.D.	41

4. Conclusions

In the present study, two novel and fully validated methods, LC–MS/MS and IC–HRMS, were successfully developed for the detection of HPPs in honeybees. These methods were validated through the collaborative efforts of three official laboratories of the Italian National Health System (Istituti Zooprofilattici Sperimentali), which are actively involved in the development of innovative analytical methods and consequent risk assessment to ensure food and environmental safety. The validation was carried out according to the criteria established by the SANTE guidelines, confirming the accuracy, sensitivity, and repeatability of both methods.

This study represents the first successful validation of the methods for detecting polar pesticides specifically in honeybees, and it establishes a fundamental step toward more comprehensive environmental monitoring. In the future, these validated methodologies could be useful as a reference for the detection of polar pesticides in other insect matrices and unconventional substrates, thus contributing significantly to food and environmental monitoring and risk assessment.

The LC–MS/MS and IC–HRMS methods are complementary due to the analytical challenges associated with detecting certain compounds. Phosphonic acid and N-acetyl-glyphosate, which are difficult to detect via LC–MS/MS, were effectively quantified through IC–HRMS, while HEPA and AMPA showed higher responsiveness and sensitivity in LC–MS/MS, with HEPA presenting high peak interferences in IC–HRMS. This complementary approach increases the robustness and reliability of the combined methods, enabling accurate and comprehensive detection of a wide range of polar pesticides.

Previously published data regarding the evaluation of different chromatographic columns [25] were used to choose the best-performing column and optimize the chromatographic performances for this unique matrix.

Furthermore, the exploitation of these innovative methods opens several future challenges and opportunities. One key challenge will be to establish new maximum residue limits (MRLs) for polar pesticides in honeybees and other pollinators, an initiative that could be undertaken by the EFSA and other international regulatory authorities. This effort could pave the way for adopting more sustainable agricultural practices that minimize pesticide exposure to pollinators and ultimately protect consumer health. Additionally, future research should focus on improving detection capabilities for a broader range of polar pesticides and their metabolites, as well as expanding these methods to cover other pollinator species. Addressing the ecological impact of polar pesticide contamination on bee health and pollinator populations remains a critical priority. Moreover, fostering collaboration between research institutions and regulatory bodies will be crucial in harmonizing methodologies and establishing new regulatory standards.

Finally, implementing these validated methods will significantly enhance environmental monitoring programs, facilitate data-driven risk assessments, and promote the adoption of healthier and more sustainable agricultural practices to protect pollinator health and environmental quality.

Supplementary Materials: The following supporting information can be downloaded at: https://www.mdpi.com/article/10.3390/jox15040095/s1, Table S1: Target polar analytes in the standard mix and relative Isotopically Labeled Internal Standards; Table S2: Chromatographic LC conditions; Table S3: Mass Spectrometry Conditions; Table S4: Target polar analytes in the standard mix and relative Isotopically Labeled internal standards; Table S5: Ionic chromatography conditions; Table S6: Acquisition parameters for the mass spectrometry coupled to IC; Table S7: Validation Levels for Recovery, Repeatability and Within-Laboratory Reproducibility; Table S8: Matrix-matched points for linearity evaluation; Table S9: Validation Levels for Recovery, Repeatability and Within-Laboratory Reproducibility.

Author Contributions: Conceptualization, M.G., I.P. and K.R.; methodology, M.G., S.O. and K.R.; validation, K.R., S.O. and M.G.; formal analysis, M.B., E.V., T.P. and T.M.; investigation, M.B., E.V., T.P. and T.M.; data curation, M.G., M.B., T.P., T.M. and E.V.; writing—original draft preparation, T.P., M.B. and E.V.; writing—review and editing, I.P., S.O., K.R., M.G., T.D. and B.A.; visualization, T.P., E.V., T.M. and M.B.; supervision, M.G.; project administration, M.G., K.R. and I.P.; funding acquisition, M.G. All authors have read and agreed to the published version of the manuscript.

Funding: This work was supported by the contribution of the Ministry of Health under the current IZSLER research project PRC 2021009 (31 December 2020–30 June 2024).

Institutional Review Board Statement: Not applicable.

Informed Consent Statement: Not applicable.

Data Availability Statement: The original contributions presented in this study are included in the article/Supplementary Material. Further inquiries can be directed to the corresponding authors.

Conflicts of Interest: The authors declare no conflicts of interest.

References

1. European Commission, 2024, Regulation (EU) 2024/1991 of the European Parliament and of the Council of 24 June 2024 on Nature Restoration and Amending Regulation (EU) 2022/869. Available online: https://eur-lex.europa.eu/legal-content/EN/TXT/?uri=CELEX:32024R1991&qid=1722240349976 (accessed on 8 April 2025).
2. Tscharntke, T.; Klein, A.M.; Kruess, A.; Steffan-Dewenter, I.; Thies, C. Landscape Perspectives on Agricultural Intensification and Biodiversity—Ecosystem Service Management. *Ecol. Lett.* **2005**, *8*, 857–874. [CrossRef]
3. Lehman, P.; Cady, N.; Ghimire, S.; Shahi, S.K.; Shrode, R.L.; Lehmler, H.; Mangalam, A.K. Low-Dose Glyphosate Exposure Alters Gut Microbiota Composition and Modulates Gut Homeostasis. *Environ. Toxicol. Pharmacol.* **2023**, *100*, 104149. [CrossRef] [PubMed]
4. Campanale, C.; Massarelli, C.; Losacco, D.; Bisaccia, D.; Triozzi, M.; Uricchio, V.F. The Monitoring of Pesticides in Water Matrices and the Analytical Criticalities: A Review. *TrAC Trends Anal. Chem.* **2021**, *144*, 116423. [CrossRef]
5. Gasparini, M.; Angelone, B.; Ferretti, E. Glyphosate and Other Highly Polar Pesticides in Fruit, Vegetables and Honey Using Ion Chromatography Coupled with High Resolution Mass Spectrometry: Method Validation and Its Applicability in an Official Laboratory. *J. Mass Spectrom.* **2020**, *55*, e4624. [CrossRef]

6. Verdini, E.; Lattanzio, V.; Ciasca, B.; Fioroni, L.; Pecorelli, I. Improved Method for the Detection of Highly Polar Pesticides and Their Main Metabolites in Foods of Animal Origin: Method Validation and Application to Monitoring Programme. *Separations* **2023**, *10*, 44. [CrossRef]
7. Rampazzo, G.; Gazzotti, T.; Pagliuca, G.; Nobile, M.; Chiesa, L.; Carpino, S.; Panseri, S. Determination of Glyphosate, Glufosinate, and Metabolites in Honey Based on Different Detection Approaches Supporting Food Safety and Official Controls. *LWT Food Sci. Technol.* **2024**, *200*, 116159. [CrossRef]
8. Bargańska, Ż.; Lambropoulou, D.; Namieśnik, J. Problems and Challenges to Determine Pesticide Residues in Bumblebees. *Critical Reviews in Analytical Chemistry* **2018**, *48*, 447–458. [CrossRef]
9. European Food Safety Authority (EFSA), Review of the Existing Maximum Residue Levels for Glyphosate According to Article 12 of Regulation (EC) No 396/2005–Revised Version to Consider Omitted Data. 2019. Available online: https://www.efsa.europa.eu/en/efsajournal/pub/5862 (accessed on 8 April 2025).
10. European Commission, 2024, Commission Implementing Regulation (EU) 2024/989 of 02 April 2024 Concerning a Coordinated Multiannual Control Programme of the Union for 2025, 2026 and 2027 to Ensure Compliance with Maximum Residue Levels of Pesticides and to Assess the Consumer Exposure to Pesticide Residues in and on Food of Plant and Animal Origin and Repealing Implementing Regulation (EU) 2023/731. Available online: https://eur-lex.europa.eu/legal-content/EN/TXT/PDF/?uri=CELEX:32024R0989 (accessed on 3 April 2025).
11. Hofman-Caris, R.; Dingemans, M.; Reus, A.; Shaikh, S.M.; Sierra, J.M.; Karges, U.; der Beek, T.A.; Nogueiro, E.; Lythgo, C.; Parra Morte, J.M.; et al. Guidance Document on the Impact of Water Treatment Processes on Residues of Active Substances or Their Metabolites in Water Abstracted for the Production of Drinking Water. *EFSA J.* **2023**, *21*, 8194. [CrossRef]
12. European Food Safety Authority Insect Pollinator Health—EFSA. Available online: https://www.efsa.europa.eu/en/topics/insect-pollinator-health (accessed on 8 April 2025).
13. Adriaanse, P.I.; Arce, A.N.; Focks, A.; Ingels, B.; Jölli, D.; Lambin, S.; Rundlöf, M.; Süssenbach, D.; Del Aguila, M.; Ercolano, V.; et al. Revised Guidance on the Risk Assessment of Plant Protection Products on Bees (*Apis Mellifera*, *Bombus* spp. And Solitary Bees). *EFSA J.* **2023**, *21*, 7989. [CrossRef]
14. Bokšová, A.; Kazda, J.; Bartoška, J.; Kamler, M. Effect of Glyphosate on the Foraging Activity of European Honey Bees (*Apis Mellifera* L.). *Plant Soil Environ.* **2023**, *69*, 195–201. [CrossRef]
15. Balbuena, M.S.; Tison, L.; Hahn, M.L.; Greggers, U.; Menzel, R.; Farina, W.M. Effects of Sublethal Doses of Glyphosate on Honeybee Navigation. *J. Exp. Biol.* **2015**, *218*, 2799–2805. [CrossRef] [PubMed]
16. Ferrante, M.; Rapisarda, P.; Grasso, A.; Favara, C.; Oliveri Conti, G. Glyphosate and Environmental Toxicity with "One Health" Approach, a Review. *Environ. Res.* **2023**, *235*, 116678. [CrossRef] [PubMed]
17. European Commission, 2005, Regulation (EC) No 396/2005 of the European Parliament and of the Council of 23 February 2005 on Maximum Residue Levels of Pesticides in or on Food and Feed of Plant and Animal Origin and Amending Council Directive 91/414/EEC Text with EEA Relevance. Available online: https://eur-lex.europa.eu/eli/reg/2005/396/oj (accessed on 8 April 2025).
18. European Commission, 2021, Analytical Quality Control and Method Validation Procedures for Pesticides Residues Analysis in Food and Feedm, SANTE 11312/2021 V2, Implemented by 01/01/2024. Available online: https://food.ec.europa.eu/system/files/2023-11/pesticides_mrl_guidelines_wrkdoc_2021-11312.pdf (accessed on 8 April 2025).
19. European Commission, Health & Consumer Protection Directorate-General Guidance Document on Analytical Quality Control and Validation Procedures for Pesticide Residues Analysis in Food and Feed, SANCO/12571/2013. Available online: https://www.eurl-pesticides.eu/library/docs/allcrl/AqcGuidance_Sanco_2013_12571.pdf (accessed on 8 April 2025).
20. Vázquez, D.E.; Villegas Martínez, L.E.; Medina, V.; Latorre-Estivalis, J.M.; Zavala, J.A.; Farina, W.M. Glyphosate Affects Larval Gut Microbiota and Metamorphosis of Honey Bees with Differences between Rearing Procedures. *Environ. Pollut.* **2023**, *334*, 122200. [CrossRef] [PubMed]
21. Berg, C.J.; King, H.P.; Delenstarr, G.; Kumar, R.; Rubio, F.; Glaze, T. Glyphosate Residue Concentrations in Honey Attributed through Geospatial Analysis to Proximity of Large-Scale Agriculture and Transfer Off-Site by Bees. *PLoS ONE* **2018**, *13*, e0198876. [CrossRef]
22. Weidenmüller, A.; Meltzer, A.; Neupert, S.; Schwarz, A.; Kleineidam, C. Glyphosate Impairs Collective Thermoregulation in Bumblebees. *Science* **2022**, *376*, 1122–1126. [CrossRef]
23. Morzycka, B. Simple Method for the Determination of Trace Levels of Pesticides in Honeybees Using Matrix Solid-Phase Dispersion and Gas Chromatography. *J. Chromatogr. A* **2002**, *982*, 267–273. [CrossRef]
24. Kamel, A. Refined Methodology for the Determination of Neonicotinoid Pesticides and Their Metabolites in Honey Bees and Bee Products by Liquid Chromatography–Tandem Mass Spectrometry (LC-MS/MS)†. *J. Agric. Food Chem.* **2010**, *58*, 5926–5931. [CrossRef]

25. Kiljanek, T.; Niewiadowska, A.; Semeniuk, S.; Gaweł, M.; Borzęcka, M.; Posyniak, A. Multi-Residue Method for the Determination of Pesticides and Pesticide Metabolites in Honeybees by Liquid and Gas Chromatography Coupled with Tandem Mass Spectrometry—Honeybee Poisoning Incidents. *J. Chromatogr. A* **2016**, *1435*, 100–114. [CrossRef]
26. Balsebre, A.; Báez, M.E.; Martínez, J.; Fuentes, E. Matrix Solid-Phase Dispersion Associated to Gas Chromatography for the Assessment in Honey Bee of a Group of Pesticides of Concern in the Apicultural Field. *J. Chromatogr. A* **2018**, *1567*, 47–54. [CrossRef]
27. Tosi, S.; Costa, C.; Vesco, U.; Quaglia, G.; Guido, G. A 3-Year Survey of Italian Honey Bee-Collected Pollen Reveals Widespread Contamination by Agricultural Pesticides. *Sci. Total Environ.* **2018**, *615*, 208–218. [CrossRef]
28. Valverde, S.; Ares, A.M.; Bernal, J.L.; Nozal, M.J.; Bernal, J. Determination of Neonicotinoid Insecticides in Bee Products by Using Ultra-High-Performance Liquid Chromatography–Tandem Mass Spectrometry. *Pestic. Toxicol.* **2021**, 27–52. [CrossRef]
29. Butovskaya, E.; Gasparini, M.; Angelone, B.; Cancemi, G.; Tranquillo, V.; Prestini, G.; Bosi, F.; Menotta, S. Occurrence of Glyphosate and Other Polar Pesticides in Honey from Lombardy and Emilia-Romagna Regions in Italy: Three-Year Monitoring Results. *Foods* **2023**, *12*, 4448. [CrossRef] [PubMed]
30. Jesús, F.; Rosa García, A.; Stecconi, T.; Cutillas, V.; Rodríguez Fernández-Alba, A. Determination of Highly Polar Anionic Pesticides in Beehive Products by Hydrophilic Interaction Liquid Chromatography Coupled to Mass Spectrometry. *Anal. Bioanal. Chem.* **2023**, *416*, 675–688. [CrossRef] [PubMed]
31. Murcia-Morales, M.; Vejsnæs, F.; Brodschneider, R.; Hatjina, F.; Van der Steen, J.J.M.; Oller-Serrano, J.L.; Fernández-Alba, A.R. Enhancing the Environmental Monitoring of Pesticide Residues through Apis Mellifera Colonies: Honey Bees versus Passive Sampling. *Sci. Total Environ.* **2023**, *884*, 163847. [CrossRef]
32. Ozols, N.; Brusbārdis, V.; Filipovičs, M.; Gailis, J.; Radenkovs, V.; Rubene, B.; Zagorska, V. Pesticide Contamination of Honey-Bee-Collected Pollen in the Context of the Landscape Composition in Latvia. *Toxics* **2024**, *12*, 862. [CrossRef]
33. Rampazzo, G.; Nobile, M.; Carpino, S.; Chiesa, L.; Ghidini, S.; Gazzotti, T.; Panseri, S. Detection of Glyphosate, Glufosinate, and Their Metabolites in Multi-Floral Honey for Food Safety. *Food Addit. Contam. Part A* **2025**, *42*, 213–222. [CrossRef]
34. Schäfer, A.-K.; Vetter, W.; Anastassiades, M. Analysis of Highly Polar Anionic Pesticides in Food of Plant and Animal Origin by Ion Chromatography and Tandem Mass Spectrometry with Emphasis on Addressing Adverse Effects Caused by Matrix Co-Extractives. *Anal. Bioanal. Chem.* **2024**, *416*, 4503–4517. [CrossRef]
35. Tóth, E.; Tölgyesi, Á.; Bálint, M.; Ma, X.; Sharma, V.K. Separation of Fosetyl and Phosphonic Acid in Food Matrices with Mixed-Mode HPLC Column Coupled with Tandem Mass Spectrometric Detection and Method Application to Other Highly Polar Pesticides. *J. Chromatogr. B* **2021**, *1189*, 123083. [CrossRef]
36. Verdini, E.; Pacini, T.; Orsini, S.; Sdogati, S.; Pecorelli, I. Chromatographic Comparison of Commercially Available Columns for Liquid Chromatography in Polar Pesticide Detection and Quantification Using a Score-Based Methodology. *Foods* **2024**, *13*, 3131. [CrossRef]
37. European Parliament and Council, Regulation 2017/625, Official Controls and Other Official Activities Performed to Ensure the Application of Food and Feed Law, Rules on Animal Health and Welfare, Plant Health and Plant Protection Products. Available online: https://data.europa.eu/eli/reg/2017/625/oj (accessed on 15 March 2017).
38. Cantwell, H. *Eurachem Guide: The Fitness for Purpose of Analytical Methods—A Laboratory Guide to Method Validation and Related Topics*, 3rd ed.; European Cooperation for Accreditation: Paris, France, 2025. Available online: http://www.eurachem.org (accessed on 8 April 2025).
39. *UNI CEN EN ISO/IEC 17025:2018*; General Requirements for the Competence of Testing and Calibration Laboratories. International Organization for Standardization, ISO: Geneva, Switzerland, 2018.
40. EU Reference Laboratories for Residues of Pesticides. Quick Method for the Analysis of Numerous Highly Polar Pesticides in Food Involving Extraction with Acidified Methanol and LC-MS/MS Measurement. In *II Food of Animal Origin (QuPPE-AO-Method), Version 3.3.* Available online: https://www.eurl-pesticides.eu/userfiles/file/EurlSRM/EurlSrm_meth_QuPPe_AO_V3_3.pdf (accessed on 8 April 2025).

Journal of *Xenobiotics*

MDPI

Review

Innovative Analytical Approaches for Food Pesticide Residue Detection: Towards One Health-Oriented Risk Monitoring

Alexandra Andreea Botnaru [1], Ancuta Lupu [2], Paula Cristina Morariu [3,*], Alin Horatiu Nedelcu [4], Branco Adrian Morariu [5], Maria Luisa Di Gioia [6], Vasile Valeriu Lupu [2], Oana Maria Dragostin [7], Ioana-Cezara Caba [8], Emil Anton [5], Madalina Vieriu [9] and Ionela Daniela Morariu [1]

1 Department of Environmental and Food Chemistry, Faculty of Pharmacy, Grigore T. Popa University of Medicine and Pharmacy, 700115 Iasi, Romania; botnaru.alexandra@yahoo.com (A.A.B.); ionela.morariu@umfiasi.ro (I.D.M.)
2 Department of Pediatrics, Faculty of General Medicine, Grigore T. Popa University of Medicine and Pharmacy, 700115 Iasi, Romania; ancuta.ignat1@umfiasi.ro (A.L.); vasile.lupu@umfiasi.ro (V.V.L.)
3 Department of Internal Medicine, Faculty of General Medicine, Grigore T. Popa University of Medicine and Pharmacy, 700115 Iasi, Romania
4 Department of Morpho-Functional Science I, Faculty of General Medicine, Grigore T. Popa University of Medicine and Pharmacy, 700115 Iasi, Romania; alin.nedelcu@umfiasi.ro
5 Faculty of General Medicine, Grigore T. Popa University of Medicine and Pharmacy, 700115 Iasi, Romania; morariubranco@gmail.com (B.A.M.); emil.anton@umfiasi.ro (E.A.)
6 Dipartimento di Farmacia, Salute e Scienze della Nutrizione, Università della Calabria, Arcavacata di Rende, 87036 Cosenza, Italy; ml.digioia@unical.it
7 Research Centre in the Medical-Pharmaceutical Field, Faculty of Medicine and Pharmacy, "Dunarea de Jos" University of Galati, 800010 Galati, Romania; oana.dragostin@ugal.ro
8 Department of Toxicology, Faculty of Pharmacy, Grigore T. Popa University of Medicine and Pharmacy, 700115 Iasi, Romania; ioana-cezara.caba@umfiasi.ro
9 Department of Analytical Chemistry, Faculty of Pharmacy, Grigore T. Popa University of Medicine and Pharmacy, 700115 Iasi, Romania; madalina.vieriu@umfiasi.ro
* Correspondence: morariu.paula-cristina@email.umfiasi.ro

Abstract: The increasing use of pesticides in agricultural products raises concerns over food safety. Furthermore, uncontrolled pesticide usage on food products can lead to residual levels that exceed the maximum residue limits (MRLs) and are potentially harmful to human health. Long-term consumption of food contaminated with pesticides can contribute to the buildup of toxic substances in the body, which has negative health effects. Advanced analytical techniques are essential to ensure the accurate and effective monitoring of pesticide residues. To ensure adherence to legal requirements, it is essential to employ rapid and accurate methods for detecting these contaminants. This review outlines current advancements (2020–2025) in the assessment of pesticide residues in diverse food matrices, including sample preparation procedures and detection methods. This review provides a standardized comparative analysis of analytical methods for detecting pesticide residues, emphasizing their advantages and limitations, sensitivity, costs, and applicability to complex food matrices, while evaluating its findings through the One Health approach, linking residue evaluation to cumulative exposure and regulatory standards. This study provides practical guidelines for laboratories and regulators while delineating research requirements for more environmentally friendly, rapid, and sensitive residue analysis in accordance with One Health-oriented risk assessment.

Keywords: pesticide residues; monitoring; food safety; chromatography; sample pretreatment; analytical methods

1. Introduction

Food safety concerns are escalating, posing significant risks to human health. Food contaminants, including heavy metal ions, foodborne viruses, pesticides, mycotoxins, and antibiotics, have emerged as insidious threats to human life and health via dietary consumption and bioaccumulation [1–3].

The extensive use of pesticides has contaminated food, feed, water, air, soil, and other sources, posing serious risks to food safety and public health [4]. Approximately 0.1% of the applied pesticide reaches its intended target, while the remaining amount turns into a pollutant in soil and the ecosystem, thus compromising future food sources [5].

Pesticides, owing to their high accumulative nature and persistence, mitigate crop loss and enhance production, which is highly beneficial for farmers. Data indicate that rice production, which sustains almost 50% of the population, has tripled due to pesticide use, while wheat production has risen by approximately 160%. Pesticides allow farmers to cultivate a greater yield on reduced land, thus contributing to the mitigation of deforestation. However, given that the majority of pesticides are toxic, we cannot overlook their detrimental consequences [6].

One Health is characterized as a solid, integrative strategy that aims to sustainably harmonize and enhance the health of humans, animals, and ecosystems. It acknowledges the interconnectedness and interdependence of human health, domestic and wild animal health, plant health, and the broader ecosystem [7]. Toxic chemicals may be identified in air, water, soil, vegetation, food, and animal feed. These residues infiltrate flora and fauna, accumulating in both people and animals through the food chain. They endanger our lives and affect our overall well-being while also annihilating beneficial organisms within the ecosystem. Researchers disclosed that most residual pesticides found in soil, agricultural goods, water, and sediment, including illegal compounds still employed for various purposes, have been recognized as potentially detrimental to ecosystems and human health [8]. The alterations in soil microbial processes, soil characteristics, and enzymatic activity due to pesticide applications are significant elements that considerably influence soil production. Disruptions in microbial community composition may result in significant alterations in the cycling of essential nutrients and metals and their consequent absorption by plants. Recent discussions in both scientific and public forums have notably focused on the detrimental effects of conventional pest control practices, especially the extensive use of neonicotinoid pesticides, on key insect pollinators like bees [9,10]. Moreover, socioeconomic factors affect the exposure of at-risk people to pesticide residues through food, housing, and occupational environments, as well as their capacity to obtain safer alternatives. The One Health concept underscores the inseparable connection among human, animal, and environmental health.

Regardless of progress in organic agriculture, current studies have shown that several food products across various regions of the globe have been contaminated by pesticides [11]. An estimated 385 million cases of accidental acute pesticide poisoning occur annually, with approximately 11,000 deaths from these cases. Approximately 44% of farmers worldwide—roughly 860 million people—suffer from pesticide poisoning each year. Moreover, 20,000 people in developing nations perish from pesticide-contaminated food each year [12]. Studies indicate that several pesticides impact the neurological system, elevating the risk of neurodegenerative disorders. Exposure to pesticides is associated with various health effects, ranging from mild symptoms such as nausea and headache to more severe conditions, including endocrine disruption, cardiovascular disease, cancer, diabetes, congenital anomalies, compromised immune and reproductive systems, Alzheimer's disease, and Parkinson's disease [8,13]. Neurodevelopmental toxicity is particularly concerning, as symptoms may present gently and with a delayed beginning following early-life exposure, highlighting the considerable neurotoxic potential of pesticide residues and underscoring the necessity for their thorough consideration in food safety evaluations [14]. Therefore, the potential toxicity of pesticides is a public health issue since it is applicable in various environments, including agricultural areas, roadways, and residential, educational, and leisure institutions [15].

Various techniques have been employed for the quantification of pesticide residue analysis. Analytical technique development consists of two primary steps, which are sample preparation (including matrix pretreatment) and analytical determination of target chemicals [16]. Sample preparation is the most critical stage before instrumental analysis. The progress of extraction methods has resulted in improvements in analytical processes, reducing the complexity of sample treatment while boosting the accuracy and precision of analysis [17]. The analytes must be separated from complex matrices using appropriate extraction, cleaning, and/or preconcentration techniques [18]. Extraction can be accomplished utilizing a solid, liquid, gas, or supercritical fluid as an extractant, contingent upon the following five fundamental chemical properties: hydrophobicity, vapor pressure, solubility, molecular weight, and acid dissociation [19]. Microextraction is garnering more interest due to its sensitivity, simplicity, and efficiency in analyzing complicated matrices. It may be utilized for a wide array of structural configurations and polarity of compounds. Filtration and centrifugation may occasionally be circumvented, using only minimal quantities of organic solvents [20]. Achieving the complete separation of analytes from other sample components through flawless sample preparation is highly challenging and unfeasible in all but a select few situations [21].

Instrumental analysis and rapid detection methods are two of the currently described approaches for detecting pesticide residue. Various factors have contributed to the rapid development in the complexity of analytical equipment over the last few decades. Whereas, in the 1970s, a chromatogram was directly printed on thermal paper, yielding minimal information, today a chromatograph connected to a high-resolution mass spectrometer may gather up to 20 spectra/second. Furthermore, as instrument prices decrease, mass spectrometers have gained popularity in both academia and industry [22]. Traditional techniques used in instrumental analysis include gas chromatography (GC), high-performance liquid chromatography (HPLC), and mass spectrometry (MS) [23]. Analytical methods have recently evolved from single to multiple residue analysis, owing to the use of highly selective analytical methods like liquid chromatography tandem mass spectrometry (LC-MS/MS) and gas chromatography tandem mass spectrometry (GC-MS/MS). Such new technologies have substantially simplified analytical processes, allowing for the identification of pesticide residues at trace levels with exceptional accuracy and precision.

Multiresidue techniques enable the analysis of a large number of chemicals with excellent recovery rates while also addressing possible sample interference. Regarding rapid detection techniques, biosensors are the most promising currently employed. A possible approach is the use of sensors employing diverse transduction principles, including fluorescence, colorimetry, and electrochemistry. Biosensors will play a crucial role in detecting pesticide residues in fruits and vegetables [24]. To ensure food safety and environmental sustainability, biosensors have the potential to either replace or enhance conventional analytical methods for monitoring pesticides in agricultural products [25]. As analytical platforms, they may be handled, offering on-site data, which are particularly advantageous for identifying pesticide residues in imported food at control points (border inspections, field testing) [26].

Prior reviews of pesticide detection have mostly focused on methods or matrices. Expanding upon this literature, we present a standardized, decision-oriented comparison of analytical workflows (environmental sustainability, cost, throughput, sensitivity, and efficacy in complex food matrices) while explicitly adopting a One Health perspective by contextualizing recent advancements within regulatory frameworks and linking food residues to environmental and biomonitoring evidence.

2. Pesticide Types, Characteristics, and Regulatory Framework

Organochlorine pesticides (OCPs), carbamates, organophosphates (OPPs), pyrethroids, and neonicotinoids are among the most commonly used types of pesticides in agriculture and pest management, depending on their chemical composition. The main classification criteria for pesticides are their chemical structure and their mode of action or mode of entry, which describe how the pesticide controls or gets rid of the target pest (Figure 1). Every type of pesticide has unique properties and different ways of action [12,27–30].

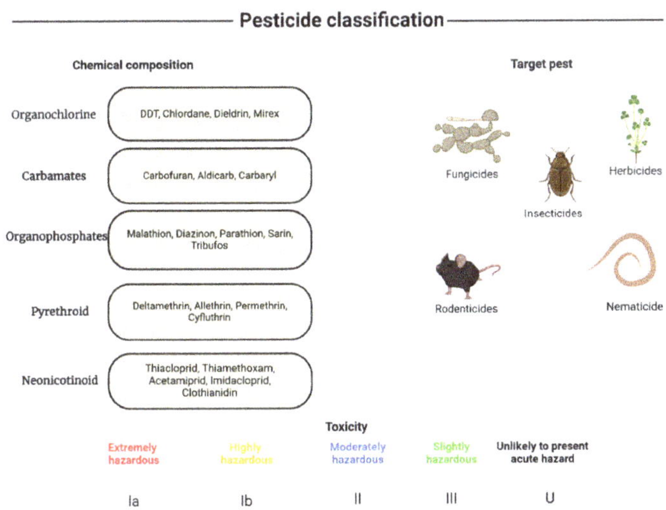

Figure 1. Pesticide classification by chemical structure, targeted pest species, and toxicity (created with BioRender.com).

2.1. OCPs

OCPs are chlorinated hydrocarbons that are extensively utilized in crops for pest control [31]. OCPs are classified as high-persistence organic pollutants in the environment. Formerly used to combat typhus and malaria, these insecticides are currently prohibited in a large number of countries [32]. Due to their persistence in the environment, ability to bioaccumulate in the food chain, and ability to accumulate in human adipose and other tissues, the extended use of organochlorines such as p,p′-dichlorodiphenyltrichloroethane (DDT), its metabolite dichlorodiphenyldichloroethylene (DDE), heptachlor epoxide, hexachlorobenzene, β-hexachlorocyclohexane, oxychlordane, and trans-nonachlor is particularly relevant to human health [33]. The chemical structure for selected OCPs is represented in Figure 2 [34].

From a toxicological perspective, OCPs are primarily linked to chronic health risks. Experimental studies provide evidence that these chemicals possess carcinogenic potential by interacting with steroid signaling pathways and promoting the growth of breast cancer cells that exhibit hormone receptors [35]. Freire et al. distinctly identified a significant correlation between blood levels of certain organochlorine pesticides and elevated total T3 in children, indicating possible thyroid-disrupting effects [36]. The GerES V research indicates a sustained reduction in plasma concentrations of organochlorine pesticides in German children and adolescents, while these enduring compounds remain detectable. Despite a minimal overall health risk, geographical disparities and food-related exposure pathways, including fish intake and breastfeeding, underscore the necessity for continuous monitoring [37].

Due to their significant persistence, bioaccumulation, and mechanisms of endocrine disruption and carcinogenicity, organochlorine pesticides (OCPs) constitute one of the most hazardous pesticides, with health effects mostly due to chronic exposure rather than acute toxicity. Moreover, soil residues continue to be a source of environmental re-emission even decades post phase-out [30,38,39].

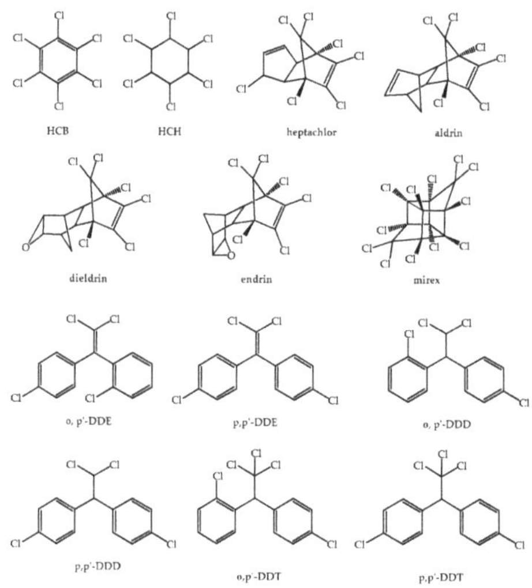

Figure 2. Chemical structure of OCPs. Modified and adapted after Adeyinka et al. [40].

2.2. OPPs

The vast quantity of OPPs that persist in food, drinking water, soil, and air can enter the human body by oral, skin, inhalation, or ocular contact, posing health risks. Chemical structures of widely used organophosphate insecticides are presented in Figure 3 [41].

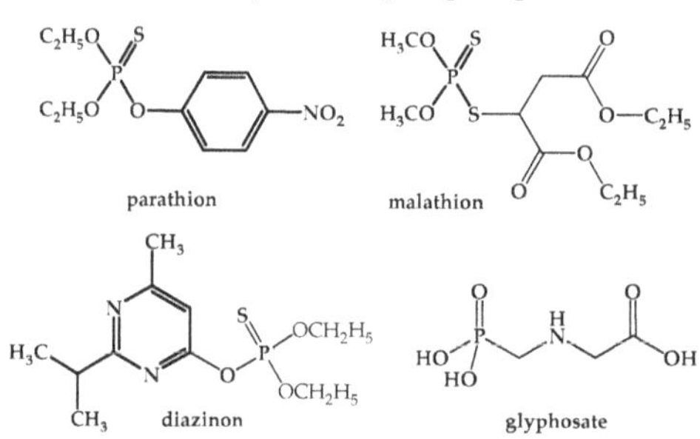

Figure 3. Chemical structures of commonly used OPPs. Modified and adapted after Mdeni et al. [42].

Only 10–15% of OPPs used in the field are effective, and a large number of residual OPPs are released into the aquatic environment because the majority of them are very soluble. This highlights a major worry about the high toxicity of parent OPPs and their metabolites [43]. Parathion, endogenous phosphorus, malathion, dimethoate, and trichlorfon are the most prevalent organophosphates employed as insecticides in the management of plant pests [44]. Ethyl parathion is one of the most extensively utilized pesticides in both agricultural and non-agricultural sectors. Conversely, according to the EPA, ethyl parathion is documented as one of the most dangerous substances in its registry [45].

2.3. Carbamates

Carbamates are a large class of compounds that include carbamic acid esters and thioesters (Figure 4). They are usually soluble in water and polar organic solvents and are available commercially as wet powders, dust granules, and emulsion concentrates. Soil microbes typically degrade these chemicals within three to five weeks [38].

Carbamates are insecticides that inhibit the enzyme AChE, allowing acetylcholine to accumulate in the nervous system. This may cause symptoms like sweating, excessive salivation, clouded vision, and, in extreme instances, paralysis and breathing difficulties [12,46].

These chemicals affect male fertility by disrupting hormonal regulation and influencing sperm production [12]. Exposure to carbamates during infancy has been associated with negative health outcomes and has garnered significant attention [47]. Furthermore, several carbamates are believed to possess carcinogenic and mutagenic properties [48].

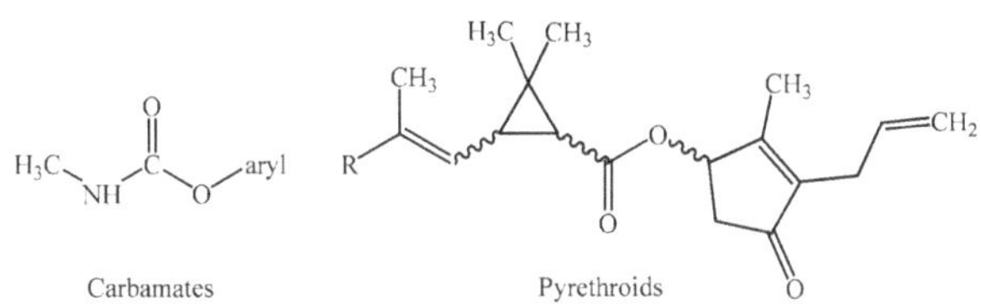

Figure 4. General structure of carbamate and pyrethroid pesticides. Modified and adapted after Hassaan and El Nemr and Tomasevic et al. [30,49].

2.4. Pyrethroids

Pyrethroids are natural insecticides obtained from the pyrethrum extracts of chrysanthemum flowers, namely pyrethrin, found in Kenya [30]. Synthetic chemical compounds (Figure 4) called pyrethroids are well-known for their ability to effectively control a variety of insect pests. Their strong insecticidal qualities make them useful in pest control and agriculture [12,26,38,50].

Pyrethroids may lead to contamination of the food chain, resulting in the bioaccumulation of these pesticides in animal-derived products, including meat, fish, milk, and honey [51]. Even while pyrethroids are usually less hazardous to people than OPPs, there are still risks associated with them, particularly when exposure is extended. Pediatric patients are a vulnerable demographic subjected to pesticide exposure through multiple pathways [15]. Recent epidemiological and longitudinal research demonstrates that prenatal exposure to pyrethroid pesticides correlates with a heightened risk of autism, developmental delays, and other developmental abnormalities in children [14].

2.5. Neonicotinoids

Recently, neonicotinoids have been the fastest-growing group of insecticides in modern agricultural protection [52]. These insecticides have characteristics comparable to those of nicotine, although they pose less of a risk to humans. Neonicotinoids and nicotine are connected chemically. After the discovery of imidacloprid, various analogs, including the 6-chloro-3-pyridylmethyl moiety, were generated, including acetamiprid, nitenpyram, and thiacloprid. Figure 5 shows the structure of imidacloprid [29,53,54].

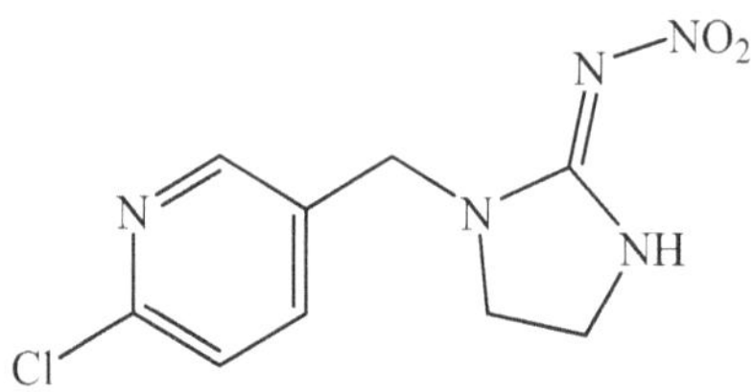

Figure 5. Chemical structure of imidacloprid. Modified and adapted after Yari et al. [55].

Neonicotinoids are effectively utilized in agriculture for crops like maize, cotton, oilseed rape, sunflower, and sugarcane, owing to their superior solubility, chemical characteristics, and selective control, which facilitate their distribution in plants through xylem and phloem transport mechanisms [56].

From a toxicological–mechanistic perspective, neonicotinoids act as agonists of nicotinic acetylcholine receptors, resulting in acute or subacute neurotoxic effects in target insects and, following sufficient exposure, in non-target species [50,52]. Their ecotoxicological impact is thoroughly documented as follows: non-target species, particularly pollinators and aquatic invertebrates, are exposed through various pathways (seed coatings, foliar sprays, contaminated dust, and soils), with field data revealing minimal plant uptake from seed treatments, resulting in the majority of the active ingredient remaining in the soil or as sowing dust. These characteristics, along with environmental persistence, can extend ecosystem-level exposures [38,57].

The environmental outcome of neonicotinoids is influenced by several parameters, such as their water solubility, adsorption to soil and sediment, input–removal dynamics, and existing abiotic conditions. Degradation processes ultimately determine their permanence, serving as the primary mechanism for removal from polluted environments. Neonicotinoids undergo degradation through a combination of abiotic transformations, including photolysis and hydrolysis, and biotic reactions facilitated by microorganisms and plants [58]. Specifically, photochemical degradation under direct sunlight generates several intermediate metabolites, some of which possess biological activity and may lead to prolonged ecotoxicological impacts in aquatic and soil environments [59].

Regarding humans, exposure data—particularly for children—remain limited, but epidemiological and longitudinal findings suggest neurodevelopmental endpoints associated with postnatal exposure to compounds such as clothianidin and imidacloprid [60,61].

2.6. Regulatory Framework

To protect human health and facilitate international commerce, the European Union and the Codex Alimentarius Commission established MRLs for pesticide residues in food products. The European Food Safety Authority (EFSA) performs thorough risk evaluations of chemical residues in food and feed within the European Union, collaborating with state agencies to deliver transparent scientific judgments that inform policy-making [62]. In the United States, regulatory authority is distributed among the EPA (Environmental Protection Agency), USDA (Department of Agriculture), and FDA (Food and Drug Administration), with the EPA responsible for pesticide registration, setting maximum residue levels (MRLs), and assessing environmental effects. The US EPA registers pesticides, establishes MRLs for raw agricultural commodities, and investigates the environmental impact of residues. The European Union has established an extensive legal framework that delineates regulations for the authorization of active chemicals under Regulation 1107/2009, their use in plant protection products, and their allowable residues in food. The MRLs denote the maximum legally permissible concentration of pesticide residue in or on food, provided that a plant protection product is utilized in compliance with its Good Agricultural Practice (GAP) [63]. The goal is to develop MRL values that decrease pesticide residues in food to the lowest possible and acceptable levels for consumers, following sensible application to preserve crops [14]. MRLs and Health-Based Guidance Values (HBGVs) have separate but complementary roles in risk assessment and food safety regulation. HBGVs, encompassing the Acceptable Daily Intake (ADI) and the Tolerable Upper Intake Level (UL), are scientifically defined parameters that specify the amount of a substance that may be consumed daily during a lifetime without posing a substantial health hazard. These numbers are generally based on toxicological data and are employed to assess total dietary exposure from all sources. Conversely, MRLs are regulatory limits set for the presence of residues from substances such as pesticides or veterinary medications in food products. They aim not to indicate safety directly but to ensure that the presence of such residues in food does not lead to consumer exposures exceeding the pertinent HBGVs. Thus, HBGVs function as toxicological benchmarks, whereas MRLs act as regulatory tools designed to ensure that actual exposure remains below permissible thresholds [64]. The cumulative risk assessment is a significant concern, as the MRLs are established for individual residues, whereas food may be contaminated with several pesticide residues [26].

Exposure modeling is an essential element of dietary risk evaluation. It assesses consumer exposure by integrating residue levels with food consumption data and contrasts the values to toxicological parameters such as the ADI and Acute Reference Dose (ARfD). The following two principal methodologies are employed: long-term (chronic) exposure assessment, which utilizes mean residue concentrations and average consumption to calculate the Estimated Daily Intake (EDI), and short-term (acute) exposure assessment, which assesses high-percentile intakes relative to the Acceptable Risk of Dose (ARfD) through the International Estimated Short-Term Intake (IESTI) model [65,66].

A wide range of programs are being conducted to assess, monitor, and reduce consumer exposure to pesticide residues in the food supply. For example, to safeguard the Brazilian populace from significant hazards linked to pesticide-contaminated food, the Brazilian National Sanitary Agency has implemented a comprehensive monitoring program for pesticide residues in fruits and vegetables since 2001. In 2009, an analysis of 20 varieties of fruits and vegetables revealed that 23.2% had pesticide residues, with 14.3% of the samples surpassing the European Union's MRLs [67]. Analytical advancements—multiresidue LC/GC-MS/MS with validated LOQ, identification criteria, and extraction efficiency control—directly guide enforceable maximum residue level decisions under Codex/EU/US regulations and facilitate One Health risk assessment by correlating residues across food, water, biota, and human biomonitoring.

3. Sample Pretreatment

The common approach for detecting and quantifying pesticide levels in samples is based on chromatographic methods (GC, LC, HPLC, UHPLC, SFC) with different detectors, although the sample preparation procedure is often matrix-specific [68]. Various sample preparation approaches have been proposed to extract pesticide residues from food products; nevertheless, we highlight the most effective and thoroughly investigated methods. Contemporary trends in analytical chemistry emphasize the simplification and miniaturization of the analytical process. The traditional liquid–liquid extraction method is hindered by time consumption, reliance on harmful volatile organic chemicals, and the generation of substantial waste solvent volumes. Consequently, to mitigate these constraints, researchers have devised many microextraction techniques [69].

3.1. MAE (Microwave-Assisted Extraction)

MAE is a technique that uses microwave radiation to heat solvents in contact with a sample, facilitating the transfer of analytes from the substrate to the solvent; hence, it provides high sample throughput with minimal solvent usage [70]. This process involves the absorption of electromagnetic waves by solid materials, which converts them into thermal energy. Subsequently, when pressure is exerted on the solid cell wall, it results in cell expansion; as the pressure escalates, the cell fractures, allowing organic pollutants inside the solid cell to leach into the organic solvent [71]. The use of MAE offers several benefits, including a high extraction rate, automation, and the potential for simultaneous sample extractions without interference. In the past few years, the use of microwaves for the extraction of constituents, primarily from plant tissues, has garnered significant study attention [70]. Only thermally stable compounds may be utilized with this technique, which poses the possibility of decomposing temperature-sensitive chemicals, and they must be dissolved in a polar solvent such as water [17,72]. However, its availability can be limited in other laboratories, as it requires an expensive instrument.

Tian et al. adopted this technique to extract mancozeb from fruits and vegetables, requiring just 50 s of pretreatment time. The mean recoveries of mancozeb varied between 81% and 112%. LOD and LOQ were 0.003 and 0.01 mg kg^{-1}, respectively [73]. Zondo et al. implemented this method effectively for the quantification of herbicides in maize crops. The recoveries of herbicides in maize ranged from 80% to 98%. The concentrations of herbicides measured varied from 2.7 to 20.4 μg L^{-1} [71].

3.2. ASE (Accelerated Solvent Extraction)

ASE is a solid–liquid organic component extraction method that operates at high temperatures (50–200 °C) and pressures (10–15 MPa), combining the advantages of high throughput, automation, and little solvent usage [74]. High temperature and high pressure are required to obtain higher extraction rates owing to lower viscosity and surface tension. However, the process also enhances the diffusion rate and solubility in the matrix [75]. This novel sample extraction technology provides various benefits, including cheap extraction costs, decreased solvent and time consumption, and streamlined extraction processes [74]. As a drawback, ASE entails considerable investment in equipment and maintenance, and the assembly and disassembly of sample extraction cells may provide challenges [76].

Currently, honey is produced in a contaminated environment. Consequently, modern honey contains a variety of chemical residues, including pesticides [77]. Upon analysis, over fifty percent of the samples revealed a combination of pesticides [78]. To separate pesticide residues from organic honey samples, two in-line ASE extraction techniques were devised and evaluated. Florisil and PSA (primary secondary amine) were used as interference retainers. The ASE with in-line clean-up is cost-effective and reduces waste creation compared to conventional procedures; by integrating extraction and clean-up in a single step, the time needed for the analysis is reduced [75]. Zhang et al. revealed that the average recovery of target compounds using ASE was 96%, with the majority of compounds falling within the confidence levels [74].

3.3. SPE (Solid Phase Extraction)

SPE techniques have been extensively utilized for the LC analysis of quaternary ammonium pesticide residues. Mixed-mode polymeric SPE cartridges can enhance analyte extraction [79]. Advantages include convenience, cost-effectiveness, simplicity, decreased organic solvent usage, the possibility for multiresidue analysis, and compatibility with different detection methods.

This method offers a variety of sorbents with distinct chemical structures, allowing for alternative extraction methods to accept pesticides with variable physicochemical qualities [80].

An examination of analytical efficiency based on cleanup techniques was carried out for Korean cabbage samples. The percentage of pesticides in these food samples that fell within the proper recovery rate range was 94–99% for SPE [81]. Chemical interferences have diminished over time owing to the enhanced selectivity attained by methods such as SPME, SPE, UHPLC, and complete two-dimensional gas chromatography (GC × GC) [21]. Numerous co-eluting chemicals make it challenging to analyze vegetable oils for pesticides. For instance, SPE with an alumina column or a C18 sorbent is often used [68].

3.4. dSPE (Dispersive Solid Phase Extraction)

The benefit of dSPE is its requirement for minimal laboratory apparatus, specifically a centrifuge and a vortex. It employs reduced sorbent material, necessitates fewer sample quantities, and provides superior interaction between the sorbent and extract compared to traditional SPE [51]. Multiple techniques have been employed to eradicate co-extracted interference from extracts, including freezing centrifugation, SPE, and dSPE. Nonetheless, except for dSPE, other clean-up methods are labor-intensive and need substantial quantities of solvent. Furthermore, certain clean-up methods may lead to the loss of pesticides due to adsorption onto sorbents or during the concentration of the extraction solution [51].

In order to obtain the removal of co-extracted substances, dSPE using various sorbent types was suggested. The hydrophobic character of the sample is maintained by applying C18, while primary–secondary amine sorbents may be utilized to achieve significant retention of fatty acids, organic acids, pigments, and sugars [68]. Clean-up is a required step for identifying pesticides in food that has a high concentration of polyphenols (apple, tea, and broccoli). In a method previously utilized in a study to enhance the precision of polyphenol quantification, the polyphenols were precipitated using d-SPE and polyvinylpolypyrrolidone as part of the sample pre-treatment step before the sample was diluted to remove the matrix impact [68].

Yun et al. examined 272 pesticide residues in food samples, including brown rice, soybeans, green peppers, mangos, and potatoes. The d-SPE technique was employed for the cleanup step, utilizing $MgSO_4$ and PSA sorbents to enhance sample purification [82].

3.5. MSPD (Matrix Solid-Phase Dispersion)

MSPD, introduced in 1989, is an adaptation of SPE that employs a sorbent functioning as an abrasive to create a modified aperture in the solid matrix, facilitating extraction [17].

This extraction has several uses as a sample preparation technique for extracting physiologically active chemicals, naturally occurring components, and other substances from complex biological matrices. MSPD requires the direct mechanical amalgamation of a sample using a sorbent mostly composed of silica bonded to octadecyl groups. The method entails the homogenization of a minimal quantity of a sample [83]. Multiresidue approaches like MSPD have been effectively used to address some of the limitations of the traditional solvent extraction of pesticide residues [84]. This approach requires decreased sample quantity and amount of organic solvent, and it may accomplish preparation, extraction, and fractionation in one step [85].

Kemmerich et al. presented a simplified alternative to the traditional MSPD method, known as balls-in-tube matrix solid-phase dispersion (BiT-MSPD). Using steel balls, this method simplifies sample preparation by carrying out every step inside a closed extraction tube. A total of 133 pesticide residues in fruits (apple, peach, pear, and plum) were successfully quantified using UHPLC-MS/MS. BiT-MSPD is a promising method for pesticide residue analysis since it is quicker, easier, and more effective than traditional MSPD [86].

3.6. SPME (Solid-Phase Microextraction)

Over three decades ago, Pawliszyn and coworkers developed SPME, a flexible sample preparation technology that is still widely used today. The technology allows samples to be extracted and preconcentrated in a single step [87]. SPME is a sample preparation technique that uses minimal quantities of an extraction phase to extract target analytes from examined sample matrices [88]. Additionally, SPME provides the concept of microextraction, but it outperforms other traditional sample preparation techniques, such as liquid extractions and SPE, proving great enrichment, high simplicity, flexibility, and, in many instances, reusability in a manner that is environmentally friendly [89].

Due to the complexity of food matrices, SPME is often conducted from the headspace. This technique allows the investigation of chemicals in solid matrices without solvents and in a shorter time [68]. SPME can be applied to gaseous, liquid, and solid matrices [90]. However, the matrix effect poses a challenge in achieving accurate quantitative results with SPME.

A primary research focus of SPME is the development of functionalized materials as extraction phases to facilitate the selective extraction of target pollutants. SPME now extensively uses covalent organic frameworks, carbon compounds, and metal–organic frameworks as coatings [91]. Hydrophobicity/hydrophilicity, pH, and matrix chemicals have a role in how effectively the analytes of interest may be sorbed onto the SPME fiber [87]. However, one of the primary limitations of adopting those technologies is acquiring pure and well-characterized materials, including the fact that they are not commercially available, which causes problems for research facilities and industries [90].

According to Agatonovic-Kustrin et al. researches used SPME combined with HPLC to analyze four pesticides applied to strawberry crops. Four SPME fibers were examined in the process of developing their methodology. When the pesticide combination was injected directly into the HPLC, the sensitivity was approximately five times lower than when an SPME fiber was used [88].

An SPME-coupled HPLC approach was investigated for determining trace amounts of dicamba, 2,4-dichlorophenoxyacetic acid (2,4-D), and picloram residues simultaneously from complex food samples. The SPME system was developed by immobilizing flavonoid moiety-incorporated carbon dots, and as an effective extractant, functional group-incorporated carbon dots demonstrated desirable sensitivity and selectivity to carboxyl-containing aromatic herbicides [92].

Herbicides containing triazine are frequently used on crops to fight weeds. Agatonovic-Kustrin et al. presented an SPME HPLC-MS technique for analyzing seven triazine herbicides [88].

3.7. SFE (Supercritical Fluid Extraction)

SFE employs a supercritical fluid, an element that displays characteristics of both a gas and a liquid when above its critical point [93]. Due to its cost-effectiveness and non-toxicity, CO_2 is commonly employed as the supercritical solvent in supercritical fluid extraction procedures [94]. SFE provides the capability to modulate the solvent power of CO_2 by altering the density of the supercritical solvent, allowing for more selectivity in the process compared to liquid CO_2 [95].

In comparison to conventional extraction techniques, SFE has advantages like reduced organic solvent usage, enhanced extract selectivity, and abbreviated processing durations. The closed system of SFE restricts contamination and inhibits oxidation and degradation of the extracted substances by excluding air and light [93]. The primary drawback of CO_2 is its non-polar nature, which results in supercritical carbon dioxide primarily extracting non-polar or low-polar molecules. In addition to the necessity for high-purity CO_2 or other supercritical solvents, the demand for substantial capital expenditure is a key issue constraining the application of this extraction method [96].

3.8. QuEChERS

The QuEChERS approach is a quick, easy, cheap, effective, rugged, and safe alternative to traditional sample preparation for the multiresidue analysis of diverse chemicals [97]. The QuEChERS method was first presented as a liquid–liquid distribution technique that included combining $MgSO_4$ and NaCl to separate an aqueous solution from an organic one [82]. This approach is versatile and may be adapted depending on the characteristics of the analytes, the composition of the sample matrix, and the available analytical equipment [98]. The QuEChERS method may be used for a broad range of analytes, including polar, semi-polar, and non-polar pesticide pollutants in various food matrices [18]. One of the most significant advantages of QuEChERS is its minimal equipment needs and cost when compared to other extraction methods. The application of QuEChERS for sample preparation in pesticide analysis has broadened to encompass fatty meals, including products of animal origin [99].

The QuEChERS technique shows great promise for the study of various types of pesticides in foods [100]. Because of how it can be coupled to chromatographic equipment (gas and liquid chromatography), it provides a broad analytical scope with a wide range of sensitivity and selectivity [98]. It is the preferred approach for food analysis since it combines many processes and increases the number of pesticides recovered compared to earlier, more time-consuming extraction procedures [100].

3.9. HF-LPME (Hollow Fiber Liquid-Phase Microextraction)

A novel LPME configuration known as HF-LPME has garnered significant research interest due to its ability to deliver a high analyte preconcentration factor for certain analytes. Moreover, it exhibits exceptional cleaning efficiency, since the HF functions effectively as a filter [101]. The technique must be repeatable and deliver an adequate signal to achieve high sensitivity [102]. This method was primarily developed for the extraction of ionic or polar analytes, including acids, bases, and metals [103]. The efficacy of an HF-LPME approach is often characterized by a pre-concentration factor (PF) or enrichment factor (EF) instead of extraction efficiency [103].

The HF-LPME techniques can be executed in two modes: three-phase and two-phase HF-LPME methods. The initial technique involves an aqueous–organic–aqueous system where the immobilized organic solvent or supported liquid membrane is subjected to two aqueous phases of the sample solution, with the aqueous acceptor phase situated within the hollow fiber. In the two-phase mode, a water-immiscible solvent is infused into the lumen of the HF, functioning as an acceptor phase [104]. Both two-phase and three-phase HF-LPME have been employed for the extraction of environmental pollutants and toxins, with this section focusing on recent advancements and uses of the approach in this domain [105].

HF-LPME is an eco-friendly sample preparation method necessitating just a few microliters of organic solvent per sample. HF-LPME facilitates significant enrichment and superior sample purification from biological and environmental components [106]. Limited publications exist about HF-LPME of foods and drinks. This is unexpected, considering that the approach is very appropriate for the extraction of pesticides. Nonetheless, the published articles unequivocally illustrate the potential of HF-LPME [106].

3.10. DLLME (Dispersive Liquid–Liquid Microextraction)

DLLME was developed in 2006 as a fast and inexpensive microextraction method with excellent analyte recovery and enrichment factors, which is particularly appropriate for extracting partially hydrophobic compounds from aqueous solutions [18].

DLLME is a tri-solvent system wherein the disperser solvent acts as an intermediary between the sample solution and the extractant, owing to its solubility and miscibility with both components. For an effective DLLME procedure, the volume of the disperser solvent must exceed that of the extractant to provide sufficient distribution throughout the sample solution. However, DLLME is less adaptable than SPE because of the restricted solvent selection criteria and the fact that it is a time-consuming method involving various stages [22].

Petrarca et al. established a more accurate and environmentally friendly approach for determining pesticide residues in soybeans. They coupled a traditional extraction approach with a DLLME stage, using a deep eutectic solvent—camphor–hexanoic acid (1:1 molar ratio)—for microextraction [107]. In a similar study, pesticides were extracted from tropical fruits, and the analytes were preconcentrated using DLLME prior to trace-level determination [18].

Zgoła-Grześkowiak et al. validated a rapid extraction technique for detecting various pollutants in chicken liver samples using LC-MS/MS. This was the first report of a DLLME approach for simultaneously determining numerous contaminants in biological chicken matrices, including aflatoxin B1, pesticides, fluoroquinolones, sulphonamides, and anthelmintics [83].

3.11. SDME (Single Drop Microextraction)

SDME was initially documented in the literature by Liu and Dasgupta in 1995, when a single drop of liquid served as an interface to capture diffusible gas components. Minimal amounts (<10 mL) of extractants are utilized in SDME processes, rendering it a very environmentally friendly analytical approach [19]. To enhance the robustness of the SDME approach, many advancements have been implemented regarding the solvent utilized as an extractant and the design of the equipment for generating or containing the micro-drop [19]. Extraction efficiency may be enhanced by selecting a suitable solvent, minimizing the volume ratio of acceptor micro-droplets to the sample, optimizing the conditions and pH of the donor and receiver stages, and employing auxiliary reagents during the extraction phase to capture the analyte [108].

This extraction technique is recognized as a simple and efficient technique for sample preparation, adeptly extracting and concentrating diverse analytes from complex sample matrices [20]. SDME was employed with GC-MS to analyze OCPs in vegetable samples [109].

3.12. CSDF-ME (Continuous Sample Drop Flow Microextraction)

CSDF-ME was devised to miniaturize and address some constraints of the continuous-flow microextraction (CFME) approach [110]. This method involves the extraction of the analyte by passing droplets of the aqueous sample solution through several microliters of an organic solvent that is immiscible with water [111]. The primary benefits of this approach are minimal utilization of extraction solvents, ease of implementation, elevated enrichment factors, and significant stability of the extraction solvents [112]. This method has been demonstrated to be repeatable and more efficient than alternative LPME (liquid-phase microextraction) methods. This improvement is attributable to a reduction in the steps of the sample preparation techniques [113].

A comparative overview of the extraction techniques discussed in this section, highlighting their main advantages, limitations, and practical considerations, is presented in Table 1.

Table 1. Advantages and limitations of sample preparation techniques in pesticide residue analysis.

Extraction	Advantages	Limitations
MAE	High extraction efficiency Green and eco-friendly Automation Low solvent consumption	Poor extraction performance for non-polar/volatile compounds Not suitable for thermally unstable analytes
ASE	Reduced solvent and time consumption Simple operation and eco-friendly	High cost of equipment and maintenance Handling of extraction cells can be challenging
SPE	Simple, cost-effective, widely available Variety of sorbents for diverse pesticide properties	Risk of co-eluting interferences
dSPE	Requires minimal equipment Efficient cleanup of fatty acids, pigments, and sugars	Risk of analyte loss during cleanup
MSPD	Low sample and solvent consumption Suitable for multi-residue analysis	Requires optimization Sorbent handling
SPME	Solvent-free/minimal solvent Simple and flexible Reusable fibers	Limited commercial availability of specialized fibers
SFE	Reduced solvent use Prevents oxidation/degradation Environmentally friendly	Mainly extracts non-polar compounds Requires high-purity CO_2 Expensive equipment
QuEChERS	Fast and simple Minimal equipment Broad analyte range	May need modifications for fatty/complex matrices
HF-LPME	Very low solvent usage Eco-friendly	Limited mainly to polar analytes
DLLME	Suitable for hydrophobic analytes Fast Low cost	Restricted solvent choices Involves multiple stages
SDME	Very eco-friendly (few μL solvents) Simple and cost-effective	Reduced reproducibility
CSDF-ME	Minimal solvent use	Limited validation

Extraction efficiency may be assessed by many metrics, including recovery rate (RR), enhancement factor (EF), and extraction recovery (ER). ER assesses the efficacy of an extraction procedure. Despite low extraction efficiency, a high enrichment factor renders the approach viable for real sample analysis [102]. The efficiency of the mentioned extraction and clean-up techniques may be examined in Table 2 by a comparative analysis based on three analytical specifications.

Table 2. Sample pretreatment for detecting pesticide residues in various food matrices (2020–2025).

Pretreatment	Food Matrix	Pesticide	EF	ER	RR (%)	Ref.
MAE	Maize	Atrazine Glyphosate Mesotrione	-	-	80–98	[71]
MAE	Fruits and vegetables	Mancozeb	-	-	81–112	[73]
MAE	Apples	Thiamethoxam	-	-	61–112	[70]
ASE	Soy products	230 pesticides	-	-	70–120	[74]
ASE	Corn flour	Glyphosate	-	-	109.19 ± 8.26	[114]

Table 2. *Cont.*

Pretreatment	Food Matrix	Pesticide	EF	ER	RR (%)	Ref.
SPE	Fruits and vegetables	Fluindapyr + metabolites	-	-	71–118	[115]
SPE	Packed fruit juice	Ametryn Chlorpyrifos Clodinafob-propargyl Fenpropathrin Oxadiazon Diniconazole Penconazole	452–751	45–75	85–101	[116]
dSPE	Orange juice	OPPs	-	-	95.35–110.75	[117]
dSPE-DLME	Strawberries	Hexaconazole Oxadiazon Tebuconazole Clodinafop-propargyl Difenoconazole	365–405	73–81	-	[118]
MSPD	Eggplant Capsicum Apple gourd Cauliflower Sponge gourd	Diafenthiuron Lufenuron Azoxystrobin Difenoconazole Chlorothalonil	-	-	88.5–116.9	[119]
MSPD	Corn	Triazines	-	-	92.6–104.7	[120]
SPME	Tea	OPPs	-	-	73.12–101.20	[121]
SPME	Fruits and vegetables	OPPs	-	-	82.6–118	[122]
HS-SPME	Grapefruit Cucumber	OPPs	-	-	85–118	[123]
SFE	Brown rice	OPPs Pyrethroid Dithiolane	-	-	96.4–105.0	[124]
SFE	Green onion	Acetamiprid Clothianidin Dinotefuran Imidacloprid Thiacloprid Thiamethoxam	-	-	70–120	[125]
QuEChERS	Mandarin Potato Green Pepper Hulled rice Soybean	Triflumezopyrim	-	-	89.7–104.3	[126]
QuEChERS	Vegetables	β-HCH, Υ-HCH Cypermethrin Profenophos 4-Nonylphenol p,p′-DDD, p,p′-DDTs	-	-	69–114	[127]
QuEChERS	Muscle chicken breast fillets	α-endosulfan Cypermethrin Endosulfan sulfate Permethrin DDT	-	-	71.2–118.80	[99]

Table 2. *Cont.*

Pretreatment	Food Matrix	Pesticide	EF	ER	RR (%)	Ref.
DLLME + dSPE	Tomato Lettuce Carrot Celery	Hexaconazole Chlorpyrifos Diazinon Tebuconazole Diniconazole	380–430	79–86	90–103	[128]
DLLME + QuEChERS	Yogurt	OCPs OPPs Dinitroanilines Carbamates and pyrethroids Triazines Chloracetamides Dicarboximides Azoles	5–16	-	70–120	[129]
HF-LPME	Canned drinks	OPPs	-	-	73.6–94.8	[102]
HF-PLM	Rice	OCPs	67.4–73.5%	76.277–81.499	86.0–92.9	[130]
CSDF-ME + UAE	Apple Strawberry Cucumber Tomato	OPPs	21–205%	-	83.0–108.0	[131]
CSDF-ME	Juice	OPPs	102–380 μg L^{-1}	17–51	83–105	[110]
CSDF-ME	Grape juice	OPPs	510–960	25.5–48.0%	90–110	[112]

UAE, ultrasound-assisted extraction; HF-PLM, hollow fiber-protected liquid-phase microextraction; HS-SPME, headspace solid-phase microextraction.

4. Detection Methods

Pesticide detection methods differ based on laboratory facilities, ensuring accessibility and relevance to varying resource levels. Laboratories with constrained infrastructure utilize quick and lab immunochemical assays, such as test strips or basic ELISA kits, which deliver prompt outcomes at a minimal cost, although with reduced sensitivity and specificity. Laboratories with moderate resources can employ improved ELISA and other immunochemical techniques, allowing for more precise detection and quantification of many chemicals concurrently, although they still need fundamental equipment and skilled workers. In adequately equipped laboratories, chromatography and mass spectrometry techniques, such as LC-MS/MS or GC-MS, provide elevated sensitivity and specificity, facilitating multi-residue analysis and advanced standardization; nonetheless, they entail significant expenses and need expert people. The selection of a detection technique is contingent upon resource availability, existing infrastructure, and testing objectives, hence assuring an extensive and adaptable application [132,133].

A comprehensive schematic illustration of the pretreatment and detection methodologies employed in pesticide residue analysis, categorized by cost-effectiveness and time requirements, is presented to facilitate comparison between rapid, economical techniques and more costly, time-intensive laboratory-based methods (Figure 6).

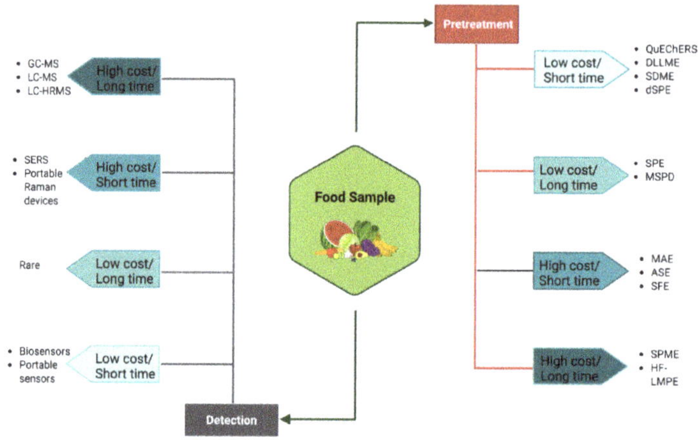

Figure 6. Overview of food sample pretreatment and detection methods classified according to cost-efficiency and time requirements (created with Biorender.com).

4.1. Chromatography Techniques

Affordable detection approaches are essential for enhancing pesticide residue monitoring in low- and middle-income countries. In several areas, access to high-performance equipment like LC–MS/MS is constrained by substantial acquisition and maintenance costs, the necessity for a specialist, and infrastructural needs. Consequently, cost-effective and accessible screening instruments provide an effective first level of surveillance. In addition to ELISA-based screening, several countries have also implemented monitoring programs and pesticide residue studies employing chromatographic techniques for confirmatory analysis (tropical fresh fruit in Brazil, fresh products in South Africa, and dairy milk in India) [134–136].

Instrumental parameters including sensitivity, selectivity, linearity, precision, accuracy, and calibration can influence the accuracy and precision of analyte concentration. Optimizing these settings is essential for achieving the most accurate results. The selection of the most suitable instrument-optimized method for quantifying analyte concentration is contingent upon several parameters, including the analyte's characteristics, sample concentration, sample matrix, solvent purity, needed sensitivity, and the desired accuracy and precision levels [137].

Alongside conventional GC-MS library matching methods, significant advancements in machine learning (ML) methodologies have emerged, facilitating the enhanced qualitative identification of chemicals and more reliable measurement of concentrations. These algorithms can identify intricate patterns in spectra, minimizing errors linked to analogous chemicals, and they can swiftly adjust to new data sets, hence providing enhanced flexibility and performance relative to traditional methods [138]. Artificial intelligence (AI) and ML have had significant progress and are swiftly gaining prominence in several predictive domains due to their capabilities, precision, and rapidity [139]. In pesticide-residue analysis, AI/ML-assisted spectrum interpretation is anticipated to enhance robustness, mitigate analyst bias, and facilitate high-throughput monitoring in intricate food matrices.

HPLC has become a key tool in pesticide analysis as the agricultural sector seeks to produce more polar pesticides that have reduced volatility and are easily degradable [22]. However, a notable drawback of HPLC is its high solvent consumption. Other chromatographic procedures need multistep sample preparation, which is laborious and time-consuming. Furthermore, these instruments are expensive and have a significant carbon footprint, rendering them unsuitable as available sensors due to calibration complications [140]. The increasing application of highly polar and ionic herbicides such as glyphosate and 2,4-D, as well as the development of HPLC coupled with MS, has made this approach more widely used in pesticide analysis. However, GC remains the usual technique for analyzing semi-volatile and non-polar pesticides [22].

To improve selectivity, MS was subsequently integrated with GC [141]. Various studies have demonstrated the efficacy of LC-MS/MS and GC-MS, respectively [142–145]. GC, GC-MS, and GC-MS/MS are frequently employed due to their superior separation efficiency, selectivity, and identification capabilities of mass spectrometry [17]. Moreover, the diverse sensitive detectors integrated with GC, including the nitrogen phosphorus detector (NPD), flame ionization detector (FID), flame photometric detector (FPD), and electron capture detector (ECD), have enhanced the detection and quantification of pesticides. The ECD is particularly effective for organochlorine pesticides, the NPD for organophosphorus and nitrogenated pesticides, and the FPD for sulfur and phosphorus pesticides [17]. The utilization of a gas chromatography flame photometric detector (GC-FPD) is very successful in identifying organophosphorus pesticides, providing enhanced accuracy and superior experimental results [146]. Despite the initial lack of uniform nomenclature among chromatographers, matrix effects have inevitably influenced several GC investigations [21].

LC–MS is extensively utilized, as it enables the simultaneous analysis of several pesticides, their metabolites, and degradation products in a single run. The primary drawback of LC–MS is the high cost of the instrument, together with its operational and maintenance expenses [141]. Numerous compounds in plant-based foods that are infrequently examined or challenging to identify, such as highly polar, non-volatile, or thermally unstable pesticides, can be swiftly and efficiently detected using the LC-MS technique, including those unsuitable for GC. Advancements in LC-MS/MS now enable the detection of pesticide residues in intricate matrices such as fruits, vegetables, cereals, and animal-derived products [17]. Its primary advantages are very high sensitivity, low LOD, and excellent selectivity [141]. Nonetheless, these techniques exhibit the following issues: The extraction and purification process entails numerous analytical procedures that are challenging and time-intensive. Recovery rates are often low and inconsistent, and the sample preparation or analysis lacks cost-effectiveness. A significant volume of hazardous solvents, including acetonitrile, methanol, and methylene chloride, is utilized as extracting agents and in liquid chromatography mobile phases, posing environmental risks [147].

Recently, high-resolution mass spectrometers (HRMS), including Orbitrap and time-of-flight (TOF) analyzer systems, have enhanced the application of mass spectrometry in analytical methods due to their exceptional capacity to detect a theoretically infinite number of compounds in full-scan mode, along with their structural information [79]. LC-HRMS provides the ability to simultaneously monitor an infinite array of compounds [148]. This approach is highly beneficial for the concurrent measurement of unidentified substances, including metabolites and transformation products in environmental samples, and it is utilized for untargeted analysis and suspect screening [79]. Moreover, LC-HRMS is an effective instrument for identifying unidentified compounds in complex matrices. HRMS is optimal for non-targeted food safety testing owing to its superior mass accuracy, resolution, scan speed, and sensitivity in full-scan mode. This has resulted in the creation of rapid and extensive multi-residue screening techniques, such as LC/Q-TOF-HRMS screening, for more than 600 multi-class compounds. In regulatory contexts, HRMS screening is facilitated by identification criteria and the Screening Detection Limit concept, establishing LC–HRMS as a versatile tool for both targeted quantification and exploratory contaminant detection in complex food matrices [149,150]. Omics-based methodologies, such as metabolomics and proteomics, provide robust tools for the untargeted assessment of pesticide effects and residue patterns, facilitating a comprehensive knowledge of exposure routes and ecotoxicological consequences [151].

Supercritical fluid chromatography (SFC) substitutes the majority of liquid mobile phases with high-density compressed gas for the separation of intricate mixtures. Carbon dioxide (CO_2) is predominantly employed as the mobile phase in SFC because of its advantageous characteristics, including a low critical point, little toxicity, and low flammability. CO_2 does not harm hardware and may be used with many liquid organic solvents [48]. In comparison to HPLC, UPLC, and GC, it diminishes the volume of organic solvents and analysis duration, aligning with the principles of environmental conservation and green chemistry [152]. The SFC-MS/MS method is commonly employed for the separation of non-volatile or thermally unstable pesticides and for quantifying chiral or achiral chemical compounds in biological samples, owing to its advantages in speed, sensitivity, and cost-effectiveness [17].

The comprehensive monitoring of pesticide exposure requires the detection of both parent chemicals and their metabolites, employing sensitive analytical methods capable of identifying both at minimal levels. Upon release into the environment, a pesticide can go through metabolic processes, resulting in the simultaneous existence of the parent molecule and its metabolites, which may have toxicological significance. European and national reference laboratories should employ suspicious screening methodologies to assess the effects of pesticide metabolites. Furthermore, authorities might provide access to metabolite standards to facilitate the accurate identification and quantification of important pesticide metabolites [153]. Global concerns over environmental contamination by fluorinated compounds, particularly per- and polyfluoroalkyl substances (PFAS), have arisen due to their established detrimental effects on human health, wildlife, and ecosystem integrity [154]. Although methods for PFAS detection are available, the analysis is challenging, requiring an intensive procedure and expensive equipment. Various analytical methods are effective tools for separation and fractionation in PFAS analysis, including GC, LC, and HPLC. GC or LC is frequently coupled with MS, or tandem mass spectrometry (LC-MS/MS, GC-MS), for the detection of both particular and non-specific per- and PFAS [155].

The primary factors that researchers analyze when evaluating the validity of a technique are the limits of detection (LOD) and limits of quantification (LOQ). LODs represent the minimum concentration of a pesticide that can be identified with acceptable precision and accuracy under defined testing circumstances. Table 3 shows current research in which various chromatographic techniques were effectively employed, together with the LODs and LOQs for the examined pesticide residues.

Table 3. Chromatography techniques used to detect pesticide residues in food products (2020–2025).

Detector	Food Matrix	Pesticide	LOD	LOQ	Ref.
1. LC-MS/MS 2. GC-MS/MS	Honey	Carbendazim Thiabendazole Azoxystrobin Chlorpyrifos Imidacloprid	10.0001–0.0004 mg/kg 0.001–0.004 mg/kg	0.0002–0.0008 mg/kg 0.002–0.008 mg/kg	[156]
LC-MS/MS	Vine leaves	512 pesticides	-	-	[142]
LC-MS/MS	Mandarins	440 pesticides	<0.01 mg kg^{-1}	-	[143]
LC-MS/MS	Pistachio	112 pesticides	0.003 mg/kg	0.01 mg/kg	[157]
LC-Q-TOF/MS	Mango	345 pesticides	-	0.5 to 20 μg/kg	[158]
HPLC	Fruit juice and white wine	Carbofuran Carbaryl Isoprocarb Diethofencarb	0.3 μg/L	-	[159]

Table 3. *Cont.*

Detector	Food Matrix	Pesticide	LOD	LOQ	Ref.
GC	Soybean	No residues of the target pesticides were detected	-	-	[107]
GC	Fruits and vegetables	Mancozeb	0.003 kg^{-1}	0.01 mg kg^{-1}	[73]
GC-MS	Oat flour	Triadimenol Flutriafol λ-cyhalothrin Difenoconazole Azoxystrobin	1.7–12.9 μg kg^{-1}	5.73–43.0 μg kg^{-1}	[145]
GC-MS	Spinach	108 pesticides	0.005–0.01 μg/g	0.01–0.025 μg/g	[160]
GC-MS	Rice	15 pesticides	0.10–1.46 μg kg^{-1}	0.390–4.85 μg kg^{-1}	[161]
GC-FID	Tomato	Spiromesifen	0.0015 μg mL^{-1}	0.006 μg mL^{-1}	[162]
GC-ECD	Milk	OCPs	3.7 to 4.8 μg L^{-1}	12–16 μg L^{-1}	[163]
GC-ECD	Melon Cucumber	Pyraclostrobin Difenoconazole Dimethomorph Azoxystrobin	-	0.01–0.05 mg/L	[164]
GC-FPD	Beet	35 pesticides	0.0047–0.0261 mg/kg	0.0143–0.0790 mg/kg	[146]
GC-MS/MS LC-MS/MS	Vegetables	80 pesticides	0.0004–0.0023 mg kg^{-1}	0.0008–0.0047 mg kg^{-1}	[165]
GC-MS/MS	Lime Lemon	45 pesticides	1.56–25.23 ng/mL	4.72–76.47 ng/mL	[166]
GC-MS/MS	Seafood	44 pesticides	2–3 ng/g	7–10 ng/g	[167]
GC-MS/MS	Rice	Triazophos Dichlorvos Chlorpyrifos Malathion	3.4–5.4 μg/kg	20 μg/kg	[168]
HPLC-MS/MS	Rice	Acetamiprid Parathion Profenofos Bixafen	10 μg/kg	-	[169]
HPLC-FL/UV GC-MS/ECD/NPD	Tomatoes	180 pesticides	5–10 μg/kg	10–20 μg/kg	[170]
HPLC-HRMS	Grapes	92 pesticides	4.88–120.16 μg L^{-1}	14.86–308.01 μg L^{-1}	[171]
UHPLC-MS/MS GC-MS/MS	Kumquat fruits	16 insecticides 7 fungicides 5 acaricides 2 plant growth modulators	-	-	[172]
UHPLC-MS/MS	Fruits and vegetables	Fluindapyr + 5 metabolites	0.0001–0.0002 mg/L	0.0003–0.0006 mg/L	[115]
UHPLC-MS/MS	Apples	Pyraclostrobin	-	0.01 mg kg^{-1}	[173]
UHPLC-MS/MS	Dates	Carbamates	0.01–0.005 μg kg^{-1}	0.003–0.04 μg kg^{-1}	[174]
SFC-MS/MS	Rice Wheat Maize	9 pesticides	0.01–42.9 μg/kg	0.4–101.8 μg/kg	[48]
SFC-MS/MS	Jujube Peach Grape Pear	Metconazole	4.30–95.9 ng/kg	10.5–143.2 ng/kg	[175]
SFC-IM-Q-TOF/MS	Yam Potato	20 pesticides	0.1–8.8 ng/mL	0.8–29.4 ng/mL	[152]
LC-HRMS/MS	Cider	18 pesticides	-	0.2 μg L^{-1}	[176]
LC-HRMS	Cereals and Grains	730 pesticides	-	5–20 μg/kg	[148]
1. GC-MS 2. HPLC-UV	Cereal (wheat, rice, corn)	323 pesticides	1. 0.0025–0.005 mg kg^{-1} 2. 0.003–0.027 mg kg^{-1}	-	[31]

Q-TOF, quadrupole-time-of-flight mass spectrometry; FL, fluorescence; UV, ultraviolet; UHPLC, ultrahigh-performance liquid chromatography; IM-Q-TOF/MS, ion mobility quadrupole time-of-flight mass spectrometry.

In mass spectrometry, quantification can be accomplished with external or internal standards. An external standard depends on a calibration curve established using a pure analyte standard assessed under equivalent experimental circumstances. The advantages include accessibility and the lack of overlap with endogenous species, as the standard is examined independently from the sample. Nonetheless, it is susceptible to variability arising from sample preparation, extraction recovery, and matrix effects, which might modify ionization responses in comparison to the calibration solution. In contrast, internal standardization entails the direct incorporation of a structurally analogous or isotopically tagged substance into the sample, facilitating the adjustment for extraction losses and ionization inconsistencies. This method provides enhanced correction for matrix-dependent variations and increases quantitative accuracy across various sample types when the internal standard closely resembles the analyte's activity [177,178]. In summary, external standardization delivers superior curve precision but is susceptible to matrix-induced variability, whereas internal standardization enables effective correction for these effects but is limited by the requirement for meticulously chosen, reliable, and frequently expensive standards. Consequently, the selection signifies a compromise between accuracy and feasibility, contingent upon the analyte–matrix combination.

Several efforts have been undertaken to mitigate matrix effects. For example, the precision of quantitative analysis may be enhanced by the utilization of internal standards in multi-residue analysis employing QuEChERS and LC-MS/MS for food products [81].

4.2. *Rapid Technologies*

4.2.1. Biosensors

Screening techniques, such as biosensors, must accomplish high throughput in tested samples and a brief analysis period while ensuring adequate detectability with limits of detection lower than MRLs. Nonetheless, pesticide residues are often removed utilizing organic solvents and extensive sample preparation procedures. This presents a significant difficulty for screening technologies, as they often involve selective biomolecules that exhibit specific tolerance to organic solvents [26].

Biosensors are analytical instruments that integrate biological sensing elements with physical and chemical transducers, enabling quantitative or semi-quantitative studies. A biosensor consists of a receptor, a transducer, and a biorecognition element that identifies specific target molecules in a medium [179]. Various biosensors possess distinct physical and chemical characteristics, which can exert particular effects on the target to facilitate identification and signal generation for analysis [180]. They provide a straightforward, economical, and dependable method for detecting pesticides in food matrices to guarantee consumer food safety.

Transducers have a key role in identifying a certain target material for biosensors. The primary role is to transform an analytical signal into an informative reading signal. Consequently, several research domains have assessed various methodologies. The most used transducers are optical (colorimetric and fluorescence), electrochemical, amperometric, and SERS (surface-enhanced Raman spectroscopy). Selecting the appropriate transducer for the biosensor is crucial, particularly in the presence of nanomaterials, since it directly affects the detection sensitivity of the target analyte [179]. For example, OPPs are detected using a variety of techniques, including chemiluminescence, electrochemical, colorimetric, and fluorescence [181]. Table 4 provides several biosensors utilized for the detection of pesticides in food products. Research on multi-residue pesticide biosensors that can simultaneously detect several pesticides is worth mentioning. Zhao et al. created an electrochemical biosensor utilizing nanogold/mercaptomethamidophos to identify 12 organophosphate pesticides in cabbage and apple samples [182].

Establishing specific research and development strategies is essential for the advancement of multi-residue sensors. The selection of detection technology must be precise; electrochemical sensors detect variations in current or potential linked to the binding of target substances, optical sensors utilize fluorescence or absorption characteristics to identify molecules, and biorecognition-based platforms, including enzymes, antibodies, or aptamers, facilitate the selective recognition of distinct residues. Additionally, the functionalization of sensors by the incorporation of specialized receptors or their integration into a multi-analytical chip enables the concurrent detection of several chemicals. Integration with microfluidics facilitates miniaturization and rapid sample analysis, enabling the manipulation of minuscule sample volumes. The utilization of genetically engineered enzymes, highly specific antibodies, microbial cells, aptamers, and molecularly imprinted polymers for biosensor design can significantly enhance sensor selectivity. Moreover, the advancement of "lab-on-a-chip" systems that integrate electrochemistry, optics, and biorecognition enables the swift and concurrent detection of many signals. The validation of sensors by testing on actual samples and comparison with conventional laboratory procedures guarantees the accuracy, sensitivity, and resilience of the system [26,183,184]. Recent research has led to portable devices for rapid pesticide detection in the field, addressing the constraints of laboratory methods. Advancements in technology have enabled smartphones and wearable gadgets with high-resolution cameras and robust image processing capabilities. These devices may now be utilized as conveniently portable and accessible tools for colorimetric analysis. The integration of colorimetric detection with smartphones and wearables has several benefits. Firstly, it obviates the necessity for cumbersome and costly laboratory apparatus, rendering it more economical and accessible, especially in resource-constrained environments or for immediate testing. Furthermore, it facilitates real-time analysis and remote monitoring, enabling rapid reaction and data dissemination [185]. Chen et al. presented a smartphone-based colorimetric sensor for on-site pesticide analysis in agricultural settings, whereas Li et al. described a portable paper-based device combined with a smartphone camera that enabled quick detection at the point of application [186,187].

Smartphone assays may provide preliminary on-site screening, eliminating the need for sample collection and transportation, while promptly producing a sample ID and delivering screening results with a specified false positive/false negative rate [26].

Nanomaterials are often employed in biosensors due to their substantial enhancement in performance, facilitating quicker, more efficient, and cost-effective detection. The enhancement in performance arises from their distinctive optical and electrical characteristics, which yield a high contact surface-to-volume ratio, elevated electrical conductivity, catalytic activity, biocompatibility, and ease of modification with functional groups, including gold nanoparticles (AuNPs), silver nanoparticles (AgNPs), silver nanowires (AgNWs), gold nanorods (AuNRs), gold nanostars (AuNSs), carbon nanotubes (CNTs), copper nanowires (CuNWs), and multi-wall carbon nanotubes (MWCNTs). In addition to pure materials, several hybrid nanostructures have also been examined [179]. Arsawiset et al. developed a paper-based analytical device utilizing CuO nanoparticle nanozymes for the quick detection of malathion in fruits and vegetables, attaining a linear detection range of 0.1–5 mg/L, a detection limit of 0.08 mg/L, and an analysis duration of roughly 10 min [194]. Qin et al. proposed an electrochemical biosensor utilizing a perovskite/AuNPs composite, which markedly improved electron transport and conductivity, enabling ultra-sensitive detection of fenitrothion with a detection limit of 0.034 μg/L in food products [195].

Electrochemical biosensors draw researchers interest due to their high sensitivity, excellent stability, ease of downsizing, low cost, and rapid detection; biosensors based on AChE are particularly promising. Nanomaterials are optimal for the fabrication of these AChE sensors [189]. The basic function of enzyme-based electrochemical biosensors is to detect and analyze changes in electrical signals. This signal is the outcome of an enzymatic reaction that produces or reduces an electroactive species [188]. Because of their high sensitivity and user-friendliness, electrochemical transducers were the most often used [179].

Table 4. Transductor-based biosensors used for pesticide detection in the food matrix.

Transductor	Food Matrix	Pesticide	Characteristics	LOD	Ref.
Electrochemical	-	Paraoxon	20 consecutive measurements → the operational stability = 94.13% Storage stability (60 days) = 70%	0.17 nM	[188]
Electrochemical	Cabbage	Dichlorvos	Good stability	0.23 nM	[189]
Electrochemiluminescence	Rape Pineapple	Methyl parathion	Highly sensitive	-	[190]
Fluorescence	Tea	Quinalphos Thiamethoxam Propargite Hexaconazole	High stability Simultaneous detection of 4 pesticides in a single sample	0.2 ng/mL	[140]
Fluorescence	Orange juice Apple juice	Ethyl parathion	High sensitivity Simpler synthetic protocol	2.40 pM	[45]
Colorimetric	Fruits Vegetables	Chlorpyrifos Profenofos Cypermethrin	High sensitivity Good linear response	0.235 mg/L 4.891 mg/L 4.053 mg/L	[191]
Colorimetric	Pear Rice Cabbage	Parathion	Wide linear range 0.01–50 $\mu g \cdot L^{-1}$	2.04 $ng \cdot L^{-1}$	[192]
SERS	-	Profenofos Acetamiprid Carbendazim	Stable and significantly short measurement time Low detection limit	0.0021 ng mL^{-1} 0.0046 ng mL^{-1} 0.0061 ng mL^{-1}	[193]
SERS	Apples	Methyl parathion	Paper-based substrate Superior reproducibility Good stability and sensitivity	0.011 $\mu g/cm^2$	[44]

SERS, surface-enhanced Raman scattering.

Fluorescence offers benefits over absorbance-based techniques, including sensitivity, selectivity, and a shorter detection time. Because of interactions between fluorophores and surface plasmons in metallic nanostructures, its sensitivity can be up to 100 times higher than that of absorbance methods. Its high selectivity is caused by the fluorescent molecule, which often has many emission spectra [196]. Fluorescence-based optical signaling of organophosphates, compared to alternative approaches, has proven to be more advantageous due to its relative simplicity, overall rapidity, and ultra-sensitivity [45].

Colorimetric transducers are more affordable, portable and lighter, have a smaller sample size, and need less equipment compared to conventional techniques [179]. Gold nanoparticles (AuNPs) have been extensively utilized as effective signal transducer elements in the creation of colorimetric pesticide sensors [5]. Regrettably, most colorimetric analytical platforms employ conventional sample preparation processes, underscoring the necessity to automate and streamline sample pretreatment to enhance the application of these approaches in practical settings [26]. Researchers utilizing this type of transducer have noted challenges, including printing reproducibility, image capture, difficulty in differentiating separate components of a combination, and stability issues [179].

SERS employs precious metal nanoparticles and the principle of electromagnetic field enhancement to induce Raman enhancement in molecules adsorbed on the surface [193]. Currently, it has been utilized in several areas, including food safety, life sciences, environmental monitoring, and the chemical industry, and it may be applicable in the quick detection of pesticide residues [180]. The uniformity and sensitivity of the signal are critical variables for assessing the SERS technique, which is associated with SERS substrates [44].

Currently, the majority of quantitative analyses of pesticide residues on the surfaces of fruits and vegetables utilizing SERS detection technology rely on the linear quantification of a singular, distinctive peak of pesticides. Nonetheless, throughout the detection process, the Raman characteristic peak of pesticide contaminants is susceptible to small shifts caused by nonlinear variables, including instrument and ambient noise [180]. Despite its promise for direct detection of pesticides at low levels in liquid samples or on solid food surfaces, the use of SERS for identifying internalized pesticides in complex solid food matrices remains challenging [141]. The integration of Raman spectroscopy with other detection methods can enhance detection capabilities. Nie et al. utilized the aptamer pesticide structure as the target for Raman spectroscopy, enabling the specific detection of malathion. Alami et al. integrated Raman spectroscopy with an enzyme inhibition technique. Consequently, the advancement of Raman signal enhancement substrates that are appropriate for detecting a broader range of pesticide residues at a reduced cost and with greater stability is essential for facilitating the application of Raman spectroscopy in pesticide residue detection [180,197,198]. SERS is a very promising method for the direct detection of pesticides at trace levels in liquid samples or on solid surfaces after straightforward extraction to enhance analyte concentration [199].

4.2.2. ELISA (Enzyme-Linked Immunosorbent Assays)

Immunoassays are generally recognized as an efficient technique for the quick detection of pesticide residues, utilizing antibodies as the recognition element. They are primarily categorized into colorimetry, optics, and electrochemistry tests based on the signal transductions. Colorimetric immunoassays have been widely favored due to their high specificity, cost-effectiveness, operational simplicity, and rapid reaction time. ELISA is considered the gold standard among colorimetric immunoassays due to its automation, high throughput, and scalability [200].

Detection methods utilizing ELISA may employ a double antibody sandwich technique or a capture approach—both of which are noncompetitive—or a competitive method. In this technique, the antibody and the target analyte generate a complex, and the detection signal is directly proportional to the concentration of the target analyte, often suitable for the detection of macromolecular antigens. Pesticides, being a category of small molecular compounds with singular antigenic characteristics, can exclusively bind to one antibody; hence, the competitive detection approach is typically employed. The detection signal exhibits a negative correlation with the concentration of the target analyte [180].

Abraxis Life Technologies™ (Los Angeles, CA, USA) delivers field and laboratory ELISA testing kits for many pesticides analyzed in diverse matrices specified in the National Environmental Methods Index. Multiple studies have demonstrated the efficacy of Abraxis pesticide kits, particularly for glyphosate analysis [201]. López Dávila et al. demonstrated that ELISA is an analytical method for the rapid identification, control, and monitoring of pesticide residues in tomatoes, sweet peppers, and cucumbers prior to chromatographic analysis (GC or LC) [202]. In another research study, the developed ELISA demonstrated an accuracy of 114% recovery with a 3% coefficient of variation for OPPs/Carbamates and 115% recovery with a 4.0% coefficient of variation for pyrethroids [201].

Researchers created very sensitive and straightforward immunoassays for detecting small quantities of imidacloprid by using phage-borne peptides as alternatives to chemically generated antigens. Two peptides were extracted from phage display libraries, competing with imidacloprid for affinity to the monoclonal antibody 3D11. Optimizations were performed on a phage–enzyme-linked immunosorbent assay (P-ELISA) and two phage time-resolved fluoroimmunoassays (P-TRFIAs), resulting in IC50 values of 0.067, 0.085, and 0.056 ng/mL, respectively, demonstrating almost four times increased sensitivity compared to prior methodologies [203].

A colorimetric ELISA test kit was employed to directly identify organophosphates and carbamates, whereas the analysis of pyrethroids was conducted using paramagnetic particles conjugated to antibodies particularly designed for pyrethroid detection. The samples were also tested by chromatography to validate the favorable results. The ELISA kits facilitated the detection of organophosphate, carbamate, and pyrethroid residues in the collected samples. The evaluated ELISA kits demonstrated quantification capabilities at levels below the detection threshold of the employed chromatographic methods. Linear correlations between the quantified values derived from the chromatographic approach and the findings obtained from the pyrethroid ELISA test kits were noted [202].

Alongside ELISA techniques, lateral flow immunoassay (LFIA) is another traditional immunochemical approach that has been extensively utilized for the on-site detection of residues. LFIA has several benefits, including user-friendliness, little time investment, simplicity, affordability, and the capability for high-throughput screening of multiple target analytes. The predominant LFIA technique utilizes gold nanoparticles (AuNPs) as reporters for colorimetric detection. A lateral flow immunoassay (LFIA) employing a widely specific anti-adamantane monoclonal antibody has been established for five adamantanes, with optical LODs ranging from 0.1 to 10 μg/kg, comparable to ELISA findings. To enhance the sensitivity of LFIA, other labels, such as fluorescent nanoparticles, have been utilized in its synthesis. An ultrasensitive fluorescent lateral flow immunoassay has been developed using a nonspecific monoclonal antibody [204].

5. Conclusions

Several studies have been conducted in recent years to evaluate the pesticide residue levels in food products. The advancement of analytical techniques has resulted in enhanced sensitivity and specificity in detection, facilitated by microextraction technologies that surpass traditional extraction approaches. Simultaneously, chromatographic techniques like HPLC and GC, when combined with high-performance detectors, provide accurate and quantitative detection of pesticide residues. Accurate detection of pesticide levels in various food types requires adequate extraction, clean-up, and enrichment of samples.

Additionally, the advancement of biosensors suggests a potential avenue, providing rapid and effective options for detecting pollutants in food items. They have the potential to revolutionize several food safety applications, enhancing regulatory compliance and safeguarding public health with increased efficacy. Intelligent biosensors will serve as a formidable instrument in enhancing risk management and guaranteeing optimal standards of food safety and quality in the present and future. Despite their advantages, biosensors encounter difficulties, including environmental interferences, signal reproducibility challenges, false positives, stability concerns, elevated material prices, and protracted baseline needs. Nonetheless, they are generally far less expensive and more portable than traditional chromatographic–mass spectrometric techniques, rendering them appealing for rapid screening.

Unfortunately, the limited availability of essential components (antibodies/enzymes for immunoassays, specialized sorbents or SPME fibers, biosensor recognition elements/nanomaterials) may restrict the replication and broad implementation of investigations among the scientific community. We support the fair dissemination of technological advances through open licensing for university research and partnerships with nonprofit laboratories to promote scientific advancement and widespread access to established methodologies.

The opportunities for enhancing food safety by reducing pesticide residues encompass the ongoing reassessment of pesticides, higher use of safer and less toxic alternatives, extensive education and training for producers, international cooperation, and the development of effective pesticide elimination strategies.

From a One Health perspective, current analytical advancements provide integrated surveillance across the food–environment–biomonitoring continuum and enhance cumulative/mixture exposure evaluation essential to biodiversity and antimicrobial resistance. Aligning these metrics with regulatory frameworks provides decisions that are legally acceptable and intervention-oriented, while accounting for socioeconomic drivers promotes effective risk-reduction approaches.

Author Contributions: Conceptualization, A.A.B. and I.D.M.; methodology, I.D.M., M.V., A.A.B., I.-C.C. and O.M.D.; formal analysis, I.D.M. and M.L.D.G.; investigation, A.A.B., B.A.M., A.L. and A.H.N.; writing—original draft preparation, A.A.B., A.L., B.A.M., V.V.L. and M.V.; writing—review and editing, P.C.M., M.L.D.G., O.M.D., I.-C.C., E.A. and I.D.M.; visualization, P.C.M., B.A.M. and A.L.; supervision, I.D.M. All authors have read and agreed to the published version of the manuscript.

Funding: This research received no external funding.

Institutional Review Board Statement: Not applicable.

Informed Consent Statement: Not applicable.

Data Availability Statement: No new data were created or analyzed in this study. Data sharing is not applicable.

Conflicts of Interest: The authors declare no conflicts of interest.

Abbreviations

The following abbreviations are used in this manuscript:

MRLs	Maximum Residue Limits
GC	Gas Chromatography
MS	Mass Spectrometry
LC	Liquid Chromatography
HPLC	High-Performance Liquid Chromatography
OCPs	Organochlorine Pesticides
OPPs	Organophosphates Pesticides
DDT	p,p′-dichlorodiphenyltrichloroethane
DDE	Dichlorodiphenyldichloroethylene
2,4-D	2,4-dichlorophenoxyacetic acid
EPA	Environmental Protection Agency
IARC	International Agency for Research on Cancer
AChE	Acetylcholinesterase
EFSA	European Food Safety Authority
GAP	Good Agricultural Practices
HBGVs	Health-Based Guidance Values
ADI	Acceptable Daily Intake
UL	Tolerable Upper Intake Level
SFC	Supercritical Fluid Chromatography
MAE	Microwave-Assisted Extraction
PFAS	Polyfluoroalkyl Substances
LOD	Limit Of Detection
LOQ	Limit Of Quantification
ASE	Accelerated Solvent Extraction
QuEChERS	Quick, Easy, Cheap, Rugged, Effective and Safe

PSA	Primary Secondary Amine
SPE	Solid Phase Extraction
dSPE	Dispersive Solid Phase Extraction
UHPLC	Ultra-High-Performance Liquid Chromatography
SPME	Solid-Phase Microextraction
MSPD	Matrix Slid-Phase Dispersion
BiT-MSPD	Balls-in-Tube Matrix Solid-Phase Dispersion
SFE	Supercritical Fluid Extraction
HF-LPME	Hollow Fiber Liquid-Phase Microextraction
PF	Pre-concentration Factor
EF	Enrichment Factor
ER	Extraction Recovery
RR	Recovery Rate
DLLME	Dispersive Liquid–Liquid Microextraction
SDME	Single Drop Microextraction
CSDF-ME	Continuous Sample Drop Flow Microextraction
CFME	Continuous Flow Microextraction
LPME	Liquid-Phase Microextraction
UAE	Ultrasound-Assisted Extraction
HF-PLM	Hollow Fiber-Protected Liquid-Phase Microextraction
G-HF-LPME	Graphene-Reinforced Hollow Fiber Liquid-Phase Microextraction
G-HC-LPME	Gas-assisted Hollow Fiber Liquid-Phase Microextraction
HS-SPME	Headspace Solid-Phase Microextraction
NPD	Nitrogen Phosphorus Detector
ECD	Electron Capture Detector
FID	Flame Ionization Detector
FPD	Flame Photometric Detector
HRMS	High-Resolution Mass Spectrometer
Q-TOF/MS	Quadrupole-Time-of-Flight Mass Spectrometry
FL	Fluorescence
SERS	Surface-Enhanced Raman Spectroscopy or Scattering
UV	Ultraviolet
ELISA	Enzyme-Linked Immune Sorbent Assays

References

1. Ouyang, M.; Liu, T.; Yuan, X.; Xie, C.; Luo, K.; Zhou, L. Nanomaterials-Based Aptasensors for Rapid Detection and Early Warning of Key Food Contaminants: A Review. *Food Chem.* **2025**, *462*, 140990. [CrossRef]
2. Scutarașu, E.C.; Trincă, L.C. Heavy Metals in Foods and Beverages: Global Situation, Health Risks and Reduction Methods. *Foods* **2023**, *12*, 3340. [CrossRef]
3. Popa, I.D.; Şchiriac, E.C.; Ungureanu, D.; Cuciureanu, R. Immune Response in Rats Following Administration of Honey with Sulfonamides Residues. *Rev. Rom. Med. Lab.* **2012**, *20*, 63–72.
4. Hashimi, M.H.; Hashimi, R.; Ryan, Q. Toxic Effects of Pesticides on Humans, Plants, Animals, Pollinators and Beneficial Organisms. *Asian Plant Res. J.* **2020**, *5*, 37–47. [CrossRef]
5. Bashir, K.; Shikha, S.; Rattu, G.; Jan, K.; Krishna, P.M.; Pattanayek, S.K. Pesticide Residues and Their Detection Techniques in Foods Using Sensors—A Review. *J. Food Sci. Technol.* **2024**, *62*, 221–239. [CrossRef] [PubMed]
6. Singhal, M.; Jadhav, S.; Sonone, S.; Singh Sankhla, M.; Kumar, R. Microalgae Based Sustainable Bioremediation of Water Contaminated by Pesticides. *Biointerface Res. Appl. Chem.* **2021**, *12*, 149–169. [CrossRef]
7. Hoffmann, V.; Paul, B.; Falade, T.; Moodley, A.; Ramankutty, N.; Olawoye, J.; Djouaka, R.; Lekei, E.; de Haan, N.; Ballantyne, P.; et al. A One Health Approach to Plant Health. *CABI Agric. Biosci.* **2022**, *3*, 62. [CrossRef]
8. Botnaru, A.A.; Lupu, A.; Morariu, P.C.; Pop, O.L.; Nedelcu, A.H.; Morariu, B.A.; Cioancă, O.; Di Gioia, M.L.; Lupu, V.V.; Avasilcai, L.; et al. Balancing Health and Sustainability: Assessing the Benefits of Plant-Based Diets and the Risk of Pesticide Residues. *Nutrients* **2025**, *17*, 727. [CrossRef]

9. Dhuldhaj, U.P.; Singh, R.; Singh, V.K. Pesticide Contamination in Agro-Ecosystems: Toxicity, Impacts, and Bio-Based Management Strategies. *Environ. Sci. Pollut. Res.* **2023**, *30*, 9243–9270. [CrossRef]
10. Falkenberg, T.; Ekesi, S.; Borgemeister, C. Integrated Pest Management (IPM) and One Health—A Call for Action to Integrate. *Curr. Opin. Insect. Sci.* **2022**, *53*, 100960. [CrossRef]
11. Gavahian, M.; Pallares, N.; Al Khawli, F.; Ferrer, E.; Barba, F.J. Recent Advances in the Application of Innovative Food Processing Technologies for Mycotoxins and Pesticide Reduction in Foods. *Trends Food Sci. Technol.* **2020**, *106*, 209–218. [CrossRef]
12. Garud, A.; Pawar, S.; Patil, M.S.; Kale, S.R.; Patil, S. A Scientific Review of Pesticides: Classification, Toxicity, Health Effects, Sustainability, and Environmental Impact. *Cureus* **2024**, *16*, e67945. [CrossRef]
13. Louppis, A.P.; Kontominas, M.G. Recent Developments (2020–23) on the Use of LC in the Determination of Food Contaminants. *Separations* **2024**, *11*, 342. [CrossRef]
14. Botnaru, A.A.; Lupu, A.; Morariu, P.C.; Jităreanu, A.; Nedelcu, A.H.; Morariu, B.A.; Anton, E.; Di Gioia, M.L.; Lupu, V.V.; Dragostin, O.M.; et al. Neurotoxic Effects of Pesticides: Implications for Neurodegenerative and Neurobehavioral Disorders. *J. Xenobiot.* **2025**, *15*, 83. [CrossRef]
15. Caba, I.-C.; Ștreangă, V.; Dobrin, M.-E.; Jităreanu, C.; Jităreanu, A.; Profire, B.; Apotrosoaei, M.; Focșa, A.-V.; Caba, B.; Agoroaei, L. Clinical Assessment of Acute Organophosphorus Pesticide Poisoning in Pediatric Patients Admitted to the Toxicology Emergency Department. *Toxics* **2022**, *10*, 582. [CrossRef] [PubMed]
16. Boneva, I.; Yaneva, S.; Danalev, D. Development and Validation of Method for Determination of Organophosphorus Pesticides Traces in Liver Sample by GC-MS/MS-Ion Trap. *Acta Chromatogr.* **2021**, *33*, 188–194. [CrossRef]
17. Tucker, S.; Dumitriu, G.-D.; Teodosiu, C. Pesticides Identification and Sustainable Viticulture Practices to Reduce Their Use: An Overview. *Molecules* **2022**, *27*, 8205. [CrossRef] [PubMed]
18. Sel, S.; Er, E.Ö.; Koyuncu, İ. Development of an Analytical Method for the Determination of Pesticides in Tropical Fruits by LC-QTOF-MS/MS after QuEChERS Extraction Sample Cleanup and DLLME Preconcentration. *Methods Appl. Fluoresc.* **2024**, *12*, 015008. [CrossRef]
19. Chormey, D.S.; Zaman, B.T.; Kasa, N.A.; Bakırdere, S. Liquid Phase Microextraction Strategies and Their Application in the Determination of Endocrine Disruptive Compounds in Food Samples. *TrAC Trends Anal. Chem.* **2020**, *128*, 115917. [CrossRef]
20. Rani, P.; Nanda, B.P.; Narang, R.K.; Bhatia, R. Advancements in Solvent Microextraction: Recent Developments and Diverse Applications in the Modern Era. *Sep. Sci. Plus* **2024**, *7*, 2300243. [CrossRef]
21. Rodríguez-Ramos, R.; Lehotay, S.J.; Michlig, N.; Socas-Rodríguez, B.; Rodríguez-Delgado, M.Á. Critical Review and Re-Assessment of Analyte Protectants in Gas Chromatography. *J. Chromatogr. A* **2020**, *1632*, 461596. [CrossRef]
22. Brinco, J.; Guedes, P.; Gomes da Silva, M.; Mateus, E.P.; Ribeiro, A.B. Analysis of Pesticide Residues in Soil: A Review and Comparison of Methodologies. *Microchem. J.* **2023**, *195*, 109465. [CrossRef]
23. Shalaby, A.A.; El-Sheikh, E.S.A.; Refaat, A.M.; Ragheb, D.A. Residue Analysis and Associated Risk Assessment of Hexythiazox and Spinosad Applied on Strawberry Plants. *Egypt. J. Chem.* **2022**, *65*, 489–498. [CrossRef]
24. Gai, T.; Nie, J.; Ding, Z.; Wu, W.; Liu, X. Progress of Rapid Detection of Pesticides in Fruits and Vegetables. *Front. Food Sci. Technol.* **2023**, *3*, 1253227. [CrossRef]
25. Moschopoulou, G.; Tsekouras, V.; Mercader, J.V.; Abad-Fuentes, A.; Kintzios, S. Development of a Portable Cell-Based Biosensor for the Ultra-Rapid Screening for Boscalid Residues in Lettuce. *Biosensors* **2024**, *14*, 311. [CrossRef]
26. Tsagkaris, A.S.; Pulkrabova, J.; Hajslova, J. Optical Screening Methods for Pesticide Residue Detection in Food Matrices: Advances and Emerging Analytical Trends. *Foods* **2021**, *10*, 88. [CrossRef]
27. Ansari, I.; Magdy El-Kady, M.; Muniyan, S. A Review on the Fatal Impact of Pesticide Toxicity on Environment and Human Health 16. In *Global Climate Change*; Elsevier: Amsterdam, The Netherlands, 2021.
28. Dawson, A.H.; Eddleston, M.; Senarathna, L.; Mohamed, F.; Gawarammana, I.; Bowe, S.J.; Manuweera, G.; Buckley, N.A. Acute Human Lethal Toxicity of Agricultural Pesticides: A Prospective Cohort Study. *PLoS Med.* **2010**, *7*, e1000357. [CrossRef] [PubMed]
29. Buszewski, B.; Bukowska, M.; Ligor, M.; Staneczko-Baranowska, I. A Holistic Study of Neonicotinoids Neuroactive Insecticides—Properties, Applications, Occurrence, and Analysis. *Environ. Sci. Pollut. Res.* **2019**, *26*, 34723–34740. [CrossRef] [PubMed]
30. Hassaan, M.A.; El Nemr, A. Pesticides Pollution: Classifications, Human Health Impact, Extraction and Treatment Techniques. *Egypt. J. Aquat. Res.* **2020**, *46*, 207–220. [CrossRef]
31. Kardani, F.; Jelyani, A.Z.; Hashemi, M.; Rashedinia, M.; Shariati, S.; Mirzaei, R.; Mahdavinia, M.; Noori, S.M.A. Determination of 323 Pesticide Residues in Iran's Cereal by GC-MS and HPLC-UV Combined with QuEChERS Extraction and Mixed-Mode SPE Clean-up Method. *J. Food Compos. Anal.* **2023**, *124*, 105670. [CrossRef]
32. Pany, B.K.; Sahu, G.; Pattnaik, M.; Pany, K.; Jena, D.; Kumar Pal, A. Effect of Organochlorine Pesticides on Living Organisms and Environment Chemical Science Review and Letters Effect of Organochlorine Pesticides on Living Organisms and Environment.

Chem. Sci. Rev. Lett. **2020**, *9*, 682–686. Available online: https://chesci.com/wp-content/uploads/2020/09/10_CS2051063_p682-686.pdf (accessed on 13 August 2025).
33. Madrigal, J.M.; Sargis, R.M.; Persky, V.; Turyk, M.E. Multiple Organochlorine Pesticide Exposures and Measures of Sex Steroid Hormones in Adult Males: Cross-Sectional Findings from the 1999–2004 National Health and Nutrition Examination Survey. *Int. J. Hyg. Environ. Health* **2021**, *231*, 113609. [CrossRef]
34. Parada, H.; Sun, X.; Tse, C.K.; Engel, L.S.; Olshan, A.F.; Troester, M.A. Plasma Levels of Dichlorodiphenyldichloroethene (DDE) and Dichlorodiphenyltrichloroethane (DDT) and Survival Following Breast Cancer in the Carolina Breast Cancer Study. *Environ. Int.* **2019**, *125*, 161–171. [CrossRef]
35. Ugalde-Resano, R.; Gamboa-Loira, B.; Mérida-Ortega, Á.; Rincón-Rubio, A.; Flores-Collado, G.; Piña-Pozas, M.; López-Carrillo, L. Exposure to Organochlorine Pesticides and Female Breast Cancer Risk According to Molecular Receptors Expression: A Systematic Review and Meta-Analysis of Epidemiological Evidence. *Curr. Environ. Health Rep.* **2023**, *10*, 442–458. [CrossRef]
36. Freire, C.; Koifman, R.J.; Sarcinelli, P.; Rosa, A.C.; Clapauch, R.; Koifman, S. Long Term Exposure to Organochlorine Pesticides and Thyroid Function in Children from Cidade Dos Meninos, Rio de Janeiro, Brazil. *Environ. Res.* **2012**, *117*, 68–74. [CrossRef]
37. Bandow, N.; Conrad, A.; Kolossa-Gehring, M.; Murawski, A.; Sawal, G. Polychlorinated Biphenyls (PCB) and Organochlorine Pesticides (OCP) in Blood Plasma—Results of the German Environmental Survey for Children and Adolescents 2014–2017 (GerES V). *Int. J. Hyg. Environ. Health* **2020**, *224*, 113426. [CrossRef] [PubMed]
38. Rapini, R.; Marrazza, G. Biosensor Potential in Pesticide Monitoring. In *Comprehensive Analytical Chemistry*; Elsevier: Amsterdam, The Netherlands, 2016; pp. 3–31.
39. Zhao, W.; Lu, J.; Lai, Y.; Hou, Y.; Zhao, X.; Wei, Q.; Zou, X.; Gou, Z. Occurrences, Possible Sources, and Risk Impacts of Organochlorine Pesticides in Soil of Changchun Central Urban Area, Northeast China. *Sustainability* **2023**, *15*, 16801. [CrossRef]
40. Adeyinka, G.C.; Moodley, B.; Birungi, G.; Ndungu, P. Evaluation of Organochlorinated Pesticide (OCP) Residues in Soil, Sediment and Water from the Msunduzi River in South Africa. *Environ. Earth Sci.* **2019**, *78*, 223. [CrossRef]
41. Zhu, S.; Zhou, Y.; Chao, M.; Zhang, Y.; Cheng, W.; Xu, H.; Zhang, L.; Tao, Q.; Da, Q. Association between Organophosphorus Insecticides Exposure and Osteoarthritis in Patients with Arteriosclerotic Cardiovascular Disease. *BMC Public Health* **2024**, *24*, 1873. [CrossRef]
42. Mdeni, N.L.; Adeniji, A.O.; Okoh, A.I.; Okoh, O.O. Analytical Evaluation of Carbamate and Organophosphate Pesticides in Human and Environmental Matrices: A Review. *Molecules* **2022**, *27*, 618. [CrossRef]
43. Wu, G.; Shi, W.; Zheng, L.; Wang, X.; Tan, Z.; Xie, E.; Zhang, D. Impacts of Organophosphate Pesticide Types and Concentrations on Aquatic Bacterial Communities and Carbon Cycling. *J. Hazard Mater.* **2024**, *475*, 134824. [CrossRef]
44. Xie, J.; Li, L.; Khan, I.M.; Wang, Z.; Ma, X. Flexible Paper-Based SERS Substrate Strategy for Rapid Detection of Methyl Parathion on the Surface of Fruit. *Spectrochim. Acta A Mol. Biomol. Spectrosc.* **2020**, *231*, 118104. [CrossRef]
45. Sharma, D.; Wangoo, N.; Sharma, R.K. Sensing Platform for Pico-Molar Level Detection of Ethyl Parathion Using Au–Ag Nanoclusters Based Enzymatic Strategy. *Talanta* **2021**, *221*, 121267. [CrossRef]
46. Colović, M.B.; Krstić, D.Z.; Lazarević-Pašti, T.D.; Bondžić, A.M.; Vasić, V.M. Acetylcholinesterase Inhibitors: Pharmacology and Toxicology. *Curr. Neuropharmacol.* **2013**, *11*, 315–335. [CrossRef]
47. Zhang, J.; Guo, J.; Wu, C.; Qi, X.; Jiang, S.; Zhou, T.; Xiao, H.; Li, W.; Lu, D.; Feng, C.; et al. Early-Life Carbamate Exposure and Intelligence Quotient of Seven-Year-Old Children. *Environ. Int.* **2020**, *145*, 106105. [CrossRef] [PubMed]
48. Xie, T.; Huang, J.; Wu, J.; Zhang, Q. Evaluation of Supercritical Fluid Chromatography Coupled to Tandem Mass Spectrometry for the Analysis of Pesticide Residues in Grain. *J. Sep. Sci.* **2024**, *47*, e2300623. [CrossRef] [PubMed]
49. Tomasevic, A.; Mijin, D.; Marinkovic, A.; Cvijetic, I.; Gasic, S. Photocatalytic Degradation of Carbamate Insecticides: Effect of Different Parameters. *Pestic. I Fitomedicina* **2019**, *34*, 193–200. [CrossRef]
50. Gu, M.; Chen, S.; Jiang, J.; Liao, J.; Long, T.; Xie, Y. Do Pyrethroid Pesticide Residues in Chinese Mitten Crab Aquaculture Areas Have an Impact on the Ecological Environment?—A Case Study of Yangcheng Lake. *J. Hazard Mater.* **2024**, 135920. [CrossRef] [PubMed]
51. Jia, Q.; Liao, G.; Chen, L.; Qian, Y.; Yan, X.; Qiu, J. Pesticide Residues in Animal-Derived Food: Current State and Perspectives. *Food Chem.* **2024**, *438*, 137974. [CrossRef]
52. Jeschke, P.; Nauen, R. Neonicotinoids—From Zero to Hero in Insecticide Chemistry. *Pest Manag. Sci.* **2008**, *64*, 1084–1098. [CrossRef]
53. Ensley, S.M. Neonicotinoids. In *Veterinary Toxicology*; Elsevier: Amsterdam, The Netherlands, 2018; pp. 521–524.
54. Fairbrother, A.; Purdy, J.; Anderson, T.; Fell, R. Risks of Neonicotinoid Insecticides to Honeybees. *Environ. Toxicol. Chem.* **2014**, *33*, 719–731. [CrossRef]
55. Yari, K.; Rahmani, A.; Asgari, G.; Azarian, Q.; Bhatnagar, A.; Leili, M. Degradation of Imidacloprid Pesticide in Aqueous Solution Using an Eco-Friendly Electrochemical Process. *Desalination Water Treat.* **2017**, *86*, 150–157. [CrossRef]

56. Malhotra, N.; Chen, K.H.-C.; Huang, J.-C.; Lai, H.-T.; Uapipatanakul, B.; Roldan, M.J.M.; Macabeo, A.P.G.; Ger, T.-R.; Hsiao, C.-D. Physiological Effects of Neonicotinoid Insecticides on Non-Target Aquatic Animals—An Updated Review. *Int. J. Mol. Sci.* **2021**, *22*, 9591. [CrossRef]
57. Basley, K.; Goulson, D. Neonicotinoids Thiamethoxam and Clothianidin Adversely Affect the Colonisation of Invertebrate Populations in Aquatic Microcosms. *Environ. Sci. Pollut. Res.* **2018**, *25*, 9593–9599. [CrossRef]
58. Gautam, P.; Dubey, S.K. Biodegradation of Neonicotinoids: Current Trends and Future Prospects. *Curr. Pollut. Rep.* **2023**, *9*, 410–432. [CrossRef]
59. Li, Y.; Li, Y. Photodegradation of Neonicotinoid Insecticides Nitenpyram, Thiacloprid, and Acetamiprid in Water and Soil Environments. *Bull. Environ. Contam. Toxicol.* **2024**, *113*, 66. [CrossRef]
60. Zhao, Y.; Zhu, Z.; Xiao, Q.; Li, Z.; Jia, X.; Hu, W.; Liu, K.; Lu, S. Urinary Neonicotinoid Insecticides in Children from South China: Concentrations, Profiles and Influencing Factors. *Chemosphere* **2022**, *291*, 132937. [CrossRef] [PubMed]
61. Wang, P.-W.; Huang, Y.-F.; Fang, L.-J.; Chen, M.-L. Prenatal and Childhood Neonicotinoid Exposure and Neurodevelopment: A Study in a Young Taiwanese Cohort. *Sci. Total Environ.* **2024**, *946*, 174232. [CrossRef] [PubMed]
62. D'Amore, T.; Smaoui, S.; Varzakas, T. Chemical Food Safety in Europe Under the Spotlight: Principles, Regulatory Framework and Roadmap for Future Directions. *Foods* **2025**, *14*, 1628. [CrossRef] [PubMed]
63. Carrasco Cabrera, L.; Di Piazza, G.; Dujardin, B.; Marchese, E.; Medina Pastor, P. The 2023 European Union Report on Pesticide Residues in Food. *EFSA J.* **2025**, *23*, e9398. [CrossRef]
64. More, S.; Bampidis, V.; Benford, D.; Bragard, C.; Halldorsson, T.; Hougaard Bennekou, S.; Koutsoumanis, K.; Machera, K.; Naegeli, H.; Nielsen, S.; et al. Statement on the Derivation of Health-Based Guidance Values (HBGVs) for Regulated Products That Are Also Nutrients. *EFSA J.* **2021**, *19*, e06479. [CrossRef]
65. Brancato, A.; Brocca, D.; Ferreira, L.; Greco, L.; Jarrah, S.; Leuschner, R.; Medina, P.; Miron, I.; Nougadere, A.; Pedersen, R.; et al. Use of EFSA Pesticide Residue Intake Model (EFSA PRIMo Revision 3). *EFSA J.* **2018**, *16*, e05147. [CrossRef]
66. Damalas, C.A.; Eleftherohorinos, I.G. Pesticide Exposure, Safety Issues, and Risk Assessment Indicators. *Int. J. Environ. Res. Public Health* **2011**, *8*, 1402–1419. [CrossRef]
67. El-Sheikh, E.S.A.; Ramadan, M.M.; El-Sobki, A.E.; Shalaby, A.A.; McCoy, M.R.; Hamed, I.A.; Ashour, M.B.; Hammock, B.D. Pesticide Residues in Vegetables and Fruits from Farmer Markets and Associated Dietary Risks. *Molecules* **2022**, *27*, 8072. [CrossRef]
68. Drabińska, N.; Marcinkowska, M.A.; Wieczorek, M.N.; Jeleń, H.H. Application of Sorbent-Based Extraction Techniques in Food Analysis. *Molecules* **2023**, *28*, 7985. [CrossRef]
69. Ezati, M.; Moinfar, S.; Mohammadi, S.; Khayatian, G. A Continuous Sample Drop Flow-Based Microextraction Method for Spectrophotometric Determination of Cobalt with 1-(2-Pyridylazo)-2-Naphthol in Water Samples. *J. Anal. Chem.* **2021**, *76*, 172–179. [CrossRef]
70. Kalogiouri, N.P.; Papadakis, E.-N.; Maggalou, M.G.; Karaoglanidis, G.S.; Samanidou, V.F.; Menkissoglu-Spiroudi, U. Development of a Microwave-Assisted Extraction Protocol for the Simultaneous Determination of Mycotoxins and Pesticide Residues in Apples by LC-MS/MS. *Appl. Sci.* **2021**, *11*, 10931. [CrossRef]
71. Zondo, S.; Mahlambi, P. Comparison of Soxhlet and Microwave-assisted Extractions Efficiency for the Determination of Herbicides in Soil and Maize Crop: Cumulative and Health Risks Assessment. *eFood* **2024**, *5*, e177. [CrossRef]
72. Barchańska, H.; Czalicka, M.; Giemza, A. Simultaneous Determination of Selected Insecticides and Atrazine in Soil by Mae–GC–ECD. *Arch. Environ. Prot.* **2013**, *39*, 27–40. [CrossRef]
73. Tian, Q.; Li, H.; Chen, L.; Han, B. Microwave-Assisted "One-Pot" Acidolysis and Extraction for the Rapid Determination of Mancozeb in Fruit and Vegetable Samples. *J. Food Qual.* **2024**, *2024*, 2577585. [CrossRef]
74. Zhang, H.; Ren, Y.; Wei, J.; Ji, Y.; Bai, X.; Shao, Y.; Li, H.; Gao, R.; Wu, Z.; Peng, Z.; et al. Optimization of the Efficient Extraction of Organic Components in Atmospheric Particulate Matter by Accelerated Solvent Extraction Technique and Its Application. *Atmosphere* **2022**, *13*, 818. [CrossRef]
75. Chiesa, L.M.; Labella, G.F.; Panseri, S.; Britti, D.; Galbiati, F.; Villa, R.; Arioli, F. Accelerated Solvent Extraction by Using an 'in-Line' Clean-up Approach for Multiresidue Analysis of Pesticides in Organic Honey. *Food Addit. Contam. Part A* **2017**, *34*, 809–818. [CrossRef]
76. Horvat, T.; Jakovljević, I.; Sever Štrukil, Z.; Pehnec, G. Optimisation of ASE for Determination of Organic Compounds Bound to Particulate Matter. *Kem. U Ind.* **2024**, *73*, 507–513. [CrossRef]
77. Morariu, I.D.; Avasilcai, L.; Cioanca, O.; Morariu, B.-A.; Vieriu, M.; Tanase, C. The Effects of Honey Sulfonamides on Immunological and Hematological Parameters in Wistar Rats. *Medicina* **2022**, *58*, 1558. [CrossRef]
78. Morariu, I.D. Immunochemical Assay of Chloramphenicol in Honey. *Farmacia* **2019**, *67*, 235–239. [CrossRef]

79. Belmonte, I.d.S.; Pizzolato, T.M.; Gama, M.R. Quaternary Ammonium Pesticides: A Review of Chromatography and Non-Chromatography Methods for Determination of Pesticide Residues in Water Samples. *Trends Environ. Anal. Chem.* **2022**, *35*, e00171. [CrossRef]
80. Avellar, Á.; Prates, G.; Alves, L.; Prestes, O.; Adaime, M.; Zanella, R. Multiresidue and Multiclass Determination of Current-Use Pesticides (Cups) in Water Samples by SPE and UHPLC-MS/MS. *Quim Nova* **2024**, *47*, e20240051. [CrossRef]
81. Lee, H.; Cho, Y.; Jung, G.; Kim, H.; Jeong, W. Comparison of Recovery Efficiency and Matrix Effect Reduction in Pesticide Residue Analysis: QuEChERS with d-SPE, SPE, and FaPEx in Apples and Korean Cabbage. *Anal. Methods* **2023**, *15*, 3709–3716. [CrossRef]
82. Yun, D.Y.; Bae, J.Y.; Kang, Y.J.; Lim, C.U.; Jang, G.H.; Eom, M.O.; Choe, W.J. Simultaneous Analysis of 272 Pesticides in Agricultural Products by the QuEChERS Method and Gas Chromatography with Tandem Mass Spectrometry. *Molecules* **2024**, *29*, 2114. [CrossRef]
83. Zgoła-Grześkowiak, A.; Grześkowiak, T.; Ligor, M.; Frankowski, R. Achievements and Challenges of Matrix Solid-Phase Dispersion Usage in the Extraction of Plants and Food Samples. *Processes* **2024**, *12*, 1146. [CrossRef]
84. Zuin, V.G.; Yariwake, J.H.; Lanças, F.M. Analysis of Pesticide Residues in Brazilian Medicinal Plants: Matrix Solid Phase Dispersion versus Conventional (European Pharmacopoeia) Methods. *J. Braz. Chem. Soc.* **2003**, *14*, 304–309. [CrossRef]
85. Souza, M.R.R.; Jesus, R.A.; Costa, J.A.S.; Barreto, A.S.; Navickiene, S.; Mesquita, M.E. Applicability of Metal–Organic Framework Materials in the Evaluation of Pesticide Residues in Egg Samples of Chicken (Gallus Gallus Domesticus). *J. Fur Verbraucherschutz Und Leb.* **2021**, *16*, 83–91. [CrossRef]
86. Kemmerich, M.; Demarco, M.; Bernardi, G.; Prestes, O.D.; Adaime, M.B.; Zanella, R. Balls-in-Tube Matrix Solid Phase Dispersion (BiT-MSPD): An Innovative and Simplified Technique for Multiresidue Determination of Pesticides in Fruit Samples. *J. Chromatogr. A* **2020**, *1612*, 460640. [CrossRef]
87. Nolvachai, Y.; Amaral, M.S.S.; Herron, R.; Marriott, P.J. Solid Phase Microextraction for Quantitative Analysis—Expectations beyond Design? *Green Anal. Chem.* **2023**, *4*, 100048. [CrossRef]
88. Agatonovic-Kustrin, S.; Gegechkori, V.; Kobakhidze, T.; Morton, D. Solid-Phase Microextraction Techniques and Application in Food and Horticultural Crops. *Molecules* **2023**, *28*, 6880. [CrossRef] [PubMed]
89. Lancioni, C.; Castells, C.; Candal, R.; Tascon, M. Headspace Solid-Phase Microextraction: Fundamentals and Recent Advances. *Adv. Sample Prep.* **2022**, *3*, 100035. [CrossRef]
90. Rosendo, L.M.; Brinca, A.T.; Pires, B.; Catarro, G.; Rosado, T.; Guiné, R.P.F.; Araújo, A.R.T.S.; Anjos, O.; Gallardo, E. Miniaturized Solid Phase Extraction Techniques Applied to Natural Products. *Processes* **2023**, *11*, 243. [CrossRef]
91. Duan, Y.; Huang, Y.; Huang, L.; Xiang, Z.; Liu, J.; Liu, S.; Chen, Z. In Vivo Tracing of Triazole Pesticides in Chinese Cabbage via a Novel Solid-Phase Microextraction Fiber. *Food Control* **2024**, *156*, 110143. [CrossRef]
92. Li, W.; Zhang, J.; Wang, S.; Ma, Z.; Feng, J.; Pei, H.; Liu, Y. Simultaneous Determination of Three Herbicide Residues in Wheat Flour Based on the Hollow Fiber Supported Carbon Dots. *J. Food Compos. Anal.* **2022**, *108*, 104426. [CrossRef]
93. Mungwari, C.P.; King'ondu, C.K.; Sigauke, P.; Obadele, B.A. Conventional and Modern Techniques for Bioactive Compounds Recovery from Plants: Review. *Sci. Afr.* **2025**, *27*, e02509. [CrossRef]
94. Edo, G.I.; Nwachukwu, S.C.; Ali, A.B.M.; Yousif, E.; Jikah, A.N.; Zainulabdeen, K.; Ekokotu, H.A.; Isoje, E.F.; Igbuku, U.A.; Opiti, R.A.; et al. A Review on the Composition, Extraction and Applications of Phenolic Compounds. *Ecol. Front.* **2025**, *45*, 7–23. [CrossRef]
95. Baldino, L.; Scognamiglio, M.; Reverchon, E. Green and Selective Supercritical Fluid Extraction of Essential Oil and Cannabidiol from *Cannabis sativa* L. *Can. J. Chem. Eng.* **2025**, *103*, 3637–3646. [CrossRef]
96. Yousefi, M.; Rahimi-Nasrabadi, M.; Mirsadeghi, S.; Pourmortazavi, S.M. Supercritical Fluid Extraction of Pesticides and Insecticides from Food Samples and Plant Materials. *Crit. Rev. Anal. Chem.* **2020**, *51*, 482–501. [CrossRef] [PubMed]
97. Morariu, I.D.; Avasilcai, L.; Vieriu, M.; Panainte, A.D.; Bibire, N. Novel Multiresidue Method for the Determination of Eight Trichothecene Mycotoxins in Pollen Samples Using QuEChERS-Based GC-MS/MS. *Rev. De Chim.* **2017**, *68*, 304–306. [CrossRef]
98. Abdulra'uf, L.B.; Junaid, A.M.; Lawal, A.R.; Ibrahim, H.B.; Tan, G.H. Determination of Pesticide Residues in Beans Using QuEChERS Technique Coupled to Gas Chromatography-Mass Spectrometry: Multivariate Optimization of CEN and AOAC Methods. *Food Chem.* **2025**, *463*, 141464. [CrossRef] [PubMed]
99. Tasic, A.M.; Ninković, M.; Pavlović, I. Validation and Application of a Method for Determination of Multi-Class Pesticides in Muscle Chicken Breast Fillets Using QuEChERS Extraction and GC/MS. *J. Vet. Res.* **2024**, *68*, 223–232. [CrossRef]
100. Wilkowska, A.; Biziuk, M. Determination of Pesticide Residues in Food Matrices Using the QuEChERS Methodology. *Food Chem.* **2011**, *125*, 803–812. [CrossRef]
101. Ma, X.; Wang, J.; Wu, Q.; Wang, C.; Wang, Z. Extraction of Carbamate Pesticides in Fruit Samples by Graphene Reinforced Hollow Fibre Liquid Microextraction Followed by High Performance Liquid Chromatographic Detection. *Food Chem.* **2014**, *157*, 119–124. [CrossRef]

102. Kaenjun, T.; Tangtreamjitmun, N. Spectrophotometric Determination of O-Phenylphenol in Canned Drinks Using Three-Phase Hollow-Fiber Liquid Phase Microextraction. *Food Chem.* **2025**, *463*, 141204. [CrossRef]
103. Madikizela, L.M.; Pakade, V.E.; Ncube, S.; Tutu, H.; Chimuka, L. Application of Hollow Fibre-Liquid Phase Microextraction Technique for Isolation and Pre-Concentration of Pharmaceuticals in Water. *Membranes* **2020**, *10*, 311. [CrossRef]
104. Moema, D.; Makwakwa, T.A.; Gebreyohannes, B.E.; Dube, S.; Nindi, M.M. Hollow Fiber Liquid Phase Microextraction of Fluoroquinolones in Chicken Livers Followed by High Pressure Liquid Chromatography: Greenness Assessment Using National Environmental Methods Index Label (NEMI), Green Analytical Procedure Index (GAPI), Analytical GREEnness Metric (AGREE), and Eco Scale. *J. Food Compos. Anal.* **2023**, *117*, 105131. [CrossRef]
105. Lee, J.; Lee, H.K.; Rasmussen, K.E.; Pedersen-Bjergaard, S. Environmental and Bioanalytical Applications of Hollow Fiber Membrane Liquid-Phase Microextraction: A Review. *Anal. Chim. Acta* **2008**, *624*, 253–268. [CrossRef] [PubMed]
106. Khan, W.A.; Arain, M.B.; Yamini, Y.; Shah, N.; Kazi, T.G.; Pedersen-Bjergaard, S.; Tajik, M. Hollow Fiber-Based Liquid Phase Microextraction Followed by Analytical Instrumental Techniques for Quantitative Analysis of Heavy Metal Ions and Pharmaceuticals. *J. Pharm. Anal.* **2020**, *10*, 109–122. [CrossRef] [PubMed]
107. Petrarca, M.H.; Cunha, S.C.; Fernandes, J.O. Determination of Pesticide Residues in Soybeans Using QuEChERS Followed by Deep Eutectic Solvent-Based DLLME Preconcentration Prior to Gas Chromatography-Mass Spectrometry Analysis. *J. Chromatogr. A* **2024**, *1727*. [CrossRef]
108. Pourhossein, M.; Khadem, M.; Omidi, F.; Heravizadeh, O.R.; Shahtaheri, S.J. Development of a Green Single Drop Microextraction Based on Deep Eutectic Solvent and HPLC-UV for Trace Residue Analysis of Three Frequent-Used Pesticides. *Iran. J. Public Health* **2023**, *52*, 2440–2449. [CrossRef]
109. Wahab, S.; Muzammil, K.; Nasir, N.; Khan, M.S.; Ahmad, M.F.; Khalid, M.; Ahmad, W.; Dawria, A.; Reddy, L.K.V.; Busayli, A.M. Advancement and New Trends in Analysis of Pesticide Residues in Food: A Comprehensive Review. *Plants* **2022**, *11*, 1106. [CrossRef]
110. Moinfar, S.; Jamil, L.A.; Sami, H.Z. Determination of Organophosphorus Pesticides in Juice and Water by Modified Continuous Sample Drop Flow Microextraction Combined with Gas Chromatography–Mass Spectrometry. *Food Anal. Methods* **2020**, *13*, 1050–1059. [CrossRef]
111. Mohammed, A.M.; Moinfar, S. Determination of Trimethoprim in Milk, Water and Plasma Using Protein Precipitation Combined with Liquid Phase Microextraction Method. *J. Food Compos. Anal.* **2023**, *118*, 105224. [CrossRef]
112. Moinfar, S.; Jamil, L.A.; Sami, H.Z.; Ataei, S. An Innovative Continuous Sample Drop Flow Microextraction for GC–MS Determination of Pesticides in Grape Juice and Water Samples. *J. Food Compos. Anal.* **2021**, *95*, 103695. [CrossRef]
113. Jamil, L.A. Optimization of New Sample Preparation Technique for the Determination of Methadone and Codeine in Plasma Sample by GC-FID. *J. Braz. Chem. Soc.* **2020**, *31*, 580–588. [CrossRef]
114. Méndez-Barredo, L.H.; Monribot-Villanueva, J.L.; Bojórquez-Velázquez, E.; Elizalde-Contreras, J.M.; Guerrero-Analco, J.A.; Ruiz-May, E. Comparative Evaluation of Different Extraction Methods for Identification and Quantification of Glyphosate in Fortified Corn Flour. *J. Mex. Chem. Soc.* **2023**, *67*, 213–226. [CrossRef]
115. Han, W.; Tang, H.; Zhao, L.; Li, Y.; Men, D.; Dong, M.; Han, L.; Wang, W. A Modified Method for the Simultaneous Determination of Fluindapyr and Its Metabolites Residues in Vegetables and Fruits by SPE and UHPLC MS/MS. *J. Food Compos. Anal.* **2025**, *137*, 106972. [CrossRef]
116. Bakhshizadeh Aghdam, M.; Farajzadeh, M.A.; Afshar Mogaddam, M.R. Partially Carbonized Cellulose Filter Paper as a Green Adsorbent for the Extraction of Pesticides from Fruit Juices. *J. Chromatogr. A* **2021**, *1648*, 462220. [CrossRef]
117. Yin, Y.; Fan, C.; Cheng, L.; Shan, Y. Efficient and Sensitive Detection of Organophosphate Pesticides in Orange Juice Using Dispersed Solid-Phase Extraction Based on Amorphous UiO-66. *J. Sep. Sci.* **2025**, *48*, e70066. [CrossRef]
118. Nazari Koloujeh, M.; Iranifam, M.; Fathi, A.A.; Farajzadeh, M.A.; Afshar Mogaddam, M.R. Development of COF@MOF Nanocomposite-Based Dispersive Solid-Phase Microextraction for the Extraction of Pesticides from Strawberries. *Food Anal. Methods* **2025**, *18*, 682–693. [CrossRef]
119. Mujahid, M.; Latif, S.; Ahmed, M.; Shehzadi, W.; Imran, M.; Ahmad, M.; Asari, A.; Jehangir, M.; Mahmud, Z. Modified Matrix Solid Phase Dispersion-HPLC Method for Determination of Pesticide Residue in Vegetables and Their Impact on Human Health: A Risk Assessment. *Front. Chem.* **2022**, *10*, 1084350. [CrossRef]
120. Yin, R.; Gao, L.; Qin, D.; Chen, L.; Niu, N. Preparation of Porous Carbon-Based Molecularly Imprinted Polymers for Separation of Triazine Herbicides in Corn. *Microchim. Acta* **2022**, *189*, 23. [CrossRef]
121. Fang, Y.; Wang, W.; Xu, Y.; Chen, Q.; Jiao, T.; Wei, J.; Chen, Q.; Chen, X. Development of a Hydrophilic-Lipophilic-Balanced Copolymer@zirconium-Based Metal-Organic Framework-Based Solid-Phase Microextraction Probe for the Trace Determination of Organophosphorus Pesticides in Tea Infusions. *Talanta* **2025**, *281*, 126823. [CrossRef] [PubMed]

122. Pang, Y.; Zang, X.; Li, H.; Liu, J.; Chang, Q.; Zhang, S.; Wang, C.; Wang, Z. Solid-Phase Microextraction of Organophosphorous Pesticides from Food Samples with a Nitrogen-Doped Porous Carbon Derived from g-C3N4 Templated MOF as the Fiber Coating. *J. Hazard Mater.* **2020**, *384*, 121430. [CrossRef] [PubMed]
123. Delińska, K.; Yavir, K.; Kloskowski, A. Head-Space SPME for the Analysis of Organophosphorus Insecticides by Novel Silica IL-Based Fibers in Real Samples. *Molecules* **2022**, *27*, 4688. [CrossRef]
124. Nakamura, K.; Otake, T.; Hanari, N. Quantitative Determination of Organophosphorus, Pyrethroid, and Dithiolane Pesticide Residues in Brown Rice Using Supercritical Fluid Extraction and Liquid Chromatography–Tandem Mass Spectrometry. *J. AOAC Int.* **2023**, *106*, 1532–1541. [CrossRef] [PubMed]
125. Nakamura, K.; Otake, T.; Hanari, N. Evaluation of Supercritical Fluid Extraction for the Determination of Neonicotinoid Pesticides in Green Onion. *J. Environ. Sci. Health Part B* **2020**, *55*, 604–612. [CrossRef] [PubMed]
126. Cho, S.M.; Lee, H.S.; Park, J.S.; Lee, S.J.; Shin, H.S.; Chung, Y.M.; Choi, H.N.; Jung, Y.H.; Oh, J.H.; Yun, S.S. Determination of Residual Triflumezopyrim Insecticide in Agricultural Products through a Modified QuEChERS Method. *Foods* **2021**, *10*, 2090. [CrossRef]
127. Muiambo, S.L.; Chaúque, E.F.C.; Gulamussen, N.J.; Chimuka, L.; Morifi, E.; Nyambe, I. Modified QuEChERS Method for the Extraction and Quantification of Persistent Organic Compounds in Vegetables from Mozambican Local Markets. *J. Hazard. Mater. Adv.* **2023**, *10*, 100262. [CrossRef]
128. Abasalizadeh, A.; Sorouraddin, S.M.; Farajzadeh, M.A.; Afshar Mogaddam, M.R. Development of a Green Approach Based on DμSPE Combined with Deep Eutectic Solvent-Based DLLME for the Extraction of Some Pesticides from Vegetable Samples Prior to GC–FID and GC–MS. *J. Iran. Chem. Soc.* **2022**, *19*, 4699–4707. [CrossRef]
129. Szarka, A.; Búčiková, K.; Kostić, I.; Hrouzková, S. Development of a Multiresidue QuEChERS–DLLME—Fast GC–MS Method for Determination of Selected Pesticides in Yogurt Samples. *Food Anal. Methods* **2020**, *13*, 1829–1841. [CrossRef]
130. Raoufi, A.; Raoufi, A.M.; Ismailzadeh, A.; Soleimani Rad, E.; Kiaeefar, A. Application of Hollow Fiber-Protected Liquid-Phase Microextraction Combined with GC-MS in Determining Endrin, Chlordane, and Dieldrin in Rice Samples. *Environ. Geochem. Health* **2023**, *45*, 5261–5277. [CrossRef]
131. Jamil, L.A.; Sami, H.Z.; Aghaei, A.; Moinfar, S.; Ataei, S. Combination of Modified Ultrasound-Assisted Extraction with Continuous Sample Drop Flow Microextraction for Determination of Pesticides in Vegetables and Fruits. *Microchem. J.* **2021**, *160*, 105692. [CrossRef]
132. Sassolas, A.; Prieto-Simón, B.; Marty, J.-L. Biosensors for Pesticide Detection: New Trends. *Am. J. Anal. Chem.* **2012**, *03*, 210–232. [CrossRef]
133. Hernández-Mesa, M.; Moreno-González, D. Current Role of Mass Spectrometry in the Determination of Pesticide Residues in Food. *Separations* **2022**, *9*, 148. [CrossRef]
134. Ciscato, C.H.P.; Bertoni Gebara, A.; Henrique Monteiro, S. Pesticide Residue Monitoring of Brazilian Fruit for Export 2006–2007. *Food Addit. Contam. Part B* **2009**, *2*, 140–145. [CrossRef]
135. Mutengwe, M.T.; Chidamba, L.; Korsten, L. Pesticide Residue Monitoring on South African Fresh Produce Exported over a 6-Year Period. *J. Food Prot.* **2016**, *79*, 1759–1766. [CrossRef]
136. John, P.J.; Bakore, N.; Bhatnagar, P. Assessment of Organochlorine Pesticide Residue Levels in Dairy Milk and Buffalo Milk from Jaipur City, Rajasthan, India. *Environ. Int.* **2001**, *26*, 231–236. [CrossRef] [PubMed]
137. Ohoro, C.R.; Wepener, V. Review of Scientific Literature on Available Methods of Assessing Organochlorine Pesticides in the Environment. *Heliyon* **2023**, *9*, e22142. [CrossRef] [PubMed]
138. Prasad, S.K.; Kalpana, D. Automation in Analytical Chemistry: The Role of AI in Chromatography. *Int. J. Appl. Pharm.* **2024**, *16*, 14–21. [CrossRef]
139. Singh, Y.R.; Shah, D.B.; Kulkarni, M.; Patel, S.R.; Maheshwari, D.G.; Shah, J.S.; Shah, S. Current Trends in Chromatographic Prediction Using Artificial Intelligence and Machine Learning. *Anal. Methods* **2023**, *15*, 2785–2797. [CrossRef]
140. Sinha, N.; Ray, S. Application of Carbon Quantum Dots Derived from Waste Tea for the Detection of Pesticides in Tea: A Novel Biosensor Approach. *ACS Omega* **2024**, *9*, 50201–50213. [CrossRef]
141. Sindhu, S.; Manickavasagan, A. Nondestructive Testing Methods for Pesticide Residue in Food Commodities: A Review. *Compr. Rev. Food Sci. Food Saf.* **2023**, *22*, 1226–1256. [CrossRef]
142. Keklik, M.; Golge, O.; González-Curbelo, M.Á.; Kabak, B. Determination of Pesticide Residues in Vine Leaves Using the QuEChERS Method and Liquid Chromatography-Tandem Mass Spectrometry. *Foods* **2024**, *13*, 909. [CrossRef]
143. Gormez, E.; Golge, O.; González-Curbelo, M.Á.; Kabak, B. Pesticide Residues in Mandarins: Three-Year Monitoring Results. *Molecules* **2023**, *28*, 5611. [CrossRef]

144. Penelope Mabunda, K.; Rejoice Maseko, B.; Ncube, S. Development and Application of a New QuEChERS-Molecularly Imprinted Solid Phase Extraction (QuEChERS-MISPE) Technique for Analysis of DDT and Its Derivatives in Vegetables. *Food Chem.* **2024**, *436*, 137747. [CrossRef] [PubMed]
145. Teixeira, A.M.; De Queiroz, M.E.L.R.; Rodrigues, A.A.Z.; de Oliveira, A.F.; Libardi, V.M.; de Freitas, J.F. Determination of Pesticide Residues in Oat Flour Using Low-Temperature Partition Extraction and GC–MS Analysis. *J. Food Sci. Technol.* **2024**, *61*, 2402–2410. [CrossRef]
146. Wang, Z.; Jiang, B.; Pang, C.; Liu, L.; Zhou, Q. Determination of Multiple Pesticide Residues and Dietary Intake Risk Assessment of 35 Pesticides in Beet from Five Provinces of Northern China. *Sugar Tech.* **2024**, *27*, 393–406. [CrossRef]
147. Morariu, I.-D.; Schiriac, E.-C.; Matiut, D.; Cuciureanu, R. Method Validation for Simultaneous Determination of 12 Sulfonamides in Honey Using Biochip Array Technology. *Farmacia* **2012**, *60*, 143–154.
148. Bessaire, T.; Savoy, M.-C.; Ernest, M.; Christinat, N.; Badoud, F.; Desmarchelier, A.; Carrères, B.; Chan, W.-C.; Wang, X.; Delatour, T. Enhanced Surveillance of >1100 Pesticides and Natural Toxins in Food: Harnessing the Capabilities of LC-HRMS for Reliable Identification and Quantification. *Foods* **2024**, *13*, 3040. [CrossRef]
149. Malm, L.; Liigand, J.; Aalizadeh, R.; Alygizakis, N.; Ng, K.; Frokjær, E.E.; Nanusha, M.Y.; Hansen, M.; Plassmann, M.; Bieber, S.; et al. Quantification Approaches in Non-Target LC/ESI/HRMS Analysis: An Interlaboratory Comparison. *Anal. Chem.* **2024**, *96*, 16215–16226. [CrossRef] [PubMed]
150. Chen, Z.; Daka, Z.; Yao, L.; Dong, J.; Zhang, Y.; Li, P.; Zhang, K.; Ji, S. Recent Progress in the Application of Chromatography-Coupled Mass-Spectrometry in the Analysis of Contaminants in Food Products. *Food Chem. X* **2025**, *27*, 102397. [CrossRef]
151. Goh, M.S.; Lam, S.D.; Yang, Y.; Naqiuddin, M.; Addis, S.N.K.; Yong, W.T.L.; Luang-In, V.; Sonne, C.; Ma, N.L. Omics Technologies Used in Pesticide Residue Detection and Mitigation in Crop. *J. Hazard Mater.* **2021**, *420*, 126624. [CrossRef]
152. Shi, Y.; Jin, H.-F.; Ma, X.-R.; Cao, J. Highly Sensitive Determination of Multiple Pesticide Residues in Foods by Supercritical Fluid Chromatography Coupled with Ion Mobility Quadrupole Time-of-Flight Mass Spectrometry. *Food Res. Int.* **2024**, *175*, 113769. [CrossRef]
153. Bauer, A.; Luetjohann, J.; Rohn, S.; Jantzen, E.; Kuballa, J. Development of a Suspect Screening Strategy for Pesticide Metabolites in Fruit and Vegetables by UPLC-Q-Tof-MS. *Food Anal. Methods* **2018**, *11*, 1591–1607. [CrossRef]
154. Donley, N.; Cox, C.; Bennett, K.; Temkin, A.M.; Andrews, D.Q.; Naidenko, O.V. Forever Pesticides: A Growing Source of PFAS Contamination in the Environment. *Environ. Health Perspect.* **2024**, *132*, 75003. [CrossRef]
155. Zahra, Z.; Song, M.; Habib, Z.; Ikram, S. Advances in Per- and Polyfluoroalkyl Substances (PFAS) Detection and Removal Techniques from Drinking Water, Their Limitations, and Future Outlooks. *Emerg. Contam.* **2025**, *11*, 100434. [CrossRef]
156. Almeida, M.O.; Oloris, S.C.S.; Faria, V.H.F.; Ribeiro, M.C.M.; Cantini, D.M.; Soto-Blanco, B. Optimization of Method for Pesticide Detection in Honey by Using Liquid and Gas Chromatography Coupled with Mass Spectrometric Detection. *Foods* **2020**, *9*, 1368. [CrossRef] [PubMed]
157. Elmi, M.; Ghane, T.; Daraei, B.; Eskandari, S.; Mohammadpour, A.; Amirahmadi, M.; Mousavi Khaneghah, A. Monitoring of Pesticide Residue in Pistachio Nut Samples by LC/MS-MS. *Food Chem.* **2024**, *437*, 137848. [CrossRef] [PubMed]
158. Wu, X.; Li, J.; Wei, J.; Tong, K.; Xie, Y.; Chang, Q.; Yu, X.; Li, B.; Lu, M.; Fan, C.; et al. Multi-Residue Analytical Method Development and Dietary Exposure Risk Assessment of 345 Pesticides in Mango by LC-Q-TOF/MS. *Food Control* **2025**, *170*, 111016. [CrossRef]
159. Phichitsaenyakorn, H.; Bunkoed, O. A Porous Composite Sorbent Incorporating a Metal Organic Framework and Zinc Oxide in Cryogel to Extract Carbamate Pesticides. *Microchem. J.* **2024**, *204*, 111156. [CrossRef]
160. Mohamadi, S.; Akbari-adergani, B.; Sadighara, P.; Jannat, B.; Abdoli, N.; Mirzaei, G.; Zeinali, T. Quantitative Analysis of Multiclass Pesticide Residues in Spinach, Iran. *Appl. Food Res.* **2023**, *3*, 100368. [CrossRef]
161. Chen, G.; Shi, L.; Wang, J.; Zhu, S.; Sheng, J.; Yang, X.; Xu, H. Pesticide Residues in Rice Planted in South and Southwest China. *Food Addit. Contam. Part B* **2023**, *16*, 176–184. [CrossRef]
162. Hammood, M.K.; Arif, M.A. Optimization and Validation of a GC-FID/QuEChERS Method for Quantitative Determination of Spiromesifen Residues in Tomato Fruits, Leaves and Soil Matrices. *Anal. Bioanal. Chem. Res.* **2024**, *11*, 447–461.
163. Lobato, A.; Fernandes, V.C.; Pacheco, J.G.; Delerue-Matos, C.; Gonçalves, L.M. Organochlorine Pesticide Analysis in Milk by Gas-Diffusion Microextraction with Gas Chromatography-Electron Capture Detection and Confirmation by Mass Spectrometry. *J. Chromatogr. A* **2021**, *1636*, 461797. [CrossRef]
164. Yang, K.; Wang, J.; Gao, S.; Wei, J.; Chen, T. The Determination Methods of Pesticide Multi-Residues and the Degradation Dynamics in Bitter Melon and Cucumber. *J. Food Sci. Technol.* **2025**. [CrossRef]
165. Ramadan, M.F.A.; Abdel-Hamid, M.M.A.; Altorgoman, M.M.F.; AlGaramah, H.A.; Alawi, M.A.; Shati, A.A.; Shweeta, H.A.; Awwad, N.S. Evaluation of Pesticide Residues in Vegetables from the Asir Region, Saudi Arabia. *Molecules* **2020**, *25*, 205. [CrossRef]

166. Ramadevi, R.; Ramachandraiah, C.; Reddy, G.V.S. Development of a Multi Residue Method for the Quantification of 45 Pesticides Using Gc-Ms/Ms and Study of Peeling Effect on Pesticide Residues in Citrus Fruits. *Orient. J. Chem.* **2023**, *39*, 1145–1155. [CrossRef]
167. Kim, M.; Cho, M.; Kim, S.-H.; Lee, Y.; Jo, M.-R.; Moon, Y.-S.; Im, M.-H. Monitoring and Risk Assessment of Pesticide Residues in Fishery Products Using GC–MS/MS in South Korea. *Toxics* **2024**, *12*, 299. [CrossRef]
168. Zheng, K.; Zheng, H.; Yu, Y.; Su, J.; Chen, L.; Zheng, M.; Liu, L.; Wu, X.; Chen, D.; Meng, X. Simultaneous Determination of Four Pesticides Residues in Rice by Modified QuEChERS Coupled with GC-MS/MS. *J. Food Compos. Anal.* **2024**, *133*, 106396. [CrossRef]
169. Bustamante, C.M.; Bravo, N.; Ruiz, P.; Grimalt, J.O.; Garí, M. Method Optimization for a Simultaneous Determination of Neonicotinoid, Carbamate/Thiocarbamate, Triazole, Organophosphate and Pyrethroid Pesticides and Their Metabolites in Urine Using UPLC-MS/MS. *J. Chromatogr. A* **2024**, *1730*, 465054. [CrossRef] [PubMed]
170. Elgueta, S.; Valenzuela, M.; Fuentes, M.; Ulloa, P.; Ramos, C.; Correa, A.; Molinett, S. Analysis of Multi-Pesticide Residues and Dietary Risk Assessment in Fresh Tomatoes (*Lycopersicum esculentum*) from Local Supermarkets of the Metropolitan Region, Chile. *Toxics* **2021**, *9*, 249. [CrossRef]
171. Crocoli, L.C.; Ramires, N.; Moura, S. Determination of Pesticide Residues in Grapes Consumed in Natura and for Juice and Wine Production by High-Performance Liquid Chromatography with High Resolution Mass Spectrometry (HPLC-HRMS). *Anal. Lett.* **2023**, *56*, 1454–1464. [CrossRef]
172. Zhang, Y.; Li, Z.; Jiao, B.; Zhao, Q.; Wang, C.; Cui, Y.; He, Y.; Li, J. Determination, Quality, and Health Assessment of Pesticide Residues in Kumquat in China. *Foods* **2023**, *12*, 3423. [CrossRef]
173. Wang, B.; Shi, L.; Ren, P.; Qin, S.; Li, J.; Cao, J. Dissipation and Dietary Risk Assessment of the Fungicide Pyraclostrobin in Apples Using Ultra-High Performance Liquid Chromatography–Mass Spectrometry. *Molecules* **2024**, *29*, 4434. [CrossRef] [PubMed]
174. Morsi, R.; Ghoudi, K.; Meetani, M.A. Determination and Health Risk Assessment of Carbamate Pesticide Residues in Date Palm Fruits (*Phoenix dactylifera*) Using QuEChERS Method and UHPLC-MS/MS. *Sci. Rep.* **2024**, *14*, 13064. [CrossRef]
175. Diao, Z.; Di, S.; Qi, P.; Liu, Z.; Wang, Z.; Zhao, H.; Wang, M.; Zhang, C.; Wang, X. Stereoselective Study on Chiral Fungicide Metconazole in Four Kinds of Fruits: Absolute Configuration, SFC-MS/MS Enantioseparation, Degradation and Risk Assessment. *Food Chem.* **2024**, *438*, 137944. [CrossRef]
176. Zušťáková, V.; Dušek, M.; Jandovská, V.; Olšovská, J. Screening and Quantification of Pesticide Residues in Ciders by Liquid Chromatography-High Resolution Mass Spectrometry. *Czech J. Food Sci.* **2023**, *41*, 29–35. [CrossRef]
177. Wang, M.; Wang, C.; Han, X. Selection of Internal Standards for Accurate Quantification of Complex Lipid Species in Biological Extracts by Electrospray Ionization Mass Spectrometry—What, How and Why? *Mass Spectrom. Rev.* **2017**, *36*, 693–714. [CrossRef]
178. Rimayi, C.; Odusanya, D.; Mtunzi, F.; Tsoka, S. Alternative Calibration Techniques for Counteracting the Matrix Effects in GC–MS-SPE Pesticide Residue Analysis—A Statistical Approach. *Chemosphere* **2015**, *118*, 35–43. [CrossRef]
179. Mirres, A.C.d.M.; Silva, B.E.P.d.M.d.; Tessaro, L.; Galvan, D.; de Andrade, J.C.; Aquino, A.; Joshi, N.; Conte-Junior, C.A. Recent Advances in Nanomaterial-Based Biosensors for Pesticide Detection in Foods. *Biosensors* **2022**, *12*, 572. [CrossRef] [PubMed]
180. Xu, L.; Abd El-Aty, A.M.; Eun, J.B.; Shim, J.H.; Zhao, J.; Lei, X.; Gao, S.; She, Y.; Jin, F.; Wang, J.; et al. Recent Advances in Rapid Detection Techniques for Pesticide Residue: A Review. *J. Agric. Food Chem.* **2022**, *70*, 13093–13117. [CrossRef]
181. Dong, J.; Yang, H.; Li, Y.; Liu, A.; Wei, W.; Liu, S. Fluorescence Sensor for Organophosphorus Pesticide Detection Based on the Alkaline Phosphatase-Triggered Reaction. *Anal. Chim. Acta* **2020**, *1131*, 102–108. [CrossRef] [PubMed]
182. Zhao, G.; Zhou, B.; Wang, X.; Shen, J.; Zhao, B. Detection of Organophosphorus Pesticides by Nanogold/Mercaptomethamidophos Multi-Residue Electrochemical Biosensor. *Food Chem.* **2021**, *354*, 129511. [CrossRef] [PubMed]
183. Liu, S.; Zheng, Z.; Li, X. Advances in Pesticide Biosensors: Current Status, Challenges, and Future Perspectives. *Anal. Bioanal. Chem.* **2013**, *405*, 63–90. [CrossRef]
184. Sekhwama, M.; Mpofu, K.; Sivarasu, S.; Mthunzi-Kufa, P. Applications of Microfluidics in Biosensing. *Discov. Appl. Sci.* **2024**, *6*, 303. [CrossRef]
185. Kant, T.; Shrivas, K.; Tejwani, A.; Tandey, K.; Sharma, A.; Gupta, S. Progress in the Design of Portable Colorimetric Chemical Sensing Devices. *Nanoscale* **2023**, *15*, 19016–19038. [CrossRef] [PubMed]
186. Wu, H.; Chen, J.; Yang, Y.; Yu, W.; Chen, Y.; Lin, P.; Liang, K. Smartphone-Coupled Three-Layered Paper-Based Microfluidic Chips Demonstrating Stereoscopic Capillary-Driven Fluid Transport towards Colorimetric Detection of Pesticides. *Anal. Bioanal. Chem.* **2022**, *414*, 1759–1772. [CrossRef]
187. Peng, S.; Wang, A.; Lian, Y.; Zhang, X.; Zeng, B.; Chen, Q.; Yang, H.; Li, J.; Li, L.; Dan, J.; et al. Smartphone-Based Molecularly Imprinted Sensors for Rapid Detection of Thiamethoxam Residues and Applications. *PLoS ONE* **2021**, *16*, e0258508. [CrossRef] [PubMed]

188. Akdag, A.; Işık, M.; Göktaş, H. Conducting Polymer-Based Electrochemical Biosensor for the Detection of Acetylthiocholine and Pesticide via Acetylcholinesterase. *Biotechnol. Appl. Biochem.* **2021**, *68*, 1113–1119. [CrossRef]
189. Hu, H.; Wang, B.; Li, Y.; Wang, P.; Yang, L. Acetylcholinesterase Sensor with Patterned Structure for Detecting Organophosphorus Pesticides Based on Titanium Dioxide Sol-gel Carrier. *Electroanalysis* **2020**, *32*, 1834–1842. [CrossRef]
190. Zhang, X.; Chen, Y.; Ling, J.; Li, J. An Electrochemiluminescence Biosensor for Detection of Methyl Parathion with a Novel Probe Ru-MIL-100 via the Acetylcholinesterase Inhibition Coupled with the Reaction Between HAc and Triethylamine. *Luminescence* **2025**, *40*, e70073. [CrossRef]
191. Zhu, J.; Yin, X.; Zhang, W.; Chen, M.; Feng, D.; Zhao, Y.; Zhu, Y. Simultaneous and Sensitive Detection of Three Pesticides Using a Functional Poly (Sulfobetaine Methacrylate)-Coated Paper-Based Colorimetric Sensor. *Biosensors* **2023**, *13*, 309. [CrossRef]
192. Chen, G.; Jin, M.; Yan, M.; Cui, X.; Wang, Y.; Zheng, W.; Qin, G.; Zhang, Y.; Li, M.; Liao, Y.; et al. Colorimetric Bio-Barcode Immunoassay for Parathion Based on Amplification by Using Platinum Nanoparticles Acting as a Nanozyme. *Microchim. Acta* **2019**, *186*, 339. [CrossRef]
193. Lu, Y.; Tan, Y.; Xiao, Y.; Li, Z.; Sheng, E.; Dai, Z. A Silver@gold Nanoparticle Tetrahedron Biosensor for Multiple Pesticides Detection Based on Surface-Enhanced Raman Scattering. *Talanta* **2021**, *234*, 122585. [CrossRef]
194. Arsawiset, S.; Sansenya, S.; Teepoo, S. Nanozymes Paper−based Analytical Device for the Detection of Organophosphate Pesticides in Fruits and Vegetables. *Anal. Chim. Acta* **2023**, *1267*, 341377. [CrossRef]
195. Qin, J.; Chen, Y.; Jia, L.; Chang, T.; Chen, J.; Li, Y.; Ma, Z.; Zhan, Y.; Yang, H. An Electrochemical Biosensor with Perovskite/AuNPs Composite for Sensitive Determination of Fenitrothion. *Microchem. J.* **2025**, *216*, 114593. [CrossRef]
196. Tessaro, L.; Aquino, A.; de Almeida Rodrigues, P.; Joshi, N.; Ferrari, R.G.; Conte-Junior, C.A. Nucleic Acid-Based Nanobiosensor (NAB) Used for Salmonella Detection in Foods: A Systematic Review. *Nanomaterials* **2022**, *12*, 821. [CrossRef] [PubMed]
197. El Alami, A.; Lagarde, F.; Tamer, U.; Baitoul, M.; Daniel, P. Enhanced Raman Spectroscopy Coupled to Chemometrics for Identification and Quantification of Acetylcholinesterase Inhibitors. *Vib. Spectrosc.* **2016**, *87*, 27–33. [CrossRef]
198. Nie, Y.; Teng, Y.; Li, P.; Liu, W.; Shi, Q.; Zhang, Y. Label-Free Aptamer-Based Sensor for Specific Detection of Malathion Residues by Surface-Enhanced Raman Scattering. *Spectrochim. Acta A Mol. Biomol. Spectrosc.* **2018**, *191*, 271–276. [CrossRef] [PubMed]
199. Xu, M.-L.; Gao, Y.; Han, X.X.; Zhao, B. Detection of Pesticide Residues in Food Using Surface-Enhanced Raman Spectroscopy: A Review. *J. Agric. Food Chem.* **2017**, *65*, 6719–6726. [CrossRef]
200. Chen, H.; An, L.; Li, M.; Liu, H.; Jin, Z.; Ma, H.; Ma, J.; Zhou, J.; Duan, R.; Zhang, D.; et al. A Self-Assembled 3D Nanoflowers Based Nano-ELISA Platform for the Sensitive Detection of Pyridaben. *Food Chem.* **2024**, *445*, 138756. [CrossRef]
201. Dávila, E.L.; Romero, O.R.; Du Laing, G.; Spanoghe, P. Du Use of Enzyme-Linked ImmunoSorbent Assay Technique to Monitor Pesticide Residues in Horticultural Crops. *J. Chem. Eng. Theor. Appl. Chem.* **2021**, *78*, 199–209.
202. López Dávila, E.; Houbraken, M.; Gil Unday, Z.; Romero Romero, O.; Du Laing, G.; Spanoghe, P. ELISA, a Feasible Technique to Monitor Organophosphate, Carbamate, and Pyrethroid Residues in Local Vegetables. Cuban Case Study. *SN Appl. Sci.* **2020**, *2*, 1487. [CrossRef]
203. Li, H.; He, S.; Liu, G.; Li, C.; Ma, Z.; Zhang, X. Residue and Dissipation Kinetics of Toosendanin in Cabbage, Tobacco and Soil Using IC-ELISA Detection. *Food Chem.* **2021**, *335*, 127600. [CrossRef]
204. Jia, M.; E, Z.; Zhai, F.; Bing, X. Rapid Multi-Residue Detection Methods for Pesticides and Veterinary Drugs. *Molecules* **2020**, *25*, 3590. [CrossRef] [PubMed]

Journal of
Xenobiotics

MDPI

Article

Biomimetic Chromatography as a High-Throughput Tool for Screening Bioaccumulation and Acute Aquatic Toxicity of Pesticides

Krzesimir Ciura

Department of Physical Chemistry, Faculty of Pharmacy, Medical University of Gdansk, Aleja Generała Józefa Hallera 107, 80-416 Gdansk, Poland; krzesimir.ciura@gumed.edu.pl

Abstract

Modern pesticide risk assessment relies on data on bioaccumulation and acute aquatic toxicity, yet generating such data is labour-intensive and animal-demanding. This study evaluated whether phospholipid affinity of pesticides, quantified by the chromatographic hydrophobicity index CHI_{IAM} obtained from high-throughput gradient biomimetic chromatography, can serve as a surrogate descriptor of these endpoints. Nineteen pesticides representing different chemical and functional classes were analyzed on IAM.PC.DD2 columns, and CHI_{IAM} values were determined. Bioconcentration factors (BCF) in fish and acute toxicity data (96 h LC_{50} for fish, 48 h EC_{50} for *Daphnia magna*) were retrieved from the Pesticide Properties DataBase. CHI_{IAM} ranged from −12.1 to 54.8 and correlated strongly with $\log_{10}$BCF (r = 0.84) and $\log_{10}LC_{50}$ in fish (r = −0.84), and moderately with $\log_{10}EC_{50}$ for *Daphnia* (r = 0.76). Highly lipophilic pesticides with high CHI_{IAM} showed elevated BCF and low LC_{50}/EC_{50} values, whereas polar compounds with low CHI_{IAM} exhibited negligible bioconcentration and low acute toxicity. Deviations from these trends, for compounds with specific modes of action, highlighted the contribution of mechanisms beyond membrane toxicity. Overall, CHI_{IAM} measured under high-throughput conditions retains prognostic value for ecotoxicological assessment and may serve as a rapid experimental descriptor to support preliminary screening.

Keywords: IAM chromatography; phospholipophilicity; bioaccumulation; aquatic toxicity; fish LC_{50}; *Daphnia magna*; pesticides

1. Introduction

Modern agriculture relies on the routine and systematic use of numerous plant protection products to maintain and improve crop productivity [1]. As a consequence, pesticides, together with other biologically active contaminants such as antibiotics and hormones, are now among the most frequently detected pollutants in aquatic and terrestrial ecosystems [2–4]. These compounds may pose significant risks to non-target organisms [5]. In regulatory environmental risk assessment of plant protection products, substance characterization typically begins with key physicochemical properties such as water solubility, vapour pressure and volatility, dissociation constants, and octanol–water partitioning, which are then integrated with environmental fate and behaviour data including degradation and photolysis in soil and in water–sediment systems, as well as adsorption–desorption related mobility. In the aquatic compartment, core ecotoxicological endpoints include acute toxicity to fish and aquatic invertebrates, reported as 96 h LC_{50} values for fish and 48 h EC_{50} values for *Daphnia magna*, together with effects on primary producers such as algal growth inhibition and tests with aquatic plants, and where relevant chronic and sediment-dweller studies as well as the bioconcentration factor (BCF) [6–9]. These parameters are used within regulatory frameworks to identify chemicals of concern. Nevertheless, generating experimental toxicity and bioaccumulation data for the growing number of active substances is labour-intensive, costly and heavily dependent on animal testing [10]. Consequently, there is a strong incentive to develop alternative in vitro and in silico methods capable of predicting bioaccumulation and aquatic toxicity and thereby supporting environmental risk assessment [11].

Among the various molecular properties that influence these endpoints, lipophilicity has long been recognized as a key determinant of bioaccumulation and baseline toxicity in aquatic organisms [12]. Traditionally, the octanol–water partition coefficient (logKow, often denoted as logP in medicinal chemistry) has served as a surrogate descriptor of a chemical's lipophilicity and its potential to bioaccumulate and it also strongly affects the distribution of pesticides between the aqueous phase, suspended particulate matter and organic matrices, with important consequences for their extraction efficiency, recovery and apparent stability in analytical determinations [2,4]. Numerous studies have reported apparent linear relationships between logKow and log BCF [13]. However, these simple models exhibit well-known limitations. In particular, highly hydrophobic compounds often deviate from the expected linear trend and display lower bioaccumulation than predicted, for example due to reduced bioavailability, sorption to particulate matter or enhanced biotransformation [14]. From an experimental standpoint, the classical shake-flask determination of logKow is also relatively labour-intensive, time-consuming and costly, which further motivates the search for more efficient and biorelevant approaches to characterizing lipophilicity.

One such approach is biomimetic chromatography, originally introduced by Valkó and co-workers in pharmaceutical research for the rapid characterization of bio-physicochemical properties of drug candidates [15–17]. Biomimetic chromatographic systems employ stationary phases coated with biomolecules to mimic key biological interfaces. Among them, immobilized artificial membrane (IAM) chromatography has played a particularly important role. In IAM stationary phases, phosphatidylcholine is covalently bonded to a silica support to form a monolayer that mimics the polar–apolar interface of a cell membrane. Chromatographic indices derived from IAM retention, such as phospholipid affinity parameters obtained under isocratic or gradient conditions, have been successfully used to describe membrane interactions and to support predictions of pharmacokinetic behaviour and nonspecific toxicity.

Several studies have highlighted the potential of IAM chromatography and related systems in environmental and ecotoxicological applications. Fernández-Pumarega et al. compared conventional reversed-phase chromatography, IAM chromatography and micellar electrokinetic chromatography to mimic nonspecific aquatic toxicity and soil sorption of selected organic compounds [18]. Tsopelas et al. demonstrated that IAM retention factors for pharmaceuticals correlate with BCF values in fish, enabling improved bioconcentration predictions for ionizable and polar compounds that are often mispredicted by logKow based models [19]. Stergiopoulos and co-workers showed that IAM retention data for pesticides can be linearly related to acute toxicity values across multiple aquatic species, including fish and *Daphnia*, and that these relationships frequently outperform models relying solely on logKow [14]. In subsequent studies, micellar chromatographic systems were also shown to predict aquatic toxicity of pesticides reasonably well, albeit with slightly

https://doi.org/0.3390/jox16010004

lower performance compared with IAM [20]. Furthermore, gradient IAM protocols and chromatographic hydrophobicity indices (CHI_{IAM}) have been applied to drugs and UV filters to support the prediction of their aquatic toxicity [21].

Despite these advances, the systematic evaluation of structurally diverse pesticides using high-throughput gradient IAM protocols and their direct linkage to standard ecotoxicological endpoints, such as 96 h LC_{50} for fish, 48 h EC_{50} for *Daphnia magna* and BCF in fish, remains limited. The working hypothesis of the present study is that the phospholipid affinity of pesticides, quantified by IAM chromatography, constitutes a major determinant of their bioaccumulative behaviour and acute aquatic toxicity. To address this hypothesis, gradient IAM chromatography was applied to a set of 19 pesticides from different chemical and functional classes, including insecticides, herbicides and fungicides. Using literature-derived ecotoxicological parameters, the ability of IAM-derived descriptors to capture variability in bioaccumulation and acute toxicity across species was assessed, and the usefulness of IAM chromatography as a potential screening tool to support environmental risk assessment of pesticides was evaluated.

2. Materials and Methods

2.1. Analytes

A set of 19 pesticides was selected to cover a broad range of chemical classes and lipophilicities. The compounds included organophosphate and carbamate insecticides (azinphos-methyl, chlorpyrifos, diazinon, fenitrothion, carbaryl), pyrethroid insecticides (bifenthrin, fenpropathrin, permethrin), neonicotinoid insecticides (imidacloprid, acetamiprid), triazine herbicides (atrazine, terbuthylazine), other herbicides (dicamba, imazapyr) and fungicides (azoxystrobin, boscalid, fludioxonil, propiconazole, tebuconazole). Analytical standards were purchased from Supelco (fenitrothion; Steinheim, Germany), VWR Chemicals (azinphos-methyl, chlorpyrifos, diazinon; Radnor, PA, USA), and Sigma-Aldrich (carbaryl, bifenthrin, fenpropathrin, permethrin, imidacloprid, acetamiprid, atrazine, terbuthylazine, dicamba, imazapyr, azoxystrobin, boscalid, fludioxonil, propiconazole, tebuconazole; Steinheim, Germany).

2.2. Biochromatographic Analysis

All biochromatographic experiments were performed using the gradient IAM protocols developed by Valkó and co-workers and previously implemented in this laboratory. Analyses were carried out on a Prominence-i LC-2030C 3D HPLC system (Shimadzu, Tokyo, Japan) controlled by LabSolutions software (v6.81, Kyoto, Japan) and equipped with an IAM.PC.DD2 column (10 × 4.6 mm, 10.0 μm) packed on the newer 'Type B' silica support and protected by a guard column (Regis Technologies, Morton Grove, IL, USA).

Mobile phase A consisted of 50 mM ammonium acetate (VWR International, Leuven, Belgium) in water, adjusted to physiological pH 7.4 with concentrated ammonia solution (Avantor Performance Materials Poland S.A., Gliwice, Poland). Ultrapure water with a resistivity of 18.2 MΩ was obtained from a Milli-Q water purification system (Merck Millipore, Darmstadt, Germany). For phospholipid affinity measurements, mobile phase B was HPLC-grade acetonitrile (Chempur, Piekary Śląskie, Poland). A linear gradient from 0 to 85% B was applied over 5.25 min, followed by an isocratic hold at 85% B for 0.5 min. The mobile phase flow rate was set at 1.5 mL/min, and the column temperature was maintained at 30 °C.

The reference substances used to calculate CHI_{IAM} values were obtained as follows: acetanilide, butyrophenone and octanophenone from Alfa Aesar (Haverhill, MA, USA); acetophenone and paracetamol from Sigma-Aldrich (Steinheim, Germany); and heptanophe-

 https://doi.org/0.3390/jox16010004

none, hexanophenone, propiophenone and valerophenone from Acros Organics (Pittsburgh, PA, USA).

Dimethyl sulfoxide (Avantor Performance Materials Poland S.A., Gliwice, Poland) was used to dissolve the analytes and to prepare stock solutions at a concentration of 200 µg/mL. UV detection was performed in the range 190–300 nm. The injection volume was 5 µL. Each compound was analyzed in duplicate, and the variation in retention times was lower than 2%. Raw retention times for the pesticides and reference standards are available in Zenodo platform https://doi.org/10.5281/zenodo.17672582.

2.3. Toxicity Data

Aquatic toxicity and bioaccumulation data for the pesticides were obtained from the Pesticide Properties DataBase (PPDB; University of Hertfordshire). Three endpoints were considered: (i) 96 h LC_{50} values for rainbow trout (*Oncorhynchus mykiss*), expressed in mg/L (LC_{50} fish); (ii) 48 h EC_{50} values for *Daphnia magna* (mg/L); and (iii) bioconcentration factors (BCF) in fish (dimensionless). The values for each pesticide are summarized in Table 1.

Table 1. Overview of 96 h LC_{50} (fish, mg/L), 48 h EC_{50} (*Daphnia magna*, mg/L), bioconcentration factors (BCF) and IAM chromatographic hydrophobicity indices (CHI_{IAM}) for the studied pesticides.

Pesticide	LC_{50} Fish	EC_{50} *Daphnia*	BCF	CHI_{IAM}
Imidacloprid	83	85	0.61	14.2
Acetamiprid	100	49.8	N.A.	13.4
Atrazine	4.5	85	4.3	27.4
Azinphos-methyl	0.02	0.0011	40	33.2
Azoxystrobin	0.47	0.23	N.A.	31.6
Bifenthrin	0.00026	0.00011	1703	54.4
Boscalid	2.7	5.33	107	38.0
Carbaryl	2.6	0.0064	44	28.80
Chlorpyrifos	0.025	0.0001	1374	48.1
Diazinon	3.1	0.0010	500	39.3
Dicamba	100	41	15	8.2
Fenitrothion	1.3	0.0086	29	38.1
Fenpropathrin	0.0023	0.00053	1100	48.6
Fludioxonil	0.23	0.4	336	43.5
Imazapyr	100	100	2.54	−12.1
Permethrin	0.0125	0.0006	300	54.8
Propiconazole	2.6	10.2	116	38.7
Tebuconazole	4.4	2.79	78	39.5
Terbuthylazine	2.2	21.2	34	34.3

N.A.—data not available.

3. Results and Discussion

Previous studies have demonstrated that IAM chromatography is a powerful tool for linking phospholipid affinity with bioaccumulation and baseline toxicity of various xenobiotics [22,23], including bisphenols [24], UV filters [21], pharmaceuticals [19,23] and selected pesticides [14]. These investigations have provided both local models for structurally related compounds and more global models covering a chemically diverse library of substances [25]. However, most of the available data were generated using earlier IAM stationary phases, such as IAM.MG.PP, which are no longer commercially available. Additionally, analyses were usually performed under isocratic conditions, which do not provide high-throughput capability.

https://doi.org/0.3390/jox16010004

In this context, the present study was designed to evaluate whether IAM indices obtained with a high-throughput gradient protocol on IAM.PC.DD2 columns can be successfully applied to ecotoxicological profiling of pesticides. To achieve this goal, a structurally diverse set of 19 pesticides was analyzed, and the resulting CHI_{IAM} values were related to standard aquatic toxicity and bioaccumulation endpoints.

The obtained CHI_{IAM} values are presented in Table 1 and span a very wide range, reflecting substantial differences in phospholipid affinity among tested pesticides. The most hydrophilic compound in the set was the herbicide imazapyr, which showed a negative CHI_{IAM} (−12.1). Such a negative value indicates that imazapyr eluted almost immediately, essentially before any significant amount of organic modifier was added; this is consistent with its highly polar, anionic nature and low lipophilicity, leading to minimal retention on the phospholipid column. In contrast, the highest CHI_{IAM} values were observed for hydrophobic insecticides such as permethrin (54.8) and bifenthrin (54.4), belonging to the pyrethroid family. Many other insecticides (e.g., chlorpyrifos, fenpropathrin) and certain hydrophobic fungicides (e.g., fludioxonil) also showed high CHI_{IAM} values in the 40 s to 50 s. Moderately lipophilic compounds, including atrazine, carbaryl, boscalid and the triazole fungicides (propiconazole, tebuconazole), exhibited intermediate or high CHI_{IAM} values in the range of approximately 30–50. Meanwhile, polar or moderately polar pesticides such as the neonicotinoids (imidacloprid, CHI_{IAM} 14.2; acetamiprid, 13.4) and the herbicide dicamba (8.2) had low CHI_{IAM} values, signifying low phospholipid affinity consistent with their high water solubility. These results confirm that CHI_{IAM} successfully captures the relative lipophilicity of the pesticides in a biologically relevant, membrane-mimetic context.

Similarly to the wide range of experimentally determined phospholipid affinities, the selected pesticides exhibited a broad spectrum of toxicity towards aquatic organisms. Fish LC_{50} values covered almost five orders of magnitude, from highly toxic pyrethroid insecticides such as bifenthrin ($LC_{50} \approx 0.00026$ mg/L) and fenpropathrin ($\approx$0.0023 mg/L) to practically non-toxic herbicides such as imidacloprid, acetamiprid and imazapyr ($LC_{50} \geq$ 83–100 mg/L). *Daphnia magna* 48 h EC_{50} values revealed an even more pronounced sensitivity range: several insecticides were extraordinarily toxic, with chlorpyrifos and diazinon showing EC_{50} values around 0.0001–0.001 mg/L (sub-ppb levels), whereas herbicides such as atrazine and imazapyr were much less toxic ($EC_{50} \approx 85$ mg/L and $\approx$100 mg/L, respectively), presumably due to their plant-specific mode of action and low bioavailability to the animals.

Fish bioconcentration factors (BCFs) also varied widely and generally reflected lipophilicity. Pesticides with low CHI_{IAM} values (e.g., imidacloprid, CHI_{IAM} 14.2) showed negligible bioconcentration (BCF $\approx$ 0.6, effectively no accumulation in fish tissues), whereas highly lipophilic insecticides such as bifenthrin and chlorpyrifos (CHI_{IAM} > 48) had BCF values on the order of 10^3 (1703 and 1374, respectively), indicating strong bioaccumulation. Intermediate BCFs were observed for moderately hydrophobic compounds (e.g., boscalid, CHI_{IAM} 38.0, BCF $\approx$ 107; terbuthylazine, CHI_{IAM} 34.3, BCF $\approx$ 34). Overall, the data collated in Table 1 illustrate clear qualitative links between membrane affinity (CHIIAM), bioconcentration potential and acute aquatic toxicity. These relationships were further quantified by calculating Pearson correlation coefficients between CHI_{IAM} and $\log_{10}$-transformed endpoints: strong correlations were observed for fish acute toxicity (r = −0.84 for CHI_{IAM} vs. $\log_{10}LC_{50}$), *Daphnia magna* toxicity (r = −0.76 for CHI_{IAM} vs. $\log_{10}EC_{50}$) and bioaccumulation (r = 0.84 for CHI_{IAM} vs. $\log_{10}$BCF), with higher membrane affinity being associated with lower LC_{50}/EC_{50} values and higher BCF (Figure 1).

https://doi.org/0.3390/jox16010004

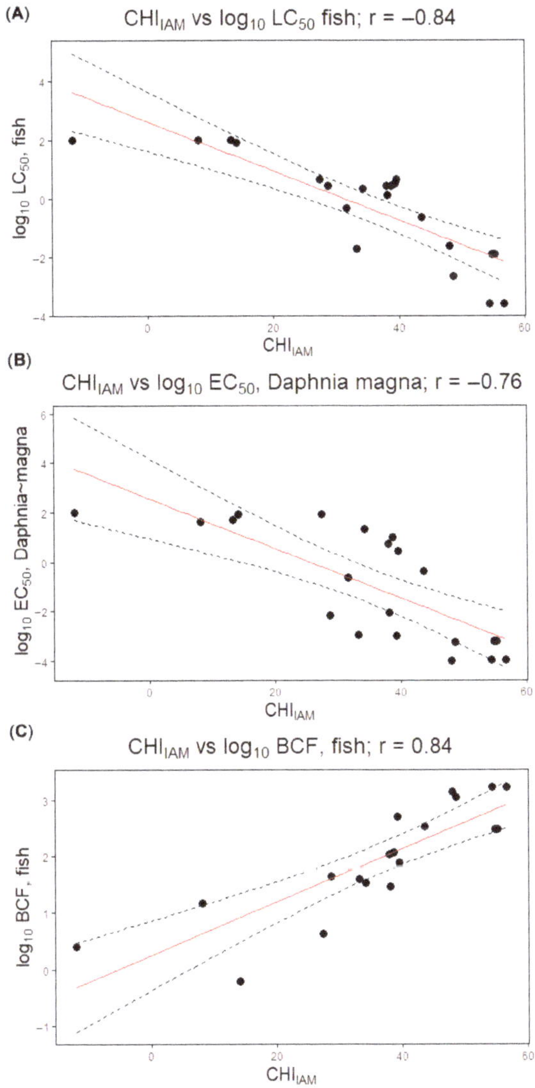

Figure 1. Relationships between CHI_{IAM} and $\log_{10}$-transformed ecotoxicological endpoints for the studied pesticides: (**A**) $\log_{10}LC_{50}$, fish, (**B**) $\log_{10}EC_{50}$, *Daphnia magna*, (**C**) $\log_{10}BCF$, fish. Points represent experimental data, and dashed lines indicate linear regression fits with their 95% confidence limits. The solid red line marks the line of identity (y = x), corresponding to perfect agreement.

These findings are consistent with the concept of baseline toxicity being largely driven by membrane partitioning: hydrophobic organic compounds accumulate in cell membranes, and lethal effects often occur when a critical membrane concentration is reached. In the present data set, the most toxic chemicals to both fish and *Daphnia*, such as bifenthrin, fenpropathrin, chlorpyrifos and azinphos-methyl, combine high CHI_{IAM} values with large BCFs, reflecting their pronounced bioaccumulative and toxic potential. A few deviations from this simple pattern merit mention. Permethrin, for example, has an extremely high CHI_{IAM} (54.8) and is very toxic to fish ($LC_{50} \approx 0.01$ mg/L), but its measured BCF ($\approx$300) is

https://doi.org/0.3390/jox16010004

lower than that of bifenthrin despite comparable CHI_{IAM}, which can plausibly be attributed to faster biotransformation or elimination. Conversely, carbaryl, a moderately hydrophobic carbamate ($CHI_{IAM} \approx 28.8$), is far more toxic to *Daphnia* ($EC_{50} \approx 0.0064$ mg/L) than would be predicted from lipophilicity alone, reflecting its specific neurotoxic mode of action in aquatic invertebrates.

The weaker correlation observed between CHI_{IAM} and *Daphnia* EC_{50} can be rationalized by physiological differences between fish and cladocerans. In fish, the presence of a blood–brain barrier means that neurotoxic effects often depend on a compound's ability to cross this barrier, a process strongly influenced by lipophilicity. In *Daphnia*, the absence of such protective barriers and the open circulatory system allow even relatively hydrophilic toxicants to access neural tissues more directly, reducing the dependence of toxicity on membrane partitioning. As a result, specific modes of action can produce extreme sensitivity irrespective of CHI_{IAM}, leading to outliers and an overall weaker CHI_{IAM}–toxicity relationship.

Nevertheless, the results of this study provide clear evidence that a pesticide's affinity for phospholipid membranes, as quantified by IAM chromatography, is an important determinant of its environmental bioaccumulation and acute toxicity profile. Numerous environmental monitoring studies have shown that pesticides are among the most frequently detected micropollutants in aquatic ecosystems, which underlines the need for efficient tools to prioritize substances for more detailed ecotoxicological and fate assessments [2–4].

One practical implication is therefore the potential use of IAM chromatography as part of an integrated, tiered strategy for environmental risk screening and chemical prioritization. Because CHI_{IAM} can be generated rapidly using short analytical runs, IAM provides a high-throughput, experimentally grounded descriptor that can be combined with conventional physicochemical and fate information to support grouping, ranking, and weight-of-evidence reasoning at early tiers. This direction is consistent with the broader transition in chemical safety assessment towards new approach methodologies, NAMs, and integrated approaches that aim to increase mechanistic interpretability while adhering to the principles of replacement, reduction and refinement of animal testing [26].

Importantly, the role and decision value of routine vertebrate testing in aquatic risk assessment is actively being re-examined, including acute fish toxicity testing for pesticides, in view of both scientific utility and 3R considerations. Within this landscape, rapid non-animal experimental descriptors that help identify substances most likely to drive risk can support more targeted deployment of resource-intensive guideline studies [27–29].

4. Conclusions

The present study shows that the phospholipid affinity of pesticides, quantified by the IAM-based index CHI_{IAM}, is closely related to their bioaccumulation potential and acute aquatic toxicity (LC_{50}, EC_{50}, BCF). Despite several mode-of-action-driven outliers, higher CHI_{IAM} values were generally associated with stronger bioconcentration and lower effect concentrations, supporting the role of membrane partitioning in baseline toxicity. CHI_{IAM} determined with a short high-throughput gradient on IAM.PC.DD2 columns can therefore be considered a useful experimental descriptor for preliminary ecotoxicological screening and prioritization of pesticides.

Funding: This research was funded by the National Science Centre of Poland, grant number 2022/47/D/NZ7/01043.

Institutional Review Board Statement: Not applicable.

Informed Consent Statement: Not applicable.

https://doi.org/0.3390/jox16010004

Data Availability Statement: The data presented in this study are openly available in Zenodo platform at [https://doi.org/10.5281/zenodo.17672582].

Acknowledgments: During the preparation of this manuscript, the author used ChatGPT 5.1 for the purposes of grammatical errors and improve the fluency of the English text. The author has reviewed and edited the output and take full responsibility for the content of this publication. Additionally, I would like to thank Dawid Górski for his support in performing the IAM experiments.

Conflicts of Interest: The author declares no conflicts of interest.

Abbreviations

BCF	fish bioconcentration factors
CHI_{IAM}	chromatographic index of hydrophobicity of IAM
IAM	immobilized artificial membrane

References

1. Chen, X. The Role of Modern Agricultural Technologies in Improving Agricultural Productivity and Land Use Efficiency. *Front. Plant Sci.* **2025**, *16*, 1675657. [CrossRef] [PubMed]
2. Barreca, S.; Busetto, M.; Forni, C.; Colzani, L.; Clerici, L.; Daverio, D.; Balzamo, S.; Calabretta, E.; Peleggi, M.; Dellavedova, P. Determination of Antibiotics, Pesticides, Herbicides, Fungicides and Hormones in Water Bodies in Italy in Occurrence with European Watch List Mechanism by Using an UHPLC-MS/MS System: Method Validation, Quantification and Evaluations. *Pollutants* **2021**, *1*, 207–216. [CrossRef]
3. Colzani, L.; Forni, C.; Clerici, L.; Barreca, S.; Dellavedova, P. Determination of Pollutants, Antibiotics, and Drugs in Surface Water in Italy as Required by the Third EU Water Framework Directive Watch List: Method Development, Validation, and Assessment. *Environ. Sci. Pollut. Res.* **2024**, *31*, 14791–14803. [CrossRef] [PubMed]
4. Barreca, S.; Forni, C.; Colzani, L.; Clerici, L.; Daverio, D.; Dellavedova, P. Study on the Stability of Antibiotics, Pesticides and Drugs in Water by Using a Straightforward Procedure Applying HPLC-Mass Spectrometric Determination for Analytical Purposes. *Separations* **2021**, *8*, 179. [CrossRef]
5. Madesh, S.; Gopi, S.; Sau, A.; Rajagopal, R.; Namasivayam, S.K.R.; Arockiaraj, J. Chemical Contaminants and Environmental Stressors Induced Teratogenic Effect in Aquatic Ecosystem—A Comprehensive Review. *Toxicol. Rep.* **2024**, *13*, 101819. [CrossRef]
6. Sánchez-Bayo, F.; Tennekes, H.A. *Toxicity and Hazard of Agrochemicals*; IntechOpen: Rijeka, Croatia, 2015. [CrossRef]
7. Constantine, L.A.; Burden, N.; Davidson, T.; Dolan, D.G.; Janer, G.; Häner, A.; Lee, M.R.; Maynard, S.K.; Nfon, E.; Perkins, A.N.; et al. Evaluation of the EMA Log Kow Trigger for Fish BCF Testing Based on Data for Several Human Pharmaceuticals. *Regul. Toxicol. Pharmacol.* **2024**, *151*, 105651. [CrossRef]
8. Jonsson, C.M.; de Queiroz, S.C.d.N. Concepts on Accumulation of Pesticides and Veterinary Drugs in Fish: A Review with Emphasis in Tilapia. *Animals* **2023**, *13*, 2748. [CrossRef]
9. Li, J.; Huang, H.; Yang, Y.; Wang, C.; Zhang, Q. Contamination Remediation and Risk Assessment of Four Typical Long-Residual Herbicides: A Timely Review. *Sci. Total Environ.* **2025**, *997*, 180169. [CrossRef]
10. Jaeschke, H.; Ramachandran, A. Are New Approach Methodologies (NAMs) the Holy Grail of Toxicology? *Toxicol. Sci.* **2025**, *208*, 1–8. [CrossRef]
11. Sheng, Q.-S.; Liu, B.; Wang, X.; Hua, L.; Zhao, S.-C.; Sun, X.-Z.; Li, M.-Y.; Zhang, X.-Y.; Wang, J.-X.; Hu, P.-L. Revolutionizing Toxicological Risk Assessment: Integrative Advances in New Approach Methodologies (NAMs) and Precision Toxicology. *Arch. Toxicol.* **2025**, *99*, 4697–4707. [CrossRef]
12. Shavalieva, G.; Papadokonstantakis, S.; Peters, G. Prior Knowledge for Predictive Modeling: The Case of Acute Aquatic Toxicity. *J. Chem. Inf. Model.* **2022**, *62*, 4018–4031. [CrossRef] [PubMed]
13. Fujikawa, M.; Nakao, K.; Shimizu, R.; Akamatsu, M. The Usefulness of an Artificial Membrane Accumulation Index for Estimation of the Bioconcentration Factor of Organophosphorus Pesticides. *Chemosphere* **2009**, *74*, 751–757. [CrossRef] [PubMed]
14. Stergiopoulos, C.; Makarouni, D.; Tsantili-Kakoulidou, A.; Ochsenkühn-Petropoulou, M.; Tsopelas, F. Immobilized Artificial Membrane Chromatography as a Tool for the Prediction of Ecotoxicity of Pesticides. *Chemosphere* **2019**, *224*, 128–139. [CrossRef] [PubMed]
15. Valko, K.L. Biomimetic Chromatography—A Novel Application of the Chromatographic Principles. *Anal. Sci. Adv.* **2022**, *3*, 146–153. [CrossRef]

 https://doi.org/0.3390/jox16010004

16. Valkó, K.L. Lipophilicity and Biomimetic Properties Measured by HPLC to Support Drug Discovery. *J. Pharm. Biomed. Anal.* **2016**, *130*, 35–54. [CrossRef]
17. Valkó, K.L.; Nunhuck, S.B.; Hill, A.P. Estimating Unbound Volume of Distribution and Tissue Binding by in Vitro HPLC-based Human Serum Albumin and Immobilised Artificial Membrane-binding Measurements. *J. Pharm. Sci.* **2011**, *100*, 849–862. [CrossRef]
18. Fernández-Pumarega, A.; Amézqueta, S.; Fuguet, E.; Rosés, M. Tadpole Toxicity Prediction Using Chromatographic Systems. *J. Chromatogr. A* **2015**, *1418*, 167–176. [CrossRef]
19. Tsopelas, F.; Stergiopoulos, C.; Tsakanika, L.-A.; Ochsenkühn-Petropoulou, M.; Tsantili-Kakoulidou, A. The Use of Immobilized Artificial Membrane Chromatography to Predict Bioconcentration of Pharmaceutical Compounds. *Ecotoxicol. Environ. Saf.* **2017**, *139*, 150–157. [CrossRef]
20. Bermúdez-Saldaña, J.M.; Escuder-Gilabert, L.; Medina-Hernández, M.J.; Villanueva-Camañas, R.M.; Sagrado, S. Modelling Bioconcentration of Pesticides in Fish Using Biopartitioning Micellar Chromatography. *J. Chromatogr. A* **2005**, *1063*, 153–160. [CrossRef]
21. Stergiopoulos, C.; Tsopelas, F.; Ochsenkühn-Petropoulou, M.; Valko, K. Predicting the Acute Aquatic Toxicity of Organic UV Filters Used in Cosmetic Formulations. *ADMET DMPK* **2024**, *12*, 781–796. [CrossRef]
22. Tsopelas, F.; Stergiopoulos, C.; Tsantili-Kakoulidou, A. Immobilized Artificial Membrane Chromatography: From Medicinal Chemistry to Environmental Sciences. *ADMET DMPK* **2018**, *6*, 225–241. [CrossRef]
23. Stergiopoulos, C.; Tsopelas, F.; Valko, K. Prediction of hERG Inhibition of Drug Discovery Compounds Using Biomimetic HPLC Measurements. *ADMET DMPK* **2021**, *9*, 191–207. [CrossRef]
24. Russo, G.; Capuozzo, A.; Barbato, F.; Irace, C.; Santamaria, R.; Grumetto, L. Cytotoxicity of Seven Bisphenol Analogues Compared to Bisphenol A and Relationships with Membrane Affinity Data. *Chemosphere* **2018**, *201*, 432–440. [CrossRef]
25. Sobańska, A.W. Affinity of Compounds for Phosphatydylcholine-Based Immobilized Artificial Membrane—A Measure of Their Bioconcentration in Aquatic Organisms. *Membranes* **2022**, *12*, 1130. [CrossRef]
26. Stokes, W.S. Animals and the 3Rs in Toxicology Research and Testing: The Way Forward. *Hum. Exp. Toxicol.* **2015**, *34*, 1297–1303. [CrossRef]
27. Ball, N.; Bars, R.; Botham, P.A.; Cuciureanu, A.; Cronin, M.T.D.; Doe, J.E.; Dudzina, T.; Gant, T.W.; Leist, M.; van Ravenzwaay, B. A Framework for Chemical Safety Assessment Incorporating New Approach Methodologies within REACH. *Arch. Toxicol.* **2022**, *96*, 743–766. [CrossRef]
28. Westmoreland, C.; Bender, H.J.; Doe, J.E.; Jacobs, M.N.; Kass, G.E.N.; Madia, F.; Mahony, C.; Manou, I.; Maxwell, G.; Prieto, P.; et al. Use of New Approach Methodologies (NAMs) in Regulatory Decisions for Chemical Safety: Report from an EPAA Deep Dive Workshop. *Regul. Toxicol. Pharmacol.* **2022**, *135*, 105261. [CrossRef]
29. Schmeisser, S.; Miccoli, A.; von Bergen, M.; Berggren, E.; Braeuning, A.; Busch, W.; Desaintes, C.; Gourmelon, A.; Grafström, R.; Harrill, J.; et al. New Approach Methodologies in Human Regulatory Toxicology–Not If, but How and When! *Environ. Int.* **2023**, *178*, 108082. [CrossRef] [PubMed]

https://doi.org/0.3390/jox16010004

Journal of
Xenobiotics

MDPI

Article

Monte Carlo Simulation of Pesticide Toxicity for Rainbow Trout (*Oncorhynchus mykiss*) Using New Criteria of Predictive Potential

Alla P. Toropova *, Andrey A. Toropov and Emilio Benfenati

Istituto di Ricerche Farmacologiche Mario Negri IRCCS, Department of Environmental Health Science, Via Mario Negri 2, 20156 Milan, Italy; andrey.toropov@marionegri.it (A.A.T.); emilio.benfenati@marionegri.it (E.B.)

* Correspondence: alla.toropova@marionegri.it; Tel.: +39-02-3901-4595

Abstract: *Background*: The toxicity of pesticides for fish in general and Rainbow Trout (*Oncorhynchus mykiss*) in particular is an important ecological indicator required by regulations, and it implies the use of a large number of fish. The number of animals needed would be even higher to evaluate metabolites and pesticide impurities. Considering ethical issues, the costs, and the necessary resources, the use of in silico models is often proposed. *Aim of the study*: We explore the use of advanced Monte Carlo methods to obtain improved results for models testing Rainbow Trout (*Oncorhynchus mykiss*) acute toxicity. Several versions of the stochastic Monte Carlo simulation of pesticide toxicity for Rainbow Trout, carried out using CORAL software, were studied. The set of substances was split into four subsets: active training, passive training, calibration, and validation. Modeling was repeated five times to enable better statistical evaluation. To improve the predictive potential of models, the index of ideality of correlation (IIC), correlation intensity index (CII), and coefficient of conformism of correlation prediction (CCCP) were applied. *Main results and novelty*: The most suitable results were observed in the case of the CCCP-based optimization for SMILES-based descriptors, achieving an R^2 of 0.88 on the validation set, in all five random splits, demonstrating consistent and robust modeling performance. The relationship of information systems related to QSAR simulation and new ideas is discussed, assigning a key role to fundamental concepts like mass and energy. The study of the mentioned criteria of predictive potential during the conducted computer experiments showed that even though they are all aimed at improving the predictive potential, their values do not correlate, except for the CII and the CCCP. This means that, in general, the information impact of the considered criteria has a different nature, at least in the case of the simulation of toxicity for Rainbow Trout (*Oncorhynchus mykiss*). The applicability domain of the model is specific for pesticides; the software identifies potential outliers by looking at rare molecular fragments.

Keywords: CORAL software; Monte Carlo method; pesticide toxicity; QSAR; rainbow trout

1. Introduction

Modern agriculture is a multifunctional system of supplying consumers with numerous needs, from citrus fruits [1–3] to seafood [4–10]. Chemical pollution caused by the agricultural industry significantly impacts customers' health. Hence, technologies that are able to at least reduce the harm of dangerous chemical pollutants are necessary. For instance, owing to the valorization of agro-wastes, citrus by-products can be a precious source of energy and natural beneficial compounds used for manifold applications from biofuels to dietary supplements [11]. A group of toxins produced by Alternaria fungi frequently contaminate tomatoes and tomato products [12], which should be controlled using corresponding fungicides. In relation to seafood, it becomes really necessary to maintain the bio-economy of proteins used in food and medicine applications [13].

Chemical pollution caused by the agricultural industry significantly impacts customers' health. Hence, technologies that are able to at least reduce the harm of dangerous chemical pollutants are necessary. In general, pesticides are widely used in agricultural activities, and there is a debate because they, while improving the economic performance of agricultural processes, nevertheless pollute the environment. The continuing growth of chemical pollution requires a comprehensive assessment of the hazard and risk of pesticides used to increase crop yields; indeed, pesticides may severely affect aquatic and other ecosystems [9,10,14,15]. In relation to aquatic environments, toxicity to fish is perhaps the most informative and used environmental indicator. A reliable assessment of toxicity to fish and the effects of substances that are pesticides, fungicides, or other xenobiotics is critical to the protection and maintenance of ecosystem stability. Pesticides may accumulate in fish, producing adverse effects related to a number of mechanisms, such as endocrine disruption, neurotoxicity, and others; in addition, the adverse effect may be indirect, since fish is the last level of the trophic chain, and thus pesticides affecting the previous levels may have consequences on fish, such as in the case of herbicides affecting algae [16–21]. In this regard, environmental risk analysis plays a key role in adequately assessing possible xenobiotic side effects and, therefore, maintaining the ecological balance [22–24].

Since ecology is essentially a changing fragment of universal human reality, researchers need to measure their own research activities and observations against the problems, taking advantage of the achievements of specialists working in related and even distant areas [9,10]. The information received by the researcher must be convenient. The convenience of the information is linked to its heuristic value. In other words, the user (researcher or assessor) must not just receive some numerical values, but also the ability to perceive the phenomenon being studied (the endpoint) in a complex manner, considering the relationship with other phenomena.

The toxicity of organic compounds for fish is an important ecological indicator. It can be used to identify reliable substances and predict safer alternatives for agricultural applications. Thus, the information on the aquatic toxicity of organic compounds can be useful not only for the assessment of existing chemicals but also for the implementation of appropriate strategies for bringing safer chemicals to market. Obviously, some kind of information flow should be discussed here, since the list of compounds expected to be future pesticides is growing rapidly. Such an evolution should be considered to increase the efficacy of efforts using the existing list of substances aimed at moving toward greener pesticides. Indeed, there is a need to develop tools for modeling the ecological, biochemical, and physicochemical behavior of generally arbitrary substances that can find practical applications, such as pesticides, biocides, or substances in other sectors [25–27]. While there are guidelines to conduct experimental studies on adult fish and related endpoints, established by the Organization for Economic Co-operation and Development (OECD), the same agency encourages the development of quantitative structure–activity relationships (QSAR) or computer-aided risk assessment of different compounds as potential toxic agents for fish [28–31]. Naturally, this leads to the development of new and diverse computational approaches to solve the problems of modeling the toxicity of various compounds for fish [32–34]. Ideally, these new models should be related to and benefit from established scientific domains. Indeed, environmental toxicology is complex and multidisciplinary. It must incorporate physical, chemical, and biological understanding to resolve present and future environmental problems caused by toxic chemicals [35].

The aim of this work is to evaluate the effectiveness of new criteria for predictive potential simulation of the acute toxicity of organic compounds used as pesticides for Rainbow Trout (*Oncorhynchus mykiss*). These kinds of substances represent the applicability domain of the model, which is further refined considering the presence of rare molecular fragments, as discussed below. From a methodological point of view, the basic idea is that the Monte Carlo method can be a tool for modeling various endpoints in general and specifically for modeling Rainbow Trout toxicity. Thus, it is possible that the statistical quality of models of toxicity for Rainbow Trout can be improved by using new criteria of predictive potential, applying the Monte Carlo approach and its recent developments, using the index ideality correlation, correlation intensity index, and coefficient of conformity of correlative prediction. In our case, the Monte Carlo method is applied within the CORAL software (https://www.insilico.eu/coral, accessed on 27 May 2025) to develop in silico models, which use SMILES as the method to represent the chemical structure, extracting relevant molecular features from this simple representation.

2. Materials and Methods

2.1. Data

Acute toxicity values (expressed as lethal concentration 50%—LC50) of 311 organic pesticides for Rainbow Trout (*Oncorhynchus mykiss*) were taken from the literature [36]. The endpoints were expressed as the negative logarithm of lethal concentration (mM/L). The data were obtained according to the OECD guideline [28]. This guideline indicates several fish species. We used data on Rainbow Trout because they were more abundant.

2.2. Simulation Scheme

The basic steps of building a model within the framework of the used method are as follows: (1) Splitting into active and passive training sets. (2) Optimizing the correlation weights of molecular features extracted from SMILES. (3) Building a regression model linking the descriptor calculated by the correlation weights with the endpoint under study (toxicity for Rainbow Trout). (4) Validating the predictive potential of the model.

In practice, within step (1), the set of all compounds was randomly divided into four subsets of approximately the same number of substances: (i) an active training set ($\approx$25%), (ii) a passive training set ($\approx$25%), (iii) a calibration set ($\approx$25%), and (iv) a validation set ($\approx$25%). Within step (2), the active training set was used to build the model, and then the passive training set was used as an inspector of the model being built. Within step (3), the calibration set was used to determine the overall parameters of the model, avoiding overtraining. Finally, within step (4), the external validation set was used to evaluate the predictive potential of the resulting model (Figure 1). The same process was repeated five times to obtain a more robust statistical basis.

2.3. Optimal Descriptors

Modeling based on optimal descriptors involves several levels of modeling activities [37,38]. The first is the list of structural parameters of the molecules of the training set. The second level is based on the assigned threshold of the list of rare and the list of active parameters. The third level is the list of parameters that are apparent promoters of an increase or apparent promoters of a decrease in the endpoint.

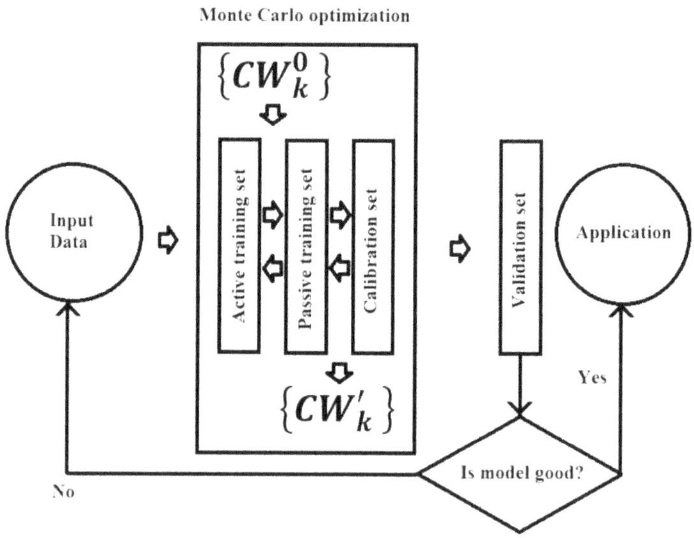

Figure 1. General scheme for building a model.

The selected list of molecular features extracted from the dataset involves the so-called SMILES atoms, which are one symbol or a group of symbols that cannot be considered separately. In addition, pairs of atoms that are neighbors in the SMILES line are recruited in the simulation. Thus, in our modeling approach, the molecular features are directly extracted from the SMILES without using additional software to calculate complex molecular descriptors. Examples of these simple molecular features are presented below in Section 3.

The optimal descriptors considered here are calculated as follows:

$$\mathrm{DCW(T,N)} = \sum \mathrm{CW}(\mathrm{S_k}) + \sum \mathrm{CW}(\mathrm{SS_k}) \quad (1)$$

CW(x) are correlation weights for molecular features extracted from SMILES. S_k is a SMILES atom; SS_k is a pair of SMILES atoms that are neighbors in the SMILES line. It should be noted that the inclusion of the above-listed features in the calculation scheme is more like conducting a public survey than strictly measuring "molecular individuality". Such surveys often contain questions that are incorrect in several directions at once. For example, the question may be unclear to the respondent. In addition, the answers "Yes" or "No" may ambiguously relate to the goals of the survey. Nevertheless, sociological surveys are practiced, and often, the results of such surveys are ultimately useful.

2.4. Optimization of Correlation Weights

The correlation weights of SMILES attributes are calculated using CORAL software (https://www.insilico.eu/coral, accesed on 27 May 2025). The optimization process applied here involves special components termed index of ideality of correlation (IIC) [39] and correlation intensity index (CII) [40]. There is an analogy between the action of the above indices and the action of catalysts. However, if the catalyst should accelerate the chemical reaction, the IIC and CII should increase the "system's attention" to that part of the training set whose position in the statistical probability aspect is closer to the status of the external validation set. In other words, the effect of these "catalysts" of the stochastic information process is aimed at improving the model for the calibration set. Unfortunately, this effect is useful for the calibration set, but, quite often, it reduces the performance observed on the initial (active and passive) training sets. As a result, the forecast is somewhat improved for the calibration set but worsens for the training sets. The influence of these factors can be regulated by using the weighting coefficients for the IIC and CII. The selection of these coefficients is carried out empirically—that is, based on the results of preliminary observations of the stochastic optimization system, with different weights for the IIC and CII. Having numerical data on the correlation weights, one can calculate the long-term toxicity under consideration with this equation:

$$pLC_{50} = C_0 + C_1 \times DCW(T, N) \tag{2}$$

C_0 and C_1 are regression coefficients; T is the threshold used to define rare and non-rare SMILES attributes. SMILES attribute is considered rare if its frequency in the active training set is less than T (rare attributes are not considered; their correlation weights are equal to zero). N is the number of epochs of Monte Carlo optimization. In this study, T = 5 and N = 15 obtained from the computational experiments within the modeling optimization.

To obtain the correlation weights necessary for calculating the descriptors by Equation (1), the following target functions were used, where IIC, CII, and the coefficient of conformism of correlation prediction (CCCP) [40] were considered:

$$TF_0 = R_{AT} + R_{PT} - |R_{AT} - R_{PT}| \times 0.1 \tag{3}$$

$$TF_1 = TF_0 + IIC \times 0.3 \tag{4}$$

$$TF_2 = TF_0 + CII \times 0.3 \tag{5}$$

$$TF_3 = TF_0 + CCCP \times 0.3 \tag{6}$$

R_{AT} and R_{PT} are correlation coefficients between observed and calculated values of pLC_{50} observed for the active training set and the passive training set, respectively.

2.5. Applicability Domain

The applicability domain for the described model, calculated with Equation (2), defines the so-called statistical defects of SMILES attributes. These defects can be calculated as follows [37,38]:

$$d_k = \frac{|P(A_k) - P'(A_k)|}{N(A_k) + N'(A_k)} + \frac{|P(A_k) - P''(A_k)|}{N(A_k) + N''(A_k)} + \frac{|P'(A_k) - P''(A_k)|}{N'(A_k) + N''(A_k)} \tag{7}$$

where $P(A_k)$, $P'(A_k)$ $P''(A_k)$ are the probability of A_k in the active training set, passive training set, and calibration set, respectively; $N(A_k)$, $N'(A_k)$, and $N''(A_k)$ are frequencies of A_k in the active training set, passive training set, and calibration set, respectively. The statistical SMILES-defects (D_j) are calculated as

$$D_j = \sum_{k=1}^{NA} d_k \tag{8}$$

where NA is the number of non-blocked SMILES attributes in the SMILES.

A SMILES falls in the domain of applicability if

$$D_J < 2 \times \overline{D} \tag{9}$$

2.6. Mechanistic Interpretation

The developed QSAR models allow for mechanical interpretation of the studied phenomena. With numerical data on the correlation weights of features that take place in several runs of the Monte Carlo optimization, one can extract three categories of these features [39,40]: (1) features that have a positive value of the correlation weight in all runs. These are promoters of endpoint increase; (2) features that have a negative value of the correlation weight in all runs. These are promoters of endpoint decrease; (3) features that have both negative and positive values of the correlation weight in different runs of the optimization. These are features with unclear roles (one cannot classify these features as promoters of an increase or decrease for the endpoint).

3. Results

Table 1 contains the statistical characteristics of models obtained with target functions TF_1, TF_2, and TF_3. Table 1 shows the complete results obtained with the different subsets. The relevant results are those observed on the calibration set, which corresponds to the final model, and, of course, the results on the validation set, which are substances not used in the model building phase.

Table 1. The statistical quality of models obtained with target functions TF_1, TF_2, and TF_3.

						TF_1					
Split	Set *	n	R^2	CCC	IIC	CII	Q^2	CCCP	RMSE	F	N_A
1	A	79	0.476	0.645	0.549	0.746	0.443	0.024	1.18	70	
	P	77	0.544	0.652	0.545	0.750	0.524	0.159	1.32	89	
	C	78	0.576	0.748	0.759	0.775	0.541	0.301	0.76	103	
	V	77	0.844	-	-	-	-	-	0.66	-	61
2	A	78	0.409	0.581	0.608	0.738	0.377	−0.019	1.28	53	
	P	78	0.409	0.542	0.608	0.701	0.377	−0.235	1.43	53	
	C	77	0.687	0.826	0.829	0.808	0.664	0.386	0.52	165	
	V	78	0.805	-	-	-	-	-	0.57	-	60
3	A	78	0.419	0.590	0.555	0.723	0.388	−0.107	1.41	55	
	P	77	0.381	0.543	0.589	0.710	0.348	−0.209	1.32	46	
	C	78	0.785	0.874	0.883	0.856	0.772	0.536	0.47	277	
	V	78	0.837	-	-	-	-	-	0.57	-	58
4	A	79	0.361	0.531	0.586	0.784	0.328	0.164	1.33	44	
	P	77	0.362	0.493	0.460	0.717	0.325	−0.182	1.49	43	
	C	78	0.738	0.839	0.859	0.842	0.715	0.474	0.53	214	
	V	77	0.831	-	-	-	-	-	0.51	-	61
5	A	78	0.453	0.624	0.640	0.765	0.425	0.177	1.37	63	
	P	78	0.414	0.615	0.499	0.740	0.384	−0.092	1.29	54	
	C	78	0.633	0.792	0.795	0.809	0.609	0.413	0.56	131	
	V	77	0.827	-	-	-	-	-	0.57	-	62
						TF_2					
1	A	79	0.554	0.713	0.656	0.769	0.527	0.240	1.09	96	
	P	77	0.594	0.678	0.585	0.779	0.576	0.140	1.30	110	
	C	78	0.366	0.576	0.573	0.760	0.310	0.063	1.08	44	
	V	77	0.632	-	-	-	-	-	1.10	-	61

Table 1. *Cont.*

Split	Set *	n	R^2	CCC	IIC	CII	Q^2	CCCP	RMSE	F	N_A
						TF_1					
2	A	78	0.498	0.665	0.706	0.751	0.469	0.090	1.18	75	
	P	78	0.497	0.640	0.689	0.740	0.473	−0.014	1.32	75	
	C	77	0.659	0.800	0.623	0.837	0.636	0.466	0.62	145	
	V	78	0.810	-	-	-	-	-	0.70	-	60
3	A	78	0.497	0.664	0.705	0.757	0.472	0.203	1.31	75	
	P	77	0.497	0.623	0.675	0.733	0.467	0.005	1.20	74	
	C	78	0.753	0.837	0.623	0.868	0.733	0.591	0.57	232	
	V	78	0.826	-	-	-	-	-	0.64	-	58
4	A	79	0.482	0.651	0.677	0.761	0.4537	0.218	1.19	72	
	P	77	0.515	0.656	0.628	0.768	0.4897	0.223	1.28	79	
	C	78	0.642	0.801	0.790	0.835	0.6105	0.465	0.65	136	
	V	77	0.720	-	-	-	-	-	0.78	-	61
5	A	78	0.529	0.692	0.657	0.768	0.505	0.097	1.27	85	
	P	78	0.521	0.706	0.704	0.776	0.496	0.179	1.21	83	
	C	78	0.583	0.742	0.545	0.832	0.543	0.484	0.66	106	
	V	77	0.778	-	-	-	-	-	0.72	-	62
						TF_3					
1	A	79	0.440	0.611	0.646	0.744	0.406	0.052	1.22	60	
	P	77	0.458	0.629	0.547	0.741	0.429	0.093	1.41	63	
	C	78	0.696	0.823	0.741	0.827	0.675	0.530	0.64	174	
	V	77	0.886	-	-	-	-	-	0.62	-	61
2	A	78	0.437	0.608	0.661	0.752	0.405	0.079	1.25	59	
	P	78	0.437	0.586	0.651	0.710	0.405	−0.140	1.40	59	
	C	77	0.799	0.890	0.803	0.899	0.785	0.749	0.44	297	
	V	78	0.886	-	-	-	-	-	0.47	-	60
3	A	78	0.456	0.627	0.610	0.740	0.429	0.080	1.36	64	
	P	77	0.456	0.567	0.611	0.726	0.425	−0.134	1.25	63	
	C	78	0.867	0.919	0.869	0.921	0.858	0.796	0.37	495	
	V	78	0.887	-	-	-	-	-	0.46	-	58
4	A	79	0.397	0.568	0.614	0.759	0.368	0.064	1.29	51	
	P	77	0.397	0.544	0.549	0.730	0.362	−0.039	1.43	49	
	C	78	0.858	0.906	0.777	0.926	0.847	0.827	0.40	459	
	V	77	0.883	-	-	-	-	-	0.42	-	61
5	A	78	0.470	0.639	0.651	0.764	0.442	0.208	1.35	67	
	P	78	0.453	0.652	0.614	0.749	0.424	−0.043	1.27	63	
	C	78	0.781	0.872	0.617	0.898	0.765	0.776	0.43	271	
	V	77	0.881	-	-	-	-	-	0.48	-	62

* A = active training set; P = passive training set; C = calibration set; V = validation set; R^2 = determination coefficient; CCC = concordance correlation coefficient; IIC = index of ideality of correlation; CII = correlation intensity index; Q^2 = cross-validated R^2; RMSE = root mean squared error; MAE = mean absolute error; F = Fischer F-ratio; N_A = the number of active (non-rare) SMILES attributes.

The number of outliers according to the statistical defect dk for the considered splits is 6, 5, 15, 4, and 9 for splits 1–5, respectively.

Table 2 contains an example of a mechanistic interpretation for a model observed in TF_3- optimization. One can see that aromaticity, double bonds, chlorine, and sulfur atoms are promoters of an increase in pLC50 for Rainbow Trout, whereas nitrogen and oxygen atoms are promoters of a decrease in toxicity for Rainbow Trout. This is observed for this dataset. However, additional computational experiments are needed to assess how true these data will be for other compounds (which are absent in considered the dataset).

Table 2. Mechanistic interpretation for the model based on TF_3 optimization with split 1.

No.	SMILES Attributes	CWs Probe 1	CWs Probe 2	CWs Probe 3	NA	NP	NC	d_k
1	1...........	0.353	1.271	0.414	70	65	61	0.0011
2	=...........	0.460	0.181	0.482	62	65	62	0.0006
3	c...1.......	0.823	0.266	0.290	57	44	56	0.0019
4	c...c.......	0.484	0.355	0.005	55	44	50	0.0017
5	(...(.......	0.904	0.393	0.512	39	33	36	0.0012
6	Cl..........	1.433	0.096	0.788	38	30	32	0.0018
7	S...........	1.643	1.467	1.151	22	28	21	0.0027
1	O...(.......	−0.128	−0.600	−0.306	59	55	56	0.0004
2	N...........	−0.885	−0.481	−0.966	51	39	45	0.0021
3	N...(.......	−0.685	−0.089	−0.094	37	28	33	0.0021
4	C...=.......	−0.243	−0.087	−0.460	31	37	28	0.0025
5	1...(.......	−0.506	−0.122	−0.522	29	24	24	0.0015
6	n...........	−0.063	−0.402	−0.277	26	13	21	0.0053
7	n...c.......	−0.807	−0.276	−0.770	24	11	21	0.0057

Table 3 contains the statistical characteristics of models for toxicity for Rainbow Trout suggested in the literature. A comparison of the results observed here in the case of the TF_3 optimization for split 1 confirms that the statistical characteristics of the model obtained here are comparable or even better than those from the literature.

Table 3. Comparison of models of toxicity for Rainbow Trout suggested in the literature.

The Number of Compounds in the Training Set	Determination Coefficient for Training Set	The Number of Compounds in Validation Set	Determination Coefficient for Validation Set	Reference
249	0.80	62	0.81	[36]
233	0.67	78	0.86	[39]
66	0.80	22	0.74	[41]
234	0.53	77	0.89	In this work

4. Discussion

In Table 1, one can see that the best predictive potential of the model of acute toxicity for Rainbow Trout was observed in the case of the optimization with target function TF_3. As we commented in the previous section, the relevant values are those related to the calibration and validation sets. Indeed, the results on the active and passive training sets correspond to the initial steps of the model building. The model's initial steps can be attracted by some peculiar substances, forcing the predictions of these substances, which may be relatively unusual and rare. Thus, these preliminary models are quite local, and in most cases, they cannot extract general rules. The use of IIC, CII, or CCCP leads to the attention of molecular features inherent in the calibration set, which "correctly" poses the attention to general characteristics, assigning a lower statistical relevance to the rare substances. This may lead to the above-mentioned paradox: the system acts in such a way as to preserve the statistical quality of the model on the calibration set, even if this "spoils" the model for the training sets.

However, since all three criteria of forecast potential considered here are aimed at the same goal, it seems reasonable to compare their values for all five splits, with three types of optimizations (target functions TF_1, TF_2, and TF_3). Figure 2 graphically shows the proportionality of these values. Considering the results obtained with IIC (Figure 2A), high improvements were obtained using CII (Figure 2B) and CCCP (Figure 2C). It should be noted, nevertheless, that the observed situation is inherent in the conditions considered. Perhaps for other endpoints, this proportionality will have a different form.

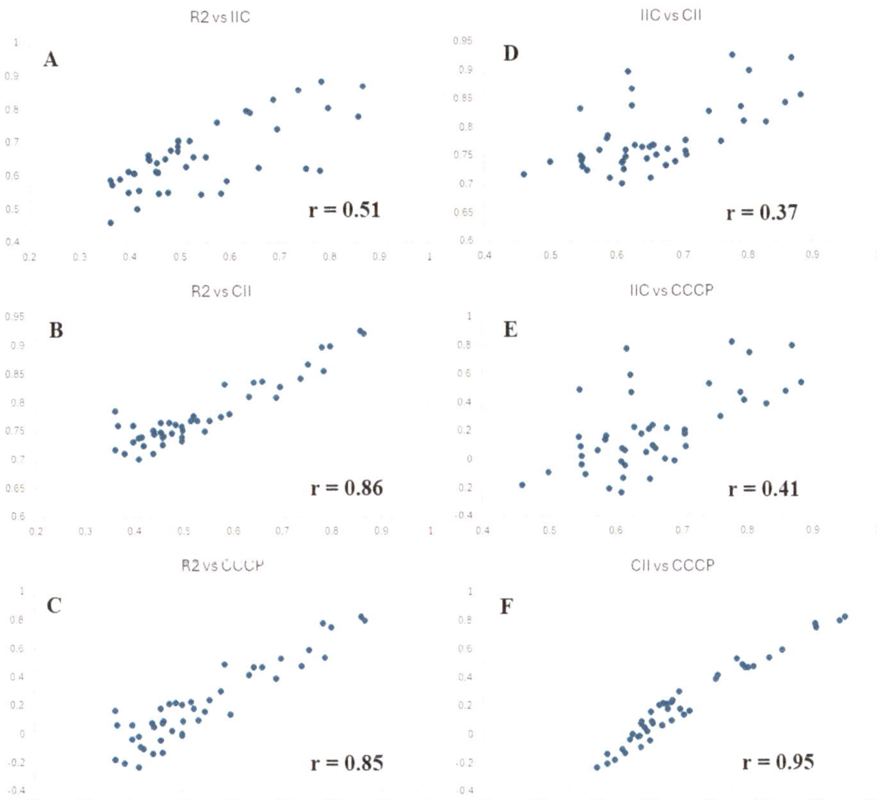

Figure 2. Comparison of the values of IIC, CII, and CCCP observed when developing toxicity models with three different target functions for five random splits into training and validation sets. The correlations of R^2 and IIC (**A**); R^2 and CII (**B**); R^2 and CCCP (**C**); IIC and CII (**D**); IIC and CCCP (**E**); CII and CCCP (**F**) are presented.

Chronologically, the first criterion of predictive potential was tested using IIC. As can be seen in Figure 2, it weakly correlates with the value of the coefficient of determination (Figure 2A). However, the most successful in this study as a "catalyst of predictive potential", CCCP, correlates with the value of the coefficient of determination much better (Figure 2C). It is important to notice, in the data presented in Table 1, that all three above criteria contribute to the improvement in predictive potential. There are correlations between the parameters that have been used. The most highly correlated values are CII and CCCP (Figure 2F). This is probably because these values have a similar origin. They are calculated based on the contributions of the correlation of opponents and supporters. However, CII is calculated based on the contributions of supporters only, while CCCP is calculated considering the contributions of both supporters and opponents of the correlation. Finally, CCCP has quite a good correlation with the determination coefficient (Figure 2C).

At present, the idea of information dynamics, which has some analogy with thermodynamics, is gaining more and more popularity [42]. The principles of infodynamics have been formulated, which, at first glance, are like the principles of thermodynamics. However, there is a certain inconsistency or even contradiction between the indicated ideologies. The first principle of thermodynamics can be formulated as "energy cannot be created or destroyed in an isolated system". The first law of information dynamics can be formulated as "information cannot be created or destroyed but only transmitted or changed". This principle is considered an all-encompassing one and therefore does not in any way imply the isolation of the system, unlike the first law of thermodynamics. The second law of thermodynamics states that the heat of a hotter system only flows toward a colder one. This is an irreversible process that always goes in the direction of increasing entropy (increasing chaos). In contrast to the second law of thermodynamics, the second law of information dynamics states that the information entropy (chaos) of systems containing information states must remain constant or decrease over time, reaching a certain minimum value at equilibrium.

According to the above discussion, it can be concluded that information systems exist similarly to thermal systems. In particular, the "technical" and "statistical" state of the model on the training set and those on the external validation set can be qualified as information systems. One would like to assume that the predictive ability of the model on the training set is a measure of the information content or information value, which can be indirectly, through the corresponding equations, extended to external information systems, i.e., to the validation set. Experience shows, however, that the value or reliability of the model on the validation set is rarely transferred from the training set without losses. One can say that the model is a message generated based on the analysis of the training set. Information entropy in this case can be defined as a measure of the uncertainty of the message source.

Despite the optimism of the second law of infodynamics, one cannot help but notice that in relation to research activities, information entropy (chaos) is far from a rare phenomenon. The desire for tools convenient for practical use often turns out to be futile, and Internet users, instead of the expected comfort, encounter a series of problems of both a bureaucratic and informational nature. If mass and energy can exist outside of human perception, then the information that one would like to talk about must be for human perception. From this perspective, it would be interesting to understand whether and, if so, how the ideas of infodynamics can be useful for QSAR analysis.

Information convenient for perception should be structured according to the user's needs. This may be contrary to the second law of infodynamics, but structured information should not be complex, i.e., requiring many steps for practical application (understanding). An increase in the number of steps is clearly capable of increasing information entropy (chaos). However, the new ideology proposed in the field of any research work should be considered and assessed in terms of its prospects for solving research problems. Whether the laws of infodynamics will help in solving the problems of the QSAR can be shown by relevant experiments and observations.

Here, we consider a system of optimal descriptors extracted from SMILES [39,40]. This is a quite convenient approach to use databases where molecules are presented in a very compact form, containing as much information as possible, which can be sorted and, if necessary, shortened. They reduce the time of computer calculation while still allowing enough chemical representation. The representation of molecular structure can be performed in many ways. The most common ways to accomplish this for QSAR analysis are SMILES and InChI. However, InChI is designed to be used "behind the scenes" by computers [43]. It is usually derived from structure representations by software, whereas SMILES conveniently supports communication between humans and molecular architecture.

In philosophy, the concept of a "dialectical pair" is known, implying a pair of concepts capable of directing thought in some rationally constructive direction, for the formation of useful hypotheses for research work. Examples of dialectical pairs, often mentioned in the general philosophical context, are essence and phenomenon, quality and quantity, space and time, cause and effect, necessity and chance, reality and possibility, matter and consciousness, and information and uncertainty. The practice of working with optimal descriptors leads to the formulation of the following dialectical pair: randomness and prediction [44].

It is obvious that the stochastic processes considered are the personification of randomness. However, due to the objective functions used in the Monte Carlo optimization processes considered here, randomness becomes capable of creating a prediction based on the information provided by SMILES. This gives rise to a number of concerns related to the practice of using SMILES. Firstly, the SMILES used must be homogeneous in their origin, i.e., in the algorithms of their creation. Second, how informative are the selected SMILES? In other words, does the selected SMILES version provide adequate molecular structure information sufficient to identify the molecular features needed for a given endpoint? The structures can be aromatic, with 3D architecture, chirality, and other features. This issue is practically resolved based on the rule: all SMILES participating in the construction of the model must be constructed by the same algorithm. The same applies to SMILES included in the external verification.

Following the principle of "QSPR/QSAR are random events" [39,40], any conclusions related to QSPR/QSAR analysis must be confirmed based on the observation of not one but several distributions of training and validation samples.

Thus, the use of the SMILES format is convenient and adequate for modeling purposes, provided that the same algorithm to generate SMILES is used. Since more and more databases contain the SMILES structure, its use is facilitated. An advantage of our approach is that there is no need to calculate molecular descriptors, since the software we used only needs the SMILES, and the molecular features used for modeling purposes and components, which are alphanumerical symbols derived from the SMILES, as shown in Table 2. This largely simplifies the modeling process and, at the same time, provides an easy representation of the molecular components that may be associated with the adverse effect. Examples of molecular components associated with adverse effects are the presence of chlorine, sulfur, benzene rings, and molecular branching. Indeed, these SMILES attributes extracted by the software have a positive coefficient (CWs), which is consistent in all the cases. Conversely, other SMILES attributes have a negative coefficient, which means that substances containing these components are relatively less toxic. Examples of these mitigating factors are the presence of oxygen, nitrogen, or oxygen in a molecule with branching and the presence of aromatic rings with nitrogen in the ring. Thus, this information may represent useful suggestions when planning safer pesticides. However, it is obvious that SMILES is limited both by the ability to convey fine details of the molecular structure (for example, the possibility of rotational isomerism in dynamics) and complex interactions with the organisms. As we have seen above, Table 2 contains quite general components, without entering into more complex features. In addition, stochastic algorithms cannot be highly accurate. Therefore, the approach under consideration has limitations in accuracy and depends on the selected SMILES variety.

Still, we notice that the approach presented here is highly valuable. Table 3, which compares the results presented here with those obtained with other in silico models, clearly shows that the models presented here for toxicity for Rainbow Trout are among the best, in particular considering the values when predicting new substances: our model gave an R^2 value of 0.89 on the validation sets, higher than those from the other models. The comparison with the values on the training set in Table 3, with all the substances, may be misleading, as discussed above, since it includes the values obtained in the initial phases of the model preparation using the active and training set; it is more correct to compare the results on the calibration set, as shown in Table 1: values range from 0.70 to 0.87 for the calibration set. Thus, the method presented here is convenient for a quick preliminary assessment of acute toxicity for Rainbow Trout (*Oncorhynchus mykiss*). Supplementary Materials section contains technical details on the model for split-1.

The study reported here is aligned with the efforts to use New Approach Methodologies (NAMs). NAMs cover a series of approaches, such as omics and in vitro methods, and include in silico models. We notice that the European Food Safety Authority (EFSA), which in Europe is dealing with pesticides, is exploring the use of NAMs [45]. This perspective aims to increase the sources of information to be used for hazard (in our case) and risk assessment, in general terms. Thus, it is consistent with the efforts to reduce the use of animal studies for ethical reasons, costs, and resources for laboratory in vivo experiments. It also introduces a positive vision, adding novel insights into toxicology. The complexity of the toxicological processes requires multiple tools, and thus initiatives in this direction are welcome. Regulatory considerations complement the scientific point of view. We mentioned that EFSA is actively moving in this direction. The different jurisdictions have different premises and contexts. For instance, in vivo studies are not allowed in Europe for cosmetics, and there is a trend to follow this in other countries. There is an intense dialogue between the scientific community and the authorities regarding a sound assessment of the hazard and risk properties of substances, and this depends on the different sectors, since different conditions apply [46]. Thus, our study is relevant to pesticides, as declared, but it is useful for other cases, too. Biocides are the closest case, since there is a large overlap between these two kinds of substances. Moreover, there are commonalities between some pesticides and some pharmaceuticals. For instance, antimycotic substances are used both as pesticides and pharmaceuticals. Furthermore, from a general point of view, pesticides as pharmaceuticals have a quite complex structure and have been developed to be active through different modes of action, and for this reason, it is possible that our model for pesticides may be applicable, at least partially, to pharmaceuticals.

5. Conclusions

Three criteria of predictive potential for models for acute fish toxicity of Rainbow Trout (*Oncorhynchus mykiss*) are presented here. The models used the Monte Carlo optimization of correlation weights of SMILES attributes. The best results were observed when using the CCCP (coefficient of conformism of correlation prediction), which gave statistical values similar or better than those obtained with other models for the same endpoint. The reliability and predictive potential of the approach considered is tested with five splits of the available data into training samples (active and passive), calibration sets, and external validation sets. The approach used is quite simple, since it only needs the SMILES structure, without the calculation of molecular descriptors. Information on the role of certain molecular features is provided by the model, increasing or decreasing the adverse effects. This information can be applied in the planning of safer pesticides. These models, as with all statistical models, have the limitation of being related to the information present in the training set. The meaning of these models is applicable to the general case represented by the population of substances in the training set and may fail if a quite different molecular skeleton is addressed. Improvements can be achieved using additional experimental data on new substances. It is suggested that the QSAR analysis can be enriched by the ideology of infodynamics if information is understood as something related to human perception. These models will be implemented on the VEGAHUB website (www.vegahub.eu, accessed on 27 May 2025) for free use.

Supplementary Materials: The following supporting information can be downloaded at: https://www.mdpi.com/article/10.3390/jox15030082/s1, Table S1: Technical details for split 1.

Author Contributions: A.P.T.: conceptualization, methodology, resources; writing—original draft preparation; visualization; formal analysis; A.A.T.: conceptualization, methodology, resources; writing—original draft preparation; visualization; formal analysis; writing—review and editing; E.B.: project administration; writing—review and editing. All authors have read and agreed to the published version of the manuscript.

Funding: We acknowledge EFSA for the financial contribution within the project sOFT-ERA, OC/EFSA/IDATA/2022/02.

Institutional Review Board Statement: Not applicable.

Informed Consent Statement: Not applicable.

Data Availability Statement: The original contributions presented in this study are included in the article/Supplementary Material. Further inquiries can be directed to the corresponding author.

Conflicts of Interest: The authors declare no conflicts of interest.

References

1. Auñón-Calles, D.; Pinciroli, M.; Nicolás, E.; Gil-Izquierdo, A.; Gabaldón, J.A.; Sánchez-Iglesias, M.P.; Carbonell-Barrachina, A.A.; Ferreres, F.; García, C.J.; Romero-Trigueros, C. Agronomic and metabolic responses of citrus clementina to long-term irrigation with saline reclaimed water as abiotic factor. *Int. J. Mol. Sci.* **2025**, *26*, 3450. [CrossRef]
2. Arias, A.; Feijoo, G.; Moreira, M.T. Agri-food waste to phenolic compounds: Life cycle and eco-efficiency assessments. *J. Environ. Manag.* **2025**, *380*, 124935. [CrossRef] [PubMed]
3. Avioz, D.; Linker, R.; Raveh, E.; Baram, S.; Paz-Kagan, T. Multi-scale remote sensing for sustainable citrus farming: Predicting canopy nitrogen content using UAV-satellite data fusion. *Smart Agric. Technol.* **2025**, *11*, 100906. [CrossRef]
4. Heshmati, A.; Sadati, R.; Ghavami, M.; Mousavi Khaneghah, A. The concentration of potentially toxic elements (PTEs) in muscle tissue of farmed Iranian rainbow trout (*Oncorhynchus mykiss*), feed, and water samples collected from the west of Iran: A risk assessment study. *Environ. Sci. Pollut. Res.* **2019**, *26*, 34584–34593. [CrossRef] [PubMed]
5. Barrientos, C.; Tapia, J.; Bertrán, C.; Peña-Cortés, F.; Hauenstein, E.; Fierro, P.; Vargas-Chacoff, L. Is eating wild rainbow trout safe? The effects of different land-uses on heavy metals content in Chile. *Environ. Pollut.* **2019**, *254*, 112995. [CrossRef]
6. Tveterås, S.; Asche, F.; Bellemare, M.F.; Smith, M.D.; Guttormsen, A.G.; Lem, A.; Lien, K.; Vannuccini, S. Fish is food—The FAO's fish price index. *PLoS ONE* **2012**, *7*, e36731. [CrossRef]
7. Føre, M.; Frank, K.; Norton, T.; Svendsen, E.; Alfredsen, J.A.; Dempster, T.; Eguiraun, H.; Watson, W.; Stahl, A.; Sunde, L.M.; et al. Precision fish farming: A new framework to improve production in aquaculture. *Biosyst. Eng.* **2018**, *173*, 176–193. [CrossRef]
8. LeDoux, M. Analytical methods applied to the determination of pesticide residues in foods of animal origin. A review of the past two decades. *J. Chromatogr. A* **2011**, *1218*, 1021–1036. [CrossRef]
9. Friedman, M.; Juneja, V.K. Review of antimicrobial and antioxidative activities of chitosans in food. *J. Food Prot.* **2010**, *73*, 1737–1761. [CrossRef]

10. Gephart, J.A.; Pace, M.L. Structure and evolution of the global seafood trade network. *Environ. Res. Lett.* **2015**, *10*, 125014. [CrossRef]
11. D'Amore, T.; Chaari, M.; Falco, G.; De Gregorio, G.; Zaraî Jaouadi, N.; Ali, D.S.; Sarkar, T.; Smaoui, S. When sustainability meets health and innovation: The case of citrus by-products for cancer chemoprevention and applications in functional foods. *Biocatal. Agric. Biotechnol.* **2024**, *58*, 103163. [CrossRef]
12. Pacini, T.; D'Amore, T.; Sdogati, S.; Verdini, E.; Bibi, R.; Caporali, A.; Cristofani, E.; Maresca, C.; Orsini, S.; Pelliccia, A.; et al. Assessment of alternaria toxins and pesticides in organic and conventional tomato products: Insights into contamination patterns and food safety implications. *Toxins* **2025**, *17*, 12. [CrossRef] [PubMed]
13. Smaoui, S.; Hosseini, E.; Tsegay, Z.T.; D'Amore, T.; Varzakas, T. Sustainable valorisation of seafood-derived proteins: Current approaches for recovery and applications in biomedical and food systems. *Food Biosci.* **2024**, *62*, 105450. [CrossRef]
14. Tulcan, R.X.S.; Ouyang, W.; Gu, X.; Lin, C.; Tysklind, M.; Wang, B. Typical herbicide residues, trophic transfer, bioconcentration, and health risk of marine organisms. *Environ. Int.* **2021**, *152*, 106500. [CrossRef]
15. Khoshnood, Z. A review on toxic effects of pesticides in Zebrafish, *Danio rerio* and common carp, *Cyprinus carpio*, emphasising Atrazine herbicide. *Toxicol. Rep.* **2024**, *13*, 101694. [CrossRef]
16. Khoshnood, Z. Effects of environmental pollution on fish: A short review. *Transylv. Rev. Syst. Ecol. Res.* **2017**, *19*, 49–60. [CrossRef]
17. Stehle, S.; Schulz, R. Agricultural insecticides threaten surface waters at the global scale. *Proc. Natl. Acad. Sci. USA* **2015**, *112*, 5750–5755. [CrossRef]
18. Damalas, C.A.; Eleftherohorinos, I.G. Pesticide exposure, safety issues, and risk assessment indicators. *Int. J. Environ. Res. Public Health* **2011**, *8*, 1402–1419. [CrossRef]
19. Horzmann, K.A.; Lin, L.F.; Taslakjian, B.; Yuan, C.; Freeman, J.L. Anxiety-related behavior and associated brain transcriptome and epigenome alterations in adult female zebrafish exposed to atrazine during embryogenesis. *Chemosphere* **2022**, *308*, 136431. [CrossRef]
20. Kucka, M.; Pogrmic-Majkic, K.; Fa, S.; Stojilkovic, S.S.; Kovacevic, R. Atrazine acts as an endocrine disrupter by inhibiting cAMP-specific phosphodiesterase-4. *Toxicol. Appl. Pharmacol.* **2012**, *265*, 19–26. [CrossRef]
21. Olsvik, P.A.; Larsen, A.K.; Berntssen, M.H.G.; Goksøyr, A.; Karlsen, O.A.; Yadetie, F.; Sanden, M.; Kristensen, T. Effects of agricultural pesticides in aquafeeds on wild fish feeding on leftover pellets near fish farms. *Front. Genet.* **2019**, *10*, 794. [CrossRef] [PubMed]
22. Takata, M.; Lin, B.-L.; Xue, M.; Zushi, Y.; Terada, A.; Hosomi, M. Predicting the acute ecotoxicity of chemical substances by machine learning using graph theory. *Chemosphere* **2020**, *238*, 124604. [CrossRef] [PubMed]
23. Austin, T.; Denoyelle, M.; Chaudry, A.; Stradling, S.; Eadsforth, C. European chemicals agency dossier submissions as an experimental data source: Refinement of a fish toxicity model for predicting acute LC50 values. *Environ. Toxicol. Chem.* **2015**, *34*, 369–378. [CrossRef] [PubMed]
24. Raimondo, S.; Barron, M.G. Application of interspecies correlation estimation (ICE) models and QSAR in estimating species sensitivity to pesticides. *SAR QSAR Environ. Res.* **2020**, *31*, 1–18. [CrossRef]
25. Veldman, M.B.; Lin, S. Zebrafish as a developmental model organism for pediatric research. *Pediatr. Res.* **2008**, *64*, 470–476. [CrossRef] [PubMed]
26. Lessman, C.A. The developing zebrafish (*Danio rerio*): A vertebrate model for high-throughput screening of chemical libraries. *Birth Defects Res. Part C Embryo Today Rev.* **2011**, *93*, 268–280. [CrossRef]
27. Hill, A.J.; Teraoka, H.; Heideman, W.; Peterson, R.E. Zebrafish as a Model Vertebrate for Investigating Chemical Toxicity. *Toxicol. Sci.* **2005**, *86*, 6–19. [CrossRef]
28. OECD. *Test No. 203: Fish Acute Toxicity Testing*; Guideline for the Testing of Chemicals, Section 2; OECD Publishing: Paris, France, 2019. [CrossRef]
29. OECD. *Test No. 210: Fish, Early-Life Stage Toxicity Test*; OECD Publishing: Paris, France, 2009; pp. 1–24. Available online: https://www.oecd.org/en/publications/test-no-210-fish-early-life-stage-toxicity-test_9789264203785-en.html (accessed on 27 May 2025).
30. OECD. *Test No. 212: Fish, Short-Term Toxicity Test on Embryo and Sac-Fry Stages*; Guidelines for the Testing Chemicals, Section 2: Effects on Biotic Systems; OECD Publishing: Paris, France, 1998; pp. 1–20. Available online: https://www.oecd.org/en/publications/test-no-212-fish-short-term-toxicity-test-on-embryo-and-sac-fry-stages_9789264070141-en.html (accessed on 27 May 2025).
31. OECD. *Test No. 236: Fish Embryo Acute Toxicity (FET) Test*; OECD Guidelines for the Testing of Chemicals, Section 2; OECD Publishing: Paris, France, 2013; pp. 1–22. Available online: https://www.oecd.org/en/publications/test-no-236-fish-embryo-acute-toxicity-fet-test_9789264203709-en.html (accessed on 27 May 2025).

32. Pulido-Reyes, G.; Moreno-Martín, G.; Gómez-Gómez, B.; Navas, J.M.; Madrid, Y.; Fernández-Cruz, M.L. Fish acute toxicity of nine nanomaterials: Need of pre-tests to ensure comparability and reuse of data. *Environ. Res.* **2024**, *245*, 118072. [CrossRef]
33. Zercher, M.; Coral, J.A.; Nabb, D.; Powers, G.; Jones, A.; Johanning, K. A Comparison of in vitro metabolic clearance of various regulatory fish species using hepatic S9 fractions. *Environ. Toxicol. Chem.* **2024**, *43*, 1390–1405. [CrossRef]
34. Bueso-Bordils, J.I.; Antón-Fos, G.M.; Martín-Algarra, R.; Alemán-López, P.A. Overview of computational toxicology methods applied in drug and green chemical discovery. *J. Xenobiotics* **2024**, *14*, 1901–1918. [CrossRef]
35. McKim, J.M. Physiological and Biochemical Mechanisms that Regulate the Accumulation and Toxicity of Environmental Chemicals in Fish. In *Bioavailability: Physical, Chemical, and Biological Interactions*; CRC Press: Boca Raton, FL, USA, 2024; pp. 179–201. [CrossRef]
36. Jia, Q.; Liu, T.; Yan, F.; Wang, Q. Norm index–based QSAR model for acute toxicity of pesticides toward Rainbow Trout. *Environ. Toxicol. Chem.* **2020**, *39*, 352–358. [CrossRef] [PubMed]
37. Toropova, A.P.; Toropov, A.A. Hybrid optimal descriptors as a tool to predict skin sensitization in accordance to OECD principles. *Toxicol. Lett.* **2017**, *275*, 57–66. [CrossRef] [PubMed]
38. Toropova, A.P.; Toropov, A.A. Does the Index of Ideality of Correlation Detect the Better Model Correctly? *Mol. Inform.* **2019**, *38*, 1800157. [CrossRef]
39. Toropov, A.A.; Toropova, A.P.; Benfenati, E. QSAR model for pesticides toxicity to Rainbow Trout based on "ideal correlations". *Aquat. Toxicol.* **2020**, *227*, 105589. [CrossRef]
40. Toropova, A.P.; Toropov, A.A.; Roncaglioni, A.; Benfenati, E. Monte Carlo simulation of aromatic molecule adsorption on multi-walled carbon nanotube surfaces using coefficient of conformism of a correlative prediction (CCCP). *C-J. Carbon Res.* **2025**, *11*, 7. [CrossRef]
41. Khan, K.; Khan, P.M.; Lavado, G.; Valsecchi, C.; Pasqualini, J.; Baderna, D.; Marzo, M.; Lombardo, A.; Roy, K.; Benfenati, E. QSAR modeling of Daphnia magna and fish toxicities of biocides using 2D descriptors. *Chemosphere* **2019**, *229*, 8–17. [CrossRef]
42. Vopson, M.M. The second law of infodynamics and its implications for the simulated universe hypothesis. *AIP Adv.* **2023**, *13*, 105308. [CrossRef]
43. Drefahl, A. CurlySMILES: A chemical language to customize and annotate encodings of molecular and nanodevice structures. *J. Cheminformatics* **2011**, *3*, 1. [CrossRef]
44. Iovine, N.; Toropova, A.P.; Toropov, A.A.; Roncaglioni, A.; Benfenati, E. Simulation of the long-term toxicity towards Bobwhite Quail (*Colinus virginianus*) by the Monte Carlo method. *J. Xenobiotics* **2025**, *15*, 3. [CrossRef]
45. Usmani, S.M.; Bremer-Hoffmann, S.; Cheyns, K.; Cubadda, F.; Dumit, V.I.; Escher, S.E.; Fessard, V.; Gutleb, A.C.; Léger, T.; Liu, Y.-C.; et al. Review of new approach methodologies for application in risk assessment of nanoparticles in the food and feed sector: Status and challenges. *EFSA Support. Publ.* **2024**, *21*, EN-8826. [CrossRef]
46. Bearth, A.; Roth, N.; Jansen, T.; Holden, L.; Čavoški, A.; Di Consiglio, E.; Hauzenberger, I.; Lee, R.; Mombelli, E.; Tcheremenskaia, O.; et al. New approach methodologies in human health risk assessment across European regulatory frameworks: Status quo, barriers and drivers for regulatory acceptance and use. *Environ. Int.* **2025**, *196*, 109279. [CrossRef] [PubMed]

Journal of
Xenobiotics

MDPI

Perspective

Advancing the One Health Framework in EU Plant Protection Product Regulation: Challenges and Opportunities

Maura Calliera [1,2,*], Ettore Capri [1,2], Nicoleta Alina Suciu [2] and Marco Trevisan [2]

[1] Opera Research Centre, Università Cattolica del Sacro Cuore, 29122 Piacenza, Italy; ettore.capri@unicatt.it
[2] Department for Sustainable Food Process (DISTAS), Università Cattolica del Sacro Cuore, 29122 Piacenza, Italy; nicoleta.suciu@unicatt.it (N.A.S.); marco.trevisan@unicatt.it (M.T.)
* Correspondence: maura.calliera@unicatt.it

Abstract: This paper explores the evolving integration of the One Health framework into the European regulatory landscape for plant protection products, focusing on key scientific and procedural challenges. The analysis addresses three fundamental components of risk evaluation—regulatory complexity, hazard identification, and characterisation—and exposure assessment, while providing an up-to-date overview of emerging policies and challenges affecting the sustainable use of plant protection products in Europe. Addressing these issues requires interdisciplinary collaboration among toxicologists, epidemiologists, ecologists, regulatory authorities, industry stakeholders, and public health experts, working synergistically to tackle complex risks. It emphasises that transitioning to more sustainable and resilient agricultural systems in line with One Health principles requires critically reviewing existing policies. By integrating evolving scientific knowledge with communication and agricultural production needs across diverse European contexts, this approach offers valuable insights to inform future policy development and risk management innovation.

Keywords: One Health; plant protection products; European regulatory framework; risk evaluation

1. Introduction

The "One Health" approach is based on understanding that human, animal, and environmental health are interconnected. It promotes collaboration between scientific disciplines and sectors such as food safety, environmental risk management, and combatting antimicrobial resistance [1].

European (ECDC, ECHA, EEA, EFSA, EMA) and international (FAO, OIE, UNEP, WHO) agencies work together to address global health challenges from a transdisciplinary perspective. This approach is also crucial for the achievement of the UN Sustainable Development Goals, particularly those related to health, environment, and partnerships [2,3].

The European Union has some of the most advanced and stringent legislation on food safety and environmental protection. Among the emerging priorities, reducing the use of Plant Protection products (PPPs) is an objective shared by many institutions, partly in response to the growing public awareness of the associated risks. However, risk assessment and risk management remain a central issue, especially in light of the new challenges posed by climate change, biodiversity loss, and the need to ensure global food security. By recognising the interdependence of human, animal, plant, and environmental health, One Health promotes integrative and sustainable strategies that jointly mitigate climate impacts while preserving ecosystem integrity [4]. For example, adopting sustainable agricultural practices aligned with One Health can enhance ecosystem resilience, reduce greenhouse gas emissions, and safeguard biodiversity. Furthermore, coordinated disease and pest surveillance that integrates data from human, animal, plant health, and environmental monitoring, enables early detection and response to emerging threats exacerbated by climate change, helping to mitigate risks holistically.

Risk analysis is a scientific and objective process aimed at estimating the likelihood of adverse effects on human health and the environment from exposure to xenobiotics. It involves the definition of a safe use dose and involves a multidisciplinary community of experts, constituting one of the pillars of European regulation. [5] However, critical issues persist that limit its effectiveness and full adherence to "One Health" principles. The ambitious objective of reducing reliance on PPPs requires more than top-down mandates; it demands leveraging the specialised expertise and practical knowledge of all stakeholders and allowing adequate time for adaptation across diverse production contexts. The policy shift in Sri Lanka—which started in 2019 and was applied in 2021, forcing a rapid transition to organic farming—led to a severe 20% drop in rice production, accompanied by economic turmoil and increased poverty among farmers. Agronomists had warned that the swift implementation without sufficient preparation or socio-economic safeguards risked undermining food security and rural livelihoods [6,7]. This experience underscores the need for carefully planned, context-sensitive strategies that balance production demands with risk management and include inclusive stakeholder engagement to support resilient food systems. While the goal of achieving a 50% reduction in pesticide use aligns with sustainability and One Health principles, it involves complex trade-offs relating to agricultural productivity, economic stability, and food security.

The implementation of the "One Health" approach in the regulation and management of PPPs in Europe is therefore a complex but unavoidable challenge for the protection of public health and the environment. A critical analysis of available policies and knowledge is essential to guide the transition towards more sustainable and resilient agricultural systems [8].

This paper critically examines the key challenges and limitations in the risk evaluation process of plant protection products (PPPs), beginning with the regulatory framework and emphasising the challenges and opportunities associated with integrating the "One Health" principle.

2. Materials and Methods

Study Design and Scope. The study follows a qualitative narrative synthesis approach aimed at critically evaluating regulatory complexity, scientific advancements in hazard identification and exposure assessment, risk characterisation methodologies, and risk communication strategies. The scope includes evaluation of EU legislation; scientific assessment process challenges, including innovative methodologies; and stakeholder engagement within a One Health context. The methodology is grounded in a structured literature and policy review, synthesising scientific, regulatory, and procedural insights from multiple authoritative sources. References published as articles, book chapters, conference proceeding, systematic review, technical reports, and project website documents were screened to ensure the relevancy with respect to the research aim and after using screening criteria as follows: to be relevant at International and European level; to be the results of a project committed by European or international and well-recognised entity such as, for example, FAO, OECD, EFSA; elements of the overall PPPs risk evaluation procedural and policy process; and relevant and/or updated to respect the on-going debate on PPPs use, sustainability, and the One Health framework.

Data were systematically collected from the following:

- Official European regulatory documents and directives, including Regulation (EC) No 178/2002 [9], Regulation (EC) No 1107/2009 [10], Sustainable Use Directive 2009/128/EC [11], and associated EU policy reforms (e.g., REFIT, Green Deal).
- Reports, guidance, and scientific publications from relevant EU agencies: Science Advice for Policy by European Academies (SAPEA), EFSA, ECHA, ECDC, and EMA.
- International health and environmental organisations such as FAO, WHO, and UNEP.
- Peer-reviewed articles and pilot project case studies.

Analytical Framework. The analysis was organised into thematic sections reflecting the critical pillars of PPP risk evaluation and communication under the One Health paradigm: regulatory framework complexity (Section 3.1), scientific methodologies for hazard and exposure evaluation, risk characterisation challenges (from Sections 3.2–3.4), communication strategies incorporating public and stakeholder perceptions (Section 3.5), and trust-building in chemical risk governance (Section 3.6).

Cross-references between legislative texts, scientific findings, and implementation case studies were employed to identify gaps, inconsistencies, and future opportunities. Emphasis was placed on the bidirectional interaction between risk assessors, risk managers, and societal actors, reflecting the complexity of One Health operationalisation.

3. Results

To facilitate readers' comprehension of the logical progression and thematic structure of Section 3, Figure 1 presents a flowchart summarising the main steps and topics discussed, starting from regulatory framework complexity to risk communication and subjective components in the chemical risk assessment.

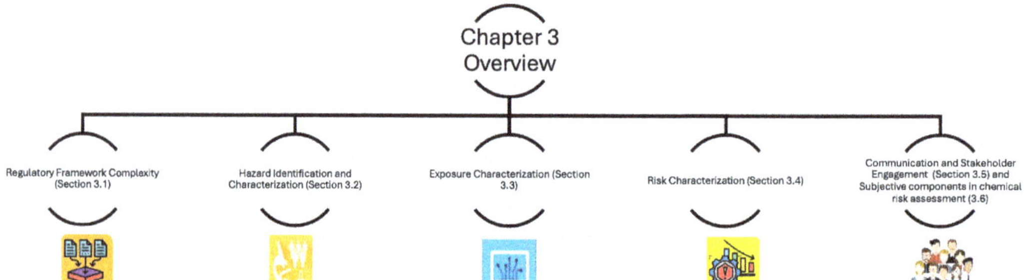

Figure 1. Flow of information, main steps, and key topics discussed in Section 3.

3.1. Complexity of the Legislative Framework

The European Union has developed a robust legislative framework aimed at jointly protecting human, animal, and environmental health. EU chemical legislation has developed over time through the adoption of specific directives and regulations, differentiated according to market types and application sectors. Thus, biocides, industrial chemicals, plant protection products, and human and veterinary medicines are regulated independently. Regulation (EC) No 178/2002 established overarching food safety principles, including risk analysis and transparency, laying the foundation for harmonised pesticide evaluation across the EU. Regulation (EC) No 1107/2009 introduced stricter risk assessment protocols, enhanced transparency and precautionary principles, and explicit criteria for approval, focusing on minimising risks to humans, animals, and ecosystems. The Sustainable Use Directive (2009/128/EC) complements product regulation by promoting integrated pest management (IPM) and reducing pesticide reliance, aiming to balance agricultural productivity with health and environmental sustainability.

This section presents the complex, multilayered European legislative framework governing plant protection products, highlighting how this regulatory complexity poses both challenges and opportunities for effectively implementing a coherent One Health approach.

3.1.1. Regulatory Framework for Plant Protection Products

PPPs, commonly known also as "pesticide," fall within the broader framework of European food legislation. In response to several food incidents in the late 1990s [12], the European Parliament and the Council adopted Regulation (EC) No 178/2002, which lays down the general principles and requirements of food law. This regulation covers all stages of the food chain: from primary production to processing, storage, transport, and retail. The main objective is to ensure a high level of protection of human and animal health, based on a scientific approach that is based on risk analysis. This analysis assesses the likelihood and severity of effects resulting from the presence of hazards in food or feed. The regulation also sets out principles of transparency, independence, and cooperation among authorities involved in food safety, and includes provisions for rapid alert systems and emergency measures in food safety crises [13]. A key feature of this regulation is the establishment of the European Food Safety Authority (EFSA), which provides independent scientific advice and risk assessments to support risk management decisions by the European Commission and Member States. EFSA plays the central role in scientific risk assessment, while the European Commission is responsible for risk management, taking the necessary preventive and control measures [14]. The risk assessment methodologies are continuously updated to address the evolving challenges, such as cumulative risk assessment, transparency, and emerging contaminants. EFSA collaborates closely with national agencies to ensure scientific thoroughness and policy relevance. This framework is structured through a combination of horizontal general rules and vertical specific substance regulations, which together form a harmonised and cooperative system across Member States [15]. Before substances such as food additives, allergens, genetically modified organisms, novel foods (e.g., insects), and PPPs can be used, they must undergo rigorous safety assessments and be authorised according to harmonised European regulations [16].

The plant protection product regulation (Regulation (EC) No 1107/2009) complements this with rigorous risk assessment protocols, ensuring only substances meeting strict safety criteria are approved, incorporating precautionary principles, environmental and operator protections, and periodic re-evaluation. Together, these regulations form a robust legislative framework for chemical food safety and PPPs in the EU, and support the broader One Health approach by addressing chemical food safety risks throughout the entire food production chain, from primary production to consumption, thus safeguarding human health while promoting sustainable agricultural practices.

3.1.2. Evaluation and Authorisation of Plant Protection Products (PPPs) and Risk Management

Regulation (EC) No 1107/2009 primarily regulates PPPs, setting strict criteria to ensure that active substances have no adverse effects on human or animal health or the environment [17]. Currently, the evaluation process for the active substances contained in PPPs involves a dual-level system: at the EU level, the European Food Safety Authority (EFSA), in cooperation with Member States, carries out a peer review of risk assessments and drafts conclusions on active substances, which are then approved or rejected at the Union level by the European Commission. Concurrently, Member States perform evaluations and authorisation of individual PPPs at the national level, based on the approved active substances and considering national agricultural and environmental contexts [18].

This system allows for rigorous, centralised scientific peer review of substances, ensuring high safety and environmental standards; however, it also introduces variability in the implementation of risk management measures, due to differing national conditions, agricultural practices, and regulatory interpretations. Harmonisation remains both possible and essential within the One Health context in other critical areas, such as scientific risk assessment methodologies and standardised data requirements. In summary, while practical regulatory implementation adapts to local contexts, harmonisation at the scientific, methodological, and procedural levels is vital to integrate One Health principles effectively across the EU PPP regulatory system [19].

On the other side, risk management is also regulated during the marketing and use of PPPs by an extensive body of legislation. Directive 2009/128/EC on the sustainable use of pesticides (SUD) is an important tool to promote responsible use and reduce environmental and health risks. The SUD represents a significant step towards One Health by advocating integrated pest management and reduced PPP reliance, contributing to ecosystem health and public safety. Nevertheless, uneven adoption and adaptation among countries highlights ongoing challenges to fully achieving its objectives. Other relevant regulations cover machinery used for applying PPPs, classification and labelling, and the collection of statistical data on agricultural inputs [20–22].

3.1.3. Evaluation of Regulatory Effectiveness and Improvement Initiatives

The effectiveness of regulations is evaluated regularly. After the review of Regulation (EC) No 1107/2009 within the framework of the European Commission's Regulatory Fitness and Performance Programme (REFIT) [23], it was found that, despite progress, implementation of the regulation is still not satisfactory. Member States must do more to reduce dependence on chemical pesticides and improve transparency to increase trust in the system [24–26]. In response, the Commission proposed an amendment to the General Food Law, culminating in Regulation (EU) 2019/1381 [27] on transparency and sustainability of risk assessment, which was applicable from March 2021. These reforms reflect clear One Health objectives by aiming to reduce chemical-related risks, promote sustainable agricultural practices, and protect human, animal, and environmental health concurrently.

In parallel, several legislative proposals have been adopted to coordinate the hazard and risk assessments of EU chemicals that could have horizontal impacts on regulatory regimes, including REACH (Registration, Evaluation, Authorisation, and Restriction of Chemicals) [28] and regulations on biocides and plant protection products. As part of the EU's "zero pollution", a key commitment of the European Green Deal, on 14 October 2020, the European Commission published the Chemical Strategy for Sustainability (CSS) [29], under which the European Commission committed to publishing a proposal for REACH reform, with the aim of better protecting citizens and the environment and promoting innovation for safe and sustainable chemicals. However, despite progress in reducing some types of pollution, achieving the targets is still a challenge. The framework is still under development and all updates can be found on the European Commission's dedicated website [30].

The Commission is also proposing to simplify current assessment methods, improve the quality and consistency of safety assessments across legislation and ensure that resources are used more efficiently to move to a 'one substance, one assessment' process for chemical safety assessments. Within this framework and in response to political and societal ambitions, EFSA has launched a strategic initiative called "European Partnership for Environmental Risk Assessment," based on Next Generation Systems (PERA) [31], which aims for a more coherent and harmonised approach in the characterisation of the environmental risks of chemicals (starting with pesticides), given that there are many commonalities (e.g., common parameters in hazard assessment).

Critical issues in the application of the SUD [32–35] have also been highlighted, following European citizens' initiatives [36] and parliamentary questions. As part of the European Green Deal and the Farm to Fork strategy [37,38], a proposal was made to revise the SUD into a Regulation on the Sustainable Use of Plant Protection Products (SUR). This proposal, which is currently on hold due to geopolitical reasons and economic concerns, aims to reduce the use and risk associated with chemical pesticides by 50 per cent by 2030, with monitoring through risk indicators and harmonised statistical data [39].

The European Green Deal prioritises sustainability with ambitious targets through the Farm to Fork and Biodiversity strategies, fostering resilient agroecosystems and minimising health risks holistically in alignment with One Health. However, practical socio-economic, agricultural, and policy constraints must be addressed for these goals to translate into consistent, measurable outcomes [40].

In conclusion, while these reforms are progressively aligned with One Health ambitions and provide a strong foundation for integrated health protection, further harmonisation, cross-sectoral collaboration, and adaptive governance are crucial to fully realise the One Health vision within EU PPP policy.

3.1.4. Harmonised Risk Indicators and Data Criticality

Recently, two harmonised risk indicators (HRIs) have been introduced, based on the quantities of active substances placed on the market and the granted emergency authorisations. Regulation (EU) 2022/2379 and the subsequent Regulation (EU) 2023/1537 lay down the rules for the collection of harmonised statistical data on agricultural inputs, which have been mandatory since 2025 [41].

On 15 July 2025, the European Commission published updated progress towards pesticide reduction targets for the period 2011–2023 that shows an overall decrease in the use and risk of chemical pesticides and in the use of more hazardous pesticides (https://food.ec.europa.eu/plants/pesticides/sustainable-use-pesticides/pesticide-reduction-targets-progress/eu-trends_en, accessed on 1 November 2025). However, Article 15 of the regulation, concerning indicators, has been the subject of debate due to its ambiguity. An important criticism is that the indicators reflect the hazard rather than the actual risk, i.e., the combination of hazard and exposure, potentially misleading risk management decisions. This limitation can inadvertently encourage substitution with compounds that are less hazardous but not necessarily less risky—as in the case of neonicotinoids and their replacements. Indeed, these indicators provide a simplified metric and do not account for actual exposure levels, dose–response relationships, or the complex interactions that determine real-world risk. In a One Health context—where simultaneous consideration of human, animal, and environmental health is essential—relying solely on hazard-based approaches can result in an incomplete or misleading understanding of risks and limits the ability to realistically assess the effectiveness of mitigation measures, such as buffer strips for environmental protection or precision application techniques. Furthermore, data on actual pesticide use are not yet available at EU level but will be compulsorily available from 2028. Currently, harmonised data on pesticide sales have been collected since 2011 (Eurostat), which show an overall stability between 2011 and 2020, with Germany, France, Spain, and Italy as the main markets [42]. Despite the availability of these data, there remain limitations regarding completeness, level of detail, and national differences, which call for caution in the analysis [43]. Recent studies also indicate that global data on pesticide use are often incomplete and underestimated, with significant differences between countries [44].

Chemical risk assessment integrates hazard identification with exposure characterisation and dose–response analysis to estimate the likelihood and severity of adverse effects. This integrated approach enables a more accurate evaluation of potential impacts across interconnected health domains. The mismatch between hazard-based rankings and risk can challenge regulatory decisions, creating gaps in protecting ecosystems and public health effectively. Consequently, reliance on HRIs alone may compromise policy legitimacy and effectiveness, underscoring the need for more integrated risk metrics that better capture real-world outcomes. Therefore, advancing risk indicators that incorporate exposure data and interdisciplinary knowledge is critical for operationalising One Health principles in PPP regulation. This shift will enhance the ability to predict and mitigate combined human–environmental risks, fostering more informed, balanced, and protective management strategies [45].

To provide a clear overview of the key challenges and corresponding ongoing initiatives within the European regulatory and scientific framework for plant protection products, Table 1 summarises the major barriers to integrating One Health principles in risk assessment and governance, alongside current efforts to address these issues. This synthesis aids in framing the detailed discussion that follows on scientific methodologies, exposure assessment, communication strategies, and policy harmonisation.

Table 1. Key challenges and ongoing initiatives in EU plant protection product risk assessment, aligned with One Health principles.

Challenge	Ongoing/Emerging Initiatives
Cumulative Risk Assessment	Development of cumulative risk assessment methodologies; EFSA guidance on dietary exposure (CRA); refinement techniques such as Expert Knowledge Elicitation (EKE) for uncertainty analysis
Exposure Characterisation	Continuous updates by EFSA on exposure assessment methods; harmonisation of data collection related to exposure sources and vulnerable populations
Adoption of NAMs (New Approach Methodologies)	EFSA's Partnership for Environmental Risk Assessment (PERA); implementation of in vitro and in silico assays; omics and AI technologies; European pilot projects; regulatory validation challenges
Public Trust and Communication	Regulation (EU) 2019/1381 enhancing transparency; inclusive stakeholder engagement; risk communication strategies to build trust and manage risk perception
Methodological Complexity and Data Gaps	Initiatives for harmonised scientific assessments; simplification efforts such as "one substance, one assessment"; continuous updating of evaluation methodologies aligned with evidence

Having established the legislative context, the following section delves into critical scientific challenges, focusing first on hazard identification and characterisation processes.

3.2. Critical Issues Related to Hazard Identification and Characterisation

Despite rigorous scientific evaluation reflected in EFSA's detailed peer reviews of pesticide active substances such as glyphosate [46], several studies identify gaps in the data and the methodological limitations that impact comprehensive hazard characterisation. Missing information and residual concerns about low-dose effects, mixture toxicity, and long-term environmental exposure are frequently noted, underscoring the need for ongoing methodological advancements.

Risk assessment of PPPs is undergoing a transformation, due to the introduction of innovative methodologies, known as NAM (New Approach Methodologies). The following subsections reviews NAMs such as in vitro cell testing, in silico computational models, omics technologies, and artificial intelligence applications, which are designed to complement or replace traditional animal testing in risk assessment. European pilot initiatives showcasing practical progress and remaining challenges in NAMs integration will be presented, with examples from the Netherlands and the ongoing efforts by EFSA's Partnership for Environmental Risk Assessment (PERA). In the following subsections, we examine regulatory and institutional barriers that hinder the uptake of NAMs, including the slow pace of regulatory acceptance, validation constraints, cultural resistance, and legal uncertainties.

3.2.1. New Approach Methodologies (NAMs) Overview

These methodologies promise multiple advantages, including better relevance to human biology, higher mechanistic insights, reduction in animal use, cost-effectiveness, and faster data generation.

In vitro and in silico methods: In vitro tests on cell cultures (e.g., human hepatocytes) allow us to identify specific adverse effects, such as liver toxicity of new pesticides, reducing reliance on animal testing [47]. In parallel, QSAR (Quantitative Structure-Activity Relationship) models have been integrated by EFSA for the preliminary evaluation of active substances, demonstrating efficacy in predicting toxicological properties based on chemical structure [48].

Omics technologies: Metabolomics has revealed metabolic alterations in fish exposed to pesticides in European streams, providing biomarkers for ecotoxicological risk assessment [49]. Transcriptomics and proteomics studies, integrated into Adverse Outcome Pathways (AOPs), have elucidated toxicity mechanisms at the molecular level, as demonstrated in assessments of chronic exposure to neonicotinoids [50].

Artificial intelligence: EFSA pilot projects use machine learning algorithms to predict bioaccumulation and environmental persistence of pesticides, optimising preliminary screening [51]. These approaches analyse toxicological big data to identify complex patterns that are not detectable by traditional methods.

3.2.2. European Case Studies

The afore-mentioned PERA (Partnership for Environmental Risk Assessment) project, coordinated by EFSA, promotes the harmonisation of NAMs in the environmental assessment of pesticides [52]. In the Netherlands, regulatory pilot projects have implemented in vitro tests and in silico models in authorisation procedures, reducing the use of animals by 30% in the last three years, as evidenced by Netherlands National Committee for the protection of animals used for scientific purposes (NCad) [53].

In the context of substances with limited data, the read-across approach, which integrates information from new approach methodologies (NAMs), is regarded as a valuable tool to minimise uncertainties and address data gaps in the risk assessment of chemicals in food and feed. The concept of read-across is based on the principle that structurally or mechanistically similar molecules tend to exhibit comparable properties. This methodology involves identifying data-rich (source) substances that closely resemble a data-poor (target) substance and using data from the source to predict the toxicity of the target. To ensure regulatory consistency and provide risk assessors and applicants with a clear framework for applying this approach, EFSA published comprehensive guidance in July 2025 [54], outlining systematic and transparent procedures for conducting read-across assessments.

3.2.3. Regulatory Challenges

Current EU frameworks, such as Regulation (EC) No 1107/2009, mandate hazard identification primarily through established toxicological and ecotoxicological studies. While these frameworks set stringent standards, the continued reliance on animal testing and classical toxicology—despite recent innovations like New Approach Methodologies (NAMs)—slows adaptation and the integration of more holistic assessment approaches.

Regulatory inertia and the reliance on traditional hazard-based methods seems to limit the NAMs ability to capture the complex interactions between human, animal, and environmental health.

Zaunbrecher et al. [55] conducted a survey involving 1381 toxicologists, 75 per cent of whom were from North America, 15 per cent from Europe, and the remaining 10 per cent from other regions of the world, highlighting how regulatory acceptance represents a significant barrier to the adoption of alternative methods, such as in vitro and in silico.

This delay in adoption is largely attributed to the following:

Rigidity and slowness of regulatory frameworks: Traditional regulations are often based on established methods, such as live animal tests, which enjoy a long history of acceptance and validation. Changing these standards requires time, robust evidence, and a complex review process.

Need for validation and harmonisation: Regulatory authorities require that new methods demonstrate reliability, reproducibility, and biological relevance comparable or superior to traditional methods. This validation process is lengthy and costly, slowing down large-scale adoption.

Cultural and institutional resistance: Even among experts and decision-makers, there may be some mistrust of unconventional methods, especially when it comes to ensuring human and environmental safety, leading to a preference for 'proven' approaches.

Legal and liability implications: The use of methods that are not yet fully recognised may expose companies and regulators to legal risks or challenges, incentivising a cautious and conservative approach.

However, NAMs implementation/adoption in the evaluation process has been introduced, but for PPPs risk assessment, they should be seen as complementing rather than wholly replacing traditional methods, with careful consideration of their current gaps in human and environmental safety evaluation.

Indeed, despite their potential, NAMs have notable shortcomings in assessing human and environmental safety [56]. Critical complex endpoints like reproductive, developmental, neurotoxicity, carcinogenicity, and endocrine disruption effects remain challenging to fully capture with current NAMs. These methods are less able to completely replace animal testing when addressing systemic, multi-organ, or long-term effects, or accounting for population variability without additional supportive data. Regulatory agencies recognise that NAMs currently serve best as screening tools or hazard indicators, rather than sole sources of definitive risk conclusions. The use of NAM data may inform but not replace default safety adjustment factors in risk assessment paradigms to ensure protective outcomes.

In the context of pesticide risk assessment, NAMs are valuable for initial screening, mechanistic understanding, and supporting reduction in animal tests, potentially improving human relevance in safety conclusions. Yet, they must be applied critically by acknowledging their limitations, especially in complex toxicological endpoints and environmental interactions. Continuous efforts are underway to refine NAMs, harmonise data, and integrate emerging AI technologies to improve predictivity and regulatory acceptance while maintaining a precautionary approach toward human and environmental risk.

3.2.4. The Challenge of Integration of New Approach Methodologies (NAMs) Within a One Health Framework

The integration of New Approach Methodologies (NAMs) within a One Health framework aligns well with the holistic perspective that One Health promotes, by reducing reliance on traditional animal testing and enhancing the specificity and relevance of safety assessments, contributing to more ethical and sustainable scientific practices and facilitating cross-sectoral data sharing and interdisciplinary collaboration, bridging gaps between environmental science, toxicology, and public health. The use of in vitro, in silico, omics, and AI technologies enables earlier identification of potential hazards that could affect not only humans but also wildlife and environmental health, fostering preventive actions and integrated risk management.

However, NAMs still require further validation to fully capture complex systemic effects and whether their adoption can help embody One Health principles by promoting safer, more predictive, and more humane approaches to chemical risk assessment that consider the health of all species and ecosystems in unison.

The following section addresses the equally complex issue of exposure characterisation, highlighting multiple exposure routes and the uncertainties that impact risk evaluation.

3.3. *Critical Issues Related to Exposure Characterisation*

This section discusses the various biological and environmental pathways through which humans and animals are exposed to plant protection products, emphasising the complexity of accurately estimating the exposure levels necessary for robust risk assessments.

Humans and animals may be exposed to plant protection products through different routes [57], such as diet, application areas, and the working environment. The main source of exposure for the general population is the diet, due to residues that are mainly on fruit and vegetables, but also on animal products and in drinking water. Occupational exposure occurs mainly on farms, during handling, the application of products, and maintenance of spraying equipment. People living near treated fields, such as residents or bystanders (e.g., cyclists), may also be exposed through drift and volatilisation of pesticides, as well as through contact with surface deposits. For children, exposure may also occur through contact with contaminated surfaces and the subsequent transfer from the hand to the mouth.

Estimating exposure to plant protection products is complex and presents many difficulties. The dose, frequency, duration, and route of exposure (e.g., inhalation, dermal, oral) are not always known or controlled and vary according to individual absorption and metabolism. The exposure level may consist of high doses for short periods or prolonged exposures at low doses, with acute effects appearing within hours or days, while chronic effects may emerge, even after years. Exposure assessment makes it possible, using indicators and models, to determine the level of exposure to the pesticide of non-target organisms and humans. Regarding the latter, different categories are considered, according to the main routes of exposure: consumer, operator, worker, and resident bystander. EFSA has developed a series of guidelines and technical reports that are continuously updated to support risk assessors, including methodologies for cumulative risk assessment (CRA) of dietary exposure to pesticide residues [58]. It is important to note, however, that CRA methodologies remain under development and are currently primarily applied in scientific and methodological contexts. While CRA represents a promising tool for better understanding combined exposure to multiple pesticide residues, it is not yet fully implemented for formal regulatory risk assessment purposes, due to ongoing challenges in methodology validation, data availability, and regulatory harmonisation. The extension of CRA into routine regulatory decision-making awaits further refinement and consensus among risk managers and assessors.

3.3.1. Innovative Epidemiological Studies on Pesticide Exposure

Furthermore, behavioural factors and personal perceptions significantly influence the extent of exposure. Given the high uncertainty, the EU Regulation governing risk assessment for the placing on the market of active substances and formulations requires that where available, epidemiological studies conducted according to recognised standards must be submitted. However, many epidemiological studies on pesticide exposure and health effects have methodological limitations that undermine their use for assessment purposes. A systematic review of the epidemiological evidence conducted by EFSA [59] highlighted several shortcomings, which were also confirmed by the 2017 report of the EFSA Panel [60], which points out that the study of associations between pesticide exposure and health effects is more complex than in other areas of epidemiology. The main difficulties lie in (i) the large number of active substances on the market; (ii) the difficulty in accurately measuring exposure; and (iii) the frequent lack of quantitative and qualitative data on exposure to individual pesticides. In particular, poor exposure characterisation—i.e., the lack of detailed and direct assessments for specific pesticides—is the most significant limitation of most available studies.

To address these issues, EFSA recently published guidelines for the evaluation and integration of evidence from epidemiological studies, for use in scientific assessments [61]. These guidelines promote an approach based on the collection, evaluation, and analysis of scientific evidence (Weight of Evidence, WOE), which involves a structured process of identifying and critically selecting data from different sources (universities, research centres, industry, Eurostat, the scientific literature). WOE analysis includes critical evaluation of studies, data analysis, and transparency in the description of the procedures followed. When empirical evidence is limited, the Expert Knowledge Elicitation (EKE) method is used, following specific protocols and guidelines [62]. In summary, EFSA does not rely on a single study or type of evidence but considers the totality of available evidence to formulate robust and reliable scientific assessments.

3.3.2. Innovative Approaches in Exposure and Effect Assessment

Following the identification and characterisation of hazards, it is crucial to advance innovative approaches in exposure and effect assessment. Better approaches need to be developed that not only include effect assessments but also integrate human biomonitoring (HBM), enhanced comparison of environmental monitoring data, exposure modelling, and evaluation of innovative exposure assessment methods. Assessing these aspects from a legislative perspective is very challenging, both in terms of cost and technical capacity. Overcoming these regulatory barriers is key to accelerating the transition to a more ethical, efficient, and sustainable risk assessment that reduces the use of laboratory animals and exploits the full potential of emerging technologies. This requires continuous dialogue between scientists, regulators, and stakeholders, investment in research and training, and the constant updating of guidelines and regulatory acceptance criteria. This is where the concept of partnership between risk assessors, authorities, and the scientific community can make a difference and help implement innovations in effect tests and exposure models and regulation [63].

We are therefore in a transition phase towards what is called Next Generation Risk Assessment (NGRA). The EU PARC project [64] can provide further insights. The project promotes the development and application of advanced methods, including in vitro, in silico, and integrated approaches based on biological data and computational models, as a continuation and extension of previous projects such as EU-ToxRisk [65] and HBM4EU [66]. The project is closely linked to the European Chemical Strategy for Sustainability and the Green Deal, aiming to better protect health and the environment while reducing the use of animal tests and integrating scientific innovation and regulatory dialogue.

The following section describes how risks are characterised, with an emphasis on cumulative and mixture effects that are relevant to One Health risk assessment.

3.4. Critical Issues Related to Risk Characterisation

The final step is the cumulative risk characterisation (CRA). Various methodologies have been formulated for the purpose of characterising the combined risk posed by PPPs that act in a similar way [67]. One such methodology involves the utilisation of reference values, such as health-based guidance values, for the determination of risk [68]. The concept of dose/concentration addition (DA/CA) involves the normalisation of the potency of each compound to an index compound (IC). Subsequently, to determine the potency of each component relative to the IC, the relative potency factor (RPF) for each compound is utilised to calculate the cumulative residue level of each sample, expressed as IC equivalents. The cumulative risk values for a given common toxic effect are then calculated and combined. The final step in the process is to characterise the risk, which includes describing the variability and the main sources of uncertainty. This approach facilitates the identification of risk contributors from PPPs and food commodities, as well as the affected subpopulations.

Another methodology is the Margin Of Exposure (MOE). The MOE is a measure of a chemical's safety, used to compare its Point Of Departure (POD) and exposure, in which critical effect levels are compared to estimates of exposure, considering uncertainties in the available exposure and toxicological databases. Some regulatory bodies use 'Margin of Safety' (MOS) interchangeably with MOE, while others define it differently. To address inconsistencies in interpretation, and in line with the One Health approach, EFSA recently published a statement with the aim to standardise its terminology, establishing MOE as a primary metric for safety assessments across human and animal health evaluations [69].

For the PPPs mixture, the total MOE (MOE_T) calculates the MOE for a group of chemicals by adding the reciprocal of individual chemicals' MOEs and then taking its reciprocal [70]. Based on this approach and a regulatory threshold of MOE_T, set at 100, EFSA completed several pilot dietary cumulative risk characterisations of pesticide residues, supported by an uncertainty analysis based on Expert Knowledge Elicitation (EKE).

One CRA considered the acute effects on the nervous system [71], while the other focused on two chronic effects on the thyroid system [72]. A third CRA considered the pesticides' effects on cranio-facial alterations [73]. In all three CRAs, at least thirty-one sources of uncertainty affecting the input data, model assumptions, and assessment methods were identified.

Despite the considerable advancements in the CRA of multiple pesticide residues in food that have been made in the past decade, thanks to robust and up to date European diet consumption data that, together with harmonised residue levels monitoring system, allow consumer exposure estimation, the harmonisation of the CRA methods used by the different regulatory agencies is still a challenge. These include methodological difficulties in integrating diverse data across sectors, ongoing data gaps regarding environmental exposures and non-target species, and institutional silos that reduce consistent cross-domain interpretation. Addressing these barriers is essential for applying the MOE concept more effectively within an integrated One Health framework.

Conversely, no such level of information is available to characterise non-dietary exposure to pesticide mixtures in occupational or residential settings. Until now, in fact, no EU requirement has been in place to collect and monitor the agricultural use of PPPs. Consequently, data has only been collected by some Member States within the remit of their national regulations, making it difficult to retrieve. Only recently, the European Commission adopted an implementation of regulation No. 1107/2009, regarding the content and format of records of PPPs kept by professional users, which will only enter into force in 2026. As a result, methodologies or approaches to address non-dietary CRA are currently at the outset [74].

Building on risk characterisation, the following section highlights the essential role of effective risk communication in fostering transparency, trust, and public acceptance of risk management decisions.

3.5. Critical Issues Related to Risk Communication

The effective communication of risk is fundamental to bridging scientific assessment and societal perception, addressing behavioural, cultural, and socio-economic factors that influence the acceptance and implementation of mitigation measures.

Public perceptions of risk vary according to individual values and motivations, which can influence the acceptance of risk management outcomes, and risk communication plays a crucial role in ensuring a two-way dialogue between scientists, managers, and civil society, fostering transparency and trust in the decision-making process. It makes it possible to link the subjective perception of risk—influenced by cognitive and emotional factors—with the objective assessment, thus mitigating discrepancies that may generate unjustified fears or underestimations of real risks.

The human sciences have recognised that behaviour and attitudes are fundamental to proposing effective and acceptable mitigation measures [75]. In the context of the sustainable use of PPPs, it is essential to consider the socio-cultural aspects that influence the acceptance or rejection of technologies and innovations. When risk assessment results for pesticides are not effectively communicated to end-users and the wider community, the discrepancy between objective risk and perceived risk, which is influenced by human behaviour, can lead to poor adherence to the recommended good practices. Misunderstanding or lack of clear information can cause users to underestimate or overestimate risks, resulting in behaviours that deviate from safety guidelines. Therefore, transparent, accessible, and context-sensitive communication is essential to bridge this gap, ensuring, for example, that users not only comprehend the scientific assessments but also trust them enough to follow the prescribed measures that protect their health and the environment.

Proper communication can help to align perceptions with objective risks, promoting compliance and ultimately improving safe pesticide use.

The following two examples illustrate distinct aspects of the impact of risk communication and perception. The first addresses challenges related to the effective implementation and communication of mitigation measures, highlighting how discrepancies between scientific risk assessment and public understanding can hinder compliance. The second example focuses on the case of glyphosate, a widely studied molecule for which, despite extensive scientific data and institutional efforts, communication has fallen short in clearly conveying the health risks. This has resulted in public alarm and diminished trust in regulatory authorities.

These cases exemplify the critical distinction between measurable risk, where outcome probabilities are known, and uncertainty, where outcomes may be identified but their likelihood remains unclear. This distinction underscores the inherent difficulties in communicating complex risk assessments to the public and emphasises the need for transparent, nuanced communication strategies that acknowledge uncertainty and foster informed decision-making.

3.5.1. Example 1—Exposure Management and Mitigation Measures

Risks identified through the risk assessment of PPPs can be effectively mitigated by implementing appropriate measures, which are closely linked to adherence to good agricultural practices. Failure to comply with these practices can lead to unintended exposures and environmental contamination, thereby undermining the outcomes of the risk assessment. Therefore, ensuring that users are fully aware of and comply with the label instructions is essential to safeguarding both worker safety and environmental health.

Recent studies show how economic and socio-cultural factors influence farmers' decisions, which must balance safety, necessity, and cost. Some mandatory measures, such as buffer strips of considerable width, are difficult to implement in specific contexts, highlighting a 'disconnect' between those who define the rules (risk analysis) and those who enforce them (risk management) [76]. Moreover, the current Environmental Risk Assessment (ERA) framework lacks sufficient flexibility to accommodate scenarios for other types of applications (e.g., precision application techniques), active substances with particular properties (e.g., volatile chemicals), or for new technologies (e.g., RNAi). Application methods that can reduce environmental exposure, such as precision application technologies and drift reduction techniques, must be considered in the environmental risk assessment. For such scenarios, conducting an explicit problem formulation would help to develop a more fit-for-purpose environmental risk assessment [77]. A compendium listing conditions of use and mitigation measures for plant protection products was recently published [78], with the aim of reducing human and environmental exposure in the EU, providing a foundation for future mapping and technology validation.

3.5.2. Example 2—The Case of Glyphosate and Uncertainty in Risk Assessment

Risk assessment for substances such as glyphosate inherently involves a degree of uncertainty, due to inconclusive or ambiguous scientific evidence. This uncertainty, however, should not be viewed as a weakness, but rather as an essential part of the scientific process that enhances the precision and clarity of data communication. Misunderstandings often arise because the public may interpret uncertainty as a lack of knowledge or reliability [79]. Additionally, individuals vary widely in how they perceive and weigh risks against potential benefits, influenced by personal values and motivations. While some are highly risk-averse, others may prioritise the potential benefits or even disregard risks altogether. Recognising and addressing these diverse perspectives is crucial for effective communication and the acceptance of risk assessment outcomes [80].

Risk-benefit assessment (RBA) is a multidisciplinary methodology requiring a wide range of expertise and collaboration between scientific disciplines (chemistry, nutrition, toxicology, microbiology, mathematical epidemiology, statistics, etc.), which is applicable to different fields of analysis.

This aspect is taken into account by EFSA, who have recently updated the Dietary Reference Guideline, which focuses on nutrition in relation to human health and considers the risks arising from the presence of hazards in food, weighing them against the benefits derived from diet, food and/or food components [80] in detail. The perception of both risk and potential benefit is one of the elements taken into account. Several factors are reported to have a negative influence on risk perception, including a lack of confidence in communications and data provided by the industry and the order in which information is provided.

Public perception of the risk assessment of PPPs within the European Union exhibits considerable variability. As reported in the Science Advice for Policy by European Academies (SAPEA) report [81], for instance, in 2013, EFSA assessed the risks posed by neonicotinoid pesticides—specifically clothianidin, imidacloprid, and thiamethoxam—to bee populations. Based on evidence identifying these substances as harmful to European honeybees, EFSA recommended restricting their use in PPPs. This decision was broadly accepted by the public. Conversely, EFSA's assessment of glyphosate, following a similar scientific approach, concluded that it is unlikely to be carcinogenic to humans, although small uncertainties remain. Despite transparent communication of this conclusion, the public response was overwhelmingly negative, sparking intense debate. This reaction appears to be unique against the backdrop of more than 8000 assessments published by EFSA over the past 15 years and remains difficult to rationalise fully.

Further complicating public understanding were the contrasting hazard classifications issued by organisations such as the International Agency for Research on Cancer (IARC), which relied exclusively on publicly available data subjects to independent scientific review [82]. These divergent evaluations contributed to confusion and calls from public initiatives to ban glyphosate-based herbicides, citing their links to human cancer and environmental degradation. Such initiatives also advocate for stricter regulatory policies, including the exclusive use of studies commissioned by public authorities, rather than industry stakeholders.

For glyphosate's assessment from 2019 to 2023, 2400 studies were used, more than 400 comments from public consultations were accepted, and 27 national authorities and more than 90 experts were involved. The glyphosate dossier is two to four times larger than a normal dossier produced for the authorisation application procedure. On 28 November 2023, the European Commission renewed the approval of glyphosate for 10 years. This is in line with EU legislation that obliges the Commission to adopt an implementing regulation when there is no qualified majority, for or against, in the standing and appeal committee, as in the case of glyphosate. In favour of public initiatives, it was evident that EFSA has interacted even more with stakeholders, and the transparency of data and information has increased, also thanks to the implementation of the Transparency Regulation (EU) 2019/1381 [83]. Furthermore, more studies have shown that the presence of AMPA (amino-methylphosphonic acid), the main transformation product of glyphosate, is related to other non-agricultural activities, such as weed control for urban purposes, but especially wastewater treatment plants. In fact, AMPA is known to be a transformation product not only of glyphosate, but also of aminopolyphosphonates used in many applications, including laundry detergents. A recent meta-analysis on glyphosate concentrations in river waters in the US and Europe points to the possibility that the main source of glyphosate in Europe may not be herbicide applications, but rather wastewater [84]. This is supported by the comparison of seasonal patterns, constant concentration data even in the absence of agricultural applications, and the presence of glyphosate and AMPA in wastewater treatment plants. Applied mitigation strategies that include restriction of even urban use have not reduced concentrations in surface water, and the concentration patterns of AMPA and glyphosate are strikingly similar, despite the different input pathways. These results indicate the need to review the sources and management strategies for glyphosate in European waters.

3.5.3. One Health Approach and PPPs Public Perception of Risk

Balancing the requirements of chemical risk assessment, therefore, with the social expectations of risk analysis is, in an OH approach, one of the main challenges to be faced for PPPs.

In addition, we must take into account the subjectivity that is inherent in chemical risk assessment, often termed "intuitive toxicology," that has been recognised in the literature since at least 1992 through the pioneering studies of Klaus and Slovic [85]. This concept highlights a natural element of uncertainty stemming from varied interpretations and perceptions of risk among toxicologists, which are influenced by differences in their professional backgrounds—such as regulation, industry, and academia—as well as factors like gender and worldview. In fact, the study found considerable divergence among toxicologists: 41% agreed or strongly agreed with the statement that "if a scientific study produces evidence that a chemical causes cancer in animals, then we can be reasonably certain that it will cause cancer in humans," whereas 58% disagreed or strongly disagreed (with 1% uncertain or neutral). Although scientific methodologies necessarily involve a degree of subjectivity—through assumptions and interpretations, particularly in extrapolation—disagreements among experts on core toxicological principles significantly affect both the perception and credibility of risk assessment results. Such expert disagreements can strongly influence public confidence, often heightening concerns around chemical risks [86], undermining trust in scientific expertise, and complicating public discourse—phenomena vividly illustrated by the controversy surrounding the herbicide glyphosate.

In alignment with the One Health approach, effectively addressing the public perception of risk requires more than scientific risk assessment alone. One Health promotes interdisciplinary collaboration and active engagement among diverse stakeholders, including policymakers, farmers, local communities, and environmental experts, to collectively pursue shared goals of health and sustainability. By integrating local knowledge and socio-cultural factors through participatory methods, understanding and acceptance of risk assessments are improved, thereby enhancing the efficacy of risk management strategies [87]. Such collaborative efforts support the development of sustainable and resilient agricultural systems, embodying the core One Health objective of optimising health outcomes across the human, animal, and ecosystem domains.

Following communication challenges, Section 3.6 discusses subjective components influencing chemical risk assessment and the critical need to rebuild public and stakeholder confidence.

3.6. Subjective Components in Chemical Risk Assessment and the Need to (Re)Build Confidence in the Assessment Process

This final section highlights the subjective nature of risk interpretation among both professionals and the public, underscoring transparency, trust, and stakeholder engagement as fundamental pillars of effective risk governance aligned with One Health principles.

Tensions between industry innovation, regulatory conservatism, and public distrust complicate the implementation of One Health principles in PPP regulation. Regulatory frameworks often lag behind scientific advances, which can slow the adoption of innovations when being included or considered in the panorama of mitigation measures. Meanwhile, public mistrust stemming from perceived opacity or conflicts of interest challenges the effectiveness of risk communication and compliance. Addressing these tensions requires fostering transparent, participatory governance models and interdisciplinary collaboration.

The examples previously discussed emphasise that public trust in the risk assessment process and the expertise involved is a critical determinant of acceptance of the information produced and the perceived balance between risks and benefits. Unfortunately, trust is fragile and often easier to erode than to build, with recent scientific disagreements regarding risk severity tending to undermine public confidence.

Focusing here on the social and cognitive dimensions:

- Risk perception plays a vital role in societal risk management and strongly influences how uncertainty and ambiguity are addressed [88].
- Transformation within these processes requires trust, which in turn depends on transparency—ensuring stakeholders can understand system functioning, identify weaknesses, and see how these are or can be addressed [89].
- Transparency must be coupled with a shared commitment among diverse stakeholders who align on common visions and values that are central to decision-making and strategy [89].
- Behavioural and cultural factors are essential to the exposure assessment and risk management of plant protection products [90].
- Clear communication of uncertainties inherent in exposure and effect assessment is paramount [91].

Table 2 summarises key challenges identified within the EU regulatory and scientific landscape for PPPs and highlights the ongoing initiatives and efforts documented in the paper to address these challenges. The aim is to provide a clear overview of the multifaceted barriers and corresponding opportunities for advancing integrated One Health approaches in risk assessment, governance, and communication.

Table 2. Challenges and opportunities in integrating One Health principles into Plant Protection Product risk assessment.

Key Point	Brief Description	Challenge for One Health Integration
Complex Legislative Framework	The EU has a robust, multilayered legislative framework protecting human, animal, and environmental health through regulations and directives for PPPs. Directive 2009/128/EC promotes integrated pest management and pesticide reduction, balancing productivity and sustainability.	Regulatory inertia slows cross-sector coordination, hindering integrated One Health responses and limiting advancements in agricultural practice.
Harmonised Risk Indicators	Harmonised risk indicators and data collections rely on hazard-based metrics that miss real exposure and risks.	Hazard-based focus limits capturing complex human–environment interactions that are essential to understanding One Health impacts.
Innovative Methodologies (NAMs)	New approaches (in vitro, in silico, omics, AI) promise better hazard identification and reduced animal testing but require validation.	Slow regulatory acceptance and validation delay the integration of innovative science that is necessary for One Health.
Exposure Characterisation	PPP exposure pathways are complex with uncertainties, requiring improved cumulative risk assessment methods.	Scientific and data gaps hamper accurate, holistic exposure assessment across human, animal, and environmental health domains.
Toward One Health	Legislative and scientific landscapes are evolving toward One Health, requiring multidisciplinary collaboration and adaptive governance.	Institutional silos, resource constraints, and policy fragmentation impede fully integrated transdisciplinary One Health action.
Communication	Clear and accessible communication supports the One Health approach, enhancing awareness, collaboration, and transparency in PPP risk assessment.	Effective, transparent communication across diverse stakeholders remains challenging, especially in conveying scientific findings to foster trust and coordinated action in One Health contexts.

4. Discussion

The regulation of Plant Protection Products (PPPs) in the European Union is situated within a broader chemical food safety framework that is increasingly evolving toward integrated approaches. These products hold a crucial role in the EU's food law and Chemical Food Safety framework and are essential for crop protection, ensuring food availability and quality. The interrelation with the One Health approach is increasingly recognised: pesticide management is not only about chemical safety in food but also about protecting environmental health and the interconnected human–animal–ecosystem health interface. One Health encourages integrated risk assessment and regulation that accounts for cumulative exposures, environmental impacts such as biodiversity loss, and indirect effects on human health. The collaboration of EU bodies such as EFSA, ECHA, and ECDC under One Health frameworks aims to operationalise this holistic perspective, enhancing the capacity to anticipate and mitigate combined risks from pesticides, thereby harmonising chemical safety, environmental protection, and public health goals in a sustainable way [92]. But, although the EU has a robust regulatory framework for PPPs, our analysis shows that improvements to the overall process are needed to achieve the Green Deal's ambitious goals.

PPPs are regulated within the EU food safety framework to ensure human, consumer, and non-target organism protection from harmful residues, while the One Health approach fosters the enhanced integration of environmental and health risk assessment, promoting more comprehensive and sustainable pesticide management. To date, the "One Health" perspective has primarily emphasised the interconnectedness of environmental, animal, and human health, with relatively less focus on PPPs, integrated pest management, and agricultural practices, as evidenced by Falkenberg T et al. (2022) [8]. These latter aspects, which are inherently linked to plant health, represent a complex yet indispensable challenge for effectively safeguarding both public health and the environment.

Building on this foundation are the following:

- Current regulatory systems for risk assessment face several challenges that must be addressed. These include the need to evaluate the cumulative risk of an increasing variety of chemicals and mixtures, the complexity of assessing diverse health effects such as endocrine disruption and neurotoxicity, and the emergence of new materials like nanoparticles in product formulations.
- To address ethical and economic concerns, innovative methods [47–51]—including in vivo and in silico methods (e.g., quantitative structure–activity relationships, omics technologies, artificial intelligence, and automated processes)—have also been adopted for PPPs in toxicological assessments. While these advances enhance evaluation, they also demand new skills from assessors, increase the decision-making complexity, and introduce uncertainty, all of which impact the credibility of the process.
- Ethical considerations are central to this evolution. The reduction demand in animal testing and the adoption of more humane, transparent, and precautionary approaches is aligned with the principles of sustainability and One Health [93].
- To meet the ambitious EU objectives, the role of exposure science must also be strengthened. Currently, pesticide risk assessment in Europe is mainly hazard-driven, although guidelines and tools for chemical exposure assessment have evolved to meet the needs of specific policy areas.
- Expert disagreement and scientific subjectivity influence perceptions of risk assessment credibility, particularly when inconsistent interpretations raise public concerns, as exemplified by controversies such as glyphosate, presented in Section 3.5.2 Example 2, for the case of glyphosate and uncertainty in risk assessment.

- Public trust plays a pivotal role: confidence in science reduces perceived risks and boosts perceived benefits, while distrust in industry or conflicting information heightens risk perception. Confusion, conflicting messages, or lack of knowledge further amplify risk concerns. Although PPPs undergo rigorous risk evaluation, public opinions often emphasise environmental and health risks more than the benefits related to food safety and security [82,83].
- To meet the EU's ambitious risk reduction targets, it is essential to assess the effectiveness of mitigation measures, including the potential impact of new technologies and innovations. Improving how the effectiveness of these measures is addressed and calculated in specific contexts could help to minimise human and environmental exposure to PPPs.
- To promote the effective use of PPPs, enhance communication, and bolster confidence in the entire risk assessment process, it is imperative to examine the behavioural and cultural factors, as well as risk perception assessments at contextual and territorial levels. Involving qualified professionals from the agricultural, forestry, livestock, and environmental sectors is essential to improve policy implementation and ensure an integrated approach according to the "One Health" principles and the UN Sustainable Development Goals. Agronomists, farmers, and rural communities play a pivotal role in managing sustainable agricultural practices and safeguarding the environment and rural livelihoods [81]. Their participation is vital to the effectiveness of environmental policies and crucial to preventing rural depopulation and the higher social and environmental costs associated with land management, as highlighted in the most recent paper of Vos et al. (2025) [44], addressing the unresolved challenges between the "One Health" approach and the agricultural sector.

In summary, an objective risk assessment of PPPs requires greater multidisciplinary integration, as well as the continuous updating of methodologies to reflect new scientific evidence and meet the requirements of public health and environmental protection.

A "One Health" approach, which fosters dialogue between science, policy, and civil society, is essential to overcoming the current critical issues and promoting more sustainable and resilient agricultural systems. Collaboration among key EU agencies exemplified by the Cross-Agency One Health Task Force reflects a commitment to operationalise One Health in regulatory science. This joint framework fosters strategic coordination, research synergy, and capacity-building to streamline scientific advice and risk assessments across sectors, underpinning more harmonised, transparent, and sustainable regulatory policies. Consequently, the role of PPPs within the EU chemical food safety framework is progressively oriented toward serving as a vital component of an integrated One Health approach, which is essential for a resilient and balanced protection of global health.

5. Conclusions

This article presents a perspective on the evolving integration of the One Health framework into the European regulatory landscape for PPPs. While grounded in a thorough review of the scientific literature, regulatory policies, and emerging methodologies, the primary aim is to provide a critical interpretative outlook, rather than a conventional systematic review. This perspective fosters a holistic understanding of the interplay between human, animal, and environmental health in the context of chemical risk assessment and regulation.

The study acknowledges the inherent limitations: notably, the reliance on existing data and public policy documents without original experimental research. Furthermore, the rapidly changing regulatory environment and scientific advancements necessitate continuous updates to maintain relevance. Despite these constraints, this perspective offers valuable insights by synthesising information streams and highlighting emerging challenges and opportunities from a multidisciplinary standpoint.

The strength of the One Health approach lies in its integrative and collaborative ethos, promoting cross-sectoral partnerships and inclusive stakeholder engagement. Applying this framework to plant protection product regulation enables more robust, harmonised strategies that simultaneously safeguard human health, animal welfare, and environmental integrity. This is especially critical in addressing complex issues such as chemical mixtures, climate change impacts, and societal expectations regarding transparency and risk communication. To facilitate effective One Health implementation, the establishment of pilot programmes, as proposed at the EU level, is vital. Such initiatives enable the capture of complex interactions across human, animal, plant, and ecosystem health, thereby guiding adaptive management. Governance models must emphasise inclusivity, transparency, and active stakeholder participation to build trust and support dynamic decision-making. Additionally, revising PPP authorisation criteria to explicitly incorporate ecosystem services and sustainability principles is essential for fully aligning regulatory outcomes with the One Health objectives. The Sri Lanka case serves not only as a cautionary example but also as a lens through which to analyse equity, trade dependencies, and disparities in technological access in global pesticide governance. A comprehensive One Health perspective must incorporate global justice dimensions, recognising differential impacts on vulnerable populations and the necessity for equitable access to sustainable technologies and support systems. This is particularly critical, given the transboundary nature of pesticide environmental and health impacts.

Ultimately, this perspective supports strategic innovation in risk assessment methodologies and governance practices, underscoring the indispensable role of the One Health principles. Embracing this approach is essential for advancing sustainable, resilient agricultural systems and comprehensive health protection policies at the European level and beyond.

Author Contributions: Conceptualisation, M.C.; investigation, M.C., N.A.S., E.C. and M.T.; resources, E.C. and M.T.; writing—original draft preparation, M.C.; writing—review and editing, M.C., N.A.S., E.C. and M.T.; supervision, E.C.; project administration, E.C.; All authors have read and agreed to the published version of the manuscript.

Funding: This paper was produced in the framework of a PhD in Agro-Food Systems (AGRISYSTEM)-XXXVIII Cycle of the Università Cattolica del Sacro Cuore (Italy). PNRR Grants Innovative PhDs that meet the innovation needs of companies and D3.2: Research Projects of Strategic Interest at Università Cattolica del Sacro Cuore, (2023), Verso una salute globale (One Health) impatto della conoscenza scientifica in credenze e culture collettive.

Institutional Review Board Statement: Not applicable.

Informed Consent Statement: Not applicable.

Data Availability Statement: No new data were created or analysed in this study.

Conflicts of Interest: The authors declare no conflicts of interest.

References

1. Coghlan, S.; Coghlan, B.J.; Capon, A.; Singer, P. A bolder One Health: Expanding the moral circle to optimise health for all. *One Health Outlook* **2021**, 3, 21. [CrossRef] [PubMed]
2. Joint Statement by European Union Agencies. Cross-Agency Knowledge for One Health Action. 2023. Available online: https://www.efsa.europa.eu/sites/default/files/2023-11/one-health-2023-joint-statement.pdf (accessed on 20 September 2025).

3. World Health Organization. *Joint Tripartite (FAO, OIE, WHO) and UNEP Statement on OHHLEP's Definition of "One Health"*; World Health Organization: Geneva, Switzerland, 2021. Available online: https://www.who.int/news/item/01-12-2021-tripartite-and-unep-support-ohhlep-s-definition-of-one-health (accessed on 20 September 2025).
4. Liu, C.; Wang, D.; Rietjens, I.M.; Zheng, L. Current and Emerging Issues in Chemical Food Safety. *Curr. Opin. Food Sci.* **2025**, *62*, 101284. [CrossRef]
5. Wilks, M.F.; Roth, N.; Aicher, L.; Faust, M.; Papadaki, P.; Marchis, A.; Calliera, M.; Ginebreda, A.; Andres, S.; Kühne, R.; et al. White paper on the promotion of an integrated risk assessment concept in European regulatory frameworks for chemicals. *Sci. Total. Environ.* **2015**, *521–522*, 211–218. [CrossRef] [PubMed]
6. Kueh, A.C. Sri Lanka: Impact Assessment Study of 2021 Ban on Conventional Pesticides and Fertilizers. 2022. Available online: https://eu-asean.eu/sri-lanka-impact-assessment-study-of-2021-ban-on-conventional-pesticides-and-fertilizers-2022/ (accessed on 1 November 2025).
7. Nordhaus, T.; Shah, S. Sri Lanka, Organic Farming Went Catastrophically Wrong. *Foreign Policy Magazine*, 5 March 2022. Available online: https://foreignpolicy.com/2022/03/05/sri-lanka-organic-farming-crisis/ (accessed on 1 November 2025).
8. Falkenberg, T.; Ekesi, S.; Borgemeister, C. Integrated Pest Management (IPM) and One Health—A call for action to integrate. *Curr. Opin. Insect. Sci.* **2022**, *53*, 100960. [CrossRef]
9. European Union. Regulation (EC) No 178/2002 of the European Parliament and of the Council of 28 January 2002 Laying Down the General Principles and Requirements of Food Law, Establishing the European Food Safety Authority and Laying Down Procedures in Matters of Food Safety. *Off. J. L* **2002**, *31*, 1–24. Available online: https://eur-lex.europa.eu/eli/reg/2002/178/oj/eng (accessed on 26 November 2025).
10. European Union. Regulation (EC) No 1107/2009 of the European Parliament and of the Council of 21 October 2009 Concerning the Placing of Plant Protection Products on the Market and Repealing Council Directives 79/117/EEC and 91/414/EEC. *Off. J. L* **2009**, *309*, 1–50. Available online: https://eur-lex.europa.eu/eli/reg/2009/1107/oj/eng (accessed on 26 November 2025).
11. European Union. Directive 2009/128/EC of the European Parliament and of the Council of 21 October 2009 Establishing a Framework for Community Action to Achieve the Sustainable Use of Pesticides. *Off. J. L* **2009**, *309*, 71–86. Available online: https://eur-lex.europa.eu/eli/dir/2009/128/oj/eng (accessed on 26 November 2025).
12. Fung, F.; Wang, H.-S.; Menon, S. Food Safety in the 21st Century. *Biomed. J.* **2018**, *41*, 88–95. [CrossRef]
13. Food Safety–EUR-Lex. Available online: https://eur-lex.europa.eu/summary/chapter/30.html (accessed on 26 November 2025).
14. European Food Safety Authority (EFSA). Available online: https://european-union.europa.eu/institutions-law-budget/institutions-and-bodies/search-all-eu-institutions-and-bodies/european-food-safety-authority-efsa_en (accessed on 18 November 2025).
15. Van der Meulen, B.M.J. The Structure of European Food Law. *Laws* **2013**, *2*, 69–98. [CrossRef]
16. D'Amore, T.; Smaoui, S.; Varzakas, T. Chemical Food Safety in Europe Under the Spotlight: Principles, Regulatory Framework and Roadmap for Future Directions. *Foods* **2025**, *14*, 1628. [CrossRef]
17. EUR Lex: Commission Regulation (EU) No 283/2013 of 1 March 2013 Setting Out the Data Requirements for Active Substances, in Accordance with Regulation (EC) No 1107/2009 of the European Parliament and of the Council Concerning the Placing of Plant Protection Products on the Market. Available online: https://eur-lex.europa.eu/eli/reg/2013/283/oj/eng (accessed on 18 November 2025).
18. EUR Lex: Commission Regulation (EU) No 284/2013 of 1 March 2013 Setting Out the Data Requirements for Plant Protection Products, in Accordance with Regulation (EC) No 1107/2009 of the European Parliament and of the Council Concerning the Placing of Plant Protection Products on the Market. Available online: https://eur-lex.europa.eu/eli/reg/2013/284/oj/eng (accessed on 18 November 2025).
19. Bullón Caro, C. Legal elements to operationalise One Health: From principles to practice. *Res. Dir. One Health* **2025**, *3*, e4. [CrossRef]
20. European Union. Directive 2009/127/EC of the European Parliament and of the Council of 21 October 2009 amending Directive 2006/42/EC with regard to machinery for pesticide application. *Off. J. L* **2009**, *310*, 29–33.
21. European Union. Regulation (EC) No 1272/2008 of the European Parliament and of the Council of 16 December 2008 on Classification, Labelling and Packaging of Substances and Mixtures, Amending and Repealing Directives 67/548/EEC and 1999/45/EC, and Amending Regulation (EC) No 1907/2006. *Off. J. L* **2008**, *353*, 1–1355. Available online: https://eur-lex.europa.eu/eli/reg/2008/1272/oj/eng (accessed on 26 November 2025).
22. European Union. Regulation (EU) 2022/2379 of the European Parliament and of the Council of 23 November 2022 on statistics on agricultural input and output, amending Commission Regulation (EC) No 617/2008 and repealing Regulations (EC) No 1165/2008, (EC) No 543/2009 and (EC) No 1185/2009 of the European Parliament and of the Council and Council Directive

96/16/EC PE/37/2022/REV/1. *Off. J. L* **2022**, *315*, 1–29. Available online: https://eur-lex.europa.eu/eli/reg/2022/2379/oj/eng (accessed on 26 November 2025).

23. European Commission. REFIT—Making EU Law Simpler, More Efficient and Future-Proof. Available online: https://commission.europa.eu/law/law-making-process/evaluating-and-improving-existing-laws/refit-making-eu-law-simpler-less-costly-and-future-proof_en (accessed on 20 September 2025).
24. European Union. Report from the Commission to the European Parliament and the Council. Evaluation of Regulation (EC) No 1107/2009 on the Placing of Plant Protection Products on the Market and of Regulation (EC) No 396/2005 on Maximum Residue Levels of Pesticides COM (2020) 208 Final. Available online: https://eur-lex.europa.eu/legal-content/EN/TXT/?uri=celex:52020DC0208 (accessed on 26 November 2025).
25. European Parliament Resolution of 13 September 2018 on the Implementation of the Plant Protection Products Regulation (EC) No 1107/2009 (2017/2128(INI)). Available online: https://www.europarl.europa.eu/doceo/document/TA-8-2018-0356_EN.html (accessed on 26 November 2025).
26. European Parliament Resolution of 16 January 2019 on the Union's Authorisation Procedure for Pesticides (2018/2153(INI)). Available online: https://www.europarl.europa.eu/doceo/document/TA-8-2019-0023_EN.html (accessed on 26 November 2025).
27. European Union. Regulation (EU) 2019/1381 of the European Parliament and of the Council of 20 June 2019 on the Transparency and Sustainability of the EU Risk Assessment in the Food Chain and Amending Regulations (EC) No 178/2002, (EC) No 1829/2003, (EC) No 1831/2003, (EC) No 2065/2003, (EC) No 1935/2004, (EC) No 1331/2008, (EC) No 1107/2009, (EU) 2015/2283 and Directive 2001/18/EC. *Off. J. L* **2019**, *231*, 1–28. Available online: https://eur-lex.europa.eu/eli/reg/2019/1381/oj/eng (accessed on 26 November 2025).
28. Your Europe. Registering Chemicals (REACH). Available online: https://europa.eu/youreurope/business/product-requirements/chemicals/registering-chemicals-reach/index_en.htm (accessed on 20 September 2025).
29. Caldeira, C.; Farcal, R.; Moretti, C.; Mancini, L.; Rasmussen, K.; Rauscher, H.; Riego Sintes, J.; Sala, S. *Safe and Sustainable by Design Chemicals and Materials—Review of Safety and Sustainability Dimensions, Aspects, Methods, Indicators, and Tools*; EUR 30991 EN; Publications Office of the European Union: Luxembourg, 2022; ISBN 978-92-76-47560-6. [CrossRef]
30. European Commission. Safe and Sustainable by Design. Available online: https://research-and-innovation.ec.europa.eu/research-area/industrial-research-and-innovation/chemicals-and-advanced-materials/safe-and-sustainable-design_en#documents (accessed on 20 September 2025).
31. European Food Safety Authority (EFSA); Devos, Y.; Auteri, D.; de Seze, G.; Fabrega, J.; Heppner, C.; Rortais, A.; Hugas, M. Building a European Partnership for next generation, systems-based Environmental Risk Assessment (PERA). *EFSA Support. Publ.* **2022**, *19*, e200503. [CrossRef]
32. EUR Lex: Report from the Commission to the European Parliament and the Council of October 2017 on Member State National Action Plans and on Progress in the Implementation of Directive 2009/128/EC on the Sustainable Use of Pesticides—COM(2017) 587 Final. Document 52017DC0587. Available online: https://eur-lex.europa.eu/legal-content/EN/TXT/?uri=CELEX:52017DC0587 (accessed on 26 November 2025).
33. European Parliament. Directive 2009/128/EC on the Sustainable Use of Pesticides European Implementation AssessementEPRS | European Par-liamentary Research Service. External Study. 2018. Available online: https://www.europarl.europa.eu/RegData/etudes/STUD/2018/627113/EPRS_STU(2018)627113_EN.pdf (accessed on 26 November 2025).
34. European Court of Auditor. Special Report 05/2020: Sustainable Use of Plant Protection Products: Limited Progress in Measuring and Reducing Risks. Available online: https://www.eca.europa.eu/en/publications?did=53001 (accessed on 1 November 2025).
35. Council of European Union. Council Conclusions on the Report from the Commission to the European Parliament and the Council on the Experience Gained by Member States on the Implementation of National Targets Established in Their National Action Plans and on Progress in the Implementation of Directive 2009/128/EC on the Sustainable Use of Pesticides. 13441/20 AGRI 447 PESTICIDE 41 SEEDS 16 AGRILEG 157. Available online: https://data.consilium.europa.eu/doc/document/ST-13441-2020-INIT/en/pdf (accessed on 1 November 2025).
36. Friends Earth. Joint Open Letter on the Need for a Strong Proposal on an EU Legislative Framework for Sustainable Food Systems. 2023. Available online: https://friendsoftheearth.eu/publication/open-letter-commission-president-must-stick-to-sustainable-food-system-law-timeline/ (accessed on 26 November 2025).
37. EUR-Lex Communication from the Commission to the European Parliament, the European Council, the Council, the European Economic and Social Committee and the Committee of the Regions, 'The European Green Deal' COM/2019/640 Final. Available online: https://eur-lex.europa.eu/legal-content/EN/TXT/?uri=celex:52019DC0640 (accessed on 20 September 2025).
38. EUR-Lex Communication from the Commission to the European Parliament, the Council, the European Economic and Social Committee and the Committee of the Regions, 'A Farm to Fork Strategy for a Fair, Healthy and Environmentally-Friendly Food

System', COM/2020/381 Final. Available online: https://eur-lex.europa.eu/legal-content/EN/TXT/?uri=celex:52020DC0381 (accessed on 20 September 2025).
39. Eurostat. 2021 Methodology for Calculating Harmonised Risk Indicators for Pesticides Under Directive 2009/128/EC. Available online: https://ec.europa.eu/eurostat/web/products-manuals-and-guidelines/-/ks-gq-21-008 (accessed on 26 November 2025).
40. One Health Governance in the European Union, Publications Office of the European Union. 2024. Available online: https://data.europa.eu/doi/10.2777/8697309 (accessed on 26 November 2025).
41. Commission Implementing Regulation (EU) 2023/1537 of 25 July 2023 Laying Down Rules for the Application of Regulation (EU) 2022/2379 of the European Parliament and of the Council as Regards Statistics on the Use of Plant Protection Products to be Transmitted for the Reference Year 2026 During the Transitional Regime 2025–2027 and as Regards Statistics on Plant Protection Products Placed on the Market (Text with EEA Relevance). Available online: https://eur-lex.europa.eu/legal-content/IT/TXT/?uri=CELEX%3A32023R1537 (accessed on 1 November 2025).
42. Eurostat. 'Pesticide Sales by Categorisation of Active Substances', Eurostat Data Browser. Available online: https://ec.europa.eu/eurostat/databrowser/view/AEI_PESTSAL_RSK/default/table?lang=en&category=agr.aei.aei_pes (accessed on 5 June 2025).
43. Shattuck, A.; Werner, M.; Mempel, F.; Dunivin, Z.; Galt, R. Global pesticide use and trade database (GloPUT): New estimates show pesticide use trends in low-income countries substantially underestimated. *Glob. Environ. Change* **2023**, *81*, 102693. [CrossRef]
44. Vos, J.; Alessandrini, M.; Trevisan, M.; Pii, Y.; Mazzetto, F.; Orzes, G.; Cesco, S. One Health approach: Addressing data challenges and unresolved questions in agriculture. *Sci. Total Environ.* **2025**, *977*, 179312. [CrossRef]
45. Bennekou, S.H. Moving towards a holistic approach for human health risk assessment—Is the current approach fit for purpose? *EFSA J.* **2019**, *17*, e170711. [CrossRef]
46. EFSA (European Food Safety Authority); Alvarez, F.; Arena, M.; Auteri, D.; Binaglia, M.; Castoldi, A.F.; Chiusolo, A.; Crivellente, F.; Egsmose, M.; Fait, G.; et al. Peer review of the pesticide risk assessment of the active substance glyphosate. *EFSA J.* **2023**, *21*, 1–52. [CrossRef]
47. Blaauboer, B.J.; Boobis, A.R.; Bradford, B.; Cockburn, A.; Constable, A.; Daneshian, M.; Edwards, G.; Garthoff, J.A.; Jeffery, B.; Krul, C.; et al. Considering new methodologies in strategies for safety evaluation. *Toxicol. Vitr.* **2016**, *32*, 198–204.
48. Toropov, A.A.; Toropova, A.P.; Rallo, R.; Leszczynska, D.; Leszczynski, J.; Kabir, M.M.; Benfenati, E. Optimal descriptors-based QSAR analysis of toxicity of organic compounds to Tetrahymena pyriformis. *SAR QSAR Environ. Res.* **2020**, *31*, 1–20. [CrossRef] [PubMed]
49. Pelkonen, O.; Abass, K.; Parra Morte, J.M.; Panzarea, M.; Testai, E.; Rudaz, S.; Louisse, J.; Gundert-Remy, U.; Wolterink, G.; Jean-Lou, C.M.D.; et al. Metabolites in the regulatory risk assessment of pesticides in the EU. *Front. Toxicol.* **2023**, *5*, 1304885. [CrossRef] [PubMed]
50. Bakker, R.; Ellers, J.; Roelofs, D.; Vooijs, R.; Dijkstra, T.; van Gestel, C.A.M.; Hoedjes, K.M. Combining time-resolved transcriptomics and proteomics data for Adverse Outcome Pathway refinement in ecotoxicology. *Sci. Total Environ.* **2023**, *869*, 161740. [CrossRef]
51. EFSA Scientific Committee. New approach methodologies for risk assessment using deep learning. *EFSA J.* **2024**, *22*, e221105. [CrossRef]
52. Sousa, J.P.; Aldrich, A.; Axelman, J.; Backhaus, T.; Brendel, S.; Dorronsoro, B.; Duquesne, S.; Focks, A.; Holz, S.; Knillmann, S.; et al. Building a European Partnership for next generation, systems-based Environmental Risk Assessment (PERA). *EFSA Support. Publ.* **2022**, *19*, 7546E. [CrossRef]
53. Netherlands National Committee for the Protection of Animals Used for Scientific Purposes (NCad). Transition Programme for Innovation Without the Use of Animals (TPI). 2022. Available online: https://www.animalfreeinnovationtpi.nl (accessed on 1 November 2025).
54. EFSA Scientific Committee. Guidance on the use of read-across for chemical safety assessment in food and feed. *EFSA J.* **2025**, *23*, e9586. [CrossRef]
55. Zaunbrecher, V.; Beryt, E.; Parodi, D.; Telesca, D.; Doherty, J.; Malloy, T.; Allard, P. Has toxicity testing moved into the 21st century? A survey and analysis of perceptions in the field of toxicology. *Environ. Health Perspect.* **2017**, *125*, 087024. [CrossRef]
56. Deepika, D.; Bharti, K.; Sharma, S.; Kumar, S.; Pathak, R.K.; Biosca Brull, J.; Sabuz, O.; García Vilana, S.; Kumar, V. Advancing human health risk assessment: The role of new approach methodologies. *Front. Toxicol.* **2025**, *7*, 1632941. [CrossRef]
57. EEA. How Pesticides Impact Human Health and Ecosystems in Europe. 2023. Available online: https://www.eea.europa.eu/publications/how-pesticides-impact-human-health/how-pesticides-impact-human-health (accessed on 30 June 2025).
58. EFSA (European Food Safety Authority); Di Piazza, G.; Dujardin, B.; Levorato, S.; Medina, P.; Mohimont, L.; Solazzo, E.; Costanzo, V. Prioritisation of pesticides and target organ systems for dietary cumulative risk assessment based on the 2019–2021 moni-toring cycle. *EFSA J.* **2024**, *22*, e8554. [CrossRef]

59. Ntzani, E.E.; Chondrogiorgi, M.; Ntritsos, G.; Evangelou, E.; Tzoulaki, I. Literature review on epidemiological studies linking exposure to pesticides and health effects. *EFSA Support. Publ.* **2013**, *EN-497*, 159. [CrossRef]
60. EFSA Panel on Plant Protection Products and Their Residues (PPR). Scientific Opinion of the PPR Panel on the follow-up of the findings of the External Scientific Report 'Literature review of epidemiological studies linking exposure to pesticides and health effects'. *EFSA J.* **2017**, *15*, 5007. [CrossRef]
61. EFSA Scientific Committee; More, S.; Bampidis, V.; Benford, D.; Bragard, C.; Hernandez-Jerez, A.; Bennekou, S.H.; Koutsoumanis, K.; Lambr, C.; Machera, K.; et al. Scientific Committee guidance on appraising and integrating evidence from epidemiological studies for use in EFSA's scientific assessments. *EFSA J.* **2024**, *22*, e8866. [CrossRef] [PubMed]
62. EFSA (European Food Safety Authority). Guidance on Expert Knowledge Elicitation in Food and Feed Safety Risk Assessment. *EFSA J.* **2014**, *12*, 3734. [CrossRef]
63. Tkalec, Z.; Antignac, J.-P.; Bandow, N.; Béen, F.M.; Belova, L.; Bessems, J.; Le Bizec, B.; Brack, W.; Cano-Sancho, G.; Chaker, J.; et al. Innovative analytical methodologies for characterizing chemical exposure with a view to next-generation risk assessment. *Environ. Int.* **2024**, *186*, 108585. [CrossRef]
64. Marx-Stoelting, P.; Rivière, G.; Luijten, M.; Aiello-Holden, K.; Bandow, N.; Baken, K.; Cañas, A.; Castano, A.; Denys, S.; Fillol, C.; et al. A walk in the PARC: Developing and implementing 21st century chemical risk assessment in Europe. *Arch. Toxicol.* **2023**, *97*, 893–908. [CrossRef]
65. EU-ToxRisk. Available online: https://cordis.europa.eu/project/id/681002/reporting (accessed on 26 November 2025).
66. European Human Biomonitoring (HBM4EU). Available online: https://www.hbm4eu.eu (accessed on 26 November 2025).
67. Yang, M.; Wang, Y.; Yang, G.; Wang, Y.; Liu, F.; Chen, C. A review of cumulative risk assessment of multiple pesticide residues in food: Current status, approaches and future perspectives. *Trends Food Sci. Technol.* **2024**, *144*, 10434. [CrossRef]
68. EFSA; Bennekou, S.H.; Allende, A.; Bearth, A.; Casacuberta, J.; Castle, L.; Coja, T.; Crépet, A.; Halldorsson, T.; Hoogenboom, R.; et al. Statement on the use and interpretation of the margin of exposure approach. *EFSA J.* **2025**, *23*, e9606. [CrossRef]
69. EFSA; Crivellente, F.; Hart, A.; Hernandez-Jerez, A.F.; Hougaard Bennekou, S.; Pedersen, R.; Terron, A.; Wolterink, G.; Mohimont, L. Establishment of cumulative assessment groups of pesticides for their effects on the nervous system. *EFSA J.* **2019**, *17*, e05800. [CrossRef]
70. EFSA; Hardy, A.; Benford, D.; Halldorsson, T.; Jeger, M.J.; Knutsen, K.H.; More, S.; Mortensen, A.; Naegeli, H.; Noteborn, H.; et al. Update: Use of the benchmark dose approach in risk assessment. *EFSA J.* **2017**, *15*, e04658. [CrossRef]
71. EFSA; Craig, P.S.; Dujardin, B.; Hart, A.; Hernandez-Jerez, A.F.; Hougaard Bennekou, S.; Kneuer, C.; Ossendorp, B.; Pedersen, R.; Wolterink, G.; et al. Cumulative dietary risk characterisation of pesticides that have acute effects on the nervous system. *EFSA J.* **2020**, *18*, e06087. [CrossRef]
72. EFSA; Craig, P.S.; Dujardin, B.; Hart, A.; Hernandez-Jerez, A.F.; Hougaard Bennekou, S.; Kneuer, C.; Ossendorp, B.; Pedersen, R.; Wolterink, G.; et al. Cumulative dietary risk characterisation of pesticides that have chronic effects on the thyroid. *EFSA J.* **2020**, *18*, e06088. [CrossRef]
73. EFSA; Anagnostopoulos, C.; Anastassiadou, M.; Castoldi, A.F.; Cavelier, A.; Coja, T.; Crivellente, F.; Dujardin, B.; Hart, A.; Hooghe, W.; et al. Retrospective cumulative dietary risk assessment of craniofacial alterations by residues of pesticides. *EFSA J.* **2022**, *20*, e07550. [CrossRef]
74. Tosti, L.; Marazzini, M.; Kanadil, M.; Metruccio, F. Cumulative risk assessment methodology applied to non-dietary exposures: Developmental alterations in professional agricultural settings. *J. Occup. Med. Toxicol.* **2025**, *20*, 7. [CrossRef]
75. SCHER (Scientific Committee on Health and Environmental Risks). Improvement of Risk Assessment in View of the Needs of Risk Managers and Policy Makers. 2011. Available online: https://ec.europa.eu/health/scientific_committees/environmental_risks/docs/scher_o_154.pdf (accessed on 26 November 2025).
76. Calliera, M.; Marchis, A.; Sacchettini, G.; Capri, E. Stakeholder consultations and opportunities for integrating socio behavioural factors into the pesticide risk analysis process. *Environ. Sci. Pollut. Res.* **2016**, *23*, 2937–2947. [CrossRef] [PubMed]
77. European Commission, Directorate-General for Health and Food Safety. Guidance to Provide Justifications as Referred in Point 1.5 of the Introduction of the Annexes of Regulations (EU) No 283/2013 and No 284/2013: Problem Formulation for Environmental Risk Assessment in the Context of Regulation (EC) No 1107/2009. 2024. Available online: https://food.ec.europa.eu/document/download/c4d6b7df-b7f9-4b3b-8ce5-b823ccdcf98c_en?filename=pesticides_ppp_app-proc_guide_horizontal_problem_formulation_era_en.pdf (accessed on 26 November 2025).
78. European Commission, Directorate-General for Health and Food Safety. Compendium of Conditions of Use to Reduce Exposure and Risk from Plant Protection Products. 2024. Available online: https://food.ec.europa.eu/document/download/cdd9b6c4-29dc-4077-a118-3e51c0abeb80_en?filename=pesticides_ppp_app-proc_guide_horizontal_compendium_conditions_of_use_en.pdf (accessed on 26 November 2025).

79. Calliera, M.; Di Guardo, A.; L'Astorina, A.; Polli, M.; Finizio, A.; Capri, E. Integrating Environmental and Social Dimensions with Science-Based Knowledge for a Sustainable Pesticides Management. *Sustainability* **2023**, *15*, 7843. [CrossRef]
80. EFSA Scientific Committee; More, S.J.; Benford, D.; Hougaard Bennekou, S.; Bampidis, V.; Bragard, C.; Halldorsson, T.I.; Hern ndez-Jerez, A.F.; Koutsoumanis, K.; Lambr, C.; et al. Guidance on risk-benefit assessment of foods. *EFSA J.* **2024**, *22*, e8875. [CrossRef] [PubMed]
81. SAPEA (Science Advice for Policy by European Academies). *Workshop Report No.3: Risk Perception and the Acceptability of Human Exposure to Pesticides*; SAPEA: Berlin, Germany, 2017.
82. International Agency for Research on Cancer; World Health Organization. IARC Monographs on the Evaluation of Carcinogenic Risks to Humans, Volume 112. 2015. Available online: https://www.iarc.who.int/wp-content/uploads/2018/07/MonographVolume112-1.pdf (accessed on 26 November 2025).
83. Initiative 'Ban Glyphosate and Protect People and the Environment from Toxic Pesticides' Called on the Commission, Under Its Third Aim, 'to Set EU-Wide Mandatory Reduction Targets for Pesticide Use, with a View to Achieving a Pesticide-Free Future' European Citizen's Initiative. Available online: https://europa.eu/citizens-initiative/_en (accessed on 1 November 2025).
84. Schwientek, M.; Rgner, H.; Haderlein, S.B.; Schulz, W.; Wimmer, B.; Engelbart, L.; Huhn, C. Glyphosate contamination in European rivers not from herbicide application? *Water Res.* **2024**, *263*, 122140. [CrossRef]
85. Kraus, N.; Malmfors, T.; Slovic, P. Intuitive toxicology: Expert and lay judgments of chemical risks. *Risk Anal.* **1992**, *12*, 215–232. [CrossRef]
86. Bearth, A.; Roth, N.; Wilks, M.F.; Siegrist, M. Intuitive toxicology in the 21st century-Bridging the perspectives of the public and risk assessors in Europe. *Risk Anal.* **2024**, *44*, 2348–2359. [CrossRef]
87. Campling, P.; Joris, I.; Calliera, M.; Capri, E.; Marchis, A.; Kuczyńska, A.; Vereijken, T.; Majewska, Z.; Belmans, E.; Borremans, L.; et al. A multi-actor, participatory approach to identify policy and technical barriers to better farming practices that protect our drinking water sources. *Sci. Total Environ.* **2021**, *755*, 142971. [CrossRef]
88. Slovic, P. Perception of Risk. *Science* **1987**, *236*, 280–285. [CrossRef]
89. WHO—European Health Observatory. Trust and Transformation: Five Policy Briefs by the Observatory and WHO/Europe in Support of the Tallinn Conference. 2023. Available online: https://eurohealthobservatory.who.int/themes/observatory-programmes/governance/tallinn-conference-2023 (accessed on 8 September 2024).
90. World Health Organization, Regional Office for Europe. Advancing Behavioural and Cultural Insights for Health Through Engagement with Experts: Report of the Meeting of the Technical Advisory Group on Behavioural and Cultural Insights, Copenhagen, Denmark, 8–9 February 2024. Available online: https://iris.who.int/handle/10665/376929 (accessed on 26 November 2025).
91. European Food Safety Authority (EFSA). Guidance on Uncertainty Analysis in Scientific Assessments. *EFSA J.* **2018**, *16*, 5123. [CrossRef]
92. Bronzwaer, S.; de Coen, W.; Heuer, O.; Marnane, I.; Vidal, A. The framework for action of the Cross-agency One Health Task Force. *One Health* **2024**, *19*, 100925. [CrossRef]
93. Lockwood, A.H. Human testing of pesticides: Ethical and scientific considerations. *Am. J. Public Health* **2004**, *94*, 1908–1916. [CrossRef]

Journal of *Xenobiotics*

MDPI

Review

Scientific and Regulatory Perspectives on Chemical Risk Assessment of Pesticides in the European Union

Fabio Buonsenso

Department of Chemistry "Ugo Schiff", University of Florence, Via della Lastruccia 3-13, Sesto Fiorentino, 50019 Florence, Italy; fabio.buonsenso@unifi.it

Abstract: People are exposed to pesticides daily through food, drinking water, and the environment, both in urban and rural settings. These chemicals, while offering economic and agricultural benefits through pest control and increased productivity, may pose a growing risk to human health and ecosystem biodiversity. While the European regulatory framework offers a robust foundation for risk assessment, significant limitations persist, especially in addressing cumulative exposure, low-dose effects, and chemical mixtures. This review focuses on selected scientific and regulatory challenges by reviewing recent *European Food Safety Authority* (EFSA) conclusions, *Organization for Economic Co-operation and Development* (OECD) test guidelines updates, and current European legislative approaches. Particular attention is given to the regulation of endocrine-disrupting and reprotoxic substances, highlighting progress and remaining gaps in implementation. A brief mention will also be made of immuno-toxic substances, for which no specific hazard class has yet been established. Building on official reports and peer-reviewed literature, this review provides a structured evaluation of the scientific and regulatory landscape, including underexplored issues like the transition to animal-free toxicology and integration of biomonitoring with health data. The goal is to propose realistic, evidence-based improvements to current frameworks using integrated, interdisciplinary approaches that connect toxicology, policy, and implementation science. A shift to a holistic, systems-based, and precautionary paradigm is vital to address emerging challenges and ensure strong protection of health and environment, as well as supporting the needs of the agricultural sector.

Keywords: climate change scenarios; CLP regulation; REACH; endocrine disruptors; immunotoxics; EU regulation; cumulative exposure; human health and environmental protection; plant protection

1. Introduction

Plant protection products (PPPs), including herbicides, fungicides, and insecticides, play a crucial role for sustaining agricultural productivity and ensuring global food security [1]. However, their widespread and prolonged use continues to raise significant concerns about their long-term effects on human health and ecosystems, in particular in the context of chronic, aggregate, and cumulative exposure [2,3]. Risk assessment associated with the use of PPPs represents an extremely complex challenge, reflecting the variety and dynamic nature of the scenarios in which these risks arise. It is essential to adopt a rigorous scientific approach, based on sustainable and reliable procedures, capable of ensuring the highest possible level of safety, in light of current knowledge. Consequently, is also essential an exhaustive regulatory control, aimed at protecting public health and the environment [4]. This need motivates increasing attention and a continuous investment of resources to ensure a constant evolution of the risk assessment and applicable mitigation measures. In the European Union (EU), pesticide regulation (Figure 1) is primarily governed by Regulation (EC) No 1107/2009, together with broader legislative instruments such as the *Registration, Evaluation, Authorisation and Restriction of Chemicals* (REACH) regulation and the *Classification, Labelling and Packaging* (CLP) regulation [5–8]. REACH and CLP provide standardized hazard classification, labeling, and exposure assessment criteria that complement the specific requirements of pesticide regulation.

Figure 1. Overview of European legislation on plant protection products. Green box: legislation in force; red box: legislation no longer in force.

Although this framework is designed to be comprehensive and well-structured, scientific and operational limitations continue to hinder its effective implementation. These include the limited integration of realistic exposure scenarios, insufficient attention to mixture toxicity, and regulatory uncertainty regarding low-dose effects and non-monotonic responses [9,10]. Despite some recent progress, the evaluation and management of endocrine-disrupting and reprotoxic substances remain fragmented and unevenly implemented across Member States [11,12].

Building on this historical and regulatory context, the focus now turns to emerging challenges posed by these substances. The persistent gaps in their assessment and management highlight the need for integrated, multidimensional approaches that account for cumulative exposure, low-dose effects, and environmental persistence. From an operational standpoint, regulatory flexibility tools, such as emergency authorizations for non-approved active substances, are frequently employed in EU countries. In some cases, these authorizations are issued repeatedly or extended beyond standard timeframes. This trend reflects the practical challenges of reconciling regulatory objectives with the demands of agricultural production [13,14]. Similarly, notable variability in pesticide use can be observed across different crops and Member States, influenced by national policies, enforcement approaches, and specific agronomic conditions, and the occurrence of so-called "minor uses", cases where pesticides are applied to less common crops or for specific pest problems not widely covered by existing authorizations [15].

Moreover, cumulative risk assessments show that a small number of active substances disproportionately influence overall risk estimates under conservative assumptions [16,17]. These findings highlight the urgency of refining methodological approaches, improving prioritization criteria, and enhancing transparency in decision-making processes. The rapid evolution of scientific and technological tools further complicates the regulatory landscape. *New Approach Methodologies* (NAMs), such as in silico modeling, machine learning, and high-throughput omics technologies, are increasingly being proposed to complement or replace conventional toxicological testing. While promising, these tools raise significant challenges regarding validation, standardization, and legal acceptance [18–20].

This review provides a perspective on emerging challenges in pesticide risk assessment, focusing on European regulation while also referencing global disparities, emerging methodological directions, and implementation barriers. In this context, integrated toxicological assessment is defined as a multidimensional approach combining classical toxicology, biomonitoring data, high-throughput in vitro and in silico testing, and environmental exposure assessments to evaluate cumulative and low-dose effects of pesticides on human and ecological health. By integrating insights from pesticide regulation and complementary chemical safety frameworks such as REACH and CLP, this approach provides a holistic, evidence-based evaluation of risk, extending beyond traditional single-substance toxicity assessments, and can inform regulatory decision-making, policy development, and prioritization of substances for further study or restriction. By analyzing the most recent opinions of the *European Food Safety Authority* (EFSA), policy documents, and scientific literature, the review aims to identify priority areas for reform and outline evidence-based strategies for a more adaptive, integrated, and precautionary risk governance model.

In an era marked by increasing ecological challenges, climate change, and rapid scientific innovation, strengthening the scientific foundations and institutional coherence of the European regulatory system is a crucial step toward reconciling agricultural productivity, public health, and ecological sustainability.

2. Evidence and Regulation: State of the Art

As highlighted above (Figure 1), in the European Union, the placing on the market of PPPs is regulated under Regulation (EC) No. 1107/2009 [5], amended for microorganisms by Regulation (EU) 2022/1438 [21], and is based on a careful pre-marketing risk assessment process [22]. A substance active can only be approved when, following an exhaustive examination, no unacceptable risks are established for human health, animal health, or the environment. The assessment is founded on large dossiers submitted by producers, which have to include experimental studies conducted according to internationally accepted and standardized guidelines, as stated by Regulations (EU) No. 283/2013 [23] and No. 284/2013 [24] (amended by Regulations (EU) 2022/1439 [25] and 2022/1440 [26] for microorganisms). The decision-making criteria used by the authorities are set in the *Uniform Principles* (Regulations (EU) No. 546/2011 [27], and 2022/1441 [28] for microorganisms).

International Uniform Chemical Information Database (IUCLID), developed by the *European Chemicals Agency* (ECHA) in association with the *Organisation for Economic Co-operation and Development* (OECD), facilitates handling of data and allows firms to organize and submit dossiers in a harmonized format, which can be easily assessed by the *Zonal Rapporteur Member State* (zRMS) and other Member States. IUCLID ensures consistency, traceability, and comparability of ecotoxicological and toxicological data, allowing integrated and multidimensional risk assessment strategies [29,30].

Toxicological assessment involves hazard characterization and identification, *No Observed Adverse Effect Level* (NOAEL), *Lowest Observed Adverse Effect Level* (LOAEL) and *Benchmark-dose* (BMD) estimation, and application of safety factors to compute reference values such as the *Acceptable Daily Intake* (ADI) and the *Acute Reference Dose* (ARfD). Concurrently, exposure assessment considers residues in food products and allows for the setting of *Maximum Residue Limits* (MRLs), as regulated by Regulation (EC) No. 396/2005, precautionary values intended to ensure safety without necessarily indicating toxicological risk [31,32]. Ecotoxicological analysis is interested in the effect of active ingredients and formulated products on non-target organisms, terrestrial or aquatic. Acute and chronic effects are studied through laboratory, semi-field, or field tests depending on product toxicity. The no-effect concentration for each species is compared to exposure levels, taking into account the *Uniform Principles* safety factors, for obtaining ecological risk characterization. Environmental destiny of chemicals is assessed considering persistence, mobility, degradation, volatility, photolysis, hydrolysis, and metabolites, and including harmonized European modeling in order to predict concentrations in soil, surface and groundwater, sediments, and air [22,27–29].

Despite this rigorous regulatory framework, recent research has shown severe problems with cumulative exposure and combinations of pesticides. As shown by the 2021 EFSA monitoring report, residues were detected in 96.1% of the food samples analyzed, and 28.9% of them contained more than one active substance [17]. While these figures are formally compliant with regulatory limits, they highlight the persistent challenge of multiple exposure scenarios, both in terms of aggregate exposure (i.e., exposure to the same chemical through different routes) and cumulative exposure (i.e., exposure to multiple chemicals with similar effects or modes of action), aspects that remain insufficiently considered in current risk assessment models. In particular, cumulative exposure is becoming increasingly relevant not only through diet but also via environmental and occupational pathways, as confirmed by the widespread detection of pesticide residues in surface water [33], soils [34], and human biological fluids [35]. New evidence suggests that simultaneous exposure to low doses of different pesticides may result in additive or synergistic effects [36,37]. While additive effects are much easier to address with existing data and the current risk assessment paradigm, based on chemical assessments for single chemicals, the potential for synergy, greater than additive effects, requires a deeper understanding, particularly from a mechanistic perspective, of how chemicals interact (e.g., whether the interaction occurs at the same receptor or at different receptors). Furthermore, while most studies still focus on single compounds, simultaneous exposure to multiple pesticides or to pesticides combined with other xenobiotic substances ("cocktail effects"; Figure 2) represents a much more realistic scenario [38].

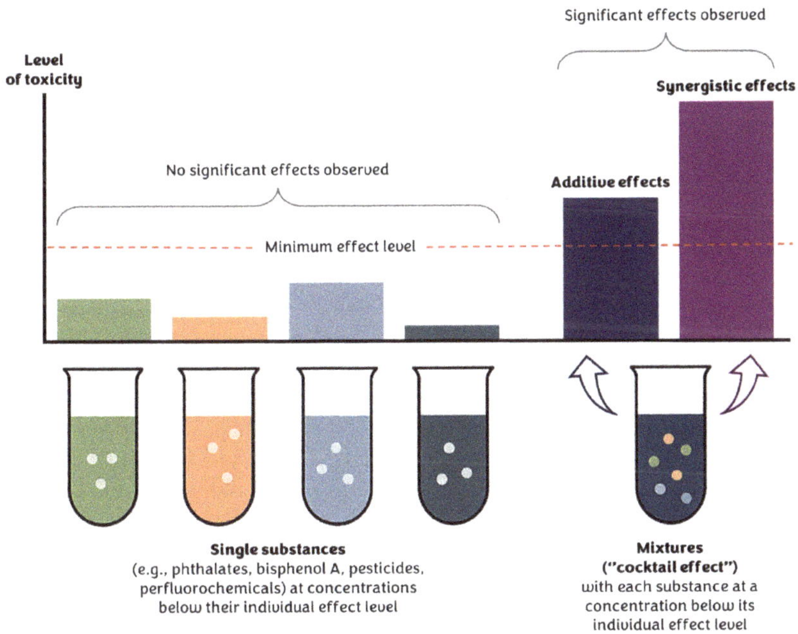

Figure 2. Single substances, when tested individually, produce no measurable toxic effect at concentrations below their respective effect levels. In contrast, mixtures of the same substances, each present at sub-threshold concentrations, can interact additively (blue bar) or synergistically (purple bar) ("cocktail effect"), leading to toxic effects that exceed the minimum effect level.

Recent experimental studies further support these concerns. Developmental exposure of rats to mixtures of pesticides, such as cypermethrin (CYP) and endosulfan (END), produces effects that are significantly different from exposure to the individual pesticides. Synergistic, preventive, or antagonistic interactions were observed depending on the neurochemical or behavioral parameters analyzed, indicating that deviations from additivity are frequent [39]. In particular, the mixture induced synergistic effects on spatial memory in males and associative learning in females, while some endpoints, such as motor coordination, showed antagonistic interactions, likely due to the prevention of γ-Aminobutyric acid (GABA) increase in the cerebellum by the mixture. Moreover, neurochemical and behavioral effects were strongly sex-dependent, with differential alterations of inflammatory markers and N-methyl-D-aspartate (NMDA) receptor subunits in males and females. These findings emphasize the importance of considering combined effects of pesticides rather than single-compound effects in health risk assessment [39,40].

Additionally, the recent EFSA cumulative risk assessment published in 2024 confirmed the relevance of considering combined effects of pesticide residues targeting the thyroid, defining two *Cumulative Assessment Groups* (CAGs) for hypothyroidism and C-cell hypertrophy, hyperplasia, and neoplasia. The report highlights that a limited number of active substances disproportionately drive cumulative risk and that mixture effects should be considered in regulatory decision-making [41,42].

Despite initiatives such as the CAGs developed by EFSA, the practical incorporation of mixture toxicity into routine regulatory evaluations remains limited, often constrained by data gaps, methodological complexity, and the lack of harmonized guidance across Member States.

A critical area of ongoing debate is the identification and regulation of endocrine disruptors (EDs). Although formal identification criteria were adopted in 2018, their application in risk management decisions has been highly inconsistent [43]. Substances evaluated as EDs are not automatically subjected to substitution or phase-out unless supported by additional data, and classification under the CLP Regulation often lags behind hazard identification. While Commission Delegated Regulation (EU) 2023/707, introducing the *Persistent, Mobile and Toxic/very Persistent and very Mobile* (PMT/vPvM) categories, represent an evolution toward a more systemic hazard-based approach, its practical implementation requires the development of operational guidance, clear threshold criteria for classification, prioritization tools, and cross-authority coordination to integrate PMT/vPvM measures into risk management decisions [44].

Another important aspect concerns the use of emergency authorizations under Article 53 of Regulation (EC) No. 1107/2009 [5]. Initially intended to address specific and time-limited threats to crop protection, this regulatory tool remains a key instrument for managing unforeseen phytosanitary challenges. Between 2017 and 2021, Member States granted over 600 such authorizations annually, with a significant proportion renewed across multiple years [13,14]. While these authorizations respond to practical agricultural needs, their frequency highlights the complex interplay between pest control demands, availability of effective alternatives, and the pace of regulatory and agronomic innovation. Strengthening the interface between risk assessment, comparative evaluation, and long-term transition strategies may help ensure that emergency measures continue to fulfill their original purpose while supporting more sustainable plant protection approaches.

Despite the scientific and political recognition of these issues, the regulatory paradigm remains dominated by a model based on the assessment of individual substances, which struggles to reflect the complexity of the real world, including the toxicity of mixtures, low-dose effects, and the dynamic interactions between environmental and human health factors. Indeed, current European regulations, such as the *Mixture Assessment Factor* (MAF) approach, do not fully address the risks arising from cumulative exposures and mixtures of substances [45,46]. Furthermore, the guidelines for their assessment are not yet harmonized across the various regulatory sectors, resulting in a significant gap in risk management. Currently used modeling tools also fail to adequately represent the ecological complexity of the real world, limiting the ability to reliably estimate overall environmental impacts. Despite increasing calls for a shift towards integrated assessment frameworks, including cumulative risk models, high-throughput screening tools and population-level biomonitoring programs, such approaches are not yet structurally embedded in EU practice. Comparative insights from other regulatory systems, such as the *Environmental Protection Agency* (EPA—United States) or the *Pest Management Regulatory Agency* (PMRA—Canada), may offer useful contrasts in how cumulative risk and endocrine disruption are operationally addressed [47–50]. In particular, the EPA provides a structured approach to pesticide risk assessment that can serve as a useful comparison. The agency evaluates both ecological and human health risks through multi-step processes, including hazard identification, dose–response assessment, exposure analysis, and risk characterization. Cumulative risk assessment is explicitly incorporated, considering simultaneous exposures to multiple pesticides with common mechanisms of toxicity across dietary, environmental, and residential pathways. Additionally, the EPA employs standardized models and technical guidelines to predict environmental concentrations and potential effects on non-target organisms [51].

Closing these gaps will require not only scientific advancement, but also institutional reform, more open implementation paths, and more investment in data and knowledge infrastructures with the potential to support integrated, multidimensional risk analysis. To take decisive steps in this direction, a more visionary, concerted, and ambitious approach is needed. First, monitoring protocols must be coordinated and strengthened at the regional, national, and EU levels, including not only all active substances but also the most significant metabolites and transformation products. To this end, it is necessary to develop harmonized risk assessment methodologies capable of addressing the combined and additive/synergistic effects of pesticide mixtures, going beyond the traditional approach, applied in the REACH Regulation, which focuses almost exclusively on individual compounds. To this end, it is necessary to implement existing regulatory provisions with workplace exposure scenarios. In parallel, research efforts must be intensified to investigate the cumulative, chronic, and sub-lethal impacts of pesticides, particularly on non-target organisms, pollinators, and soil microbiota, which play a key role in ecosystem stability but are often overlooked in conventional assessments. Integrating advanced analytical techniques, such as biomonitoring, omics approaches, and remote sensing, could substantially enhance the accuracy and temporal resolution of environmental data. At the same time, it is necessary to ensure environmental surveys, to obtain an accurate and comprehensive photography of the contamination situation. This is accompanied by the need to strengthen resistance monitoring programs to quickly identify any emerging outbreaks and specify more resilient management and rotation practices. Finally, a tangible measure to promote more responsible use would be the enforcement of mandatory phytosanitary prescriptions prepared by certified experts, could be complemented by farmer training, public awareness campaigns, and incentive schemes for integrated pest management practices. This measure would enable the regulation and rationalization of the employment of plant protection products, as well as ensuring increased safety for the environment and public health.

3. Emerging Critical Issues in Risk Assessment

Current regulatory frameworks for pesticide risk assessment remain grounded in a reductionist approach, relying heavily on standardized testing of individual active substances. While this has historically facilitated harmonization and comparability, it falls short in addressing the systemic, dynamic, and context-specific nature of real-world exposure [52,53].

Several European agencies are responsible for assessing environmental and health risks associated with both occupational and non-occupational exposure to pesticides. The EFSA provides scientific evaluations of risks to human health, animals, and the environment, including residue levels in food and feed. The European Commission, based on EFSA assessments, authorizes active substances and establishes maximum residue limits, while Member States implement these regulations at the national level and monitor pesticide use. The ECHA supports safe chemical management, including data relevant to pesticide risk assessment. Despite these roles, gaps remain, such as limited integration of cumulative exposure scenarios, inconsistent risk management of endocrine disruptors, and fragmented coordination across Member States. Italy represents an excellent model of virtuous management and coordination of PPPs. In fact, in Italy, several authorities are involved in pesticide regulation. The *Ministry of Health* is the competent authority for the authorization of PPPs and for the monitoring and enforcement of regulations relating to human health, while the *Ministry of Agriculture, Food Sovereignty and Forests*, particularly through the *National Phytosanitary Service*, supervises the use of pesticides in crop protection. The *Ministry of the*

Environment, Italian *Institute for Environmental Protection and Research* (ISPRA) and regional health authorities are responsible for environmental monitoring and risk assessment. All these bodies cooperate with the European Commission, EFSA and other European organizations, such as the *European and Mediterranean Plant Protection Organization* (EPPO) and *European Minor Uses Coordination Facility* (MUCF), to ensure compliance with EU standards.

This section discusses three interconnected challenges, cumulative exposure, low-dose effects, and environmental persistence, that illustrate the limits of current paradigms and the need for integrated adaptive methodologies.

3.1. *Cumulative Exposure and Chemical Mixtures*

In both occupational and environmental settings, exposure to pesticides occurs through complex chemical mixtures, yet current risk assessment procedures still prioritize single-compound evaluations [54]. These regulatory and operational challenges intersect with methodological complexities in risk assessment. The development of CAGs by EFSA represents an important conceptual advance, particularly for organophosphates and triazole fungicides, but these methods remain largely non-binding and inconsistently applied in regulatory decision-making. The derivation of CAGs relies on grouping substances based on shared mechanisms of toxicity, chemical structure, or target organs, with thresholds set to reflect combined risk potential. However, the operational application of these thresholds is limited by the availability of co-exposure data, the need for validated models for interaction effects, and uncertainties in extrapolating to different populations or exposure scenarios [3,16,55].

Integrating cumulative exposure into EU regulatory practice would require not only adoption of group-based thresholds, but also explicit guidance on threshold calculation, monitoring of high-risk substance clusters, and implementation of post-market surveillance to verify compliance and effectiveness. Agencies like the U.S. EPA have developed cumulative risk frameworks based on shared mechanisms of toxicity, albeit with operational limitations. The EU could benefit from a more explicit adoption of group-based regulatory thresholds, combined with targeted monitoring of high-risk substance clusters. A shift towards this model would also require greater data sharing, interoperable exposure databases, and post-market surveillance mechanisms. Overall, clearer methodological guidance and enhanced data infrastructure are essential to translate CAG-based thresholds into actionable regulatory decisions.

3.2. *Low-Dose Effects and Non-Monotonic Dose–Response Curves*

A key principle of chemical risk assessment is the protection of populations from adverse effects, rather than the suppression of all measurable biological changes, which are part of normal organismal variability. Therefore, the observation of biological responses below established reference points, such as NOAEL, LOAEL and BMD, is not in itself indicative of regulatory failure. However, such sub-threshold responses acquire toxicological relevance in contexts of aggregate or cumulative exposure, where multiple simultaneous exposures, to the same substance via different routes or to different substances sharing similar modes of action, may combine such that the net response exceeds the threshold into adversity. Evidence of low-dose biological activity for certain classes of pesticides (e.g., some neonicotinoids, organophosphates, and pyrethroids) highlights the need to integrate mechanistic information, like *Mode of Action* (MOA) and *Adverse Outcome Pathways* (AOPs), weight-of-evidence evaluation, and exposure context into assessment frameworks, rather than discarding the classical toxicological paradigm outright. The 2021 EFSA Scientific Committee Opinion [56,57] on *non-monotonic dose responses* (NMDRs) provides an operational analysis of when apparent NMDRs warrant further consideration in human health risk assessment. The Opinion offers practical criteria, including evaluation of biological plausibility, proximity to adversity, and availability of mechanistic MOA evidence, to guide assessors. When these criteria indicate potential regulatory relevance, targeted additional low-dose analyses or a refined uncertainty evaluations are recommended.

Furthermore, some pesticides exhibit a NMDR where lower doses give rise to a stronger or qualitatively different response than higher doses [58]. For example, for endocrine disruptors, non-monotonicity can be observed through mechanisms such as receptor down-regulation, selective activation of signaling pathways in response to varying doses, or feedback loops in hormonal systems. These non-monotonic dose–response relationships complicate the definition of health reference values because they undermine the original toxicological notion of monotonicity and hence well-defined effect thresholds. NMDRs, therefore, represent a significant challenge in risk assessment as they inhibit the reliable determination of NOAELs and margins of safety and illustrate the limitations of the current *Organization for Economic Co-operation and Development* (OECD) Test Guidelines in capturing all these complexities in their assessments and thus limit their ability to inform hazard identification. In other words, this implies that current OECD guidelines, which are primarily designed to identify monotonic dose–response relationships, may miss adverse effects that occur only at low or intermediate doses, resulting in a likely underestimation of the hazard potential.

To address these gaps, EU regulatory science would benefit from incorporating low-dose and NMDR data into risk assessments, revising testing protocols, and embracing mechanistic and omics-based evidence. Specifically, as will be discussed in greater detail in the following paragraphs, omics technologies, including genomics, transcriptomics, proteomics, metabolomics, and epigenomics, can provide mechanistic insights into molecular pathways affected by pesticide exposure, identify early biomarkers of effect, and reveal systemic or transgenerational impacts that classical approaches cannot detect. Integration of these datasets into risk assessment frameworks allows for the identification AOPs, supports probabilistic modeling, and strengthens predictive capabilities for human and ecological health. Moreover, multi-omics approaches can be combined with computational tools, such as *quantitative structure–activity relationship* (QSAR) and *physiologically based pharmacokinetic* (PBPK) models, to improve mechanistic understanding and guide the design of safer chemicals.

Shifting from fixed-threshold models toward probabilistic and systems-based risk evaluation approaches could substantially improve the realism and reliability of pesticide regulation [59].

3.3. *Environmental Persistence and Bioaccumulation Under Climate Change Scenarios*

Many pesticides are highly persistent in the environment and prone to bioaccumulation, impacting soils, aquatic ecosystems, and entire food webs. Their persistence and lipophilicity facilitate accumulation through dermal, respiratory, and dietary exposure, often leading to internal concentrations exceeding those in the surrounding environment, although this does not necessarily indicate a health risk, further assessment is needed considering additional factors such as toxicity and duration of exposure [60] (Figure 3).

Despite these risks, current monitoring and assessment programs remain fragmented and unevenly applied across EU Member States. Ecotoxicological indicators, such as shifts in macroinvertebrate populations or changes in soil microbiomes, are still underutilized, even though they can serve as early warning signals of ecological disruption [61,62].

In addition, a growing body of research demonstrates that climate change interacts with chemical pollution, affecting pesticide persistence, bioavailability, and toxicity. Changes in temperature, precipitation, and extreme weather events can modify chemical occurrence, alter uptake and elimination in organisms, and influence trophic transfer within ecosystems. For example, higher temperatures can increase chemical volatility and degradation rates, while also modulating bioaccumulation and toxicity through oxidative stress, metabolic interference, and disruption of species adaptation mechanisms [63–65]. Extreme weather events may further exacerbate chemical risks by breaching biological adaptation thresholds, while temperature fluctuations can produce nonlinear or sex-dependent effects on organisms. These factors introduce significant uncertainty into ecological risk assessment, highlighting the need for frameworks that quantitatively integrate climate-driven changes in chemical occurrence, bioaccumulation, and toxicity [66].

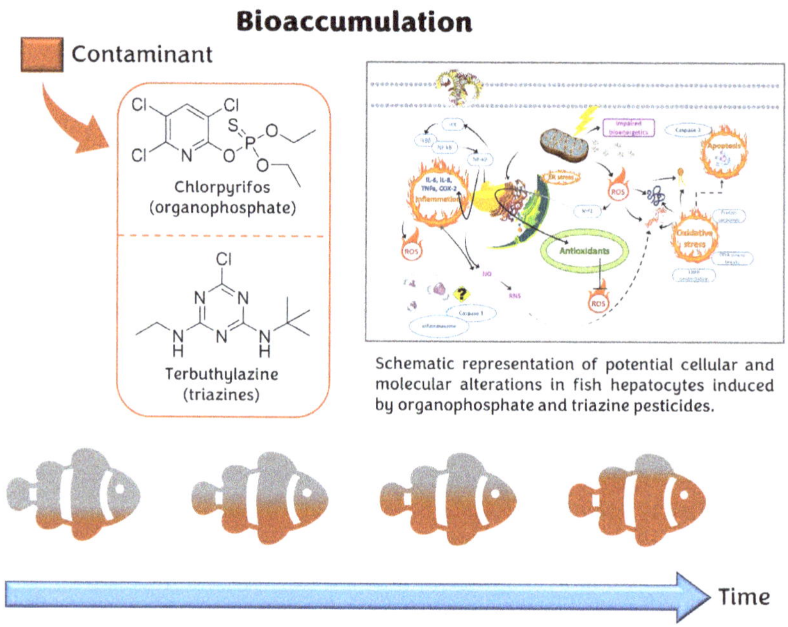

Figure 3. Progressive bioaccumulation of contaminant over time due to their persistence and lipophilicity. Fish, as key inhabitants of aquatic ecosystems, are a major focus in the assessment of the chronic toxicity of pesticides, particularly because of their potential for bioaccumulation and bioconcentration. These processes may occur through dermal, respiratory, or oral routes, with pesticide residues entering the organism either by direct diffusion from water or through ingestion of contaminated prey, leading to biomagnification. Among the pesticides most frequently detected in aquatic environments worldwide, including Central and Eastern Europe, are organophosphates (e.g., chlorpyrifos, glyphosate), phenylureas (e.g., chlorotoluron), triazines (e.g., terbuthylazine), and neonicotinoids. Fish can accumulate these compounds more efficiently than mollusks, and once accumulated, they may impair redox balance and cause toxic effects such as oxidative stress, cellular damage, inflammation, and altered immune and neuroendocrine functions [67,68]. The diagram of potential cellular and molecular alterations shown in the figure is adapted from [67].

Climate change profoundly influences pest and pathogen dynamics, reshaping their geographic distribution, life cycles, population growth, and interactions with crops. Rising temperatures, altered precipitation patterns, and higher atmospheric CO_2 levels can intensify pest pressure and extend growing seasons. At the same time, these factors may reduce pesticide efficacy, often requiring more frequent or larger applications. Sustainable crop management strategies, such as crop rotations, remain essential to mitigate pest prevalence and decrease reliance on chemical controls. Tools like CropRota integrate agronomic criteria with historical crop data to optimize rotations at field, farm, or regional scales, supporting long-term planning [69].

In parallel, process-based crop–pest models such as Pest-EPIC allow the simulation of crop–pest interactions in both conventional and organic systems, helping assess yield gaps and pesticide requirements under current and projected climatic conditions [70]. The use of such models not only strengthens the robustness of risk assessments but also contributes to the design of more effective adaptation strategies under evolving agro-ecological scenarios.

Addressing the uncertainties inherent in climate–crop–pest interactions requires holistic, scenario-based approaches that combine predictive models with empirical data on pest populations, pesticide use, and crop rotation practices. This integration enables researchers, farmers, and policymakers to develop adaptive strategies that maximize crop production, minimize environmental impacts, and enhance system resilience. Nonetheless, important knowledge gaps persist, particularly regarding pesticide efficacy under changing climate conditions, shifts in pest populations, and the interactions between climate change and pesticide persistence and degradation. Understanding how elevated temperatures, increased CO_2 levels, and altered precipitation influence pest life cycles, chemical resistance, and the effectiveness of control measures, including biological control agents, is critical. It is equally important to evaluate whether current regulatory frameworks are flexible enough to accommodate climate-related uncertainties and support adaptive agricultural practices. Through targeted research and strong collaboration among scientists, farmers, and policymakers, sustainable, climate-resilient, and environmentally conscious pest management strategies can be achieved, ensuring safe and effective food production in the face of global change [69,70].

The introduction of hazard categories for PMT (*Persistent, Mobile and Toxic*) substances under Commission Delegated Regulation (EU) 2023/707 offers a new opportunity to address these challenges [44]. However, fully operationalizing these classifications in EU risk assessment requires improved methodological frameworks, integration of environmental fate data, adoption of passive sampling technologies, and bioindicator-based monitoring. Implementing these measures would enhance the ecological relevance of pesticide risk evaluation and support more sustainable regulatory decisions.

3.4. Endocrine Disruptors and Reprotoxicants

In recent years, growing attention has been directed toward *endocrine disruptors* (EDs) and *reprotoxic* (Repr.) substances found in plant protection products [71–73]. Considering these concerns, it is essential to map which substances used as active ingredients in plant protection products have been associated with endocrine-disrupting effects [74,75]. Table 1 identifies the pesticides with endocrine-disrupting effects according to the *Endocrine Disruptor List* [76] and most recent EFSA reports [77,78]. These evidence-based lists of (potential) EDs have been compiled from existing sources and published by the national authorities of Belgium (*Federal Public Service—Health, Food Chain Safety and Environment* and *National Environment and Health Action Plan*), Denmark (*Danish Environmental Protection Agency*), France (*French Agency for Food, Environmental and Occupational Health & Safety—ANSES* and *French Ministry for the Ecological and Inclusive Transition*), the Netherlands (*The Ministry of Infrastructure and Water Management*), Spain (*General Directorate for Environmental Quality and Assessment—Ministry for the Ecological Transition and Demographic Challenge* and *General Directorate for Public Health—Ministry of Health*), and Sweden (*Swedish Chemicals Agency*—KEMI). The lists are updated at least twice a year. These compounds represent significant hazards to human health and ecosystems, as they can disrupt endocrine function (EDs) or cause reproductive toxicity (Repr.), even at very low concentrations, especially during sensitive developmental windows. Such effects are frequently associated with *non-monotonic dose–response* (NMDR) relationships and transgenerational outcomes, which are poorly captured by conventional toxicological models based on linear assumptions and fixed thresholds like NOAEL and LOAEL. Experimental data from both in vivo and in vitro studies show that substances such as chlorpyrifos, linuron, and vinclozolin can induce endocrine or developmental toxicity at exposure levels within current regulatory safety margins, indicating a consistent underestimation of risk in existing frameworks [79–81].

Table 1. The substances are divided into three categories according to the *Endocrine Disruptor List*. **List I**—Substances identified as EDs that have completed the full evaluation process for endocrine disruption under EU legislation, including the Plant Protection Products Regulation (PPPR), the Biocidal Products Regulation (BPR), or REACH (Candidate and Authorization Lists); **List II**—Substances currently under assessment in an EU legislative process due to explicit concerns about potential endocrine-disrupting properties; **List III**—Substances considered, by the evaluating National Authority, to have endocrine-disrupting properties based on scientific evidence. Abbreviations: Biocidal Products Committee (BCP); Community rolling action plan (CoRAP).

Active Substance	Chemical Class (Type)	Health Effects	Environmental Effects	Status (Status Year)	Regulatory Field
List I					
Asulam	Carbamate (herbicide)	No	Yes	Legally adopted (2024)	PPPR
Benthiavalicarb	Carbamate (fungicide)	Yes	No	Legally adopted (2024) Concluded ED in EFSA opinion (2021)	PPPR
Clofentezine	Tetrazine (acaricide)	Yes	No	Legally adopted (2024) Concluded ED in EFSA opinion (2021)	PPPR
Cyanamide	Nitrile (herbicide)	Yes	Yes	Legally adopted (2024) Concluded ED in EFSA opinion (2021)	BPR
Dimethomorph	Morpholine (fungicide)	Yes	Yes	Legally adopted (2024) Concluded ED in EFSA opinion (2023)	PPPR
Mancozeb	Dithiocarbamate (fungicide)	Yes	Yes	Legally adopted (2021) Concluded ED in EFSA opinion (2020)	PPPR
Mepanipyrim	Aminopyrimidines (fungicide)	Yes	Yes	Legally adopted (2024) Concluded ED in EFSA opinion (2023)	PPPR
Metiram	Dithiocarbamate (fungicide)	Yes	No	Legally adopted (2024) Concluded ED in EFSA opinion (2023)	PPPR
Metribuzin	1,2,4-triazines (herbicide)	Yes	No	Legally adopted (2024) Concluded ED in EFSA opinion (2023)	PPPR
Propiconazole	Triazole (fungicide)	Yes	Yes	Legally adopted (2024) Concluded ED in BPC opinion (2022)	PPPR

Table 1. *Cont.*

Active Substance	Chemical Class (Type)	Health Effects	Environmental Effects	Status (Status Year)	Regulatory Field
Triflusulfuron-methyl	sulfonylurea (herbicide)	Yes	No	Legally adopted (2024) Concluded ED in EFSA opinion (2022)	PPPR
List II					
Deltamethrin	Pyrethroid (type 2) ester (insecticide)	Yes	No	Commission EDC list (2019)	Cosmetics
Ziram	Dimethyldithiocarbamate (fungicide)	Yes	Yes	CoRAP (ECHA) list (2012)	REACH
Diuron	Urea (herbicide)	Yes	Yes	Concluded ED in SEV (2024) CoRAP (ECHA) list (2012)	REACH
Thiabendazole	Benzimidazoles (fungicide)	Yes	No	Concluded ED in EFSA opinion (2022)	PPPR
Fludioxonil	Phenylpyrrole (fungicide)	Yes	Yes	Concluded ED in EFSA opinion (2024)	PPPR
Flufenacet	Thiadiazole (herbicide)	Yes	Yes	Concluded ED in EFSA opinion (2024)	PPPR
Cyprodinil	Anilinopyrimidine (fungicide)	Yes	Yes	Concluded ED in EFSA opinion (2025)	PPPR
3-iodo-2-propynylbutylcarbamate (IPBC)	Carbamate (fungicide)	No	Yes	Concluded ED in BPC opinion (2024)	BPR
List III					
Prochloraz	Imidazole (fungicide)	Yes	No	List III National Authority evaluation (2020)	-

Current scientific understanding indicates that EDs may act as agonists or antagonists of key hormone receptors, such as steroid or thyroid receptors, and can interfere at multiple levels, including hormone synthesis, secretion, transport, binding, action, and elimination. These disruptions may not manifest as acute toxicity but can result in long-term or delayed adverse effects, including reproductive disorders, certain types of cancer, metabolic conditions such as obesity and diabetes, and cardiovascular disease. Moreover, environmental exposure, even at low doses, can produce population-level impacts with potential ecological repercussions, especially in wildlife, due to the capacity of these compounds for bioaccumulation [82,83].

3.4.1. Evolving Regulatory Criteria and Implementation Gaps

Regulation (EC) No. 1107/2009 provides for the automatic exclusion from the market of active substances classified as *endocrine disruptors* or *reprotoxic* (Category 1A or 1B under the CLP Regulation), unless exposure is deemed negligible [5]. Official scientific criteria for ED identification were established with Regulation (EU) 2018/605 (plant protection products) [73] and Regulation (EU) 2017/2100 (biocidal products) [84], aligned with the *World Health Organization* (WHO) definition [85]. These criteria require three elements: (1) evidence of an adverse effect, (2) identification of an endocrine mode of action, and (3) a demonstrated causal link between the two. Despite this regulatory progress, implementation remains uneven. A significant number of substances with documented or suspected endocrine-disrupting activity are still not formally classified under the CLP Regulation, resulting in delays in substitution procedures and fragmented regulatory responses across Member States.

Some EDs are regulated because of their general toxicity, carcinogenicity, or reproductive toxicity, even though their endocrine-disrupting effects occur at doses higher than those that cause these other effects and are therefore not the primary driver of regulation. However, is still important to understand the risks posed by EDs under low-dose, long-term, and combined exposure conditions, as current toxicological models and testing strategies are insufficient to capture the full complexity of endocrine disruption, highlighting the need for precautionary and mechanistic approaches.

The recent adoption of Commission Delegated Regulation (EU) 2023/707 [44], which amends the CLP Regulation by introducing new hazard classes [86], including EDs (for both human health and the environment), *Persistent, Bioaccumulative and Toxic/very Persistent and very Bioaccumulative* (PBT/vPvB), and *Persistent, Mobile and Toxic/very Persistent and very Mobile* (PMT/vPvM), represents a significant step toward a more integrated and precautionary risk assessment framework. These classifications have become legally binding as of 1 May 2025 for newly registered substances and will become legally binding from 1 November 2026 for substances already on the market. Their implementation is expected to influence authorization re-evaluations, safety data sheet labelling, national risk management systems, and substitution procedures under the "candidates for substitution" provisions of Regulation (EC) No. 1107/2009 [5].

Evidence from EFSA conclusions, the *European Chemicals Agency* (ECHA), and peer-reviewed studies confirm that many substances previously approved in the EU exhibit endocrine or reproductive toxicity, and several remain in use outside the EU, highlighting the need for harmonized international monitoring and trade alignment.

Table 2 presents a selection of active substances with well-documented endocrine-disrupting or reprotoxic effects, Prochloraz, Chlorpyrifos, Epoxiconazole, Vinclozolin, Linuron, Mancozeb, and Carbendazim, together with their chemical class, regulatory status, mechanisms of action, and CLP classification.

Notably, several of these substances, such as Epoxiconazole, Linuron, and Carbendazim, are officially classified under the CLP Regulation as *Repr.* Category 1B and are listed as candidates for substitution, indicating a regulatory trajectory towards the adoption of safer alternatives. Furthermore, epoxiconazole is considered PBT by most data submitters in the ECHA database, although this classification has not yet been formalized under the CLP [87–89]. Chlorpyrifos, by contrast, is currently under assessment for potential PBT properties, in addition to its documented neurotoxic and suspected endocrine-disrupting effects. In 2019, EFSA identified concerns for genotoxicity and developmental neurotoxicity, supported by epidemiological evidence, and concluded that the substance does not meet the approval criteria in the EU. Following Council Decision (EU) 2021/592 of 7 April 2021 [90] on the submission, on behalf of the European Union, of a proposal for the listing of chlorpyrifos in Annex A to the Stockholm Convention on Persistent Organic Pollutants (POPs), Chlorpyrifos is officially recognised POP (Substances proposed as POPs) and has recently been added to Annex A (Elimination) of the Convention [91]. For these reasons Chlorpyrifos is no longer approved in the EU since 2020 (Commission Implementing Regulation (EU) 2020/18 of 10 January 2020 [92]) and is not currently listed as a candidate for substitution [93].

Table 2. Overview of active substances in plant protection products, mechanisms of actions in human and animals, observed effects and their regulatory status. [a] Based on scientific literature; [b] Verified on ECHA C&L Inventory—https://echa.europa.eu/information-on-chemicals/cl-inventory-database (accessed on 21 April 2025); [c] Verified on The EU Pesticides Database—https://food.ec.europa.eu/plants/pesticides/eu-pesticides-database_en (accessed on 21 April 2025). ED: endocrine disruptors; Repr: reprotoxic; ✔: Harmonized classification—Annex VI of Regulation (EC) No. 1272/2008 (CLP Regulation); ✖: not classified.

Active Substance	Chemical Class (Type)	Mechanism of Action in Human and Animals	Observed Effects	Category [a]	Regulatory Classification (EU) [b]	EU Status [c]
Carbendazim	Benzimidazole (fungicide)	Inhibition of meiosis	Malformations and spermatogenesis toxicity [94]	Repr. 1B	✔ Repr. 1B (CLP) + ✖ ED (not classified)	Not approved (Expiry: 30 November 2014)
Chlorpyrifos	Organophosphorus (insecticide)	Acetylcholinesterase inhibition, neuroendocrine alterations	Delayed brain development [95]	Suspected ED, Neurotoxic	✖ ED (not classified)	Not approved (Expiry: 16 January 2020)
Epoxiconazole	Triazole (fungicide)	Aromatase inhibition	Fetal and developmental toxicity [96]	Repr. 1B, Repr. 2, suspected ED	✔ Repr. 1B and 2 (CLP) + ✖ ED (not classified)	Not approved (Expiry: 30 April 2020)
Linuron	Phenylurea (herbicide)	Androgenic antagonism	Sexual development disorders [97]	Repr. 1B, Repr. 2, suspected ED	✔ Repr. 1B and 2 (CLP) + ✖ ED (not classified)	Not approved (Expiry: 3 March 2017)
Mancozeb	Dithiocarbamate (fungicide)	Thyroid interference	Neurotoxicity, thyroid dysfunctions [98]	Repr. 1B, suspected ED	✔ Repr. 1B (CLP) + ✖ ED (not classified)	Not approved (Expiry: 4 January 2021)
Prochloraz	Imidazole (fungicide)	Inhibition of steroidogenic enzymes	Testicular dysfunction, thyroid alterations [99]	Suspected Repr. 1B and ED	✖ Repr. 1B and ED (not classified)	Not approved (Expiry: 31 December 2021)
Vinclozolin	Dicarboximide (fungicide)	Androgen receptor antagonism	Transgenerational effects [100]	Repr. 1B	✔ Repr. 1B (CLP) + ✖ ED (not classified)	Banned

For Prochloraz, EFSA published a Reasoned Opinion in 2023 on the review of *maximum residue levels* (MRLs) under Article 12 of Regulation (EC) No. 396/2005, highlighting consumer risks that led the European Commission to lower the MRLs to the *limit of determination* (LOD) [32]. Since 31 December 2021, Prochloraz has no longer been approved in the EU and is also listed as a candidate for substitution [101].

Regarding Mancozeb, EFSA concluded in 2019 that the substance did not meet the approval criteria under Article 4 of Regulation (EC) No. 1107/2009, based on its potential as an endocrine disruptor for humans and non-target organisms, risks from non-dietary exposure, and adverse effects on terrestrial fauna. As a result, its EU approval was not renewed; however, it is not currently listed as a candidate for substitution [102].

For Vinclozolin, although no recent EFSA opinions with the same level of detail have been identified, longstanding evidence of its endocrine-disrupting properties and reproductive toxicity has already led to revocations or bans within the EU [100].

These examples clearly illustrate the gap between robust scientific evidence and formal regulatory action, where classification of endocrine disruptors often requires particularly stringent criteria. Such gaps highlight the urgent need for stronger integration of EFSA peer reviews, the joint ECHA/EFSA guidance on endocrine disruptors, and OECD testing standards into regulatory decision-making. Without this, regulatory frameworks risk lagging behind scientific advances, undermining the protection of human and environmental health.

An interesting case currently under evaluation is acetamiprid. It is a neonicotinoid agonist of nicotinic acetylcholine receptors (nAChRs) with systemic and translaminar action. Currently, through Commission Implementing Regulation (EU) 2018/113 [103], acetamiprid is approved until 28 February 2033. However, in November 2021, EFSA recommended evaluating the endocrine-disrupting properties of acetamiprid in line with the criteria established by Regulation (EU) 2018/605 [73,104]. Furthermore, in its report of 15 May 2024, EFSA highlighted gaps in available studies regarding developmental neurotoxicity [105]. Consequently, it was not possible to establish with certainty whether acetamiprid still meets the approval criteria set out in Article 4 of Regulation (EC) No. 1107/2009 [5] with regard to its developmental neurotoxicity, as well as for its endocrine-disrupting properties, for which no assessment is available. Therefore, a review of the authorization under Article 21 of Regulation (EC) No. 1107/2009 is ongoing from December 2024 to assess developmental neurotoxicity and its endocrine-disrupting properties [106]. The EFSA report of 30 September 2025 expresses a favorable opinion on the new tests on acetamiprid, with some uncertainties considering the completeness of the data set. It recommends additional tests to investigate endocrine activity in non-mammalian species and a review of the scientific literature to assess the effects on humans and non-target organisms [107].

Overall, these cases underscore the need for a dynamic, precautionary, and evidence-responsive regulatory approach, capable of rapidly incorporating new scientific insights across the EU.

3.4.2. Implications for Risk Assessment: Toward a Science-Based Paradigm Shift

The specific characteristics of EDs and Repr. substances have significant implications for the future of pesticide risk assessment, requiring a shift from traditional, linear, single-compound models to a more integrated approach. A growing body of literature shows that many EDs exhibit adverse effects at extremely low concentrations, often below conventional safety thresholds such as NOAELs [9], challenging the regulatory reliance on threshold-based toxicology and, consequently, the associated risk assessment. Non-monotonic dose–response curves, where lower doses may cause stronger or qualitatively different effects than higher ones, complicate the establishment of reliable reference values and safety factors.

Emerging data on epigenetic and transgenerational effects, as observed in substances like vinclozolin and prochloraz [92,93], indicate that even transient exposures during sensitive life stages can produce heritable molecular and physiological changes. In particular, these transgenerational outcomes have been documented across multiple generations in both animal models and epidemiological studies, demonstrating effects on reproductive health, metabolic regulation, and neurodevelopment [108–110]. Such findings underscore the importance of incorporating multi-generational endpoints into risk assessment frameworks, as standard one-generation tests may underestimate long-term hazards [111,112]. These findings suggest a need to expand both the temporal and biological scope of toxicological evaluations.

Standard regulatory testing frameworks also show critical methodological limitations: many validated protocols do not include endpoints sensitive to endocrine disruption, such as steroidogenesis, early thyroid signaling, or hypothalamic–pituitary–gonadal axis function, resulting in frequently missed receptor-mediated activity (e.g., estrogenic or anti-androgenic effects). For example, recent studies demonstrate that glyphosate may inhibit aromatase, reducing estrogen synthesis, while neonicotinoids like thiacloprid and imidacloprid can act on estrogen receptors, potentially impairing reproductive function even at low exposure levels [113,114].

Additionally, oxidative stress, exacerbated by poor diet and cumulative environmental exposures, may act as a co-factor in endocrine disruption, particularly by triggering epigenetic modifications in germ cells and developing tissues [115]. These cross-cutting mechanisms emphasize the need for an integrated, systems-level understanding that integrates biochemical, physiological, environmental, and social determinants of risk.

4. Regulatory Implementation Proposals

Based on the gaps and challenges identified in previous sections, the following operational proposals aim to strengthen the EU regulatory framework for endocrine disruptors and reproductive toxicants. These measures focus on improving scientific assessment, harmonizing regulatory practices, and enhancing transparency and precaution in pesticide governance.

- Retrospective re-evaluation of legacy substances: Substances approved before 2018, prior to the adoption of harmonized ED criteria, should undergo re-assessment under the updated scientific framework. This is essential to ensure consistency across the EU market and prevent the persistence of hazardous compounds based on outdated evaluations.
- Mandatory inclusion of ED-specific test guidelines: Regulatory submissions should require the systematic application of OECD testing guidelines specifically designed to detect endocrine and reproductive effects, such as Test Guidelines (TG) 440, 443, 456, and 458 [116–119]. These should be complemented by mechanistic and receptor-binding assays, where relevant.
- Modernized, transparent data systems: Platforms like OpenFoodTox, IUCLID, and the substance database of ECHA should be enhanced with standardized visual flags (e.g., ED, Repr. 1B, PMT), searchable endpoints, and dynamic links to scientific references. These improvements would allow regulators, researchers, and stakeholders to quickly access hazard classifications, support comparative assessments and substitution, and improve transparency for consumers.
- Regulatory harmonization across frameworks: Increased alignment is needed between Regulation (EC) No. 1107/2009 (plant protection products), REACH (industrial chemicals), and the CLP Regulation, particularly in terms of classification criteria, data requirements, hazard communication and workplace exposure. This would reduce redundancy, promote consistent decision-making, and avoid regulatory "blind spots" between sectors.
- A new hazard class is proposed for chemicals that have harmful effects on the immune system (immunotoxic substances), which would be incorporated into the CLP Regulation [7] to define their classification and labeling. In fact, various studies conducted in recent times showed that exposure to fungicides, herbicides, and insecticides is associated with an enormous number of diseases, including inflammatory diseases, rheumatoid arthritis, metabolic syndrome, cancer, respiratory diseases, nervous system diseases, and alteration in various organ systems [120,121]. These disorders are strongly associated with immunological dysfunction, one of the principle aims of the toxic activity of pesticides. Immuno-toxic substances, to which many pesticides belong, are capable of inhibiting innate and acquired immune functions, causing toxicity of varied severity to the organism. Tests conducted during the past few years on cells, animals, and human subjects have proved that pesticides substantially alter humoral and cell-mediated immunity [122–124]. These medications can specifically target single

cells of the immune system, such as T lymphocytes, B lymphocytes, natural killer (NK) cells, and macrophages, inhibiting their survival, growth, and functional capability. Effects most frequently observed are apoptosis, or programmed cell death, cell cycle inhibition, DNA damage, and alteration of intracellular signal transduction mechanisms. On the molecular level, pesticides interfere with vital biological pathways and induce oxidative stress, mitochondrial injury, and endoplasmic reticulum stress [125,126]. These pathways lead to modifications of the expression of key factors like nuclear factor kappa B (NF-κB), pro-inflammatory cytokines including inter-leukin-6 (IL-6), interleukin-8 (IL-8), and interferon-gamma (IFN-γ), and anti-inflammatory mediators like interleukin-10 (IL-10) [127,128]. It has been noted that widely used or banned pesticides, like organophosphorus compounds (OPs), carbamates, pyrethroids, and other chemicals, have been associated with reduced antibody formation, inhibition of virus-induced B-cell differentiation into plasma cells, decreased T-lymphocyte proliferation, and repressed lytic activity of NK cells [120,121,129]. Certain pesticides have caused decreased phagocytosis, oxidative injury, and reorganization of cytokine production in macrophages that are essential for anti-viral and inflammatory function. These results, both in models and in studies of exposed populations, such as farmers, confirm the high immuno-toxic potential of pesticides. It has been observed that during peak agricultural seasons, B and T lymphocytes among workers exhibit higher levels of DNA damage than unexposed populations, posing a real risk to human health [130,131]. These immune system alterations can have long-term consequences, resulting in the onset of chronic diseases and increased susceptibility to infections, cancer, and inflammatory disorders. However, pesticide exposure can affect not only adults, particularly workers, but also children. Studies demonstrate an association between residential pesticide exposure in children and adolescents and the onset of leukemia, the most common pediatric cancer. In addition to leukemia, pesticide exposure increases the risk of non-Hodgkin's lymphoma (NHL), amyotrophic lateral sclerosis, asthmatic events, type 2 diabetes, and neurodegenerative diseases such as Alzheimer's and Parkinson's [128,132–134].

On this basis, which adds to that about endocrine-disrupting and reprotoxic activities of some pesticides, it is a question of priority to improve new assessments based on immuno-toxicological markers, capable of identifying the risks early on and guiding the use of more restrictive regulation for pesticide use. At the same time, efforts must be made to encourage the development of less hazardous substitutes and integrated control methods in a manner that will protect public health without jeopardizing agricultural production.

Meeting the challenge of EDs, Repr., and immunotoxics requires a precautionary, integrated approach bridging toxicology, systems biology, environmental monitoring, and global regulatory cooperation, representing a paradigm shift toward more responsive, anticipatory, and transparent pesticide governance.

5. Future Scientific Directions: Predictive Toxicology and Innovative Approaches

Among the future outlooks, particular attention should be paid to predictive toxicology and the development of NAMs. To this end, not only financial resources, but also operational and expertise are needed to validate in vitro and in silico tools (such as omics technologies, QSARs, and machine learning-based classifiers) so that they can reliably predict EDs and reproductive toxicity more efficiently and with fewer animal studies.

Validation programs must go beyond mere technical reproducibility and be able to explicitly link molecular and cellular signals measured in vitro in order to establish biological pathways or *Adverse Outcome Pathways* (AOPs), establishing quantitative decision thresholds capable of distinguishing adaptive changes from potentially harmful ones [135,136]. This will require well-characterized reference sets, multicenter reproducibility trials, transparent performance metrics, and early dialogue with regulators and stakeholders to foster acceptance and implementation.

5.1. Human Biomonitoring and Real-World Exposure

Human Biomonitoring (HBM) represents a fundamental advancement in assessing pesticide exposure, overcoming the limitations of predictive models based solely on laboratory data and dietary intake. The analysis of biological fluids, such as blood, urine, hair, or breast milk, enables the direct detection of specific metabolites, offering concrete evidence of both the presence and concentration of contaminants in the body [137]. Importantly, these biomonitoring results can be directly linked to trade patterns and climate change impacts, as changes in pesticide use, environmental distribution, and degradation influence human exposure profiles.

Among the commonly used biomarkers for estimating population exposure are various urinary metabolites, such as 3-phenoxybenzoic acid (3PBA), a common metabolite of several pyrethroids; 4-fluoro-3-phenoxybenzoic acid (FPBA), specific to cyfluthrin; 3-(2,2-dibromovinyl)-2,2-dimethylcyclopropane carboxylic acid (DBCA), specific to deltamethrin; cis-/trans-3-(2,2-dichlorovinyl)-2,2-dimethylcyclopropane-1-carboxylic acid (DCCA); and dialkyl phosphates, including dimethyl phosphate (DMP), diethyl phosphate (DEP), O,O-dimethyl thiophosphate (DMTP), O,O-diethylthiophosphate (DETP), O,O-dimethyl dithiophosphate (DMDTP), and O,O-diethyl dithiophosphate (DEDTP), specific for organophosphate pesticides [138–142]. These biomarkers provide evidence of how climate-driven changes in chemical persistence, bioavailability, and international pesticide trade influence internal doses in human populations.

HBM allows for the identification of new chemical exposures, monitoring of temporal trends, and analysis of exposure distribution across various population groups, particularly vulnerable ones (e.g., children) [143,144]. It is a methodology capable of linking potential external exposure, often described or estimated in worst-case scenarios, to actual internal concentrations, thereby providing reliable information on the extent and intensity of exposure, to be combined with toxicological data for risk characterization [145–147].

HBM plays a decisive role in health prevention and risk management, providing essential data for regulatory agencies and policymakers. Recent research in Europe has highlighted the widespread presence of pesticide metabolites in both urban and rural populations, with particularly high exposure levels observed in children, considered a vulnerable group. In Slovenia, 74 urinary biomarkers, including banned pesticides and emerging substances, have been identified in children, revealing the presence of multiple environmental chemicals during early childhood. Some studies have reported associations between prenatal pesticide exposure, as measured in maternal and umbilical cord blood, and adverse birth outcomes such as preterm birth and low birth weight, highlighting the need for further research to clarify causal relationships [148,149].

Studies conducted in New Zealand and Ethiopia have identified varying patterns of pesticide exposure, assessed via urinary metabolites, influenced by seasonal factors, dietary habits, and occupational activities. Notably, cases of pesticide poisoning were documented among workers in floriculture enterprises [150,151]. Such geographic differences in exposure emphasize the need to consider both climate variability and trade-driven pesticide distribution when interpreting biomonitoring data.

Emerging evidence also indicates possible involvement of pesticides in molecular and epigenetic alterations. The *Parental Pesticide and Offspring Epigenome* (PaPoE) study, conducted in Morocco, linked parental pesticide exposure, measured in blood, to altered DNA methylation profiles in cord blood of offspring, suggesting potential intergenerational effects. Similarly, another study documented oxidative DNA damage in Cypriot children, assessed through urinary levels of glyphosate metabolites and 8-hydroxy-2′-deoxyguanosine (8-OHdG), reinforcing concerns about the genotoxic potential of currently used pesticide residues [152,153].

Large-scale projects such as *Human Biomonitoring for Europe* (HBM4EU) [154–156] and the *European Partnership for the Assessment of Risks from Chemicals* (PARC) [157] are significantly contributing to the development of reference databases on human exposure, useful for guiding health and environmental policies.

Despite significant progress in recent years, current systems for monitoring pesticide exposure remain largely focused on residues in agricultural products and food, failing to capture the complexity of real-world exposure. This includes non-dietary routes such as inhalation or dermal absorption, exposure to heterogeneous chemical mixtures, and impacts on particularly vulnerable life stages such as childhood and pregnancy [148,149].

To more effectively address these challenges, a structural shift is needed to systematically include HBM in both national and European surveillance programs. This approach should be complemented by integrated environmental monitoring, capable of detecting the persistence and bioaccumulation potential of certain pesticides, which may remain in ecosystems and food chains long after application. Strengthening the link between environmental indicators and public health data, through tools such as ecotoxicological biomarkers or sentinel species, would enable a more realistic, preventive, and evidence-based approach to pesticide regulation. Integrating HBM findings with trade and climate data will allow for more targeted policies that account for the dynamic environmental and economic factors affecting human exposure.

However, critical knowledge gaps remain, particularly regarding long-term health effects and the translation of biomonitoring data into risk management policies. Methodological inconsistencies persist, along with limited geographic coverage and poor representativeness of the most exposed populations. There is an urgent need to develop standardized methodologies, promote studies in low- and middle-income countries by strengthening infrastructure, and integrate biomonitoring with participatory surveillance systems and a clear focus on environmental justice. This requires greater inclusion of socially and geographically marginalized groups, enhanced stakeholder engagement, and effective risk communication.

Reconceptualizing HBM not only as a scientific endeavor but also as a socio-political tool will allow the field to contribute more meaningfully to the protection of environmental health on a global scale.

5.2. Alternative Models and Advanced Technologies

A radical shift is currently taking place in the scientific tools available for pesticide risk assessment, with the aim of eliminating approaches that rely on animal testing. Among the alternatives to in vivo experimentation, in silico methods are demonstrating great potential, driven by recent advances in computational resources and the application of artificial intelligence and machine learning to large, curated datasets [158–160].

One widely used in silico approach is the *Quantitative Structure–Activity Relationship* (QSAR), a modeling technique applied in chemistry, pharmacology, and toxicology to predict the biological activity, properties, or behavior of chemical compounds based on their molecular structure. This method can be effectively integrated with machine learning techniques. Recent studies have confirmed the reliability and utility of in silico models in predicting acute toxicity across different chemical domains. The Leadscope QSAR model was evaluated for predicting the *United Nations Globally Harmonized System* (GHS) hazard categories, using acute oral toxicity data in rats. The study included pharmaceuticals, plant protection products, pharmaceutical intermediates, and their metabolites, and demonstrated the high reliability of the model in forecasting risks across a wide variety of chemical structures, making it applicable to several industrial sectors. The findings also emphasized the importance of a weight-of-evidence approach, which increases the accuracy and appropriateness of predictive modeling [161].

Although many studies have explored acute toxicity prediction in different species through in silico methods, most of them have focused on non-human endpoints, such as LC_{50} or LD_{50} values in rodents, fish, or birds. Decision tree-based QSAR models have been applied to evaluate the toxicity of pesticides in various bird species, addressing regulatory needs related to ecological protection rather than human health. Acute toxicity syndromes in fish have also been identified through a QSAR approach, which proved useful for mechanistic categorization but revealed limitations due to species specificity. More recently, chemometric approaches have been used to predict pesticide toxicity in chickens, reflecting the increasing refinement of computational models in the field of ecotoxicology [162–165].

Nonetheless, despite these advancements, a significant research gap remains concerning the development of QSAR or hybrid models capable of predicting human-specific acute toxicity endpoints, such as the *minimum toxic dose* (or *Toxic Dose Low*, TD_{Lo}) [166]. This underlines the urgent need for a solid and interpretable computational framework that utilizes available chemical and biological data to anticipate acute toxicity risks to human health. The current study [166] addresses this unmet need by developing the *Quantitative Read Across Structure–Toxicity Relationship* (q-RASTR) model, which merges the strengths of traditional QSTR methods with the read-across technique, in compliance with OECD principles and relevant Testing Guidelines (TG 440, 443, 456, 458) [116–119].

Physiologically Based Pharmacokinetic (PBPK) models provide a detailed simulation of how chemical compounds are distributed and metabolized within the body, using realistic representations of organs and tissues. These models consist of interconnected sub-models for each physiological compartment, linked by blood flow and parameterized using known anatomical and physiological data. Unlike classical compartmental models, which rely on empirical volumes and constants, PBPK models are grounded in measurable biological processes, making them more realistic and predictive in describing chemical kinetics [167].

PBPK models can be developed using a bottom-up approach, based on in vitro studies, the literature, or known physicochemical properties, or through a top-down approach, which involves adjusting parameters based on in vivo experimental data. In the case of pyrethroids, permethrin has been used as a model compound to validate predictive tools for human exposure, including a PBPK model and the high-resolution *Stochastic Human Exposure and Dose Simulation* (SHEDS) model, taking advantage of the extensive exposure and toxicokinetic data available for this substance [168–170].

Nevertheless, one of the most critical issues is species extrapolation, since scaling results from animal models to humans introduces uncertainties that must be carefully addressed. These uncertainties are typically managed through the application of guidance documents and standardized methodologies that incorporate uncertainty factors,

PBPK modeling, and interspecies scaling approaches. Additional concerns include interindividual variability, the limited availability of high-quality in vivo data for parameterization, and the lack of harmonized validation criteria across regulatory bodies.

Reference documents now exist that address many of the barriers to regulatory adoption of PBPK models. The *International Programme on Chemical Safety/World Health Organization* (IPCS/WHO) Harmonization Report [171,172] and the U.S. EPA Guidance [173] define principles of good practice for model development, emphasizing transparency, documentation of assumptions and parameter sources, and the systematic use of sensitivity and uncertainty analyses. Both publications recommend comparing model predictions with independent experimental data as a key element of validation. The OECD Guidance [174] translates these principles into operational tools for regulatory use, introducing minimum reporting standards, templates and checklists, both qualitative and quantitative validation criteria, as well as recommendations for peer review and traceability of input data; together, these elements facilitate harmonization of assessments across jurisdictions and contribute to more consistent acceptance decisions.

In parallel, the study by McLanahan et al. [175] draws attention to essential practical aspects for regulatory adoption. In particular, the importance of code and dataset accessibility, independent validation, and transparent communication of uncertainty is emphasized. The authors call for closer collaboration between academia, industry, and regulatory authorities to ensure reproducible and fit-for-purpose models. These guidelines address typical critical issues in a complementary manner by defining procedures and templates for validating and reporting predictive performance. Furthermore, they require that interspecies scaling choices be made explicit and justified, with the possibility of parameterization through in vitro and in silico approaches and with assessment of the related uncertainties. Another element is the inclusion of population variability, such as population PBPK or Monte Carlo simulations, and the support for the use of read-across and alternative approaches to address the paucity of in vivo data, always accompanied by sensitivity and uncertainty analyses. Consistent adoption of these requirements, clear declaration of intended use, complete documentation of the structure and parameters, sensitivity and uncertainty analyses, comparison with independent data, and transparent sharing of supporting material, substantively strengthens the evidentiary basis for the use of *Physiologically Based Pharmacokinetic/Physiologically Based Toxicokinetic* (PBPK/PBTK) models in regulatory decisions.

Another interesting case involves prothioconazole (PTC), a triazole fungicide that has shown strong in vitro inhibition of important cytochrome P450 (CYP450) enzymes involved in drug metabolism. To assess the risk of in vivo inhibition, a *Physiologically Based Toxicokinetic* (PBTK) model, conceptually equivalent to a PBPK model but applied in a toxicological context, was used. Building a PBTK model for PTC has been challenging due to the limited availability of human pharmacokinetic data, often restricted for ethical reasons. Nevertheless, adherence to PBTK/PBPK modeling principles consistent with international guidance allowed to explicitly evaluate the population variability and to support the conclusion of absence of clinically relevant *Pesticide–Drug Interactions* (PDI) at exposure levels close to the *Acceptable Daily Intake* (ADI), illustrating how adherence to the guidelines improves the robustness and acceptability of the assessments. Furthermore, the model allowed us to assess interindividual variability in PDI, based on the different metabolic phenotypes of the cytochrome P450 2C9 (CYP2C9) enzyme present in the population. Overall, the results support the reliability of the current regulatory exposure limits and confirm the value of the PBTK and PBPK models as robust and predictive tools for assessing health risks associated with pesticide exposure [176,177].

At the same time, omics technologies, such as genomics, transcriptomics, proteomics, metabolomics, and epigenomics, are increasingly employed to detect early, sub-lethal, or systemic biological responses to pesticide exposure. These approaches provide a detailed and multidimensional analysis of molecular changes induced by environmental contaminants. Recent research has shown that multi-omics integration is an effective strategy for overcoming the limitations of individual technologies, enabling a more comprehensive and unified understanding of pesticide mechanisms of action at both the cellular and systemic levels.

For example, genomics and transcriptomics have been used to identify genes regulated by pesticides like dichlorodiphenyltrichloroethane (DDT) and chlorpyrifos, which are involved in neuronal apoptosis as well as cardiac and neural development. Proteomic analyses have revealed significant changes in protein expression in model organisms such as zebrafish and plaice exposed to insecticides, including alterations in key proteins like *Heat Shock Protein* 70 (HSP70), CYP450, and metallothioneins. Metabolomic techniques, using liquid chromatography coupled with mass spectrometry (LC-MS) and gas chromatography coupled with mass spectrometry (GC-MS), have shown profound shifts in metabolic profiles even at low exposure doses, pointing to disruptions in energy metabolism, changes in gut microbiota composition, and potential links to disorders such as autism spectrum conditions. Epigenomic studies have further demonstrated that pesticide exposure can induce epigenetic modifications, including DNA methylation, histone changes, and alterations in microRNA, potentially leading to transgenerational effects, as observed in research on prenatal glyphosate exposure [138,178,179].

Together, these omics technologies offer a powerful toolset for understanding the mechanisms of pesticide toxicity, identifying predictive biomarkers, and integrating environmental and health risk assessments in a more sophisticated manner.

In parallel, the *European Safe and Sustainable by Design* (SSbD) framework, outlined in the Chemicals Strategy for Sustainability of the EU, encourages the development of chemical substances that are not only effective but also inherently safer. The application of green chemistry principles and the *Design for Environment* (DfE) approach supports the creation of molecules that are more biodegradable, less persistent, and with reduced bioaccumulation potential [8,180–182]. Additionally, emerging technologies such as exposomics and blockchain-based traceability offer promising avenues to enhance pesticide governance. Exposomics allow for the comprehensive assessment of real-world human and environmental exposures to chemical mixtures, linking these exposures to early biomarkers of effect and improving cumulative risk assessment models [183–185]. Blockchain technologies, on the other hand, can increase transparency and traceability across the entire pesticide supply chain, from production to field application, ensuring compliance with regulatory standards and supporting data-driven decision-making [186,187]. The integration of these approaches with omics data and SSbD strategies could enable more proactive, precise, and resilient management of chemical risks, fostering both environmental protection and sustainable agricultural practices.

These strategies are essential to decouple agricultural innovation from ecological degradation. They contribute to building a more dynamic, ethical, and economically sustainable regulatory system, one that is better prepared to evaluate new active substances in future pesticide regulatory frameworks.

5.3. Regulatory Review and the Precautionary Principle

The evolution of the scientific landscape calls for an integrated and innovative approach, not only on the technological front but also from a regulatory perspective, to ensure effective and up-to-date management of chemical risks arising from pesticides and other substances. Foundational regulations such as Regulation (EC) No. 1107/2009 [5] and Regulation (EC) No. 1907/2006 (REACH) [6] remain essential pillars in chemical risk governance; however, they do not adequately address the complexities associated with cumulative exposure, the effects of chemical mixtures, and emerging endpoints such as endocrine disruption, reprotoxic compounds, immunotoxic substances and epigenetic toxicity, phenomena that are now widely recognized in the scientific literature.

The precautionary principle must therefore be explicitly and systematically integrated into assessment and authorization processes, with particular attention to substances suspected of having endocrine or reproductive toxicity. In this context, *Cumulative Risk Assessment* (CRA) is emerging as a paradigm that goes beyond the traditional single-substance approach. CRA, as developed and promoted by EFSA through *Cumulative Assessment Groups* (CAGs), seeks to evaluate the combined effects of substances that share similar toxicological modes of action or common target organs, enabling a more realistic estimation of population risk.

Human exposure to chemicals involves a complex mixture of compounds originating from various sources, with food consumption representing the primary route of intake for many pesticides and environmental contaminants. CRA incorporates individual dietary consumption data with chemical occurrence data from official monitoring programs—such as the *EU multi-annual control programmes* [188], using probabilistic methodologies to provide a more accurate and representative evaluation of real-world exposures. Key quantitative tools include the *Hazard Index* (HI), the *Relative Potency Factor* (RPF), and the *Toxic Equivalency Factor* (TEF), which allow for normalization of cumulative exposure in reference to a benchmark compound. These tools facilitate risk characterization and help identify potential exceedances of safety thresholds [189,190].

The tiered approach adopted in CRA enables an initial conservative assessment, followed, where necessary, by more refined and specific evaluations (Tier 2). This strategy optimizes resource use and improves the precision of risk estimation. Risk communication relies on the *Margin of Exposure Total* (MoET), with threshold values aligned with traditional toxicological safety margins (e.g., MoET $\geq$ 100 at the 99.9th percentile), which serve as guides for regulatory decision-making [191].

Although the EFSA CAG framework follows a structured methodology, a systematic quantitative evaluation of its statistical power and uncertainty is still lacking. While probabilistic approaches and RPF normalization facilitate risk characterization, the magnitude of uncertainty across chemicals, population subgroups, and model assumptions remain largely unquantified. Recognizing this limitation is essential, and further research is needed to validate assumptions, perform sensitivity analyses, and reinforce confidence in the framework for regulatory governance.

Despite methodological progress, significant challenges remain. These include the need to harmonize the regulatory framework between REACH and plant protection product legislation, reducing inconsistencies in classification, data requirements, and evaluation timelines to avoid inefficiencies and regulatory gaps. Furthermore, it is necessary to integrate the REACH regulation with workplace exposure provisions. The EU Chemicals Strategy for Sustainability proposes mechanisms for joint review of authorization procedures, promoting transparency, reducing duplication, and fostering a more coherent policy landscape.

It is also essential to incorporate *Human Biomonitoring* (HBM) data into cumulative risk assessments and to account for toxicodynamic interactions, which remain underexplored. Advanced tools such as the *Aggregate Exposure Pathway* (AEP) framework and integrated exposure and biokinetic modeling platforms, such as the U.S. EPA models *Exposure Forecasting* (ExpoCast) and SHEDS, offer promising perspectives for more comprehensive integration of multiple exposures into regulatory models [137,191,192].

6. Conclusions

Pesticide chemical risk assessment is at a decisive crossroads. Mounting evidence from human biomonitoring, mixture toxicology, and mechanistic studies reveals that current models, based largely on isolated substances and simplified exposure scenarios, are insufficient to safeguard human and environmental health.

The next generation of risk assessment must be multidimensional, anticipatory, and grounded in real-world conditions. This means embracing novel tools such as omics technologies, in vitro/in silico models, and human biomonitoring, alongside implementing green chemistry and SSbD principles. Crucially, regulatory frameworks must evolve in parallel to keep pace with scientific advancement and societal expectations.

The precautionary principle should remain central to decision-making, particularly when addressing emerging risks like endocrine disruptors, reprotoxic compounds, or substances with unclear cumulative effects, and immuno-toxic compounds. Enhanced data transparency, methodological harmonization, and regulatory adaptability are essential to reduce long-term risks, protect future generations, and support the ecological transition of European agriculture.

Delaying this transformation risks not only public and environmental health, but also the strategic competitiveness of the agricultural and agrochemical sectors of the EU. Only through an integrated, forward-looking and science-based approach can we achieve a sustainable balance between productivity, human well-being, and environmental resilience.

In conclusion, an effective chemical risk assessment and management system must combine scientific innovation, regulatory integration, and a rigorous application of the precautionary principle. This is vital for protecting public health and environmental sustainability in a context of increasing and complex exposure to chemical mixtures, while at the same time promoting solutions that concretely address the needs of the agricultural sector and support its economic and productive sustainability.

Funding: This research received no external funding.

Institutional Review Board Statement: Not applicable.

Informed Consent Statement: Not applicable.

Data Availability Statement: No new data were created or analyzed in this study.

Conflicts of Interest: The author declares no conflicts of interest.

References

1. Pretty, J.; Bharucha, Z.P. Integrated pest management for sustainable intensification of agriculture in Asia and Africa. *Insects* **2015**, *6*, 152–182. [CrossRef]
2. Mostafalou, S.; Abdollahi, M. Pesticides and human chronic diseases: Evidences, mechanisms, and perspectives. *Toxicol. Appl. Pharmacol.* **2013**, *268*, 157–177. [CrossRef] [PubMed]
3. Silva, E.; Rajapakse, N.; Kortenkamp, A. Something from "nothing"—Eight weak estrogenic chemicals combined at concentrations below NOECs produce significant mixture effects. *Environ. Sci. Technol.* **2019**, *36*, 1751–1756.
4. European Food Safety Authority (EFSA); Carrasco Cabrera, L.; Pastor, P.M. The 2019 European Union report on pesticide residues in food. *EFSA J.* **2021**, *19*, e06491. [CrossRef]
5. Consolidated Text: Regulation (EC) No 1107/2009 of the European Parliament and of the Council of 21 October 2009 Concerning the Placing of Plant Protection Products on the Market and Repealing Council Directives 79/117/EEC and 91/414/EEC. Available online: http://data.europa.eu/eli/reg/2009/1107/2022-11-21 (accessed on 3 April 2025).

6. Consolidated Text: Regulation (EC) No 1907/2006 of the European Parliament and of the Council of 18 December 2006 Concerning the Registration, Evaluation, Authorisation and Restriction of Chemicals (REACH), Establishing a European Chemicals Agency, Amending Directive 1999/45/EC and Repealing Council Regulation (EEC) No 793/93 and Commission Regulation (EC) No 1488/94 as Well as Council Directive 76/769/EEC and Commission Directives 91/155/EEC, 93/67/EEC, 93/105/EC and 2000/21/EC (Text with EEA Relevance). Available online: http://data.europa.eu/eli/reg/2006/1907/2024-12-18 (accessed on 3 April 2025).
7. Consolidated Text: Regulation (EC) No 1272/2008 of the European Parliament and of the Council of 16 December 2008 on Classification, Labelling and Packaging of Substances and Mixtures, Amending and Repealing Directives 67/548/EEC and 1999/45/EC, and Amending Regulation (EC) No 1907/2006 (Text with EEA Relevance). Available online: http://data.europa.eu/eli/reg/2008/1272/2025-02-01 (accessed on 3 April 2025).
8. European Commission. *Chemicals Strategy for Sustainability Towards a Toxic-Free Environment. COM(2020)667 Final*. 2020. Available online: https://eur-lex.europa.eu/legal-content/EN/TXT/?uri=celex:52020DC0667 (accessed on 3 April 2025).
9. Vandenberg, L.N.; Colborn, T.; Hayes, T.B.; Heindel, J.J.; Jacobs, D.R.; Lee, D.H.; Shioda, T.; Soto, A.M.; vom Saal, F.S.; Welshons, W.V.; et al. Hormones and endocrine-disrupting chemicals: Low-dose effects and nonmonotonic dose responses. *Endocr. Rev.* **2012**, *33*, 378–455. [CrossRef]
10. Carvalho, F.P. Pesticides, environment, and food safety. *Food Energy Secur.* **2017**, *6*, 48–60. [CrossRef]
11. Bergman, Å.; Heindel, J.J.; Jobling, S.; Kidd, K.A.; Zoeller, R.T.; Endocrine Disrupting Chemicals-2012. Geneva: United Nations Environment Programme and the World Health Organization 2013. Available online: https://iris.who.int/bitstream/handle/10665/78102/WHO_HSE_PHE_IHE_2013.1_eng.pdf (accessed on 3 April 2025).
12. La Merrill, M.A.; Vandenberg, L.N.; Smith, M.T.; Goodson, W.; Browne, P.; Patisaul, H.B.; Guyton, K.Z.; Kortenkamp, A.; Cogliano, V.J.; Woodruff, T.J.; et al. Consensus on the key characteristics of endocrine-disrupting chemicals as a basis for hazard identification. *Nat. Rev. Endocrinol.* **2020**, *16*, 45–57. [CrossRef]
13. Transparency International EU & PAN Europe. Emergency Authorisations and Pesticide Regulation in the EU. 2023. Available online: https://www.pan-europe.info/resources/reports/2023/01/banned-pesticides-still-use-eu?utm_source=chatgpt.com (accessed on 3 April 2025).
14. Carisio, L.; Delso, N.S.; Tosi, S. Beyond the urgency: Pesticide Emergency Authorisations' exposure, toxicity, and risk for humans, bees, and the environment. *Sci. Total Environ.* **2024**, *947*, 174217. [CrossRef]
15. EFSA (European Food Safety Authority); Carrasco Cabrera, L.; Di Piazza, G.; Dujardin, B.; Marchese, E.; Medina Pastor, P. The 2022 European Union report on pesticide residues in food. *EFSA J.* **2024**, *22*, e8753. [CrossRef]
16. Yang, M.; Wang, Y.; Yang, G.; Wang, Y.; Liu, F.; Chen, C. A review of cumulative risk assessment of multiple pesticide residues in food: Current status, approaches and future perspectives. *Trends Food Sci. Technol.* **2024**, *144*, 104340. [CrossRef]
17. EFSA (European Food Safety Authority); Carrasco Cabrera, L.; Di Piazza, G.; Dujardin, B.; Medina Pastor, P. The 2021 European Union report on pesticide residues in food. *EFSA J.* **2023**, *21*, e07939. [CrossRef] [PubMed]
18. Li, Z.; Wang, D.; Zhu, T.; Tao, Y.; Ni, C. Review of deep learning-based methods for non-destructive evaluation of agricultural products. *Biosyst. Eng.* **2024**, *245*, 56–83. [CrossRef]
19. Dedeoglu, V.; Malik, S.; Ramachandran, G.; Pal, S.; Jurdak, R. Chapter Nine—Blockchain meets edge-AI for food supply chain traceability and provenance. In *Comprehensive Analytical Chemistry*; Nelis, J.L.D., Tsagkaris, A.S., Eds.; Elsevier: Amsterdam, The Netherlands, 2023; Volume 101, pp. 251–275.
20. Wang, M.W.H.; Goodman, J.M.; Allen, T.E.H. Machine learning in predictive toxicology: Recent applications and future directions for classification models. *Chem. Res. Toxicol.* **2020**, *34*, 217–239. [CrossRef]
21. Commission Regulation (EU) 2022/1438 of 31 August 2022 Amending Annex II to Regulation (EC) No 1107/2009 of the European Parliament and of the Council as Regards Specific Criteria for the Approval of Active Substances That Are Micro-Organisms. Available online: https://eur-lex.europa.eu/legal-content/EN/TXT/PDF/?uri=CELEX:32022R1438 (accessed on 22 September 2025).
22. European Commission. Procedure to Apply for Authorisation of a PPP. Available online: https://food.ec.europa.eu/plants/pesticides/authorisation-plant-protection-products/ppp-auth_en (accessed on 20 September 2025).
23. Consolidated Text: Commission Regulation (EU) No 283/2013 of 1 March 2013 Setting Out the Data Requirements for Active Substances, in Accordance with Regulation (EC) No 1107/2009 of the European Parliament and of the Council Concerning the Placing of Plant Protection Products on the Market. Available online: https://eur-lex.europa.eu/legal-content/EN/TXT/PDF/?uri=CELEX:02013R0283-20221121&qid=1758704708346 (accessed on 22 September 2025).

24. Consolidated Text: Commission Regulation (EU) No 284/2013 of 1 March 2013 Setting Out the Data Requirements for Plant Protection Products, in Accordance with Regulation (EC) No 1107/2009 of the European Parliament and of the Council Concerning the Placing of Plant Protection Products on the Market. Available online: https://eur-lex.europa.eu/legal-content/EN/TXT/PDF/?uri=CELEX:02013R0284-20221121&qid=1758704783082 (accessed on 22 September 2025).
25. Commission Regulation (EU) 2022/1439 of 31 August 2022 Amending Regulation (EU) No 283/2013 as Regards the Information to Be Submitted for Active Substances and the Specific Data Requirements for Micro-Organisms. Available online: https://eur-lex.europa.eu/legal-content/EN/TXT/PDF/?uri=CELEX:32022R1439 (accessed on 22 September 2025).
26. Commission Regulation (EU) 2022/1440 of 31 August 2022 Amending Regulation (EU) No 284/2013 as Regards the Information to Be Submitted for Plant Protection Products and the Specific Data Requirements for Plant Protection Products Containing Micro-Organisms. Available online: https://eur-lex.europa.eu/legal-content/EN/TXT/PDF/?uri=CELEX:32022R1440&qid=1758704962525 (accessed on 22 September 2025).
27. Consolidated Text: Commission Regulation (EU) No 546/2011 of 10 June 2011 Implementing Regulation (EC) No 1107/2009 of the European Parliament and of the Council as Regards Uniform Principles for Evaluation and Authorisation of Plant Protection Products. Available online: https://eur-lex.europa.eu/legal-content/EN/TXT/PDF/?uri=CELEX:02011R0546-20221121&qid=1758705109929 (accessed on 22 September 2025).
28. Commission Regulation (EU) 2022/1441 of 31 August 2022 Amending Regulation (EU) No 546/2011 as Regards Specific Uniform Principles for Evaluation and Authorisation of Plant Protection Products Containing Micro-Organisms. Available online: https://eur-lex.europa.eu/legal-content/EN/TXT/PDF/?uri=CELEX:32022R1441&qid=1758705216629 (accessed on 22 September 2025).
29. European Commission. Approval of Active Substances, Safeners and Synergists. Available online: https://food.ec.europa.eu/plants/pesticides/approval-active-substances-safeners-and-synergists_en (accessed on 22 September 2025).
30. European Chemicals Agency (ECHA), IUCLID 6. Available online: https://iuclid6.echa.europa.eu/ (accessed on 22 September 2025).
31. Consolidated Text: Regulation (EC) No 178/2002 of the European Parliament and of the Council of 28 January 2002 Laying Down the General Principles and Requirements of Food Law, Establishing the European Food Safety Authority and Laying Down Procedures in Matters of Food Safety. Available online: https://eur-lex.europa.eu/legal-content/EN/TXT/PDF/?uri=CELEX:02002R0178-20240701 (accessed on 22 September 2025).
32. Consolidated Text: Regulation (EC) No 396/2005 of the European Parliament and of the Council of 23 February 2005 on Maximum Residue Levels of Pesticides in or on Food and Feed of Plant and Animal Origin and Amending Council Directive 91/414/EEC. Available online: https://eur-lex.europa.eu/legal-content/EN/TXT/PDF/?uri=CELEX:02005R0396-20250512 (accessed on 22 September 2025).
33. Barbieri, M.V.; Monllor-Alcaraz, L.S.; Postigo, C.; López de Alda, M. Improved fully automated method for the determination of medium to highly polar pesticides in surface and groundwater and application in two distinct agriculture-impacted areas. *Sci. Total Environ.* **2020**, *745*, 140650. [CrossRef] [PubMed]
34. Shiva, S.; Hofman, K. A worldwide review of currently used pesticides' monitoring in agricultural soils. *Sci. Total Environ.* **2022**, *812*, 152344.
35. Hussari, M.N.; Chronister, B.N.C.; Yang, K.; Tu, X.; Martinez, D.; Parajuli, R.P.; Suarez-Torres, J.; Boyd Barr, D.; Hong, S.; Suarez-Lopez, J.R. Associations Between Urinary Pesticide Metabolites and Serum Inflammatory Biomarkers in Adolescents Living in an Agricultural Region. *Expo. Health* **2025**, *17*, 887–904. [CrossRef]
36. Mostafalou, S.; Abdollahi, M. Pesticides: An update of human exposure and toxicity. *Arch. Toxicol.* **2017**, *91*, 549–599. [CrossRef]
37. Xu, L.; Abd El-Aty, A.M.; Eun, J.B.; Shim, J.H.; Zhao, J.; Lei, X.; Gao, S.; She, Y.; Jin, F.; Wang, J.; et al. Recent advances in rapid detection techniques for pesticide residue: A review. *J. Agric. Food Chem.* **2022**, *70*, 13093–13117. [CrossRef]
38. Esbaugh, A.J.; Khursigara, A.; Johansen, J. Toxicity in Aquatic Environments: The Cocktail Effect. In *Development and Environment*; Burggren, W., Dubansky, B., Eds.; Springer: Cham, Switzerland, 2018.
39. Gomez-Gimenez, B.; Llansola, M.; Cabrera-Pastor, A.; Hernandez-Rabaza, V.; Agusti, A.; Felipo, V. Endosulfan and cypermethrin pesticide mixture induces synergistic or antagonistic effects on developmental exposed rats depending on the analyzed behavioral or neurochemical end points. *ACS Chem. Neurosci.* **2018**, *9*, 369–380. [CrossRef]
40. Gordon, C.J.; Herr, D.W.; Gennings, C.; Graff, J.E.; McMurray, M.; Stork, L.; Coffey, T.; Hamm, A.; Mack, C.M. Thermoregulatory response to an organophosphate and carbamate insecticide mixture: Testing the assumption of dose-additivity. *Toxicology* **2006**, *217*, 1–13. [CrossRef]
41. European Food Safety Authority (EFSA); Crivellente, F.; Hernández-Jerez, A.F.; Lanzoni, A.; Metruccio, F.; Mohimont, L.; Nikolopoulou, D.; Castoldi, A.F. Specific effects on the thyroid relevant for performing a dietary cumulative risk assessment of pesticide residues: 2024 update. *EFSA J.* **2024**, *22*, e8672.

42. Metruccio, F.; Castelli, I.; Baroni, A.; Marazzini, M.; Mammone, T.; Tosti, L. Data collection, hazard characterisation and establishment of cumulative assessment groups of pesticides for specific effects on the thyroid: 2024 update. *EFSA Support. Publ.* **2024**, *21*, 9012E. [CrossRef]
43. European Chemicals Agency (ECHA) 2023. Annual Report 2022. Available online: https://echa.europa.eu/annual-report-2022 (accessed on 3 April 2025).
44. Commission Delegated Regulation (EU) 2023/707 of 19 December 2022 Amending Regulation (EC) No 1272/2008 as Regards Hazard Classes and Criteria for the Classification, Labelling and Packaging of Substances and Mixtures (Text with EEA Relevance). Available online: http://data.europa.eu/eli/reg_del/2023/707/oj (accessed on 3 April 2025).
45. Backhaus, T. Exploring the mixture assessment or allocation factor (MAF): A brief overview of the current discourse. *Curr. Opin. Toxicol.* **2024**, *37*, 100460. [CrossRef]
46. Treu, G.; Schulze, J.; Galert, W.; Hassold, E. Regulatory and practical considerations on the implementation of a mixture allocation factor in REACH. *Environ. Sci. Eur.* **2024**, *36*, 101. [CrossRef]
47. Environmental Protection Agency (EPA)—Pesticide Environmental Stewardship Program (PESP). Available online: https://www.epa.gov/pesp (accessed on 19 July 2025).
48. Environmental Protection Agency (EPA)—Endocrine Disruptor Screening Program (EDSP). Available online: https://www.epa.gov/endocrine-disruption (accessed on 19 July 2025).
49. Pest Management Regulatory Agency (PMRA)—Pest Control Products (Pesticides) Acts and Regulations. Available online: https://www.canada.ca/en/health-canada/services/consumer-product-safety/pesticides-pest-management/public/protecting-your-health-environment/pest-control-products-acts-and-regulations-en.html (accessed on 20 July 2025).
50. Chintada, V.; Veraiah, K.; Golla, N. Monitoring and Managing Endocrine Disrupter Pesticides (EPDS) for Environmental Sustainability. In *Biotechnology for Environmental Sustainability*; Interdisciplinary Biotechnological Advances; Verma, P., Ed.; Springer: Singapore, 2025.
51. Environmental Protection Agency (EPA)—Overview of Risk Assessment in the Pesticide Program. Available online: https://www.epa.gov/pesticide-science-and-assessing-pesticide-risks/overview-risk-assessment-pesticide-program (accessed on 22 August 2025).
52. EFSA (European Food Safety Authority); Charistou, A.; Coja, T.; Craig, P.; Hamey, P.; Martin, S.; Sanvido, O.; Chiusolo, A.; Colas, M.; Istace, F. Guidance on the assessment of exposure of operators, workers, residents and bystanders in risk assessment of plant protection products. *EFSA J.* **2022**, *20*, e07032.
53. Zaller, J.G. What is the problem? Pesticides in our everyday life. In *Daily Poison: Pesticides—An Underestimated Danger*; Springer International Publishing: Cham, Switzerland, 2020; pp. 1–125.
54. Orton, F.; Ermler, S.; Kugathas, S.; Rosivatz, E.; Scholze, M.; Kortenkamp, A. Mixture effects at very low doses with combinations of anti-androgenic pesticides, antioxidants, industrial pollutant and chemicals used in personal care products. *Toxicol. Appl. Pharmacol.* **2014**, *278*, 201–208. [CrossRef]
55. EFSA Panel on Plant Protection Products and their Residues (PPR). Scientific Opinion on the identification of pesticides to be included in cumulative assessment groups on the basis of their toxicological profile (2014 update). *EFSA J.* **2013**, *11*, 3293.
56. Beausoleil, C.; Beronius, A.; Bodin, L.; Bokkers, B.G.H.; Boon, P.E.; Burger, M.; Cao, Y.; De Wit, L.; Fischer, A.; Hanberg, A.; et al. Review of non-monotonic dose-responses of substances for human risk assessment. *EFSA Support. Publ.* **2016**, *13*, 1027E. [CrossRef]
57. EFSA Scientific Committee; More, S.; Benford, D.; Hougaard Bennekou, S.; Bampidis, V.; Bragard, C.; Halldorsson, T.; Hernandez-Jerez, A.; Koutsoumanis, K.; Lambré, C.; et al. Opinion on the impact of non-monotonic dose responses on EFSA's human health risk assessments. *EFSA J.* **2021**, *19*, e06877.
58. McMahon, T.A.; Romansic, J.M.; Rohr, J.R. Nonmonotonic and monotonic effects of pesticides on the pathogenic fungus *Batrachochytrium dendrobatidis* in culture and on tadpoles. *Environ. Sci. Technol.* **2013**, *47*, 7958–7964. [CrossRef]
59. Skinner, M.K. Endocrine disruptor induction of epigenetic transgenerational inheritance of disease. *Mol. Cell. Endocrinol.* **2014**, *398*, 4–12. [CrossRef] [PubMed]
60. Simms, A.; Robert, K.; Spencer, R.J.; Treby, S.; Williams-Kelly, K.; Sexton, C.; Korossy-Horwood, R.; Terry, R.; Parker, A.; Van Dyke, J. A systematic review of how endocrine-disrupting contaminants are sampled in environmental compartments: Wildlife impacts are overshadowed by environmental surveillance. *Environ. Sci. Pollut. Res.* **2025**, *32*, 8670–8678. [CrossRef]
61. Vischetti, C.; Casucci, C.; De Bernardi, A.; Monaci, E.; Tiano, L.; Marcheggiani, F.; Ciani, M.; Comitini, F.; Marini, E.; Taskin, E.; et al. Sub-Lethal Effects of Pesticides on the DNA of Soil Organisms as Early Ecotoxicological Biomarkers. *Front. Microbiol.* **2020**, *11*, 1892. [CrossRef]
62. Gomes, I.B. The overlooked interaction of emerging contaminants and microbial communities: A threat to ecosystems and public health. *J. Appl. Microbiol.* **2025**, *136*, lxaf064. [CrossRef]

63. Hooper, M.J.; Ankley, G.T.; Cristol, D.A.; Maryoung, L.A.; Noyes, P.D.; Kent, E. PinkertonInteractions between chemical and climate stressors: A role for mechanistic toxicology in assessing climate change risks. *Environ. Toxicol. Chem.* **2013**, *32*, 32–48. [CrossRef] [PubMed]
64. Delnat, V.; Tran, T.T.; Verheyen, J.; Van Dinh, K.; Janssens, L.; Stoks, R. Temperature variation magnifies chlorpyrifos toxicity differently between larval and adult mosquitoes. *Sci. Total Environ.* **2019**, *690*, 1237–1244. [CrossRef]
65. Wang, M.; Hou, J.; Deng, R. Co-exposure of environmental contaminants with unfavorable temperature or humidity/moisture: Joint hazards and underlying mechanisms. *Ecotoxicol. Environ. Saf.* **2023**, *264*, 115432. [CrossRef] [PubMed]
66. He, W.; Chen, C.; Yang, L.; Chen, Z.; Zhu, L.; Zhu, L.; Feng, J. Chemical Ecological Risk Assessment for Aquatic Life Under Climate Change: A Review from Occurrence, Bioaccumulation, and Toxicity Perspective. *Rev. Environ. Contam. Toxicol.* **2025**, *263*, 10. [CrossRef]
67. Khatib, I.; Rychter, P.; Falfushynska, H. Pesticide Pollution: Detrimental Outcomes and Possible Mechanisms of Fish Exposure to Common Organophosphates and Triazines. *J. Xenobiot.* **2022**, *12*, 236–265. [CrossRef]
68. Rossi, A.S.; Fantón, N.; Michlig, M.P.; Repetti, M.R.; Cazenave, J. Fish inhabiting rice fields: Bioaccumulation, oxidative stress and neurotoxic effects after pesticides application. *Ecol. Indic.* **2020**, *113*, 106186. [CrossRef]
69. Schönhart, M.; Schmid, E.; Schneider, U.A. CropRota—A crop rotation model to support integrated land use assessments. *Eur. J. Agron.* **2011**, *34*, 263–277. [CrossRef]
70. Rasche, L. Estimating Pesticide Inputs and Yield Outputs of Conventional and Organic Agricultural Systems in Europe under Climate Change. *Agronomy* **2021**, *11*, 1300. [CrossRef]
71. Mnif, W.; Hassine, A.I.H.; Bouaziz, A.; Bartegi, A.; Thomas, O.; Roig, B. Effect of Endocrine Disruptor Pesticides: A Review. *Int. J. Environ. Res. Public Health* **2011**, *8*, 2265–2303. [CrossRef]
72. Gallo, A. Reprotoxic Impact of Environment, Diet, and Behavior. *Int. J. Environ. Res. Public Health* **2022**, *19*, 1303. [CrossRef]
73. Consolidated Text: Commission Regulation (EU) 2018/605 of 19 April 2018 Amending Annex II to Regulation (EC) No 1107/2009 by Setting Out Scientific Criteria for the Determination of Endocrine Disrupting Properties (Text with EEA Relevance). Available online: http://data.europa.eu/eli/reg/2018/605/2018-04-20 (accessed on 3 April 2025).
74. Colborn, T.; Vom Saal, F.S.; Soto, A.M. Developmental effects of endocrine-disrupting chemicals in wildlife and humans. *Environ. Health Perspect.* **1993**, *101*, 378–384. [CrossRef]
75. Colborn, T. Neurodevelopment and endocrine disruption. *Environ. Health Perspect.* **2004**, *112*, 944–949. [CrossRef] [PubMed]
76. Endocrine Disruptor List. Available online: https://edlists.org/the-ed-lists (accessed on 21 August 2025).
77. EFSA (European Food Safety Authority); Álvarez, F.; Arena, M.; Auteri, D.; Leite, S.B.; Binaglia, M.; Castoldi, A.F.; Chiusolo, A.; Colagiorgi, A.; Colas, M.; et al. Peer review of the pesticide risk assessment of the active substance fludioxonil. *EFSA J.* **2024**, *22*, e9047. [CrossRef]
78. EFSA (European Food Safety Authority); Álvarez, F.; Arena, M.; Auteri, D.; Leite, S.B.; Binaglia, M.; Castoldi, A.F.; Chiusolo, A.; Colagiorgi, A.; Colas, M.; et al. Peer review of the pesticide risk assessment of the active substance flufenacet. *EFSA J.* **2024**, *22*, e8997. [CrossRef]
79. Dias, G.R.M.; Giusti, F.C.V.; de Novais, C.O.; de Oliveira, M.A.L.; Paiva, A.G.; Kalil-Cutti, B.; Mahoney, M.M.; Graceli, J.B. Intergenerational and transgenerational effects of endocrine-disrupting chemicals in the offspring brain development and behavior. *Front. Endocrinol.* **2025**, *16*, 1571689. [CrossRef]
80. Eaton, D.L.; Daroff, R.B.; Autrup, H.; Bridges, J.; Buffler, P.; Costa, L.G.; Coyle, J.; McKhann, G.; Mobley, W.C.; Nadel, L.; et al. Review of the Toxicology of Chlorpyrifos With an Emphasis on Human Exposure and Neurodevelopment. *Crit. Rev. Toxicol.* **2008**, *38* (Suppl. S2), 1–125. [CrossRef]
81. Ding, H.; Zheng, W.; Han, H.; Hu, X.; Hu, B.; Wang, G.; Su, L.; Li, H.; Li, Y. Reproductive toxicity of linuron following gestational exposure in rats and underlying mechanisms. *Toxicol. Lett.* **2017**, *266*, 49–55. [CrossRef]
82. Patisaul, H.B.; Adewale, H.B. Long-term effects of environmental endocrine disruptors on reproductive physiology and behavior. *Front. Behav. Neurosci.* **2009**, *3*, 10. [CrossRef]
83. Heindel, J.; Newbold, R.; Schug, T. Endocrine disruptors and obesity. *Nat. Rev. Endocrinol.* **2015**, *11*, 653–661. [CrossRef] [PubMed]
84. Commission Delegated Regulation (EU) 2017/2100 of 4 September 2017 Setting Out Scientific Criteria for the Determination of Endocrine-Disrupting Properties Pursuant to Regulation (EU) No 528/2012 of the European Parliament and Council (Text with EEA Relevance). Available online: http://data.europa.eu/eli/reg_del/2017/2100/oj (accessed on 3 April 2025).
85. Solecki, R.; Kortenkamp, A.; Bergman, Å.; Chahoud, I.; Degen, G.H.; Dietrich, D.; Greim, H.; Håkansson, H.; Hass, U.; Husoy, T.; et al. Scientific principles for the identification of endocrine-disrupting chemicals: A consensus statement. *Arch. Toxicol.* **2017**, *91*, 1001–1006. [CrossRef] [PubMed]
86. Kättström, D.; Beronius, A.; Boije af Gennäs, U.; Rudén, C.; Ågerstrand, M. Impact of the new hazard classes in the CLP regulation on EU chemicals legislation. *Env. Sci. Eur.* **2025**, *37*, 19. [CrossRef]

87. EFSA (European Food Safety Authority). Peer review of the pesticide risk assessment for the active substance epoxiconazole in light of confirmatory data submitted. *EFSA J.* **2015**, *13*, 4123. [CrossRef]
88. EFSA (European Food Safety Authority). Peer review of the pesticide risk assessment of the active substance linuron. *EFSA J.* **2016**, *14*, 4518. [CrossRef]
89. EFSA (European Food Safety Authority); Bellisai, G.; Bernasconi, G.; Binaglia, M.; Carrasco Cabrera, L.; Castellan, I.; Castoldi, A.F.; Chiusolo, A.; Crivellente, F.; Del Aguila, M.; et al. Updated reasoned opinion on the toxicological properties and maximum residue levels (MRLs) for the benzimidazole substances carbendazim and thiophanate-methyl. *EFSA J.* **2024**, *22*, e8569.
90. Council Decision (EU) 2021/592 of 7 April 2021 on the Submission, on Behalf of the European Union, of a Proposal for the Listing of Chlorpyrifos in Annex A to the Stockholm Convention on Persistent Organic Pollutants. Available online: https://eur-lex.europa.eu/legal-content/EN/TXT/HTML/?uri=CELEX:32021D0592 (accessed on 18 July 2025).
91. Stockholm Convention on Persistent Organic Pollutants (POPs). Available online: https://www.pops.int/TheConvention/ThePOPs/AllPOPs/tabid/2509/Default.aspx (accessed on 11 October 2025).
92. Commission Implementing Regulation (EU) 2020/18 of 10 January 2020 Concerning the Non-Renewal of the Approval of the Active Substance Chlorpyrifos, in Accordance with Regulation (EC) No 1107/2009 of the European Parliament and of the Council Concerning the Placing of Plant Protection Products on the Market, and Amending the Annex to Commission Implementing Regulation (EU) No 540/2011. Available online: https://eur-lex.europa.eu/legal-content/EN/TXT/HTML/?uri=CELEX:32020R0018 (accessed on 11 October 2025).
93. EFSA (European Food Safety Authority). Statement on the available outcomes of the human health assessment in the context of the pesticides peer review of the active substance chlorpyrifos. *EFSA J.* **2019**, *17*, e05809.
94. Rama, E.M.; Bortolan, S.; Vieira, M.L.; Gerardin, D.C.C.; Moreira, E.G. Reproductive and possible hormonal effects of carbendazim. *Regul. Toxicol. Pharmacol.* **2014**, *69*, 476–486. [CrossRef]
95. ur Rahman, H.U.; Asghar, W.; Nazir, W.; Sandhu, M.A.; Ahmed, A.; Khalid, N. A comprehensive review on chlorpyrifos toxicity with special reference to endocrine disruption: Evidence of mechanisms, exposures and mitigation strategies. *Sci. Total Environ.* **2021**, *755*, 142649. [CrossRef]
96. Liu, Y.; Wang, F.; Li, L.; Fan, B.; Kong, Z.; Tan, J.; Li, M. The potential endocrine-disrupting of fluorinated pesticides and molecular mechanism of EDPs in cell models. *Ecotoxicol. Environ. Saf.* **2025**, *289*, 117615. [CrossRef]
97. Ma, Y.; Pedersen, M.; Vinggaard, A.M. In vitro antiandrogenic effects of the herbicide linuron and its metabolites. *Chemosphere* **2024**, *349*, 140773. [CrossRef]
98. Norouzi, F.; Alizadeh, I.; Faraji, M. Human exposure to pesticides and thyroid cancer: A worldwide systematic review of the literatures. *Thyroid. Res.* **2023**, *16*, 13. [CrossRef] [PubMed]
99. Blystone, C.R.; Lambritth, C.S.; Howdeshell, K.L.; Furr, J.; Sternberg, R.M.; Butterworth, B.C.; Durhan, E.J.; Makynen, E.A.; Ankley, G.T.; Wilson, V.S.; et al. Sensitivity of fetal rat testicular steroidogenesis to maternal prochloraz exposure and the underlying mechanism of inhibition. *Toxicol. Sci.* **2007**, *97*, 512–519. [CrossRef] [PubMed]
100. Rizzo, R.; Bortolotti, D.; Rizzo, S.; Schiuma, G. Endocrine disruptors, epigenetic changes, and transgenerational transmission. In *Environment Impact on Reproductive Health*; Robert, M., Ed.; Springer: Cham, Switzerland, 2023; pp. 49–74.
101. EFSA (European Food Safety Authority). Targeted risk assessment for prochloraz. *EFSA J.* **2023**, *21*, e08231. [CrossRef] [PubMed]
102. EFSA (European Food Safety Authority); Abdourahime, H.; Anastassiadou, M.; Arena, M.; Auteri, D.; Barmaz, S.; Brancato, A.; Bura, L.; Carrasco Cabrera, L.; Chaideftou, E.; et al. Peer review of the pesticide risk assessment of the active substance mancozeb. *EFSA J.* **2020**, *18*, e05755. [CrossRef]
103. Commission Implementing Regulation (EU) 2018/113 of 24 January 2018 Renewing the Approval of the Active Substance Acetamiprid in Accordance with Regulation (EC) No 1107/2009 of the European Parliament and of the Council Concerning the Placing of Plant Protection Products on the Market, and Amending the Annex to Commission Implementing Regulation (EU) No 540/2011. Available online: https://eur-lex.europa.eu/legal-content/EN/TXT/?uri=OJ:L:2018:020:TOC (accessed on 8 October 2025).
104. EFSA PPR Panel (EFSA Panel on Plant Protection Products and their Residues); Hernandez Jerez, A.; Adriaanse, P.; Berny, P.; Coja, T.; Duquesne, S.; Focks, A.; Marinovich, M.; Millet, M.; Pelkonen, O.; et al. Statement on the active substance acetamiprid. *EFSA J.* **2022**, *20*, e07031. [CrossRef]
105. EFSA (European Food Safety Authority); Hernandez-Jerez, A.; Coja, T.; Paparella, M.; Price, A.; Henri, J.; Focks, A.; Louisse, J.; Terron, A.; Binaglia, M.; et al. Statement on the toxicological properties and maximum residue levels of acetamiprid and its metabolites. *EFSA J.* **2024**, *22*, e8759. [CrossRef]
106. Standing Committee on Plants, Animals, Food and Feed Section Phytopharmaceuticals—Legislation 4–5 December 2024. Summary Report. Available online: https://food.ec.europa.eu/document/download/bbbd95c0-c323-4a91-a534-4d496f280db2_en?filename=sc_phyto_20241204002_ppl_sum.pdf (accessed on 8 October 2025).

107. EFSA (European Food Safety Authority). Statement on the testing strategy and timelines for the assessment of developmental neurotoxicity and endocrine disruption properties of acetamiprid in the context of the review of the approval of the active substance. *EFSA J.* **2025**, *23*, e9639. [CrossRef]
108. Colopi, A.; Guida, E.; Cacciotti, S.; Fuda, S.; Lampitto, M.; Onorato, A.; Zucchi, A.; Balistreri, C.R.; Grimaldi, P.; Barchi, M. Dietary Exposure to Pesticide and Veterinary Drug Residues and Their Effects on Human Fertility and Embryo Development: A Global Overview. *Int. J. Mol. Sci.* **2024**, *25*, 9116. [CrossRef]
109. Robaire, B.; Delbes, G.; Head, J.A.; Marlatt, V.L.; Martyniuk, C.J.; Reynaud, S.; Trudeau, V.L.; Mennigen, J.A. A cross-species comparative approach to assessing multi- and transgenerational effects of endocrine disrupting chemicals. *Environ. Res.* **2022**, *204 Pt B*, 112063. [CrossRef]
110. Thorson, J.L.M.; Beck, D.; Ben Maamar, M.; Nilsson, E.E.; Skinner, M.K. Epigenome-wide association study for pesticide (Permethrin and DEET) induced DNA methylation epimutation biomarkers for specific transgenerational disease. *Environ. Health* **2020**, *19*, 109. [CrossRef]
111. Karlsson, O.; Svanholm, S.; Eriksson, A.; Chidiac, J.; Eriksson, J.; Jernerén, F.; Berg, C. Pesticide-induced multigenerational effects on amphibian reproduction and metabolism. *Sci. Total Environ.* **2021**, *775*, 145771. [CrossRef]
112. Wu, C.; He, K.; Li, H.; Zhang, L.; Mao, L.; Zhu, L.; Jiang, J.; Liu, X. Transgenerational combined toxicity effects of neonicotinoids and triazole pesticides at environmentally relevant concentrations on D. magna: From individual to population level. *J. Hazard. Mater.* **2025**, *486*, 137023. [CrossRef]
113. Stone, A.M.; Camp, O.G.; Biernat, M.M.; Bai, D.; Awonuga, A.O.; Abu-Soud, H.M. Re-Evaluating the Use of Glyphosate-based Herbicides: Implications on Fertility. *Reprod. Sci.* **2025**, *32*, 950–964. [CrossRef] [PubMed]
114. Zhang, C.; Schilirò, T.; Gea, M.; Bianchi, S.; Spinello, A.; Magistrato, A.; Gilardi, G.; Di Nardo, G. Molecular Basis for Endocrine Disruption by Pesticides Targeting Aromatase and Estrogen Receptor. *Int. J. Environ. Res. Public Health* **2020**, *17*, 5664. [CrossRef]
115. Han, X.; Huang, Q. Environmental pollutants exposure and male reproductive toxicity: The role of epigenetic modifications. *Toxicology* **2021**, *456*, 152780. [CrossRef] [PubMed]
116. OECD. Test No. 440: Uterotrophic Bioassay in Rodents: A short-term screening test for oestrogenic properties. In *OECD Guidelines for the Testing of Chemicals*; Section 4; OECD Publishing: Paris, France, 2007.
117. OECD. Test No. 443: Extended One-Generation Reproductive Toxicity Study. In *OECD Guidelines for the Testing of Chemicals*; Section 4; OECD Publishing: Paris, France, 2018.
118. OECD. Test No. 456: H295R Steroidogenesis Assay. In *OECD Guidelines for the Testing of Chemicals*; Section 4; OECD Publishing: Paris, France, 2023.
119. OECD. Test No. 458: Stably Transfected Human Androgen Receptor Transcriptional Activation Assay for Detection of Androgenic Agonist and Antagonist Activity of Chemicals. In *OECD Guidelines for the Testing of Chemicals*; Section 4; OECD Publishing: Paris, France, 2023.
120. Cestonaro, L.V.; Macedo, S.M.D.; Piton, Y.V.; Garcia, S.C.; Arbo, M.D. Toxic effects of pesticides on cellular and humoral immunity: An overview. *Immunopharmacol. Immunotoxicol.* **2022**, *44*, 816–831. [CrossRef]
121. Lee, G.-H.; Choi, K.-C. Adverse effects of pesticides on the functions of immune system. *Comp. Biochem. Physiol. Part C Toxicol. Pharmacol.* **2020**, *235*, 108789. [CrossRef] [PubMed]
122. Naidenko, O.V.; Andrews, D.Q.; Temkin, A.M.; Stoiber, T.; Uche, U.I.; Evans, S.; Perrone-Gray, S. Investigating molecular mechanisms of immunotoxicity and the utility of ToxCast for immunotoxicity screening of chemicals added to food. *Int. J. Environ. Res. Public Health* **2021**, *18*, 3332. [CrossRef]
123. Casida, J.E.; Durkin, K.A. Pesticide chemical research in toxicology: Lessons from nature. *Chem. Res. Toxicol.* **2017**, *30*, 94–104. [CrossRef]
124. Corsini, E.; Sokooti, M.; Galli, C.L.; Moretto, A.; Colosio, C. Pesticide induced immunotoxicity in humans: A comprehensive review of the existing evidence. *Toxicology* **2013**, *307*, 123–135. [CrossRef] [PubMed]
125. Cheng, B.; Zhang, H.; Jia, K.; Li, E.; Zhang, S.; Yu, H.; Cao, Z.; Xiong, G.; Hu, C.; Lu, H. Effects of spinetoram on the developmental toxicity and immunotoxicity of zebrafish. *Fish Shellfish Immunol.* **2020**, *96*, 114–121. [CrossRef] [PubMed]
126. Jiang, J.; Wu, S.; Wang, Y.; An, X.; Cai, L.; Zhao, X.; Wu, C. Carbendazim has the potential to induce oxidative stress, apoptosis, immunotoxicity and endocrine disruption during zebrafish larvae development. *Toxicol. Vitr.* **2015**, *29*, 1473–1481. [CrossRef]
127. Weis, G.C.C.; Assmann, C.E.; Cadona, F.C.; Bonadiman, B.D.S.R.; de Oliveira Alves, A.; Machado, A.K.; Duarte, M.M.M.F.; da Cruz, I.B.M.; Costabeber, I.H. Immunomodulatory effect of mancozeb, chlorothalonil, and thiophanate methyl pesticides on macrophage cells. *Ecotoxicol. Environ. Saf.* **2019**, *182*, 109420. [CrossRef]
128. Martin, T.J.; Gabure, S.; Maise, J.; Snipes, S.; Peete, M.; Whalen, M.M. The organochlorine pesticides pentachlorophenol and dichlorodiphenyltrichloroethane increase secretion and production of interleukin 6 by human immune cells. *Environ. Toxicol. Pharmacol.* **2019**, *72*, 103263. [CrossRef]

129. Mokarizadeh, A.; Faryabi, M.R.; Rezvanfar, M.A.; Abdollahi, M. A comprehensive review of pesticides and the immune dysregulation: Mechanisms, evidence and consequences. *Toxicol. Mech. Methods* **2015**, *25*, 258–278. [CrossRef]
130. Alam, R.T.; Imam, T.S.; Abo-Elmaaty, A.M.; Arisha, A.H. Amelioration of fenitrothion induced oxidative DNA damage and inactivation of caspase-3 in the brain and spleen tissues of male rats by N-acetylcysteine. *Life Sci.* **2019**, *231*, 116534. [CrossRef] [PubMed]
131. Lone, M.I.; Nabi, A.; Dar, N.J.; Hussain, A.; Nazam, N.; Hamid, A.; Ahmad, W. Toxicogenetic evaluation of dichlorophene in peripheral blood and in the cells of the immune system using molecular and flow cytometric approaches. *Chemosphere* **2017**, *167*, 520–529. [CrossRef]
132. Van Maele-Fabry, G.; Gamet-Payrastre, L.; Lison, D. Household exposure to pesticides and risk of leukemia in children and adolescents: Updated systematic review and meta-analysis. *Int. J. Hyg. Environ. Health* **2019**, *222*, 49–67. [CrossRef]
133. Luo, D.; Zhou, T.; Tao, Y.; Feng, Y.; Shen, X.; Mei, S. Exposure to organochlorine pesticides and non-Hodgkin lymphoma: A meta-analysis of observational studies. *Sci. Rep.* **2016**, *6*, 25768. [CrossRef]
134. Falzone, L.; Marconi, A.; Loreto, C.; Franco, S.; Spandidos, D.A.; Libra, M. Occupational exposure to carcinogens: Benzene, pesticides and fibers. *Mol. Med. Rep.* **2016**, *14*, 4467–4474. [CrossRef]
135. OECD. Adverse Outcome Pathways. Available online: https://www.oecd.org/en/topics/sub-issues/testing-of-chemicals/adverse-outcome-pathways.html (accessed on 20 September 2025).
136. OECD. Guidance Document on Developing and Assessing Adverse Outcome Pathways. Available online: https://one.oecd.org/document/env/jm/mono(2013)6/en/pdf (accessed on 20 September 2025).
137. de Paula Nunes, E.; Abou Dehn Pestana, B.; Pereira, B.B. Human biomonitoring and environmental health: A critical review of global exposure patterns, methodological challenges and research gaps. *J. Toxicol. Environ. Health Part B* **2025**, 1–19.
138. Birolli, W.G.; Lanças, F.M.; dos Santos Neto, Á.J.; Silveira, H.C.S. Determination of pesticide residues in urine by chromatography-mass spectrometry: Methods and applications. *Front. Public Health* **2024**, *12*, 1336014.
139. Kim, U.J.; Hong, M.; Choi, Y.H. Environmental Pyrethroid Exposure and Cognitive Dysfunction in U.S. Older Adults: The NHANES 2001–2002. *Int. J. Environ. Res. Public Health* **2021**, *18*, 12005. [CrossRef] [PubMed]
140. Hsieh, N.H.; Kwok, E.S.C. Biomonitoring-Based Risk Assessment of Pyrethroid Exposure in the U.S. Population: Application of High-Throughput and Physiologically Based Kinetic Models. *Toxics* **2025**, *13*, 216. [CrossRef] [PubMed]
141. Đuc, N.; Khanh, T.; Thang, P.; Hung, T.; Thu, P.; Chi, L.; Khuyen, V.; Tuan, N. Ultra-High Performance Liquid Chromatography–Tandem Mass Spectrometry Method Development and Validation to Quantify Simultaneously Six Urinary DIALKYL Phosphate Metabolites of Organophosphorus Pesticides. *J. Mass. Spectrom.* **2025**, *60*, e5128. [CrossRef]
142. Seo, S.H.; Batterman, S.; Karvonen-Gutierrez, C.A.; Park, S.K. Determinants of urinary dialkyl phosphate metabolites in midlife women: The Study of Women's Health Across the Nation Multi-Pollutant Study (SWAN-MPS). *J. Expo. Sci. Environ. Epidemiol.* **2025**, *35*, 590–601. [CrossRef]
143. Roca, M.; Miralles-Marco, A.; Ferré, J.; Pérez, R.; Yusà, V. Biomonitoring exposure assessment to contemporary pesticides in a school children population of Spain. *Environ. Res.* **2014**, *131*, 77–85. [CrossRef]
144. Iglesias-Gonzalez, A.; Hardy, E.M.; Appenzeller, B.M.R. Cumulative exposure to organic pollutants of French children assessed by hair analysis. *Environ. Int.* **2020**, *134*, 105332. [CrossRef]
145. Sexton, K.; Needham, L.L.; Pirkle, J.L. Human Biomonitoring of Environmental Chemicals: Measuring chemicals in human tissues is the "gold standard" for assessing people's exposure to pollution. *Am. Sci.* **2004**, *92*, 38–45. [CrossRef]
146. Angerer, J.; Ewers, U.; Wilhelm, M. Human biomonitoring: State of the art. *Int. J. Hyg. Environ. Health* **2007**, *210*, 201–228. [CrossRef]
147. Connolly, A.; Coggins, M.A.; Koch, H.M. Human Biomonitoring of Glyphosate Exposures: State-of-the-Art and Future Research Challenges. *Toxics* **2020**, *8*, 60. [CrossRef] [PubMed]
148. Tkalec, Ž.; Codling, G.; Tratnik, J.S.; Mazej, D.; Klánová, J.; Horvat, M.; Kosjek, T. Suspect and non-Targeted screening-Based Human Biomonitoring Identified 74 Biomarkers of Exposure in Urine of Slovenian Children. *Environ. Pollut.* **2022**, *313*, 120091. [CrossRef] [PubMed]
149. Jaacks, L.M.; Diao, N.; Calafat, A.M.; Ospina, M.; Mazumdar, M.; Hasan, O.S.I.; Wright, R.; Quamruzzaman, Q.; Christiani, D.C. Association of Prenatal Pesticide Exposures with Adverse Pregnancy Outcomes and Stunting in Rural Bangladesh. *Environ. Int.* **2019**, *133*, 105243. [CrossRef] [PubMed]
150. Li, Y.; Wang, X.; Feary McKenzie, J.; Mannetje, A.; Cheng, S.; He, C.; Leathem, J.; Pearce, N.; Sunyer, J.; Eskenazi, B.; et al. Pesticide Exposure in New Zealand School-Aged Children: Urinary Concentrations of Biomarkers and Assessment of Determinants. *Environ. Int.* **2022**, *163*, 107206. [CrossRef]
151. Shentema, M.G.; Kumie, A.; Bråtveit, M.; Deressa, W.; Ngowi, A.V.; Moen, B.E. Pesticide Use and Serum Acetylcholinesterase Levels Among Flower Farm Workers in ethiopia-A Cross-Sectional Study. *Int. J. Environ. Res. Public Health* **2020**, *17*, 964. [CrossRef]

152. Menouni, A.; Duca, R.C.; Berni, I.; Khouchoua, M.; Ghosh, M.; El Ghazi, B.; Zouine, N.; Lhilali, I.; Akroute, D.; Pauwels, S.; et al. The Parental Pesticide and offspring's Epigenome Study: Towards an Integrated Use of Human Biomonitoring of Exposure and Effect Biomarkers. *Toxics* **2021**, *9*, 332. [CrossRef]
153. Makris, K.C.; Efthymiou, N.; Konstantinou, C.; Anastasi, E.; Schoeters, G.; Kolossa-Gehring, M.; Katsonouri, A. Oxidative Stress of Glyphosate, AMPA and Metabolites of Pyrethroids and Chlorpyrifos Pesticides Among Primary School Children in Cyprus. *Environ. Res.* **2022**, *212*, 113316. [CrossRef]
154. HBM4EU. HBM4EU Priority Substances. *Human Biomonitoirng for Europe*. Available online: https://www.hbm4eu.eu/hbm4eu-substances/ (accessed on 6 August 2025).
155. HBM4EU. The Project | HBM4EU-Science and Policy for a Healthy Future. Available online: https://www.hbm4eu.eu/the-project/ (accessed on 6 August 2025).
156. Vorkamp, K.; López, M.E.; Gilles, L.; Göen, T.; Govarts, E.; Hajeb, P.; Katsonouri, A.; Knudsen, L.E.; Kolossa-Gehring, M.; Lindh, C.; et al. Coordination of chemical analyses under the European Human Biomonitoring Initiative (HBM4EU): Concepts, procedures and lessons learnt. *Int. J. Hyg. Environ. Health* **2023**, *251*, 114183. [CrossRef]
157. Ramhøj, L.; Svingen, T.; Vanhaecke, T. Editorial: European partnership on the assessment of risks from chemicals (PARC): Focus on new approach methodologies (NAMs) in risk assessment. *Front. Toxicol.* **2024**, *6*, 1461967. [CrossRef] [PubMed]
158. Nagy, K.; Duca, R.C.; Lovas, S.; Creta, M.; Scheepers, P.T.; Godderis, L.; Ádám, B. Systematic review of comparative studies assessing the toxicity of pesticide active ingredients and their product formulations. *Environ. Res.* **2020**, *181*, 108926. [CrossRef]
159. Villaverde, J.J.; Sevilla-Morán, B.; López-Goti, C.; Alonso-Prados, J.L.; Sandín-España, P. QSAR/QSPR models based on quantum chemistry for risk assessment of pesticides according to current European legislation. *SAR QSAR Environ. Res.* **2019**, *31*, 49–72. [CrossRef] [PubMed]
160. Goh, M.S.; Lam, S.D.; Yang, Y.F.; Naqiuddin, M.; Addis, S.N.K.; Yong, W.T.L.; Luang-In, V.; Sonne, C.; Ma, N.L. Omics technologies used in pesticide residue detection and mitigation in crop. *J. Hazard. Mater.* **2021**, *420*, 126624. [CrossRef]
161. Bercu, J.; Masuda-Herrera, M.J.; Trejo-Martin, A.; Hasselgren, C.; Lord, J.; Graham, J.; Schmitz, M.; Milchak, L.; Owens, C.; Lal, S.H. A cross-industry collaboration to assess if acute oral toxicity (Q)SAR models are fit-for-purpose for GHS classification and labelling. *Regul. Toxicol. Pharmacol.* **2021**, *120*, 104843. [CrossRef] [PubMed]
162. McKim, J.M.; Bradbury, S.P.; Niemi, G.J. Fish acute toxicity syndromes and their use in the QSAR approach to hazard assessment. *Environ. Health Perspect.* **1987**, *71*, 171–186. [CrossRef]
163. Bhhatarai, B.; Gramatica, P. Per-and polyfluoro toxicity (LC50 inhalation) study in rat and mouse using QSAR modeling. *Chem. Res. Toxicol.* **2010**, *23*, 528–539. [CrossRef]
164. Basant, N.; Gupta, S.; Singh, K.P. Predicting toxicities of diverse chemical pesticides in multiple avian species using tree-based QSAR approaches for regulatory purposes. *J. Chem. Inf. Model.* **2015**, *55*, 1337–1348. [CrossRef]
165. Das, S.; Samal, A.; Ojha, P.K. Chemometrics-driven prediction and prioritization of diverse pesticides on chickens for addressing hazardous effects on public health. *J. Hazard. Mater.* **2024**, *471*, 134326. [CrossRef]
166. Das, S.; Ojha, P.K. Revolutionizing toxicity predictions of diverse chemicals to protect human health: Comparative QSAR and q-RASAR modeling. *Toxicol. Lett.* **2025**, *411*, 16–24. [CrossRef]
167. Upton, R.N.; Foster, D.J.R.; Abuhelwa, A.Y. An introduction to physiologically-based pharmacokinetic models. *Pediatr. Anesth.* **2016**, *26*, 1036–1046. [CrossRef]
168. Zartarian, V.; Xue, J.; Glen, G.; Smith, L.; Tulve, N.; Tornero-Velez, R. Quantifying children's aggregate (dietary and residential) exposure and dose to permethrin: Application and evaluation of EPA's probabilistic SHEDS-Multimedia model. *J. Expo. Sci. Environ. Epidemiol.* **2012**, *22*, 267–273. [CrossRef]
169. Tornero-Velez, R.; Davis, J.; Scollon, E.J.; Starr, J.M.; Setzer, R.W.; Goldsmith, M.R.; Chang, D.T.; Xue, J.; Zartarian, V.; De Vito, M.J.; et al. A pharmacokinetic model of cis-and trans-permethrin disposition in rats and humans with aggregate exposure application. *Toxicol. Sci.* **2012**, *130*, 33–47. [CrossRef]
170. Isaacs, K.K.; Glen, W.G.; Egeghy, P.; Goldsmith, M.R.; Smith, L.; Vallero, D.; Brooks, R.; Grulke, C.M.; Grulke, H.O. SHEDS-HT: An integrated probabilistic exposure model for prioritizing exposures to chemicals with near-field and dietary sources. *Environ. Sci. Technol.* **2014**, *48*, 12750–12759. [CrossRef]
171. International Programme on Chemical Safety (IPCS). *Characterization and Application of Physiologically Based Pharmacokinetic Models in Risk Assessment*; World Health Organization, International Programme on Chemical Safety: Geneva, Switzerland, 2010. Available online: https://iris.who.int/server/api/core/bitstreams/9a07bd8a-bc8d-4c8c-9349-63e46fca607f/content (accessed on 15 September 2025).
172. Meek, M.B.; Barton, H.A.; Bessems, J.G.; Lipscomb, J.C.; Krishnan, K. Case study illustrating the WHO IPCS guidance on characterization and application of physiologically based pharmacokinetic models in risk assessment. *Regul. Toxicol. Pharmacol.* **2013**, *66*, 116–129. [CrossRef]

173. *Approaches for the Application of Physiologically Based Pharmacokinetic (PBPK) Models and Supporting Data in Risk Assessment (Final Report)*; NCEA Publication: S. EPA; U.S. Environmental Protection Agency: Washington, DC, USA, 2006. Available online: https://iris.epa.gov/Document/&deid=157668 (accessed on 15 September 2025).
174. OECD. Guidance document on the characterisation, validation and reporting of Physiologically Based Kinetic (PBK) models for regulatory purposes. In *OECD Series on Testing and Assessment*; No. 331; OECD Publishing: Paris, France, 2021. [CrossRef]
175. McLanahan, E.D.; El-Masri, H.A.; Sweeney, L.M.; Kopylev, L.Y.; Clewell, H.J.; Wambaugh, J.F.; Schlosser, P.M. Physiologically based pharmacokinetic model use in risk assessment—Why being published is not enough. *Toxicol. Sci.* **2012**, *126*, 5–15. [CrossRef]
176. Perovani, I.S.; Yamamoto, P.A.; da Silva, R.M.; Lopes, N.P.; de Moraes, N.V.; de Oliveira, A.R.M. Unveiling CYP450 inhibition by the pesticide prothioconazole through integrated in vitro studies and PBPK modeling. *Arch. Toxicol.* **2025**, *99*, 2845–2854. [CrossRef] [PubMed]
177. Mangas, I.; Spilioti, E.; Terron, A.; Panzarea, M.; Nepal, M.; Viviani, B.; Binaglia, M.; Crofton, K.M. European Food Safety Authority database for in vivo developmental neurotoxicity studies of pesticides. *Crit. Rev. Toxicol.* **2025**, *55*, 587–598. [CrossRef] [PubMed]
178. Munnia, A.; Bollati, V.; Russo, V.; Ferrari, L.; Ceppi, M.; Bruzzone, M.; Dugheri, S.; Arcangeli, G.; Merlo, F.; Peluso, M. Traffic-Related Air Pollution and Ground-Level Ozone Associated Global DNA Hypomethylation and Bulky DNA Adduct Formation. *Int. J. Mol. Sci.* **2023**, *24*, 2041. [CrossRef] [PubMed]
179. Cao, L.; Kang, Q.; Tian, Y. Pesticide residues: Bridging the gap between environmental exposure and chronic disease through omics. *Ecotoxicol. Environ. Saf.* **2024**, *287*, 117335. [CrossRef]
180. Manzoor, R. Toxicity in the Twenty-First Century (Tox-21): A Multidisciplinary Approach for Toxicity Prediction of Marine Organic Pollutants. In *Recent Trends in Marine Toxicological Assessment*; Chuan, O.M., Hamid, N., Ghazali, A., Eds.; Springer: Cham, Switzerland, 2025.
181. Anastas, P.T.; Warner, J.C. Principles of green chemistry. In *Green Chemistry: Theory and Practice*; Oxford University Press: Oxford, UK, 1998; Volume 29, pp. 14821–14842.
182. Fitzgerald, D.P.; Herrmann, J.W.; Sandborn, P.A.; Schmidt, L.C.; Gogoll, T.H. Design for environment (DfE): Strategies, practices, guidelines, methods, and tools. *Environ. Conscious Mech. Des.* **2007**, *1*, 1–24.
183. Jamin, E.L.; Bonvallot, N.; Tremblay-Franco, M.; Cravedi, J.P.; Chevrier, C.; Cordier, S.; Debrauwer, L. Untargeted profiling of pesticide metabolites by LC–HRMS: An exposomics tool for human exposure evaluation. *Anal. Bioanal. Chem.* **2014**, *406*, 1149–1161. [CrossRef]
184. Lefèvre-Arbogast, S.; Chaker, J.; Mercier, F.; Barouki, R.; Coumoul, X.; Miller, G.W.; David, A.; Samieri, C. Assessing the contribution of the chemical exposome to neurodegenerative disease. *Nat. Neurosci.* **2024**, *27*, 812–821. [CrossRef]
185. Kunde, H.; Ioannou, E.I.; Chovatiya, J.; Jagani, R.; Charisiadis, P.; Arora, M.; Andra, S.S.; Makris, K.C. An exposomics analysis of 125 biomarkers of exposure to food contaminants and biomarkers of oxidative stress: A randomized cross-over chrononutrition trial of healthy adults. *Environ. Int.* **2025**, *202*, 109682. [CrossRef]
186. Gozali, L.; Kristina, H.J.; Yosua, A.; Zagloel, T.Y.M.; Masrom, M.; Susanto, S.; Tanujaya, H.; Irawan, A.P.; Gunadi, A.; Kumar, V.; et al. The improvement of block chain technology simulation in supply chain management (case study: Pesticide company). *Sci. Rep.* **2024**, *14*, 3784. [CrossRef] [PubMed]
187. Santos, R.d.A.; Teles, E.O.; Freires, F.G.M. Applying the Circular Economy Framework to Blockchain Agricultural Production. *Sustainability* **2024**, *16*, 8004. [CrossRef]
188. European Commission. EU Multi-Annual Control Programmes. Available online: https://food.ec.europa.eu/plants/pesticides/maximum-residue-levels/enforcement/eu-multi-annual-control-programmes_en (accessed on 15 September 2025).
189. Nikolopoulou, D.; Ntzani, E.; Kyriakopoulou, K.; Anagnostopoulos, C.; Machera, K. Priorities and Challenges in Methodology for Human Health Risk Assessment from Combined Exposure to Multiple Chemicals. *Toxics* **2023**, *11*, 401. [CrossRef] [PubMed]
190. D'Amore, T.; Smaoui, S.; Varzakas, T. Chemical Food Safety in Europe Under the Spotlight: Principles, Regulatory Framework and Roadmap for Future Directions. *Foods* **2025**, *14*, 1628. [CrossRef] [PubMed]
191. Cattaneo, I.; Kalian, A.D.; Di Nicola, M.R.; Dujardin, B.; Levorato, S.; Mohimont, L.; Nathanail, A.V.; Carnesschi, E.; Astuto, M.C.; Tarazona, J.V.; et al. Risk Assessment of Combined Exposure to Multiple Chemicals at the European Food Safety Authority: Principles, Guidance Documents, Applications and Future Challenges. *Toxins* **2023**, *15*, 40. [CrossRef]
192. Reale, E.; Zare Jeddi, M.; Paini, A.; Connolly, A.; Duca, R.; Cubadda, F.; Benfenati, E.; Bessems, J.; Galea, K.S.; Dirven, H.; et al. Human Biomonitoring and Toxicokinetics as Key Building Blocks for next Generation Risk Assessment. *Environ. Int.* **2024**, *184*, 108474. [CrossRef] [PubMed]

Journal of *Xenobiotics*

MDPI

Review

Global Evidence on Monitoring Human Pesticide Exposure

Tatiane Renata Fagundes [1], Carolina Coradi [1], Beatriz Geovana Leite Vacario [2], Juliana Maria Bitencourt de Morais Valentim [1] and Carolina Panis [1,*]

[1] Laboratory of Tumor Biology, Western Paraná State University, Francisco Beltrão 85605-010, Parana, Brazil; tatiane.fagundes@uenp.edu.br (T.R.F.); carolina.coradi@uel.br (C.C.); juliana.mbitencourt@uel.br (J.M.B.d.M.V.)

[2] Laboratory of Mutagenesis and Oncogenetics, State University of Londrina, Londrina 86057-970, Paraná, Brazil; beatriz.geovana@uel.br

* Correspondence: carolina.panis@unioeste.br

Abstract: This study analyzes global data on human exposure to pesticides, focusing on glyphosate, POPs, carbamates, and organophosphates, which are among the most widely used in agricultural and urban environments, providing an overview of global human contamination by these substances. Current research has increasingly focused on the unintended consequences of pesticide use, including food, water, and soil contamination, biodiversity loss (especially beneficial insects such as pollinators), and the growing evidence of adverse impacts on human health (neurological, reproductive, endocrine, and carcinogenic effects). Therefore, we compiled information from several existing studies that evaluated pesticide residues in human biological samples, specifically urine, blood, and breast milk, to assess the extent of exposure. The analysis takes a global perspective, highlighting the importance of monitoring exposure in countries that demonstrate exceptionally high pesticide use (in terms of absolute volume), such as Brazil, the United States, and China, which are among the largest global consumers. The data cover both contemporary pesticides, whose consumption is driven by intensive agriculture in these and other countries, and persistent legacy compounds (POPs) that continue to circulate in nature and accumulate in the human body decades after their ban in many countries. Globally, there is a wide disparity in global regulations, and many developing countries continue to use pesticides that have been banned or severely restricted in more developed nations. Finally, it provides a critical overview of global data on human pesticide contamination. The data reinforce the critical importance of establishing preventive initiatives and strengthening surveillance and monitoring systems to detect and control pesticide residues in human populations globally, ultimately aiming to mitigate the harms of chronic pesticide exposure to human health and well-being.

Keywords: pesticide; environmental pollutants; biological monitoring

1. Introduction

The widespread use of pesticides in global agriculture has been a cornerstone of ensuring food security and increasing agricultural productivity [1]. However, this intensive reliance raises significant concerns about the potential risks to human and environmental health, given the persistence and toxicity of many of these compounds.

This exposure occurs through occupational, environmental, and incidental pathways, including unintentional contact, such as spray drift or food and water contamination, and has been linked to a variety of health problems, including cancer, diabetes, and neurodegenerative diseases. Furthermore, the presence of these substances may be associated with adverse maternal and child health outcomes, such as congenital anomalies and reproductive dysfunction [1,2].

Despite historical limitations, such as complex and costly analytical procedures and the need for specialized laboratories, there has been significant progress in the biomonitoring of human pesticide exposure. Validated methods for common pesticide metabolites are routinely applied, and international initiatives have supported standardization and quality assurance efforts. At the same time, several regulatory agencies (e.g., European Food Safety Authority (EFSA), European Chemicals Agency (ECHA), and United States Environmental Protection Agency (US EPA)) have published reference values for selected compounds, contributing to better interpretation of biomonitoring data. Although health-based reference values, such as the Acceptable Daily Intake (ADI), exist, these values vary significantly between countries, making standardization of reference values in human matrices difficult. Consequently, biological indices of exposure, defined as concentrations of pesticides or their metabolites in human samples, such as urine or blood, indicative of exposure levels, are still lacking or inconsistent for several compounds [3].

Pesticide regulations vary significantly across countries. Some impose strict standards, while others adopt more lenient policies, influenced by numerous factors, including agricultural practices, pest prevalence, and regional environmental conditions, which increase population exposure in areas with more lenient regulations. In the early 2000s, pesticide application rates were highest in Asia and the Americas, averaging 2.92 and 2.42 kg/ha, respectively. In 2018, these rates increased to 3.69 kg/ha in Asia and 3.54 kg/ha in the Americas. In contrast, pesticide use in Europe, Africa, and Oceania during 2018 was considerably lower, at 1.67, 0.39, and 2.07 kg/ha, respectively [4]. Therefore, distinct patterns of exposure may emerge among populations around the world.

According to the 2022 Food and Agriculture Organization of the United Nations (FAO) report, the Americas led global pesticide use, applying over 1 million tons, with an average intake of 1.17 kg per person [5]. Europe followed with an average of 0.64 kg/person, while Africa recorded lower use than the others, with an intake per person of 0.11 kg/person. In terms of the absolute amount applied in 2020, the top three countries were the United States (0.408 million tons), Brazil (0.377 million tons), and China (0.273 million tons) [6], countries with the most significant monitoring data due to their high rates of use of these compounds. We also present data from many European Union countries, which historically have stricter legislation on the usual limits for these compounds in food and water, and compare the regulatory differences between them.

Most pesticide biomonitoring studies in humans are conducted in high-income countries, such as those in the European Union and North America, creating a significant lack of data in low- and middle-income countries, where agricultural practices, the use of personal protective equipment, and regulations may be less stringent, exposing populations to greater and poorly documented risks, limiting the creation of public policies to address this problem. Data on pesticide exposure and its health effects are from studies that are cross-sectional and not longitudinal, limiting the understanding of the effects of long-term exposure and hindering the creation of standardized global biomonitoring programs that allow for comparison of data across countries and populations. This also impedes the creation of a globally comparable database to assess exposure trends over time and identify more vulnerable populations.

Against this backdrop, it becomes evident that human exposure to pesticides results not only from their intensive use but also from regulatory weaknesses, the insufficiency of consistent monitoring programs, and the absence of effective environmental remediation strategies. Therefore, this review is guided by the hypothesis that legislative permissiveness, the lack of systematic biomonitoring, and the irreversibility of environmental and biological contamination act together to intensify risks to human health and ecological balance. These shortcomings are further aggravated by socioeconomic inequalities, which leave groups such as rural workers, pregnant women, and children disproportionately vulnerable, particularly in countries with more lenient regulations, in contrast to the more precautionary framework.

In this context, this review presents a global overview of human pesticide contamination based on biomonitoring data from biological samples of blood, urine, and breast milk, focusing on current and past pesticides, including glyphosate and its metabolite AMPA, DDT, and its organophosphate and carbamate derivatives DDD and DDE. It provides an overview of the global landscape of environmental contamination and its currently documented effects on human health. Reviewing this topic is crucial to synthesizing existing evidence and highlighting its gaps, providing a solid basis for public health decision-making and the formulation of more effective regulatory policies. The lack of data in certain regions and the inability to assess cumulative exposure hinder the definition of safety limits and the implementation of risk mitigation strategies globally.

To make the structure of the article clearer, the authors have provided an outline of the manuscript, which is included in the Supplementary Materials.

2. Pesticide Levels in Human Samples: Primary Matrices, Methods, and Target Analytes

Blood and urine are the most commonly used biological samples for pesticide detection. Blood is considered an ideal matrix for detecting many chemical compounds, as it circulates throughout the body, promoting exchanges between tissues and organs. Thus, whole blood or serum and plasma are important to assess internal exposure to persistent pesticides, such as organochlorine insecticides, which have long biological half-lives and can be measured as the parent compound or its metabolites [7].

For pesticides with a short half-life in the human body and therefore found in low concentrations in the blood, urine is the matrix of choice, especially for the evaluation of non-persistent pesticides such as pyrethroids, organophosphate insecticides, or different classes of herbicides, such as glyphosate, which, due to its rapid metabolism, limits its detection in blood samples [8]. The compounds found in urine indicate recent exposure, as non-persistent pesticides are metabolized and eliminated from the body within a few days.

Non-invasive alternative matrices for monitoring pesticides, such as hair, meconium or breast milk, have attracted increasing interest in recent years due to their ease of acquisition. Human hair is an easy-to-collect, -transport, and -store sample that is stable and provides information on short- and long-term exposures, but it is difficult to differentiate between endogenous deposition, diffusion from blood capillaries, and exogenous environmental deposition. Several organochlorine and organophosphate insecticides have been found in hair samples, mainly by analysis of their parent compound [9].

Compounds that have lipophilic bioaccumulation characteristics are excreted in breast milk, a sample that can be used to monitor maternal and infant exposure to lipophilic organochlorine pesticides and other persistent organic pollutants (POPs). Comparison of the levels found should be adjusted according to the lipid content, since the amount of lipids in milk varies. Milk analysis requires careful preparation, with volumes between 1 and 5 mL and removal of proteins and fats [10].

Meconium is considered an appropriate matrix to measure prenatal pesticide exposure, since these compounds can accumulate in meconium from the third month of gestation until birth. This analysis provides a long-term fetal exposure profile and can be collected in large quantities [11].

There are many quantitative methods for detecting pesticides in biological samples. Sample preparation involves multiple steps that remove interfering substances or hydrolyze conjugated forms of the compounds of interest (pre-analytical step). In some cases, pretreatment consists of a simple dilution of the sample with water or formic acid to reduce the matrix effect and matrix variability between samples, as is done with urine [10,11]. Blood or breast milk samples need to undergo precipitation processes and protein denaturation, and the compounds used for this purpose include formic acid and solvents such as n-propanol, isopropanol, methanol or acetonitrile, which increase the extraction efficiency of lipophilic pesticides and avoid clogging of solid phase extraction [10,12–14].

To protect public health, the European Commission (EC) established Maximum Residue Limits (MRLs) for pesticides in food, measured in μg/kg or mg/kg [15]. MRLs vary depending on the matrix analyzed, the specific environmental occurrence, and the toxicological profile of the pesticide. A general MRL of 10 μg/kg is established when an MRL value is not specified [16]. Therefore, monitoring pesticide residues in food becomes essential to comply with these [17].

The wide variety of pesticides legally approved for use requires multi-residue analytical methods for their determination. However, determining low detection limits in accordance with the established MRLs and the complexity of some biological matrices, the different chemical classes to which these compounds belong, complicate this analysis [18].

The analysis of pesticides in human matrices still uses classical extraction and purification strategies, such as solid-phase extraction (SPE), liquid–liquid extraction (LLE), solid–liquid extraction (SLE), and solid-phase microextraction (SPME). Liquid–liquid extraction is a routine technique used for many years, but it requires a long time and solvent use and has been replaced by solid-phase extraction recently. Solid-phase extraction is an important approach for extracting and purifying pesticides from urine and milk [19,20]. For urine, several solvents are used in the extraction and purification of pesticide metabolites from urine. Oasis HLB, for example, is especially recommended for specific metabolites of organophosphates, pyrethroids and herbicides, while Strata XC (mixed-mode reversed phase/strong cation exchange) is more suitable for atrazine metabolites [21,22].

QuEChERS is a proprietary sample preparation method that has become extremely popular and is often considered an alternative or evolution for the analysis of pesticide residues in complex matrices. Its simplicity and effectiveness have allowed laboratories around the world to adopt similar protocols, which addresses the issue of the lack of standardized methods in different countries. QuEChERS (acronym for quick, easy, cheap, effective, robust, and safe)-based methods consist of analyte extraction by liquid partition using a small volume of acetonitrile, subsequent collection of the organic fraction after a salting-out step (promoting equilibrium between an aqueous and an organic layer), and cleanup of the organic extract using dispersive solid-phase extraction (d-SPE) [17]. The versatility of this approach allows the use of different buffered salts for the salting-out step, depending on the matrix and pesticide properties. This method is widely applicable in biological matrices with good extraction efficiency for a large number of organic compounds, allowing the selective extraction of multi-residue pesticides with a significant reduction in interferents in the sample and achieving high extraction efficiencies for pesticides from different chemical groups [23].

The choice of instrumental techniques for separating, detecting, and quantifying pesticides is crucial and depends on the physicochemical properties and concentrations of the target compounds. Gas chromatography–mass spectrometry (GC-MS) and high-performance liquid chromatography–tandem mass spectrometry (HPLC-MS/MS) are commonly used for pesticide detection. GC-MS is well-suited for nonpolar and volatile compounds, such as organochlorines (OC), pyrethroids (PYR), and organophosphates (OPs), or for more polar compounds, such as dialkyl phosphates (DAPs), when a prior derivatization step is used. In contrast, HPLC-MS/MS is preferable for more polar, less volatile, or thermally labile compounds, including metabolites of OPs and PYR, triazine herbicide metabolites, phenoxy acid herbicides, and neonicotinoid metabolites.

Conventional LC-MS/MS analysis determines a list of known analytes through the use of reference standards. LC-HRMS offers an alternative for broad screening of hundreds of polar contaminants in full-scan mode, enabling the detection of both known and unknown non-target compounds with high sensitivity and selectivity. However, in the absence of reference standards, quantification is generally limited to a semi-quantitative level [24,25].

Persistent pesticides are analyzed together with other POPs (polychlorinated biphenyls (PCBs) and dioxins), covered by the Stockholm Convention. The effectiveness of this treaty has led to a decrease in the concentrations of persistent pollutants in environmental and biological matrices, and more sensitive instrumental methods are needed to reliably quantify the amounts of persistent pollutants in mixtures of compounds [26]. The GC-MS method is the routine method for quantification of persistent pesticides and other POPs in various environmental and biological matrices [27].

The literature demonstrates that various regions and populations across Latin America, European Union countries, North America, Asia, and some African countries have experienced harmful health effects due to prolonged occupational or dietary exposure to pesticides, even at low doses, reinforcing that pesticide contamination is a global phenomenon. Additionally, these substances are known to quickly disperse through the atmosphere and infiltrate soil, contaminating groundwater sources such as water tables and aquifers, rendering them unsafe for human consumption [28].

Much research has been carried out mainly in the ten countries that currently use the highest volumes of pesticides worldwide: China (1806 million kg), the United States (386 million kg), Argentina (265 million kg), Thailand (87 million kg), Brazil (76 million kg), Italy (63 million kg), France (62 million kg), Canada (54 million kg), Japan (52 million kg), and India (40 million kg) [29,30]. The widespread and often uncontrolled use of pesticides in these countries is largely driven by the demand to maximize crop productivity, contain pests, and prevent crop diseases, improving harvest efficiency and ensuring significant economic returns for the agricultural sector. However, this dependence harms both the biosystem and rural communities and consumers [29].

Globally, the most widely used pesticides in countries where agriculture is the main economic activity include trifluralin, 2-methyl-4-chlorophenoxyacetic acid, glyphosate, 2,4-D, mancozeb, aldicarb, atrazine, chlorthalonil, diazinon, dicamba, diuron, malathion, metolachlor, and chlorpyrifos. Among these and other pesticides, many are banned in certain European Union countries but remain widely used in developing nations; these pesticides include chlorpyrifos, endosulfan, cypermethrin, atrazine, paraquat, glyphosate (both dimethylammonium and monoammonium salts), tebuthiuron, lambda-cyhalothrin, imidacloprid, dichlorvos, emamectin, and mancozeb [5,31,32].

While banned in countries in the Northern Hemisphere due to their harmful effects on the human body and ecosystems, they continue to be widely used in other countries in the Southern Hemisphere, where regulations are more lenient. Consequently, countries with stricter domestic policies may still expose their populations to banned pesticides through the import of products such as coffee, orange juice, and soybeans, which have not been adequately evaluated and come from countries with more lenient regulations [29]. Maggi and Cols (2021) [33] analyzed three key variables to assess and correlate pesticide use in different countries with potential risks to human health, including pesticide toxicity, gross population exposure based on geographic distribution, and human intake through air inhalation and drinking water contamination. This approach highlights how the type and scale of pesticide use, combined with local exposure pathways, contribute to varying levels of health risk across regions. The assessment revealed that critical regions common to all three variables were East Asia, mainly China, and South America, specifically Brazil. The study also highlighted that European Union countries, despite high pesticide application rates, exhibit lower average risk levels to their populations because of severe prohibitions and laws governing the application of these substances.

Therefore, there is a clear need for globally harmonized and stricter regulations on pesticide bans and maximum residue limits that are not regionally flexible. Such measures would better protect public health from risks such as cancer, endocrine disruption, neurological disorders, and environmental harm.

Among the various types of pesticides currently in use, glyphosate, POPs, carbamates, and organophosphates rank among the most utilized across agricultural and urban environments. These compounds are increasingly scrutinized due to their potential to compromise human health and disturb ecological systems. Proper assessment of their environmental fate and effects requires a thorough understanding of their physicochemical properties, as these determine how they interact with soil, water, air, and living organisms, as well as their degradation dynamics and persistence within ecosystems [34].

In the following sections, we will examine the principal pesticides used globally, the concentrations found in biological matrices, and the analytical techniques applied to measure them.

2.1. Glyphosate-AMPA

2.1.1. Chemical Composition and General Properties

Glyphosate (N-phosphonomethyl-glycine) (Figure 1) is an organophosphorus compound used as a broad-spectrum, non-selective herbicide, widely applied in agriculture for weed control. Its mechanism of action is based on the inhibition of the enzyme 5-enolpyruvylshikimate-3-phosphate synthase (EPSPS), which is involved in the shikimate pathway—essential for the biosynthesis of aromatic amino acids (phenylalanine, tryptophan, and tyrosine) in plants [35].

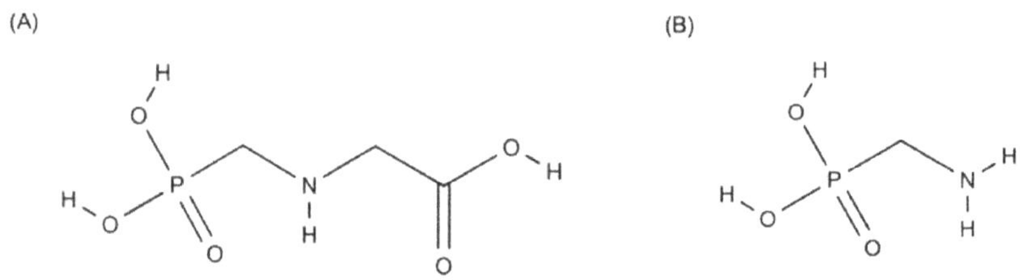

Figure 1. Chemical structure of glyphosate (**A**) and AMPA (**B**) [36].

It was in the 1970s that glyphosate was introduced to the market by the company Monsanto, under the trade name Roundup. Since then, it has become one of the most widely used pesticides worldwide. With the introduction of genetically modified crops resistant to glyphosate, its use has intensified [37].

The main method of glyphosate application in large agricultural areas is through aerial or ground spraying, which facilitates its dispersion through the air to non-target areas. After application, glyphosate can contaminate the environment through spray drift, surface runoff and leaching into groundwater. Residues of glyphosate and its metabolite, aminomethylphosphonic acid (AMPA), have already been detected in rivers, lakes, groundwater, and even in drinking water [38,39]. It is a compound moderately soluble in water (12 g/L at 25 °C) and exhibits adsorption to soil particles. Being stable to chemical and photochemical decomposition, its degradation depends on both biotic and abiotic factors and is mediated by microorganisms such as species of the genera *Pseudomonas* and *Achromobacter*, which produce AMPA and glyoxylic acid [40].

Currently, glyphosate is one of the main environmental contaminants and can affect non-target organisms such as soil microorganisms, amphibians, and insects. Reports indicate that its presence alters soil microbiota and interferes with nitrogen fixation in legumes, impacting soil fertility [41]. In aquatic environments, glyphosate can be toxic to algae, fish, and amphibians, especially in formulations that contain surfactants such as POEA (polyethoxylated tallow amine) [42].

The application of glyphosate on food crops leads to detectable residues in foods consumed by the general population. Studies have identified glyphosate and AMPA in various foods such as cereals, legumes, fruits, and vegetables, raising concerns about the effects of chronic ingestion of these residues [43].

In 2015, glyphosate was classified as probably carcinogenic to humans (Group 2A) by the International Agency for Research on Cancer (IARC), based on evidence of genotoxicity and the occurrence of tumors in laboratory animals under high levels of exposure [44]. However, this classification remains the subject of international debate by regulatory agencies, which allege insufficient evidence of carcinogenicity in humans under typical exposure conditions. Human exposure occurs mainly through occupational routes, ingestion of contaminated food and water. Rural workers involved in the preparation and application of pesticides are considered the most exposed individuals; however, women who do not work directly in the fields but are responsible for decontaminating clothing and personal protective equipment of their male partners who work in the fields show high levels of glyphosate contamination, mainly due to dermal exposure [45].

Chronic exposure to glyphosate has been associated with various conditions, such as alterations in immune system cells [46], breast cancer [45], follicular lymphoma [47], neurological disorders [48] and infertility [49].

2.1.2. Regulatory Limits in Different Countries and Health Effects in Humans

Regulation of the maximum allowable levels of glyphosate in drinking water varies significantly between countries and international organizations, reflecting different risk assessment approaches as well as the influence of economic policies and regulatory precautions. The European Union (EU) adopts one of the strictest standards worldwide, setting the limit at 0.1 μg/L for any individual pesticide, including glyphosate, in its Directive 98/83/EC on drinking water quality [50]. In contrast, countries such as the United States and Brazil have considerably more permissive limits. The U.S. Environmental Protection Agency (EPA) sets a Maximum Contaminant Level (MCL) of 700 μg/L [51]. In Brazil, the Ministry of Health, through Ordinance GM/MS No. 888/2021, establishes a limit of 500 μg/L [52].

Other countries also adopt different limits. In Canada, the maximum allowable value is 280 μg/L, according to Health Canada guidelines [53]. Australia is even more permissive, allowing up to 1000 μg/L, based on criteria established by the National Health and Medical Research Council (NHMRC) [54]. The World Health Organization (WHO), in its most recent provisional guideline, proposes a safety limit of 900 μg/L for glyphosate in drinking water [55].

This wide disparity in regulatory limits demonstrates the lack of international scientific consensus regarding the safety of glyphosate at residual levels, which hinders global risk comparisons and compromises coordinated strategies for environmental monitoring and public health protection.

Concerning human exposure, Table 1 presents the levels of glyphosate and AMPA detected in urine samples from individuals from different countries.

Table 1. Urinary levels of glyphosate and AMPA detected in samples from different countries.

Country	Sample	Methodology	LOQ	LOD	Outcome (Association)	Reference
Belgium	Urine (n = 428)	LC–MS/MS	LOQ GLY/AMPA = 0.1 μg/L	Not reported	Male adolescents living within 2000 m of agricultural areas had increased urinary AMPA levels	[56]
Canada	Urine (n = 1880)	UPLC-MS/MS	LOQ GLY = 0.26 μg/L, AMPA = 0.29 μg/L Glufosinate = 0.28 μg/L	LOD GLY = 0.08 μg/L AMPA = 0.09 μg/L Glufosinate = 0.08 μg/L	Because the population is predominantly urban, direct pesticide use and spray drift are unlikely pathways of contamination. Diet is believed to be a potential source of GLY exposure	[57]
Canada	Urine (n = 1765)	UPLC-MS/MS	LOQ GLY = 0.26 AMPA = 0.29 μg/L	LOD GLY/AMPA = 0.09 μg/L	No association was identified between urinary concentrations of GLY or AMPA and the risk of preterm birth	[58]
China	Urine (n = 134)	GC-MS/MS	Not reported	LOD GLY = 0.02 AMPA = 0.01 mg/L GLY ≤ 0.020–17.202 mg/L (median, 0.292 mg/L); AMPA ≤ 0.010 mg/L–2.730 mg/L (median, 0.068 mg/L)	The urinary concentration of GLY and AMPA of the workers was related to the TWA (time weighted average) value, which reflects the exposure of these workers	[59]
Cyprus	Urine (n = 177)	LC-MS/MS	LOQ GLY = 0.19 μg/L AMPA = 0.18 μg/L. GLY = 0.06–3.18 μg/L (mean 0.23 μg/L) AMPA = 0.06–1.44 μg/L (mean 0.24 μg/L)	Not reported	The study identified a significant association between AMPA levels found in the population and oxidative DNA damage	[60]
Democratic Republic of the Congo	Urine (n = 15)	LC-MS/MS	LOQ GLY = 80 pg/mL	Not reported	The study indicates exposure to environmental pollutants in the studied population	[61]
France	Urine (n = 1299)	UPLC-MS/MS	LOD GLY/AMPA = 0.02 μg/L	LOQ GLY/AMPA = 0.05 μg/L	The study identified higher concentrations of AMPA in children than in adults	[62]
Germany	Urine (n = 40)	GC-MS/MS	LOQ GLY = 0.1 μg/L AMPA = 0.1 μg/L. Maximum glyphosate = 2.80 μg/L; Maximum AMPA = 1.88 μg/L	Not reported	Not reported	[63]
Germany	Urine (n = 2144)	GC-MS/MS	Not reported	LOD GLY/AMPA = 0.1 μg/L	Not reported	[64]
Germany	Urine (n = 301)	LC–MS/MS	LOQ GLY/AMPA = 0.2 μg/L. M GLY = 0.16 μg/L; M AMPA = 0.20 μg/L	LOD GLY = 0.05 μg/L, LOD AMPA = 0.09 μg/L	The study identified positive correlations between urinary glyphosate excretion and legume consumption, as well as between urinary AMPA excretion and mushroom consumption	[65]

Table 1. *Cont.*

Country	Sample	Methodology	LOQ	LOD	Outcome (Association)	Reference
Ireland	Urine (n = 50)	LC-MS/MS	Not reported	LOD GLY = 0.5 μg/L. Median = 0.87 μg/L	Results suggest environmental exposure to GLY	[66]
Italy	Urine (n = 17)	UPLC-MS/MS	LOQ GLY = 0.1 μg/L AMPA = 0.5 μg/L. Median GLY = 2.30 μg/L	Not reported	Reports the absorption of xenobiotic by vineyard workers after application of GLY	[67]
Malaysia, Uganda and United Kingdom	Urine (n = 271)	LC-MS/MS	LOQ GLY/AMPA = 0.5 μg/L	Not reported	Higher GLY concentrations were associated with lower use of personal protective equipment, lower education levels, and longer duration of pesticide use	[68]
Mexico	Urine (n = 30)	HPLC	Not reported	LOD GLY = 5.0 AMPA = 15 μg/L	The study identifies, through the calculation of the external dose, that agricultural workers ingest up to 146 mg/kg/day of GLY	[69]
Puerto Rico	Urine (n = 247)	GC-MS/MS	Not reported	LOD GLY/AMPA = 0.20 μg/L	Study identifies associations between urinary GLY and AMPA concentrations and increased likelihood of preterm birth	[70]
Portugal	Urine (n = 79)	GC-MS/MS	LOQ GLY = 0.05 μg/L	LOD GLY = 0.02 AMPA = 0.013 μg/L	GLY and AMPA were detected in the study population, in both rounds of testing and in concentrations above the values of previous studies carried out in European countries	[71]
Slovenia	Urine (n = 246)	GC-MS/MS	LOQ GLY/AMPA = 0.1 μg/L	Not reported	Study shows low exposure to GLY and AMPA	[72]
Spain	Urine (n = 97)	LC-MS/MS	LOQ GLY/AMPA = 0.1 μg/L	Not reported	GLY concentration was related to the consumption of eggs and fruits in the 72 h preceding urine collection	[73]
Thailand	Urine (n = 124)	HPLC	Highest concentration = 2.92 ng/ml	LOD GLY = 0.5 μg/L	Exposure to GLY affects the entire population living in the agricultural community	[74]
Thailand	Urine (n = 180)	LC-MS/MS	LOQ GLY = 5 g/L	LOD GLY = 2.5 g/L	Workers applying GLY presented greater oxidative stress and worsening of lung function	[75]
Thailand	Urine (n = 15)	HPLC	LOQ GLY = 0.4 μg/L. M = 19.42 μg/L. Maximum level = 57.49 μg/L	LOD GLY = 0.1 μg/L	Workers are at significant risk of exposure to GLY	[76]
USA	Urine (n = 175)	LC-MS/MS	LOD GLY 1 ppb	LOD GLY 1 ppb Glyphosate concentrations $\leq$ 1–233 ppb	Farmers who did not use personal protective equipment, such as rubber gloves, had higher urinary GLY concentrations than other farmers	[77]

Table 1. *Cont.*

Country	Sample	Methodology	LOQ	LOD	Outcome (Association)	Reference
USA	Urine (n = 71)	LC-MS/MS	LOQ GLY = 0.5 ng/mL	LOD GLY = 0.1 ng/mL	Shortened gestational length was associated with high urinary GLY levels	[78]
USA	Urine (n = 187)	HPLC	LOQ GLY = 0.5 ng/mL	LOD GLY = 0.1 ng/mL	Elevated maternal GLY levels in the first trimester of pregnancy were associated with lower body weight percentiles (BWT) and increased risk of NICU admission	[79]
USA	Urine (n = 368)	IC-MS/MS	Not reported	LOD GLY = 0.2 μg/L	Study suggests that GLY exposure may be associated with oxidative stress	[80]
USA	Urine (n = 52)	LC-MS/MS	LOQ GLY = 0.03 ng/mL	LOD GLY = 0.03 ng/mL	Urinary GLY concentration was associated with increased odds of preterm birth (adjusted for urine specific gravity and maternal race)	[81]
USA	Urine (n = 480)	UPLC-MS/MS	Not reported	LOD GLY = 0.08 μg/L AMPA = 0.09 μg/L	Detection of AMPA in urine in children at 5 years of age was associated with elevated transaminases and metabolic syndrome. At 14 years of age, it was associated with metabolic syndrome. Overall, increased urinary AMPA during childhood was associated with an increased risk of elevated liver transaminases and metabolic syndrome. Individuals who lived near agricultural areas during early childhood were also associated with metabolic syndrome at age 18	[82]
USA	Urine (n = 305)	MS	LOQ GLY = 0.041 AMPA = 0.040 ng/mL	LOD GLY = 0.014 AMPA = 0.013 ng/mL	The study suggests that grains are the largest dietary source of GLY, while AMPA is found in soybeans, wine and fast food	[83]
USA	Urine (n = 62)	UPLC-MS/MS	Not reported	LOD GLY/glufosinate = 0.08 μg/L AMPA = 0.09 μg/L	Women represent a growing proportion of agricultural workers and may be uniquely vulnerable to pesticides	[84]
USA	Urine (n = 40)	IC-MS/MS	Not reported	LOD GLY = 0.1 μg/L	The study suggests that people living near agricultural areas suffer greater exposure to GLY	[85]
USA	Urine (n = 2132)	IC-MS/MS	Not reported	LOD GLY = 0.2 μg/L. Concentrations = 0.141–8.210 μg/L	Urinary GLY has been linked to decreased muscle strength and increased prevalence of physical function limitations in middle-aged and older adults	[86]

Table 1. *Cont.*

Country	Sample	Methodology	LOQ	LOD	Outcome (Association)	Reference
USA	Urine (n = 368)	MS	Not reported	LOD GLY = 0.2 µg/L	Farmers who did not wear rubber gloves when handling the pesticide had higher urinary GLY concentrations	[87]
USA	Urine (n = 2745)	IC-MS/MS	Not reported	LOD GLY = 0.2 µg/L	Increased urinary GLY concentrations were associated with a 46% increase in diabetes prevalence (after adjustment for covariates)	[88]
USA	Urine (n = 5224)	IC-MS/MS	Not reported	LOD GLY 0.2 ng/ml	Urinary GLY concentration was associated with a statistical score that was developed to predict the presence of metabolic syndrome	[89]
USA	Urine (n = 2842)	IC-MS/MS	GLY concentration 0.31 ng/mL	Not reported	Exposure to GLY may pose a risk to cardiovascular health, with elevated serum insulin levels being a key mediator in this process	[90]
USA	Urine (n = 2094)	IC-MS/MS	Not reported	LOD GLY 0.2 µg/L. M = 0.304 µg/g creatinine	Urinary GLY levels have been associated, in a dose-dependent manner, with glycemic dyshomeostasis, especially in individuals with overweight/obesity or central obesity	[91]
USA	Urine (n = 1689)	IC-MS/MS	Not reported	LOD GLY 0.2 ng/mL	GLY levels in urine correlate with the severity of arthritis, specifically its osteoarthritis subtype	[92]

Abbreviations: GLY: glyphosate, AMPA: aminomethylphosphonic acid, c, LOQ: Limit of Quantitation, USA: United States of America, GC-MS/MS: gas chromatography–tandem mass spectrometry, LC-MS/MS: liquid chromatography–tandem mass spectrometry, UPLC-MS/MS: high-performance liquid chromatography–tandem mass spectrometry, HPLC: high-performance liquid chromatography, M: median, MS: mass spectrometry, IC-MS/MS: ion chromatography–tandem mass spectrometry, GC-MS/MS: gas chromatography-electrospray ionization–mass spectrometry.

The compilation of biomonitoring studies demonstrates that human exposure to glyphosate and AMPA is a global phenomenon, with residues detected in urine samples from children, adolescents, adults, and pregnant women in both rural and urban settings. Detection levels vary among studies, reflecting differences in the intensity of herbicide use, dietary habits, proximity to agricultural areas, application methods, regulatory policies, and the analytical sensitivity of the laboratory methods employed.

The detection of glyphosate and AMPA in urine samples from various EU countries highlights widespread environmental exposure to this herbicide, especially considering the 0.1 μg/L limit for pesticides in drinking water. Available data show that both adults and children are exposed to glyphosate, with detectable levels found in rural and urban populations. Studies conducted in Belgium [56], France [62] and Germany [64], indicate higher contamination in children and adolescents, reflecting the greater susceptibility of these groups. In France, detectable levels of AMPA were higher in children than in adults. In Belgium, greater contamination was associated with proximity to agricultural areas, underscoring the importance of geographic location as a variable in environmental exposure.

The association between AMPA and oxidative stress markers observed in Cyprus [60] reinforces concerns about the biological effects of pesticide exposure, especially in developing populations such as children. This finding is consistent with previous literature indicating that glyphosate can induce oxidative and genotoxic damage in experimental models [93–95].

Furthermore, dietary patterns have proven to be an important exposure factor in adult populations. In Germany [65], positive correlations were identified between urinary glyphosate excretion and legume consumption, and between AMPA excretion and mushroom consumption. In Spain [73], a relationship was observed between the presence of urinary glyphosate in infants and recent consumption of fruits and eggs.

In rural workers, such as vineyard workers in Italy [67], detected levels were substantially higher (median of 2.30 μg/L), reflecting occupational absorption following product application. This exposure is concerning given that several studies suggest that occupational exposure to glyphosate may be linked to adverse health effects, including endocrine disruption and an increased risk of certain cancers, such as non-Hodgkin lymphoma [96].

Even in countries with lower reported agricultural use, such as Ireland [66] and Portugal [71], measurable concentrations of glyphosate and AMPA were detected in participants' urine. In Portugal, the average values detected were higher than those reported in previous studies in European countries, suggesting possible underreporting of exposure or differences in herbicide use patterns.

The presence of glyphosate and AMPA in the urine of North American populations has been detected in both rural and urban settings, across a range of age groups, with concentrations varying depending on factors such as diet, proximity to agricultural areas, and workplace safety practices.

In Mexico [69], the analysis of occupational exposure in agricultural workers revealed alarming levels of estimated intake of glyphosate and AMPA. With relatively high detection limits, extrapolation of the external dose indicated a potential intake of up to 146 mg/kg/day, an extremely high value compared to acceptable limits set by international regulatory agencies such as the EFSA and the EPA. For glyphosate, EFSA establishes an acceptable daily intake (ADI) of 0.5 mg/kg/day [97], while the EPA adopts an even more permissive limit of 1.75 mg/kg/day [98]. Compared to these reference values, the estimated dose represents a higher potential exposure than what is considered safe for chronic use.

In Puerto Rico [70], widespread exposure was detected among pregnant women. Analysis revealed a significant association between urinary levels of glyphosate and AMPA at 26 weeks of gestation and an increased risk of preterm birth. In Canada, two cohorts of pregnant women demonstrated the presence of glyphosate and AMPA in nearly all samples analyzed, with diet identified as the main route of exposure [57]. These data corroborate previous evidence that processed foods and vegetables grown with glyphosate-based herbicides contribute significantly to general population exposure [43].

In the United States, several studies have demonstrated associations between urinary levels of glyphosate and AMPA and adverse health outcomes, particularly in pregnant women and children. Higher concentrations of glyphosate have been associated with reduced gestational length [78,80,99], lower birth weight, and increased risk of neonatal intensive care unit admission [79]. Furthermore, exposure to these compounds during childhood has been associated with an increased risk of metabolic syndrome and liver abnormalities during adolescence [82]. This association was stronger among individuals who lived near agricultural areas during childhood.

On the other hand, in occupational populations such as farmers in the USA [77,87] and Mexico [69], the lack of personal protective measures, such as glove use, was associated with higher urinary glyphosate concentrations. This finding highlights the urgent need for implementation and enforcement of safety protocols in the field, especially among female farmworkers, a group that represents a growing share of the agricultural workforce and may have additional physiological vulnerabilities to pesticide exposure [45,100].

The detection of glyphosate in the urine of participants from African countries, although documented in a limited number of studies, provides relevant evidence of environmental and occupational exposure to this herbicide in the region. In the Democratic Republic of Congo [61], the detection of glyphosate in all participants suggests significant environmental exposure. A multicenter study involving participants from Malaysia, Uganda, and the United Kingdom [68], although not exclusive to the African continent, provides relevant data on the exposure context in Uganda. The results show that higher glyphosate concentrations were associated with behavioral and socioeconomic factors such as lower use of personal protective equipment (PPE), lower educational levels, and longer duration of herbicide use. These associations reinforce the role of occupational vulnerability and social inequality in amplifying health risks arising from pesticide exposure. Specifically, in Uganda, a country highly dependent on the agricultural sector with limited regulations controlling pesticide use, these findings highlight the importance of educational programs, enforcement of good agricultural practices, and access to basic protective measures [101].

In Asian countries such as China and Thailand, widespread pesticide use is evident. In China [59], a broad range of urinary concentrations of glyphosate and AMPA was observed. The concentrations of both compounds showed a correlation with the time-weighted average (TWA), confirming that urinary levels reflect the intensity of occupational exposure.

In Thailand, the three included studies also indicate significant exposure among the rural population. Glyphosate exposure was associated with oxidative stress and impaired lung function among herbicide applicators. Elevated levels of serum malondialdehyde and reduced glutathione were observed, along with worsening respiratory parameters such as forced expiratory volume in the first second (FEV1), the FEV1/FVC ratio, and peak expiratory flow [75]. It was also demonstrated that glyphosate exposure affects the entire rural community population, not just the direct applicators of the product, suggesting that spray drift, soil and water contamination, as well as domestic handling of agricultural products, may represent significant secondary sources of exposure [74,76].

In South America, Brazil stands out as one of the largest consumers of pesticides in the world, especially glyphosate. In 2023 alone, 253,301.95 tons were sold, representing 48% of the total [102]. Despite this widespread use, there are still gaps in monitoring human exposure to this compound by the standard gold method. A study by Panis and colleagues analyzed urine samples from breast cancer patients in southern Brazil by enzyme immuneassay and detected glyphosate residues at concentrations ranging from 0.25 to 2.12 ppb, particularly among those who reported washing work clothes without personal protective equipment. Additionally, women exposed to glyphosate showed a higher risk of developing breast cancer and metastasis [45].

These regional differences in study populations and exposure profiles illustrate the complexity of the problem and the need for specific surveillance and regulatory approaches that consider local contexts. The data suggest that countries with higher pesticide use and more permissive safety limits tend to have higher levels of human exposure, especially among rural workers and socioeconomically vulnerable populations.

2.2. *Persistent Organic Pollutants (POPs): Dichlorodiphenyltrichloroethane and Hexachlorocyclohexane*

2.2.1. Chemical Properties and Associated Health Effects

POPs include non-volatile organic compounds (polychlorinated dibenzo-p-dioxins (PCDDs), dibenzofurans (PCDFs), polychlorinated biphenyls (PCBs), dichlorodiphenyltrichloroethane and its metabolites (DDTs), hexachlorocyclohexane isomers (HCHs), chlordane compounds (CHLs), organochlorine pesticides (OCs), and hexachlorobenzene (HCB)), so named because of their resistance to environmental degradation and high potential for bioaccumulation in wildlife and humans, raising concern due to their cytotoxicity [103].

The 2001 Stockholm Convention is an international agreement aimed at protecting human health and the environment from POPs. The stated goal was to reduce or restrict the production and use of DDT (Figure 2) and PCBs, due to their ability to persist in the environment, bioaccumulate, and move over great distances, even to remote regions where they have never been used [104–106]. Countries that ratify the convention committed to taking measures to achieve its objectives by eliminating production and use, identifying and monitoring the presence of POPs in their territories, and promoting appropriate waste management to prevent environmental contamination [106].

DDT and HCH are part of the organochlorine pesticides (OCs), a class of synthetic chemicals effective against several insect species and known for their high environmental persistence and highly lipophilic nature, accumulating in food and fatty tissues. This category includes p,p′-dichlorodiphenyltrichloroethane (p,p′-DDT) and its metabolite p,p′-dichlorodiphenyldichloroethylene (p,p′-DDE), β-hexachlorocyclohexane (β-HCH), an isomer of HCH with a longer half-life, and hexachlorobenzenes (HCBs) [107].

In human tissues, bioaccumulation is a concern due to their estrogen-like properties and potential adverse effects in humans [108]. There is considerable evidence of carcinogenic potential in experimental animals, but epidemiological evidence in humans is insufficient, and these compounds are classified by IARC as Group 2B, possibly carcinogenic to humans [45,109].

DDT was once one of the most widely used organochlorine pesticides in agriculture, having been banned in the 1980s, and is currently used primarily to control vector-borne diseases in some developing countries. According to the WHO in 1971, the use of other pesticides, such as malathion or propoxur, as alternatives to DDT would increase the cost of malaria control by approximately 3.4 and 8.5 times, and some countries, such as Africa, would not be able to afford this transition in the coverage of their control programs [110]. Table 2 shows the levels of POPs detected in biological samples from different countries.

Table 2. Levels of POPs detected in biological samples from different countries.

Country	Sample	Methodology	LOD	LOQ	Outcome (Association)	Reference
Beijing	Blood (n = 100)	GC/MS	LOD (0.4 and 21.6 ng/mL)	LOQs = 1.4 and 71.9 ng/mL. LOQ p,p′-DDD = 0–400.7 ng/g of lipid LOQ p,p′-DDT (3.0–2534.1 ng/g of lipid) LOQ total DDT (25.4–9361.9)	The blood of the analyzed individuals indicated low exposure of the general population of Beijing to several POPs compared to other countries.	[111]
Brazil	Breast milk (n = 69)	GC	LOD p,p′-DDE (0.0040 ng/mL) LOD p,p′-DDD (0.0340 ng/mL) LOD p,p′-DDT (0.0040 ng/mL)	LOQ not reported	In a study in Brazil, all breast milk samples were contaminated with DDT and its metabolites and 8.7% of the estimated daily intake (EDI), in terms of total DDT, a value higher than the acceptable daily intake proposed by the WHO.	[112]
China	Breast milk (n = 40)	Gel permeation chromatography	LOD p,p′-DDT = 12–380 ng/g LOD p,p′-DDD ≤ 0.51–28 ng/g LOD Σ DDTs = 780–5400 ng/g	LOQ not reported	The most common contaminant found in human breast milk was DDTs, with higher levels than in other countries. An estimate of daily DDT intake by infants from human breast milk was made, and breast milk from Dalian had higher concentrations of DDTs compared to samples from Shenyang, suggesting that infants from Dalian may be at higher risk from these contaminants.	[113]
Colombia	Breast milk (n = 60)	GC	LOD PBDEs: 2 a 42 pg/g LOD PCBs: 16 a 99 pg/g LOD OCPs: 8.55 a 89.2 pg/g	LOQ PBDE: 2.5–54.7 pg/g LOQ PCBs: 39–271 pg/g LOQ OCPs: 20.4–259 pg/g	DDT concentrations were found in breast milk samples from women in Colombia at levels comparable to those found in other countries, such as Brazil, Uruguay, Chile and Asian countries.	[114]
Delhi	Blood (n = 50)	GC	LOD OCPs: 4 pg/mL	LOQ not reported	The levels of β-HCG found in the blood of the study subjects were associated with the pathogenesis of metabolic syndrome.	[115]
France	Blood (n = 386)	GC	HCB = 22.8 ng/g α- HCH = 0.74 ng/g ß-HCH = 27.0 ng/g DDT = 3.8 ng/g DDE = 104.6 ng/g LOD not reported	LOQ not reported	Serum levels of OC pesticides in French adults were higher (except for DDT and DDE) than levels found in American, Canadian, and German populations, and lower than or similar to levels found in other European countries.	[116]
India	Blood (n = 18)	GC	M = 32.61 µg/L LOD not reported	LOQ not reported	The levels of total DDT and total HCG found in the blood of the individuals were lower than previous findings from India, which may be due to the restriction on the use of these pesticides in agriculture.	[117]

Table 2. *Cont.*

Country	Sample	Methodology	LOD	LOQ	Outcome (Association)	Reference
India	Blood (n = 193)	GC/MS	LOD not reported	LOQ p,p-DDE = 330 (273–399) ng/g LOQ p,p-DDE = 579 (521–643) ng/g	The levels of p,p′-DDE in the blood of individuals from Delhi were exceptionally high, which is expected in this locality because the National Vector-borne Disease Control Programme still recommends indoor residual spraying of homes with DDT for the control of malaria, visceral leishmaniasis and other vector-borne diseases.	[118]
Israel	Breast milk (n = 52)	GC/MS	M DDT = 168 ng/g M DDE = 147 ng/g M p′p DDT 4.3 ng/g LOD not reported	LOQ not reported	Compounds of DDT, hexachlorocyclohexanes and PCBs were above detectable levels in breast milk samples from Israeli women. Although considered high, the levels were still lower than those reported in European countries included in the WHO/UNEP study.	[103]
Korea	Blood (n = 1295)	HRGC/HRMS	p,p′-DDE = M = 128.47 ng/g of lipid LOD not reported	LOQ not reported	More than 60% of the blood samples analyzed had HCB, p,p′-DDT, and p,p′-DDE at concentrations lower than or similar to those reported in Korean nationwide biomonitoring surveys.	[119]
Luxembourg	Hair analysis (n = 497)	GC/MS	M γ-HCH = 0.37 pg/mg M lindano = 0.15 pg/mg M HCB = 14.1 pg/mg LOD not reported	LOQ not reported	In unusual hair samples, 17 chemicals were found in concentrations of these chemicals comparable to those found in studies conducted in France and China. The main compounds found were γ-HCH, lindane and HCB.	[120]
Morocco	Blood (n = 59)	GC/MS	∑DDT = 237.9 ng/g. LOD not reported	LOQ not reported	Low levels of POPs were observed in human milk from Moroccan women compared to those from other countries, reflecting the effectiveness of compliance with the Stockholm Convention requirements in the country.	[121]
Portugal	Blood (n = 203)	GC	LOD not reported	LOQ p,p′ DDE = 43.5–390.25 µg/L	Blood levels of p,p′DDE ranged from undetected to 390.5 µg/L in urban samples, and lower levels (undetected to 43.5 µg/L and 171.2 µg/L) in rural samples. Portugal presented in this study one of the highest levels of contamination when compared to levels found in Europe, Asia and America.	[122]
Saudi Arabia	Umbilical cord /maternal serum (n = 1578)	GC	Umbilical cord and maternal: LOD p,p′-DDE = 0.085 µg/L LOD p,p′-DDD = 0.186 µg/L LOD p,p′-DDT = 0.202 µg/L, Placental tissues: LOD p,p DDE = 0.043 µg/kg dry weight LOD p,p-DDD = 0.093 µg/kg dry weight LOD p,p-DDT = 0.1 µg/kg dry weight	LOQ not reported	p,p′-DDE and p,p′-DDT were found in the umbilical cord and maternal serum, and placental tissues. Based on these findings, the authors report that intrauterine exposure to DDT may have caused a reduction in fetal head circumference, crown-to-heel length, and birth weight.	[123]

Table 2. *Cont.*

Country	Sample	Methodology	LOD	LOQ	Outcome (Association)	Reference
Spain	Blood (n = 953)	GC	M p,p′-DDE = 822 ng/g M β-HCH = 167 ng/g M HCB = 379 ng/g MLOD not reported	LOQ not reported	Relatively high levels of OCPs were found in the blood of the subjects in this study, mainly in those involved in agricultural practices. Serum HCB levels were particularly high in one of the regions due to industrial use. In general, serum β-HCH and HCB levels were found to be substantially higher than in most Western countries.	[105]
Tanzania	Breast milk (n = 95)	GC	LOD p,p′-DDE = 24–2400 ng/g LOD p,p′-DDT = undetected/133 ng/g	LOQ not reported	The estimated daily intake of all DDTs (parent compound and metabolites) exceeded the tolerable daily intake in this study with breast milk. Decreased fetal head circumference was associated with high levels of p,p′-DDE in female infants, suggesting that exposure to OCs during pregnancy may influence fetal growth.	[124]
Thailand	Blood (n = 97)	GC	LOD p′, p′-DDE = 1325.1–12,683.7 ng/g	LOQ not reported	The levels of p′,p′-DDE found in the blood of the research subjects were relatively higher when compared to other studies.	[125]

Abbreviations: DDTs: dichlorodiphenyltrichloroethane and its metabolites; ∑DDT (sum of o,p′-DDD, p,p′-DDD, o,p′-DDE, p,p′-DDE, o,p′-DDT and p,p′-DDT); p,p′-DDT: 1,1,1-trichloro-2,2-bis-(p-chlorophenyl)ethane); p,p′DDE: 1, 1-dichloro-2, 2-bis (p-chlorophenyl) ethylene; p,p′-DDD: 1, 1-dichloro-2, 2- bis (p-chlorophenyl) ethane; s: chlordane compounds; OCs or OCPs: organochlorine pesticides; GC: gas chromatography, HCB: hexachlorobenzene; HRGC: high-resolution gas chromatography, HRMS: high-resolution mass spectrometry, α-HCH: alpha-Hexachlorocyclohexane; β-HCH: beta-Hexachlorocyclohexane; γ-HCH: gamma-Hexachlorocyclohexane or lindane; PCBs: polychlorinated biphenyls; MS: mass spectrometry, M: average, LOD: Limit of Detection, LOQ: Limit of Quantitation.

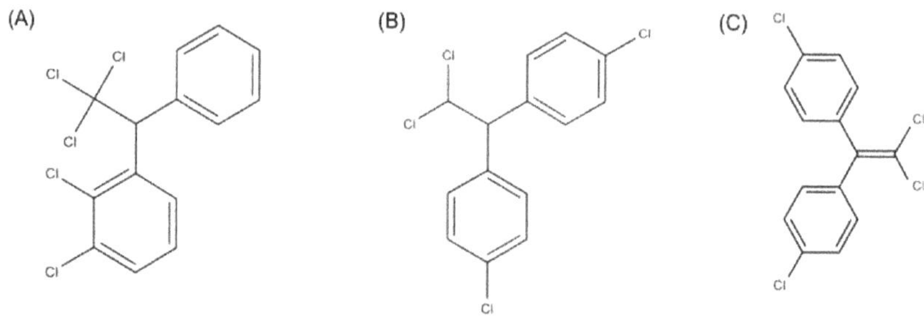

Figure 2. Chemical structure of DDT (**A**), DDD (**B**) and DDE (**C**) [36].

The first countries to ban the use of DDT were Hungary, Norway, and Sweden, followed by the United States and West Germany in the 1970s. The United Kingdom and Brazil followed later in the 1980s. Despite this, high levels of this pesticide have been observed in human samples, such as in Portugal in the 2000s, mainly in young women [122]. In some areas of India and Beijing, where the use of these pesticides in agriculture has been restricted, the detected levels were lower (in 2004 and 2019) [111,117], and levels of POPs and DDT found in individuals residing in Delhi were exceptionally high, which is expected in this location because the National Vector-Borne Disease Control Program still recommends indoor residual spraying of homes with DDT for the control of malaria, visceral leishmaniasis, and other vector-borne diseases (2018) [111].

The main reason DDT is still found in the environment and in human samples, even after being banned in many countries, is its status as a persistent organic pollutant (POP). DDT and its byproducts, such as DDE, are chemically stable and slow to degrade, with decomposition taking decades; historical sources, such as contaminated soils and water from sites where DDT was used extensively in the past, continue to release the compound into the environment, and its slow degradation allows it to persist for long periods. As DDT moves through the food chain, its concentration increases at each trophic level, a process known as bioaccumulation or biomagnification, and animals at the top of the food chain, including humans, accumulate much higher concentrations of the compound [122]. Furthermore, DDT is semi-volatile, which allows it to evaporate, travel through the atmosphere, and be deposited in other parts of the globe, including regions where it has never been used, such as the Arctic.

Biological monitoring of DDT exposure can be performed by measuring compounds or their metabolites in blood, serum, plasma, breast milk, and urine samples [3]. However, breast milk has a relatively high amount of fat, making it an ideal matrix for monitoring maternal and infant exposure to these compounds [104]. The detection of different organochlorine pesticides (PCDD/Fs, PCBs) in breast milk samples suggests that the population was exposed to multiple contaminants, which is very common; in China, DDT was predominant in breast milk, with relatively higher levels than in other countries [113]. In Brazil and Colombia, all breast milk samples analyzed showed contamination by DDT and its metabolites, higher than the acceptable daily intake proposed by the WHO [112,114].

In Israel, concentrations of DDT, hexachlorocyclohexanes, and PCBs were above detectable levels in breast milk samples; however, these levels were lower than those reported in European countries [103]. The estimated daily intake of DDT exceeded the tolerable daily intake through breast milk in Tanzania, with a negative correlation with fetal circumference growth [124]. Only Morocco had lower levels of POPs in breast milk compared to other countries, possibly due to its compliance with the requirements of the Stockholm Convention [121].

2.2.2. Regulatory Limits and Restrictions in Different Countries

It is also suspected that in some developing countries, organochlorine insecticides, such as DDT and HCH, are still in use even after the bans, and relatively high levels of these contaminants have been observed in human breast milk [117,126].

The levels detected and reported in studies may reflect past exposures, as many of these compounds have already been banned. These data are not compared with currently permitted intake doses in food or water, as their restricted and controlled use assumes that they would not accumulate in human tissues [127,128], except in countries that still use these pesticides and have WHO oversight in the implementation of malaria control programs, including the safe application of DDT, surveillance of its presence in biological samples, and assessment of its health impacts.

Regarding the health effects of DDT, it has been suggested that exposure is associated with deterioration in semen quality in men [129], early pregnancy loss [130], tumors [131], and diabetes [132]. Chronic effects are of most concern due to their bioaccumulation, and recent evidence indicates a greater risk in women exposed earlier in life [133,134]. Although DDT played an important role in preventing malaria, its risks to human health are significant, especially during infancy, and are associated with higher rates of prematurity and reduced breastfeeding, both important risk factors for infant mortality in developing countries. The WHO is currently reassessing the health risks of DDT, although progress in this assessment has been limited [135].

Biomonitoring programs, such as those conducted by the U.S. Centers for Disease Control and Prevention (CDC), show that DDE levels in the blood and urine of the general population have decreased significantly since the ban, but are still detectable in almost all samples [136]. National and international legislation does not establish "permissible levels" of DDT in the human body, focusing instead on Maximum Residue Limits (MRLs) in food and drinking water to prevent exposure and accumulation. The U.S. Environmental Protection Agency (EPA) and the European Union have strict MRLs for DDT residues in food products such as fruits, vegetables, and meat. The Codex Alimentarius, an international reference, also establishes residue limits for global trade and allows 5 mg/kg of DDT in meat. The maximum permitted concentration in fruits and vegetables in general in the European Union is 0.01 mg/kg, a generic detection limit, reflecting a zero-tolerance policy for the presence of banned pesticides [137]. In the US, an MRL of 0.05 ppm is permitted in food, but most MRLs for DDT have been revoked or reduced to a minimum to reflect its ban on domestic use in the country [51].

In Brazil, the use of DDT has been banned by a series of decrees and laws over the decades. The final and complete ban, for any use, was consolidated in 2009 by Law No. 11,936, which prohibited the production, import, export, storage, marketing, and use of DDT. Thus, DDT is no longer a registered product for use in the country. Regulatory limits for food and the environment are based on the zero-tolerance principle, meaning its presence in food is unacceptable. However, due to its environmental persistence, small amounts of DDT and its metabolites (DDE) can be detected in samples. In these cases, regulations are based on the Limit of Quantification (LQ), which is the lowest concentration of a residue that can be accurately measured through laboratory testing. The Brazilian Health Regulatory Agency (ANVISA), through its Pesticide Residue Analysis Program in Food (PARA), monitors the presence of pesticide residues, including DDT, in plant-based foods sold in the country. If DDT is detected above the LQ, the product is considered to be in violation of the legislation [52].

Hexachlorocyclohexane (HCH) is an organochlorine pesticide that, like DDT, is a POP, formed by a mixture of isomers, the best-known being gamma-HCH (lindane), a potent insecticide. Regulation of HCH is complex because it addresses not only the main substance but also the other isomers (alpha-HCH, beta-HCH) that are byproducts of its production. Beta-HCH is the most persistent and accumulates the most in fatty tissues. Lindane (gamma-HCH), despite having been the main insecticide, is less persistent than beta-HCH. Therefore, regulations often consider total HCH (the sum of all isomers) or specify limits for each of them.

The Stockholm Convention included alpha-HCH, beta-HCH, and lindane (gamma-HCH) in its list of POPs, with the aim of eliminating their production and use. The exception, in some cases, is its use for controlling lice and scabies in humans, but even this use has been gradually phased out. Countries such as the European Union and the United States banned the agricultural use of HCH decades ago, with lindane being discontinued for most uses in the US in 2006 and in the EU in 2002 [138].

Despite the ban, MRLs are established to monitor food contamination, which can still occur due to the persistence of HCH in soil and water. Limits vary by food type and country, reflecting each regulatory agency's risk assessment. The EU and the US maintain a strict approach, with very low MRLs for HCH and its isomers, typically 0.01 mg/kg, which is the standard limit for unauthorized substances, unless a higher limit is justified. In Brazil, HCH is banned; the tendency is for its limits to be as low as possible, usually below analytical detection limits, to ensure no contamination [139,140].

2.3. Carbamates

2.3.1. Chemical Characteristics and General Properties

Carbamates constitute a class of substances originating from carbamic acid, a molecule whose inherent instability contrasts with the relatively high stability of its derivatives. Structurally, these compounds feature a characteristic functional group (Figure 3), formed by the substitution of carboxyl and amino groups with organic radicals such as alkyl or aryl chains [141]. These molecules may adopt either cyclic or acyclic conformations. The interaction of the carbamate group with inorganic elements, both metallic and non-metallic, leads to the formation of compounds referred to as inorganic carbamates [142].

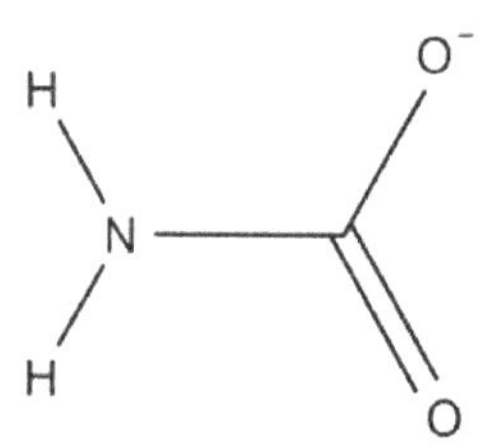

Figure 3. Chemical structure of carbamate [36].

Due to their structural and functional versatility, carbamates are employed across multiple fields. In the pharmaceutical sector, they are found in formulations approved by international regulatory agencies, acting as active ingredients or prodrugs in the therapeutic resource for neurological disorders such as myasthenia gravis, alzheimer's and glaucoma [143]. Within the chemical and agrochemical industries, carbamates assume a central function within the formulation of pest control agents, being particularly valued due to their efficacy and specificity. Furthermore, they are widely used in organic synthesis as protective groups, reactive intermediates, and ligands within the context of combinatorial methodologies [144]. In addition, they serve an important role within the materials industry, where they are incorporated into paint and polyurethane formulations as essential structural elements [142].

Although extensively employed, several carbamates are classified as potentially carcinogenic, prompting regulatory restrictions in various regions, including both the U.S. and the EU [144]. These restrictions derive not only from human health toxicological risks but also on the adverse environmental impacts associated with their use. Among the main concerns are contamination of water sources and soil, atmospheric pollution, and threats to biodiversity in affected ecosystems [145]. At the same time, international regulatory bodies have re-examined health evidence regarding these compounds. The International Agency for Research on Cancer (IARC), for instance, classified carbaryl and aldicarb in Group 3, meaning they are "not classifiable as to their carcinogenicity to humans," due to insufficient evidence in humans and animals [146,147]. While certain in vitro assays suggest genotoxic potential, such results remain inconclusive for human carcinogenesis [148].

The acute toxicity of carbamate insecticides is typically assessed using the median oral lethal dose (LD_{50}) in experiment animals, a standard toxicological parameter inversely proportional to the compound's toxicity, lower LD_{50} values indicate higher toxic potential. Regulatory authorities such as the World Health Organization (WHO), the U.S. Environmental Protection Agency (EPA), and the European Chemicals Agency (ECHA) have established classification systems based on LD_{50} thresholds to support risk assessment and regulatory decision-making. According to these frameworks, carbamate insecticides can be grouped into three general toxicity categories: high toxicity ($LD_{50} \leq 50$ mg/kg), moderate toxicity (LD_{50} between 50–500 mg/kg), and low toxicity ($LD_{50} > 500$ mg/kg) [149]. In this context, aldicarb stands out as one of the most acutely toxic carbamates, with oral LD_{50} values in rats close to 0.5–1 mg/kg, whereas carbaryl demonstrates comparatively moderate toxicity [150]. Despite the reversibility of acetylcholinesterase inhibition, which reduces the duration of effects relative to organophosphates, highly potent carbamates such as aldicarb and carbofuran can still cause intoxications of comparable severity.

According to the WHO's Recommended Classification of Pesticides by Hazard, substances are grouped as follows: Class Ia—Extremely hazardous: $LD_{50} \leq 5$ mg/kg; Class Ib—Highly hazardous: $LD_{50} > 5$–50 mg/kg; Class II—Moderately hazardous: $LD_{50} > 50$–2000 mg/kg; Class III—Slightly hazardous: $LD_{50} > 2000$ mg/kg [151].

Similarly, the EPA classifies acute oral toxicity for pesticide products into the following categories: Category I—High toxicity: $LD_{50} \leq 50$ mg/kg; Category II—Moderate toxicity: $LD_{50} > 50$–500 mg/kg; Category III—Low toxicity: $LD_{50} > 500$–5000 mg/kg; Category IV—Very low toxicity: $LD_{50} > 5000$ mg/kg [152].

2.3.2. Regulatory Residue Limits and Related Health Effects

These classification systems are widely used in toxicological evaluations and regulatory labeling, including the Globally Harmonized System (GHS) adopted by ECHA under EU Regulation (EC) No 1272/2008. The alignment of classification thresholds across these agencies facilitates international harmonization in chemical hazard communication and pesticide management [153]. However, beyond hazard-based categories, risk assessment frameworks in the European Union and other jurisdictions also establish quantitative reference values to better capture chronic and occupational exposures.

For example, the European Food Safety Authority (EFSA) set an acceptable daily intake (ADI) for carbaryl at 0.0075 mg/kg bw/day, while both the acceptable operator exposure level (AOEL) and the acute reference dose (ARfD) were defined at 0.01 mg/kg bw/day [154]. For carbofuran, the ARfD was established at a markedly lower value of 0.00015 mg/kg bw, reflecting its high neurotoxic potential [155]. Aldicarb presents extremely low no-observed-adverse-effect levels (NOAELs), underscoring its high potency [150]. Conversely, propoxur lacks harmonized values in the EU, and maximum residue limits (MRLs) were therefore adjusted to the analytical limit of detection [156].

Epidemiological studies further demonstrate that chronic exposure to carbamates, even at levels consistent with authorized agricultural use, may be associated with neurocognitive deficits in children, respiratory alterations, and possible endocrine disruption [156]. These findings underscore the importance of continuous biomonitoring, strict enforcement of reference values, and progressive restriction of highly hazardous carbamates, especially in low- and middle-income countries where regulatory oversight and enforcement remain limited.

Carbamates inhibit acetylcholinesterase (AChE) through reversible carbamylation of the active site serine residue, leading to a transient accumulation of acetylcholine in the synaptic cleft and the appearance of the cholinergic toxidrome. Unlike organophosphates, this inhibition is not permanent, and the carbamate–AChE complex is unstable and undergoes spontaneous hydrolysis, allowing full enzymatic recovery within 24–48 h without the need for new enzyme synthesis [157–159]. This reversibility explains the generally shorter duration and lower severity of carbamate poisoning compared to organophosphates, and also why the use of oximes such as pralidoxime shows inconsistent results in these cases [158,160].

Although carbamates are often described as presenting lower persistence and somewhat less irreversible toxicity compared to organophosphates, this should not be interpreted as an indication of inconsequential health effects. On the contrary, carbamate exposure has been consistently associated with significant acute and chronic outcomes, including severe intoxication and even intentional lethal use in some contexts [158,161–163]. Therefore, while mechanistic differences exist between carbamates, organophosphates, and organochlorines, all these pesticide classes pose relevant risks to human health and require careful monitoring and regulation [151,152].

Symptoms observed in acute carbamate poisoning differ according to the absorbed dose, pathway of exposure, as well as individual susceptibility. Symptoms may include muscarinic signs (such as sialorrhea, bradycardia, and bronchospasm), nicotinic signs (such as fasciculations accompanied by reduced muscle strength), along with less common central nervous system manifestations (including agitation, convulsions and coma) [161,162].

Exposure to carbamates may occur acutely or chronically, with absorption possible through multiple routes: dermal, respiratory, ocular, mucosal, or gastrointestinal. Although transdermal absorption is generally limited, it can be significantly enhanced when skin integrity is compromised or when contact occurs with compounds of high lipophilicity. Experimental rodent models indicate that maximal suppression of cholinesterase activity is observed around half an hour following ingestion [161]. Under circumstances of massive exposure, symptom onset can be almost immediate, occurring within as little as five minutes. The latency between exposure and clinical manifestation depends on the amount absorbed and the inherent toxic potential of the specific carbamate compound [164]. Compounds with elevated lipid solubility tend to rapidly partition toward adipose tissue compartments, which may transiently attenuate their initial systemic effects and delay observable toxicodynamics [165].

Carbamates undergo predominant biotransformation within the hepatic system via hydrolytic, hydroxylation, as well as conjugation reactions. Elimination occurs predominantly via the kidneys, with approximately 90% of the compound excreted within a few days [164]. There is still some debate in the literature regarding whether these agents can penetrate the blood–brain barrier and subsequently reach the cerebrospinal fluid [162,166]. Nevertheless, it is observed that while adults typically exhibit limited central neurological involvement, suppression of central nervous system activity is often a prominent clinical feature in pediatric cases [164].

It is important to emphasize that, unlike organophosphates, carbamates do not induce the irreversible enzymatic "aging" process, a phenomenon associated with the permanent phosphorylation of acetylcholinesterase [158]. In contrast, the bond between the carbamate and the enzyme is unstable and dissociates spontaneously within a few hours, generally resulting in a more favorable clinical outcome.

Administration of oximes like pralidoxime for treating intoxication caused by carbamates remains a controversial issue. While their efficacy in treating organophosphate poisoning is well established, outcomes in carbamate intoxication are inconsistent. Table 3 summarizes biomonitoring and clinical findings on human exposure to carbamates across different countries. Reported concentrations range from trace urinary levels in the general population [62] to markedly elevated values in occupational settings and acute poisoning cases [167]. Together, these data highlight the heterogeneous nature of carbamate exposure and reinforce the importance of sensitive analytical methods to support public health monitoring and risk assessment.

The analysis of the data presented in Table 3 highlights the importance of biomonitoring as an essential tool for understanding human exposure to carbamates across different geographic and sociocultural contexts. In China, for example, the use of GC-MS/MS enabled the detection of elevated concentrations of carbofuranphenol in pregnant women and children living in agricultural areas. Across several studies, concentrations detected in urine were consistently associated with adverse developmental outcomes, including reduced birth weight and length, deficits in neurobehavioral performance, lower intelligence quotient, and delayed physical growth [168–171]. These results demonstrate that even exposures at low levels, when occurring continuously, may exert deleterious effects on vulnerable populations, particularly during critical windows of child development.

In Europe, results from the French national biomonitoring program (Esteban, 2014–2016) revealed low detection frequencies of propoxur, 2-isopropoxyphenol, and carbofuranphenol in samples from adults and children [62]. Although the concentrations observed were lower than those found in populations from highly exposed agricultural regions, the detection of these compounds in the general population indicates the persistence of environmental exposure even in countries with strict regulatory policies. The sensitivity of the analytical method was crucial to uncover these low concentrations, reinforcing the need for continuous monitoring, particularly among more susceptible groups such as children and pregnant women.

In India and Israel, distinct exposure patterns were also identified using sensitive chromatographic methods. Among Indian agricultural workers, elevated levels of carbaryl and propoxur were identified in blood and urine, reflecting mainly occupational exposure [172]. In Israel, urinary quantification of 1-naphthol revealed diffuse exposure to carbaryl, detected in 100% of the samples, although no statistically significant associations were found with dietary factors such as vegetable consumption [148]. These findings highlight that carbamate exposure is not limited to agricultural environments but may also occur in urban contexts through more diffuse environmental routes.

Taken together, the evidence suggests that human exposure to carbamates transcends geographic and socioeconomic boundaries. In countries with high agricultural production, such as China and India, more intense occupational and environmental exposures predominate; in European countries, low but still measurable levels are observed in the general population. These findings underscore that carbamate exposure represents a global public health concern, paralleling the international challenges already highlighted for glyphosate and POPs.

Table 3. Carbamate exposure: analysis of biological samples and clinical implications.

Country	Sample	Methodology	Levels Found	LOD	LOQ	Outcome (Association)	Reference
China	Urine (pregnant women, $n \approx 1100$; Sheyang Birth Cohort)	GC-MS/MS	Carbofuranphenol detected in 96.6% of samples; GM ≈ 0.81 µg/L (IQR: 0.29–2.07 µg/L); values up to 395.4 µg/L	0.01 µg/L	0.05 µg/L	Urinary concentrations associated with lower birth weight and reduced length of newborns	[168]
China	Urine (children age 3, $n \approx 300$; SMBCS Cohort)	GC-MS/MS	Carbofuranphenol GM ≈ 0.70 µg/L; P95 ≈ 5.6 µg/L	0.01 µg/L	0.05 µg/L	Carbamate exposure associated with worse neurobehavioral performance in young children	[169]
China	Urine (children age 7, $n \approx 400$; SMBCS Cohort)	GC-MS/MS	Carbofuranphenol median ≈ 0.56 µg/L; P95 ≈ 4.9 µg/L	0.01 µg/L	0.05 µg/L	Early exposure to carbamates associated with reduced IQ at age 7	[170]
China	Urine (children age 7, $n \approx 420$; SMBCS Cohort)	GC-MS/MS	Carbofuranphenol median ≈ 0.62 µg/L; range 0.05–9.4 µg/L	0.01 µg/L	0.05 µg/L	Carbamate exposure associated with delayed physical development (lower height and weight)	[171]
France	Urine (children, $n = 500$)	UPLC-MS/MS	Propoxur (P95) = 0.05 µg/L; 2-isopropoxyphenol (P95) = 0.30 µg/L; carbofuranphenol < LOQ (máx = 12.2 µg/L)	Propoxur/2-IPP = 0.02 µg/L; carbofuranphenol = 1 µg/L	Propoxur/2-IPP = 0.05 µg/L; carbofuranphenol = 5 µg/L	Low frequency of quantification in the general population	[62]
France	Urine (adults, $n = 899$)	UPLC-MS/MS	Propoxur (P95) = 0.13 µg/L; 2-isopropoxyphenol (P95) = 0.19 µg/L; carbofuranphenol < LOQ (máx = 13.4 µg/L)	Propoxur/2-IPP = 0.02 µg/L; carbofuranphenol = 1 µg/L	Propoxur/2-IPP = 0.05 µg/L; carbofuranphenol = 5 µg/L	Low levels in the general population; quantification in 1.7–8.5% of adults	[62]
India	Blood/urine ($n = 42$)	LC-MS/MS	Carbaryl/Propoxur = 0.3 to 2.5 µg/L	Not reported	LOQ = 0.1 µg/L	Elevated carbamate levels were observed among agricultural workers, suggesting occupational exposure	[167]
Israel	Urine (**n** = 42)	GC-MS	1-naphthol = detected in 100% of urine samples IPPX = detectability less than 20%.	Not reported	Not reported	Higher 1-naphthol levels compared to other populations reported in the literature- No statistically significant association with vegetarian diet- IPPX with levels too low for association analyses	[148]

Abbreviations: LOD: Limit of Detection; LOQ: Limit of Quantification; GC-MS: gas chromatography–mass spectrometry; GC-MS/MS: gas chromatography coupled with tandem mass spectrometry; UPLC-MS/MS: ultra-performance liquid chromatography coupled with tandem mass spectrometry; LC-MS/MS: liquid chromatography coupled with tandem mass spectrometry.

2.4. Organophosphate Pesticides

2.4.1. Chemical Properties and Health Effects in Humans

Organophosphate pesticides (OPs) are chemical compounds found in herbicides, pesticides, and insecticides, produced to eliminate and control insects such as fleas, lice, flies, and mosquitoes. They are used in agriculture, livestock, and domestic settings. Due to their faster decomposition and consequently lower persistence in the environment, their use has become a safer and preferable option compared to organochlorines [173,174].

However, due to their high toxicity, there are frequent reports of OP poisoning through ingestion of contaminated food or dust, suicide attempts, and inhalation during occupational exposure to spraying. It acts mainly by preventing the activity of the enzyme acetylcholinesterase (AChE), which is found in the nervous system [174,175]. The toxicity class of OPs is between I and IV via the respiratory or oral route, according to the US-USEPA, and the mechanisms triggered in organisms depend on the type of exposure, quantity of the substance and its chemical structure [176].

The chemical structure of OPs (Figure 4) is essentially composed of a terminal oxygen double-bonded to a phosphoryl group. Attached to the phosphorus atom are two lipophilic groups and a leaving group, usually a halide. These compounds originate from phosphorus esters, commonly derived from thiol or amide derivatives of acids, such as thiophosphoric, and often contain side chains such as thiocyanate, phenoxy and cyanide groups. OPs are resistant to degradation by thermal, photolytic, and chemical processes [176].

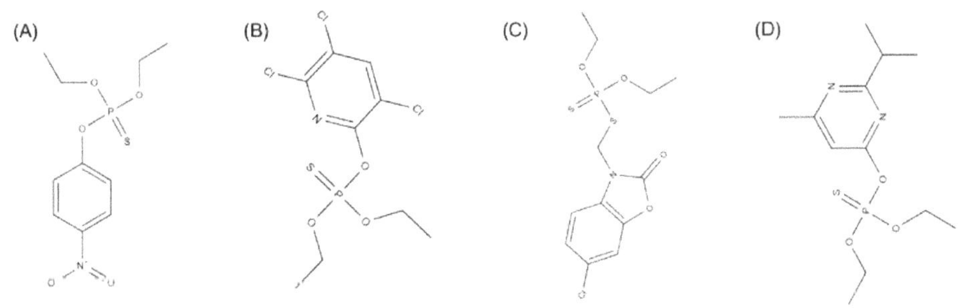

Figure 4. Chemical structure of parathion (**A**), chlorpyrifos (**B**), phosalone (**C**), diazinon (**D**) [36].

When their structure contains an aromatic group such as p-CH_3-S or p-NO_2, the electrophilicity of the phosphorus atom increases during metabolic biotransformation in humans, making these compounds more toxic to the organism. However, when the chain length of the lipophilic groups is greater, the central phosphorus atom reduces the electrophilicity, reducing the toxicity of OPs [177]. Most OPs exhibit characteristics such as low water solubility, high volatility, and elevated octanol–water partition coefficients (K_ow), which reflect their strong lipophilicity and potential for bioaccumulation in biological tissues. Certain OPs, such as diazinon, chlorpyrifos and parathion, have an affinity for fat and can bioaccumulate in human adipose tissue for long periods of time [178].

OPs are active as formulated in pesticides, while others require metabolic activation in the body, particularly those needing conversion to their active oxygen form to exert toxicity. Enzymes such as oxygenases, A-esterases, carboxylesterases, and fluorohydrolases mediate these activation processes through oxidative, reductive, or hydrolytic reactions. For detoxification and elimination, conjugation reactions are necessary, involving enzymes like glutathione S-transferase and amidase [179,180]. The paraoxonase enzyme (PON1) is an A-esterase produced in the liver that breaks the active metabolites of OP compounds that are generated during the first phase of metabolism [180–183].

Due to their lipophilic characteristics and chemical structure, OPs, like other classes of pesticides, can mimic the action of hormones such as TSH by binding to receptors in the thyroid and activating its signaling pathways. As evidenced in the work carried out in Mexico, with rural works [184], and China [185], with pregnant women. Exposure to these endocrine disruptors contributes to the increase in endocrine disorders in individuals, as well as cardiovascular diseases, neurological problems, bone disorders, and even cancer development [186,187].

OPs, being structural analogs of acetylcholine, covalently bind to the hydroxyl (OH) group of the serine residue at position 203 (Ser203) at the place where the product is transformed by AChE. During the process of adding a phosphate group, OPs phosphorylate the hydroxyl group, rendering the enzyme inactive [188]. AChE is found between motor neurons and muscle fibers and in red blood cells. This inhibition can be reversible or irreversible, as a result of the chemical structure of the OPs [189,190]. These changes can be seen in studies produced in Brazil [191], Greece [192], Thailand [193] and USA [194].

Due to their inhibitory action on AChE, individuals exposed to OPs—either through single or multiple doses—may develop delayed neuropathy, characterized by prolonged uncoordinated and clumsy movements and decreased muscle tone [195,196]. In the USA, it was pointed out that exposed children also demonstrated memory and attention problems [194].

IARC classifies OP insecticides such as malathion and diazinon as probable carcinogens (Group 2A), while dichlorvos, tetrachlorvinphos and parathion are categorized as possible carcinogens (Group 2B) [197]. In Brazil [191] and Portugal [28], analyses of biological samples of an occupationally exposed population revealed changes in genetic material. When metabolized in the body, reactive oxygen species are produced. Thus, in both subchronic and chronic exposure, the main mechanism of OP toxicity is the induction of oxidative stress, defined by increased production of free radicals [198].

The production of free radicals can cause damage to genetic material by inducing oxidation or chain breakage, both directly and indirectly. It may induce mutations in oncogenes and tumor suppressor genes during cell division, potentially leading to carcinogenesis [199,200].

There are also reports in the literature of OPs interfering in sexual reproduction and gametes, described in Spain [201], China [202], and Egypt [203] with female and male rural workers. Oogenesis and spermatogenesis depend on the hypothalamic–pituitary–gonadal axis, which is impaired by the action of OPs, compromising the production of gonadotropins and disrupting the normal regulation of oogenesis and spermatogenesis [204].

Through a literature review and selection of articles meeting established criteria, such as pesticide dosage in human samples, it was observed that studies addressing the effects of OP exposure in populations without acute toxicity symptoms are scarce (Table 4). However, the health impacts of exposure to these substances are well documented by in vitro, in vivo, and observational studies, even when OP levels were not detected in biological samples. Reported health issues include endocrine dysfunction, reproductive problems in both females and males, neurological and motor impairments, genetic damage, mainly by inhibiting the activity of the enzyme AChE, which is responsible for the characteristic symptoms of intoxication. Table 4 presents data from studies analyzing pesticide concentrations in biological samples of individuals in relation to associated health problems.

Table 4. Mean levels of organophosphate pesticides detected in biological samples from different countries.

Country	Sample	Metodology	LOQ	LOD	Outcome (Association)	Reference
Brazil	Urine (n = 188)	GC/MS	LOQ DETP = 0.078 μgL^{-1} LOQDEDTP = 0.0431 μgL^{-1}	LOD DETP = 0.023 μgL^{-1} LOD DEDTP = 0.0129 μgL^{-1}	Decrease in AChE and DNA damage, problems in cell division, changes in proliferative potential, and cell death	[191]
China	Serum (n = 325)	GC/MS	LOQ DMTP = 2.27 μg/L LOQ DEP = 11.20 μg/L LOQ DETP = 1.99 μg/L	LOD DMP = 9.81 μg/L LOD DMTP = 0.79 μg/L LOD DEP = 5.00 μg/L LOD DETP = 0.78 μg/L LOQ DMP = 21.40 μg/L	Association with FT4 and TSH. Changes in thyroid function in pregnant women	[185]
China	Semen/ Urine (n = 75)	GC	Not reported	LOD p-nitrophenol = 0.018 mg/L	When comparing exposed and unexposed works, the analysis of the sample of the first spermatozoa showed worse characteristics with decreased quantity and motility	[202]
Egypt	Follicular fluid (n = 300)	GC/MS	Not reported	LOD Chlorpyrifos = 0.098 μg/L LOD Diazinon = 0.055 μg/L LOD Malathion = 0.024 μg/L LOD Pretilachlor = 0.003 μg/L LOD DDT = 0.005 μg/L LOD Lindane = 0.083 μg/L	Decrease in the oocytes retrieved. High concentrations associated with lower implantation	[205]
Greece	Blood/hair (n = 1)	GC/MS	LOQ DMP = 1289.4 pg/mg LOQ DEP = 709.4 pg/mg, LOQ opDDE = 484.0 pg/mg, LOQ ppDDE = 526.6 pg/mg, LOQ opDDD = 448.4 pg/mg, LOQ ppDDD + opDDT = 259.9 LOQ pg/mg, ppDDT = 573.7 pg/mg	Not reported	Exposure association with motor neuron disease	[192]
Mexico	Urine/ blood (n = 148)	GC-liquid/GC	Not reported	LOD ΣDMP = 1.45 μmol/g creatinine LOD ΣDEP = 0.31 μmol/g creatinine LOD ΣDAP = 1.99 μmol/g creatinine LOD p,p-DDE = 6.22 ng/mL	OPs alter the amounts of thyroid hormone, especially in those with PON1 192RR polymorphism or even decreased PON1 enzyme activity	[184]

Table 4. *Cont.*

Country	Sample	Metodology	LOQ	LOD	Outcome (Association)	Reference
Portugal	Urine (n = 85)	colorimetric kit	Not reported	LOD 2.23 μg/mmol	Increase in DNA strand breaks and alterations in the percentage of B lymphocytes	[28]
Spain	Urine (n = 116)	GC/MS	LOQ DEDTP = 0.01 μg/L LOQ DAP = 0.1 μg/L	LOD DMP = 1.3 μg/L LOD DMTP = 1.0 μg/L LOD DMDTP = 0.08 μg/L LOD DEP = 2.6 μg/L LOD DETP = 0.94 μg/L LOD DEDTP = 0.05 μg/L LOD ∑DAP = 75.5 nmol/L LOD ∑DAP = 100 mg/L	OPs have been associated with male infertility, as they affect the biosynthesis of hormones responsible for reproduction, consequently reducing the number of sperm, as well their mobility	[206]
Thailand	Urine/ Blood (n = 150)	GC	LOQ DETP = 1 μg/L LOQ DMP = 25 μg/L	LOD DMP = 5.0 μg/g LOD DMTP = 0.4 μg/g LOD DMDTP = 0.36 μg/g LOD DEP = 0.51 μg/g LOD DEDTP = 0.4 μg/g LOD DETP = 0.26 μg/g LOD DETP = 0.1 μg/L LOD DMP = 2.5 μg/L	Muscle weakness in exposed individuals	[193]
USA	Urine (n = 26)	GC/MS	LOQ AZM = 5.30 μg/g LOQ DMP = 4.0 ng/mL LOQ DEP = 2.0 ng/mL LOQ DMTP = 2.2 ng/mL LOQ DMDTP = 1.6 ng/mL LOQ DETP = 1.6 ng/Ml	LOD DMP/DEP/DMTP/ DMDTP/DETP = 0.01 μg/gm LOD AZM = 0.10 μg/gm	Neurological and behavioral assessment using questionnaires applied to works exposed to OPs indicated worse performance compared to those who did not work with crops, therefore not exposed	[207]
USA	Urine (n = 652)	GC/FPD	Not reported	LOD MP ≥ 100 ppb	The assessment test applied to verify neurological and motor performance showed that children exposed to OPs have the worst results	[208]

Abbreviations: AZM: Azinphos-methyl, AChE: Acetylcholinesterase, DAP: ialkylphosphate, DEDTP: Diethyldithiophosphate, DEP: Diethylphosphate, DETP: Diethylthiophosphate, DMDTP: Dimethyldithiophosphate, DMP: Dimethylphosphate, DMTP: Dimethylthiophosphate, FT4: free T4, GC: gas chromatography, MS: mass spectrometry, LOD: Limit of Detection, LOQ: Limit of Quantification MP: Methyl parathion, OP: Organophosphate, p,p′-DDE: P,p′-dichlorodiphenyldichloroethene, PON1: Paraoxonase-1 enzyme, USA: United States of America.

2.4.2. Regulatory Limits and Different Countries

Because OPs persist in the environment for a relatively short period, they are sometimes considered as alternatives to organochlorines. However, studies on these pesticides worldwide, within the criteria established for this review, are limited. Larger studies have been conducted primarily in China [185,202] and the USA [194,203]. Demonstrating a scarcity of biomonitoring studies of this class worldwide. Most other countries were represented by a single study. The majority of investigations involved exposed workers over 18 years of age, with only one study in the USA including exposed children.

Globally, OPs account for approximately 40% of the total pesticide market. Countries with the highest average annual insecticide use (kg/ha) include Japan (6634), Mexico (2313), Brazil (1163), the USA (0.627), France (0.196), the United Kingdom (0.174), Canada (0.125), Germany (0.124), and India (0.09) [209]. According to the Global Pesticide Market, the compound annual growth rate (CAGR) of OPs between 2018 and 2023 was 5.5%, their use being mainly in developing countries, where the economy is centered on agriculture [210,211].

OP metabolites can be detected in various biological samples, particularly blood and urine. Specific metabolites, as well as dialkyl phosphate metabolites (DAPs), can be identified, making them extremely important for biomonitoring the exposure of rural workers or even those exposed to contaminated food and water [211]. Currently, the most commonly used OPs in agricultural practices are monocrotophos, methylparathion, diazinon, dichlorvos, chlorpyrifos, trichlorfon, and malathime. Some OPs have already been banned in Europe and the USA; however, their extensive use continues in developing countries [212].

In a study, nine metabolites of OPs and pyrethroid pesticides were measured in 322 urine samples collected across eight countries between 2006 and 2014 using HPLC-MS/MS. The highest concentrations were observed in Vietnam (28.9 ng/mL), India (14.2 ng/mL), China (13.6 ng/mL), Korea (12.5 ng/mL), Greece (12.3 ng/mL), Saudi Arabia (11.3 ng/mL), and the USA (7.9 ng/mL) [213].

Even with the increase in biomonitoring studies relating the presence of these metabolites in human samples and their pathological outcomes already observed in adults and children exposed and not exposed to rural areas, as portrayed in this study, and investigation of the amounts of residues present in food and water, there is still negligence on the part of authorities to reduce population exposure [214].

According to EFSA, the NOAEL (No Observed Adverse Effect Level) values of some organophosphates for neurochemical effects are pirimiphos-methyl (0.004–0.4 mg/kg bw), phosmet (0.01–1 mg/kg bw), ethoprofos (0.0004–0.04 mg/kg bw), chlorpyrifos (0.001–0.1 mg/kg bw), dimethoate (0.001–0.1 mg/kg bw), fenamiphos (0.0008 mg/kg bw), methiorcab (0.013–1.3 mg/kg bw) and chlorpyrifos-methyl (0.01–1 mg/kg bw) [215]. The values for the daily intake correspond to malathion (300 μg/kgbw/d), diazinon (5 μg/kgbw/d), parathion (4 μg/kgbw/d), phorate (4 μg/kgbw/d) and dimethoate (2 μg/kgbw/d). It is known, however, that the occupationally exposed population ends up having a higher daily intake of OPs, and the values are in the range of 1.08 to 19.72 μg/kgbw/d [216].

Many countries adopt MRLs for pesticides in food and water as determined by the FAO and WHO. The most commonly used OPs include chlorpyrifos, malathion, dichlorvos, profenofos, dimethoate, and AZM, with MRLs set by the FAO and the WHO at 0–0.01 mg/kg, 0.01 mg/kg, 0.01–7 mg/kg, 0.01–15 mg/kg, 0.01–20 mg/kg, 0.001–4 mg/kg, and 0–0.03 mg/kg, respectively [217]. According to the European Reference Laboratories (EURLs), residual values for diazinon were reported for products of animal origin (LOQ = 0.02 mg/kg) and for general food products (LOQ = 0.05–5 mg/kg) [218]. The EPA and WHO indicate that the maximum permitted levels of organophosphates in water and fish are 30 μg/kg [219]. Analyses on the concentrations of OP residues detected in rivers around the world were developed, and in some of the countries shown in Table 4, values were found, from lowest to highest, respectively, in Greece: 6–30 ng/L^{-1}, Spain: 9–97 ng/L^{-1}, Mexico: 33–149 ng/L^{-1}, Brazil: 535–700 ng/L^{-1}, and China: 4–1560 ng/L^{-1} [220].

In Brazil, the estimated average intake of OPs from fruits and vegetables for children and adults was calculated based on FAO and WHO recommendations and analyzed by the Pesticide Residue Analysis Program (PARA) and the National Health Surveillance Agency (ANVISA). Acceptable intake values for OPs generally range from 0.001 to 0.070 g/day [221]. For water intended for human consumption, Brazilian agencies recommend a maximum of 100 μg/L [222]. For China, the average residue threshold in food is 0.05 mg/kg for triazophos and omethoate and 0.1 mg/kg for other OPs investigated. However, residues in rice frequently exceed these limits (0.011–1.756 mg/kg) [223]. A study in Changchun, in northwest China, reported that most foods exceeded permitted levels, with vegetables showing higher contamination than fruits and roots. The associated risk index was 0.448 for adults and 0.343 for children [224]. According to China's "Agricultural Product Quality and Safety Risk Monitoring Plan," the geometric mean of OP residues in the general population's diet ranges from 0.02 to 0.04 μg/kg bw/day [225].

In the USA, the EPA sets permitted residual limits for vegetables, roots, and fruits at 0.001–0.010 ppm for chlorpyrifos, 0.001–0.040 ppm for dichlorvos, 0.001–0.050 ppm for methamidophos, and 0.001–0.020 ppm for omethoate [226]. However, as noted throughout this article, these standards are not particularly stringent and are often difficult to enforce, leaving the population exposed to the adverse effects of OPs [227]. A study by the FDA conducted between 2009 and 2017 reported that chlorpyrifos was among the pesticides most frequently detected in food above the permitted tolerance, with concentrations ranging from 0.06 to 0.08 mg/kg, exceeding the US limit of 0.05 mg/kg [228]. According to NHANES, the most recent urinary analyses for OP metabolite residues in 2018 showed geometric mean values of 2.39 μg/g for DEP, 2.45 μg/g for DMP, 0.298 μg/g for DETP, levels below the detection limit for DEDTP, and 0.481 μg/g for DMDTP [229].

In the European Union, including Spain, Portugal, and Greece, where studies on OP exposure and related human health effects have been conducted, the standards for pesticide application are stricter. Residues of OPs in vegetables ranged from 0.02 to 5 ng/g, with diazinon (0.017 ng/g) and chlorpyrifos (31 ng/g) being the most frequently detected. In fruits, levels generally ranged from 0.006 to 9.32 ng/g, mainly associated with ethoprophos and chlorpyrifos, but none exceeded the maximum permitted residue limits. The risk index was low for both vegetables (3.93) and fruits (4.26), remaining within acceptable limits. These findings highlight the effectiveness of stringent regulatory standards in reducing population exposure to OPs in these countries [230].

In Thailand, urinary levels of OP metabolites in conventional and organic rural workers were analyzed. In this analysis, conventional workers exhibited higher concentrations than organic workers. Another study in the same country examined OP metabolite residues in vegetables used in school meals and correlated them with the presence of these metabolites in children's urine samples. Analysis of the children's urine revealed DMDTP at 5258 μmol/mol creatinine and DEP at 2884 μmol/mol creatinine [231].

In Mexico, a study assessed urinary concentrations of OP metabolites in workers with varying levels of pesticide exposure, showing that the metabolite found in the highest concentration was DMTP [232]. In a study conducted in southwestern Mexico, the aim was to analyze the presence of various pesticides in fruits and vegetables classified as highly hazardous, following the criteria established by the FAO/WHO (LD50 < 2 μg/bee).

Among these, the presence of chlorpyrifos and methamidophos was detected, exceeding the MRLs published by the country, the latter being identified as highly toxic by the Rotterdam Convention [233].

In Egypt, MRLs established by European Union legislation are generally adopted, with limits of 10 ng/g for chlorpyrifos. However, a study reported that chlorpyrifos levels in buffalo milk exceeded the permitted limit of 0.02 mg/kg [234].

Given the above, it is understood that changes caused by occupational exposure or the ingestion of food and water contaminated by OPs cause numerous types of harm to human health. Therefore, a more rigorous investigation of this class of pesticides is necessary, since their use serves diverse sectors such as agriculture, domestic, industrial, commercial, and governmental use, in addition to being one of the most widely consumed classes of insecticides for these purposes. There is also a disparity in legislation regarding the use of certain OPs among countries, despite extensive scientific evidence regarding their negative effects. In those that adopt prohibitive measures, there is a lack of rigorous investigations into their illegal use in agriculture [235].

3. Discussion

The global analysis of exposure to glyphosate and its metabolite AMPA, DDT and other POPs, carbamates, and OPs reveals not only methodological differences in biomonitoring studies but also inequalities in pesticide use regulation, with populations in the Global South frequently exposed to high concentrations of these compounds [236].

In an effort to mitigate the harmful effects of pesticides on human health, the FAO established regulations for pesticide production in 1985. However, many developing countries did not adopt these standards; likewise, the WHO warnings regarding the control of highly toxic pesticides were not followed, leaving populations exposed to their detrimental effects [237].

The role of the agrochemical industry in this scenario is central. The pesticide market generated approximately USD 108.1 billion in 2024, dominated by a few multinational corporations that control around 70% of global sales [238]. Glyphosate, the most widely marketed herbicide worldwide, accounted for about USD 8.99 billion in 2024, and its market value is projected to reach USD 12.44 billion by 2034 [239].

FAOstat data from 2019 to 2025 reveal that some countries in Western Europe, South America, and Oceania, although not the largest absolute consumers (in tons), show high rates of pesticide application per hectare. This pattern reflects the intensification of use in specific agricultural areas, indicating that cultivated soils receive proportionally high chemical loads [240]. Countries with a higher intensity of use per hectare tend to present greater risks of direct and indirect exposure (occupational, environmental, and dietary); however, there is a scarcity of biomonitoring studies in these regions.

The growing expansion of globalization exerts pressure on large-scale food production for both human and animal consumption. Multinational corporations controlling seed and pesticide production have developed genetically modified crops resistant to pesticides, creating a cycle of dependency that compels producers to also use these chemical inputs [241]. This corporate dominance was further highlighted by the divergence between the IARC classification—which places glyphosate, DDT, and OP insecticides such as malathion and diazinon in Group 2A, "probably carcinogenic to humans"—and the broader body of scientific evidence demonstrating carcinogenic effects [44].

Reports produced by international organizations reveal that European Union countries manufacture and export pesticides banned for domestic use to developing countries, exposing a duality in legislation: what is deemed too hazardous for their own populations is not considered harmful enough to prevent export to poorer nations. This dynamic reinforces a trade model that prioritizes economic interests over public health [242].

The global distribution of pesticide exposure reflects not only local agricultural practices but also political and economic dynamics between producing countries, consumers, and regulatory agencies. In countries where pesticide regulations are stricter, exposure tends to be lower and primarily linked to dietary intake.

In Latin America, Asia, and Africa, more intense exposures are observed, particularly among agricultural workers. From a socioeconomic perspective, industrial pressure is disproportionately exerted on rural workers, who represent a vulnerable population. This vulnerability is further aggravated by the fact that populations with lower educational levels tend to use personal protective equipment (PPE) less frequently, thereby increasing occupational exposure, as evidenced by studies conducted in Malaysia, Uganda, and the United Kingdom [68], Mexico [69], and the USA [77,87]. Beyond rural workers, women—often overlooked during the pesticide application cycle—are consistently exposed through secondary contamination, such as when washing clothes of family members involved in spraying [45,84].

From an environmental perspective, the application of pesticides such as glyphosate, POPs, carbamates, and OPs affects soil microbiota and disrupts ecosystems, highlighting that the risks posed by pesticides impact human, animal, and environmental health. Once present in soil and water, these compounds require degradation by specific microorganisms, which are rare and often insufficient to achieve bioremediation, allowing them to persist in the environment for extended periods [176].

This reality demonstrates that pesticide exposure is not limited to the moment of application. In this context, the use of organochlorine pesticides, such as DDT and its derivatives (DDE and DDD), represents one of the greatest public and environmental health dilemmas of the 20th and 21st centuries. Although effective in eradicating pests and controlling disease vectors, their chemical properties of persistence and bioaccumulation pose serious long-term risks. Western countries banned the use of DDT and other organochlorines as early as the 1970s and 1980s; however, before this period, the widespread application of these pesticides resulted in bioaccumulation still observed today, since discussions on the problems caused by these compounds were virtually absent at that time [243]. However, DDT is still used today in some developing countries for the control of vector-borne diseases in endemic areas, mainly due to its low cost [135]. Even with its total ban in some countries, such as Brazil—where regulatory limits in food and the environment are based on the principle of zero tolerance—the presence of DDT and DDE can still be detected in biological samples due to their persistence in the environment, as evidenced by studies conducted in China [113] and in European Union countries such as Luxembourg, Portugal, and Spain [120,122,123]. The levels of these pesticides found in blood, urine, and breast milk samples [117,118] are likely associated with the use of DDT in malaria vector control.

Substances such as carbofuran and aldicarb, belonging to the carbamate group, have been banned by the European Union and the USA due to their high risk; however, countries in Latin America, Africa, and Asia continue to report high levels of agricultural use [154, 244]. This class of pesticides shares the same enzymatic target as organophosphates (acetylcholinesterase) but exhibits reversible inhibition, which in some contexts has led to the false perception that they are safer alternatives. However, carbamates do not represent a safe alternative to organophosphates but rather add to the already complex "pesticide cocktail," increasing the public health risk burden. These effects are demonstrated by studies conducted in rural populations in China [168–171], France [62], India [167] and Israel [148], where residues were detected in the blood and urine of pregnant women and children, confirming significant environmental and occupational exposure despite the compounds' rapid metabolism.

Recently, chlorpyrifos, an organophosphate, was banned in the European Union (2019) and the USA (2021) due to the health impacts associated with this class of compounds, including neurological and motor alterations, infertility, and genetic damage, as reported in studies conducted in Spain [206], Portugal [28] and Greece [192] with urine and blood samples collected from children and adults. The Health and Environmental Alliance (HEAL) reported an analysis of pesticide residues in fruits and vegetables, in which chlorpyrifos was among the five most frequently detected compounds exceeding the MRL values [245].

In China, chlorpyrifos was widely used in rice cultivation; however, following studies—such as one conducted with pregnant women linking exposure to thyroid gland alterations [185], along with other evidence of its toxicity to the human body—it was banned by the government in 2018. Nevertheless, due to its persistence in the environment and illegal use in agriculture, its negative effects continue to impact human health, highlighting the ineffectiveness of enforcement measures [246]. Despite being banned in 35 countries, chlorpyrifos continues to be widely used in developing countries such as Brazil, where the economy relies heavily on agriculture and agribusiness, and legislation favors companies and rural producers to increase productivity and profits [247], even in the face of evidence of alterations in cell proliferation and DNA damage reported in studies conducted with rural workers in the country.

Finally, differences in laboratory infrastructure between developed and developing countries impact the assessment of population exposure, leading to an underestimation of the true burden of pesticide exposure and its harmful effects [248]. Countries such as France and the United States maintain robust biomonitoring programs, such as the Esteban Study and the National Health and Nutrition Examination Survey (NHANES), which allow for continuous evaluation of residues in biological samples and support more responsive public policies [55]. Moreover, international initiatives, such as the Codex Alimentarius [217] and Human Biomonitoring for Europe (HBM4EU) [249], aim to harmonize safety parameters and standardize analytical methodologies, strengthening global comparisons and supporting regulatory decisions. In developing countries, such as China and Brazil, the lack of comparable surveillance creates a gap between law and practice, contributing to the continued invisibility of exposure from an epidemiological perspective [250].

In addition to the lack of governmental surveillance to assess the impact of exposure on population health, the high costs and complexity of biomonitoring techniques create difficulties in producing scientific research on this topic in developing countries, such as those in Africa, South Asia, and Central and South America [251]. Population monitoring is extremely relevant, as its results allow for verification of whether regulations limiting pesticide use are being effectively enforced or require adjustments to reduce their harmful effects [252].

Therefore, these points reveal a concerning scenario in which permissive legislation, the lack of systematic monitoring, the absence of remediation mechanisms, and the lack of international regulatory harmonization converge to reinforce risks to human health and environmental balance.

4. Concluding Remarks

Human exposure to pesticides remains a global concern, with concentrations in biological matrices ranging from trace levels to values exceeding regulatory limits. Diet, occupational contact, rural life, and socioeconomic conditions are consistently associated with increased exposure. Furthermore, regulatory laxity in countries like Brazil, coupled with insufficient surveillance and enforcement, exacerbates risks compared to the more precautionary European framework. The persistence and bioaccumulation of banned compounds illustrate the irreversibility of environmental and biological contamination, giving pesticide exposure a chronic and transgenerational nature. These vulnerabilities are further exacerbated by socioeconomic inequalities, which disproportionately affect rural workers, women, and children in low- and middle-income countries, where monitoring systems are often limited or nonexistent.

The development of broad-spectrum, nonspecific biomarkers is therefore crucial, along with robust analytical methods and interlaboratory harmonization to ensure data quality and comparability. Without these efforts, exposure patterns will remain poorly characterized, and the lack of standardized reference values will continue to hinder the interpretation of risks and the implementation of preventive interventions.

Finally, existing studies often focus on individual pesticides, while populations are actually exposed to complex mixtures, the so-called "cocktail effect." The lack of data and methodologies to assess cumulative and synergistic effects represents a significant gap in the understanding of health risks. Across all pesticide classes, children, pregnant women, and agricultural workers emerge as disproportionately affected, reflecting broader socio-environmental inequalities. Furthermore, regulatory fragmentation (from stricter limits in the European Union to more permissive limits in Brazil, the United States, and China) creates an uneven global risk market. Addressing these disparities requires harmonized regulation, coordinated biomonitoring, and integrative risk assessment frameworks that explicitly consider cumulative and synergistic exposures.

Supplementary Materials: The following supporting information can be downloaded at: https://www.mdpi.com/article/10.3390/jox15060187/s1. We added the index to the supplementary material.

Author Contributions: T.R.F., C.C., J.M.B.d.M.V., B.G.L.V. and C.P. contributed to the study conception and design, material preparation, data collection and analysis. The first draft of the manuscript was written by T.R.F. and C.P. All authors have read and agreed to the published version of the manuscript.

Funding: Brazilian National Council for Scientific and Technological Development (CNPq; Grants 402364/2021-0, 441017/2023-1, and 305335/2021-9) and Itaipu Binacional (grant 4500007500/2024).

Institutional Review Board Statement: Not applicable.

Informed Consent Statement: Not applicable.

Data Availability Statement: No new data were created or analyzed in this study. Data sharing is not applicable to this article.

Conflicts of Interest: The authors have no conflicts to declare.

References

1. Mostafalou, S.; Abdollahi, M. Pesticides and human chronic diseases: Evidences, mechanisms, and perspectives. *Toxicol. Appl. Pharmacol.* **2013**, *268*, 157–177. [CrossRef]
2. Luo, Y.; Zhang, M. Multimedia transport and risk assessment of organophosphate pesticides and a case study in the northern San Joaquin Valley of California. *Chemosphere* **2009**, *75*, 969–978. [CrossRef]
3. Aprea, C.; Colosio, C.; Mammone, T.; Minoia, C.; Maroni, M. Biological monitoring of pesticide exposure: A review of analytical methods. *J. Chromatogr. B* **2002**, *769*, 191–219. [CrossRef]
4. Sabzevari, S.; Hofman, J. A worldwide review of currently used pesticides' monitoring in agricultural soils. *Sci. Total Environ.* **2022**, *812*, 152344. [CrossRef]
5. Food and Agriculture Organization of the United Nations. *Pestic. Use Trade 1990–2021*; FAO: Rome, Italy, 2023. [CrossRef]

6. Rajan, S.; Parween, M.; Raju, N.J. Pesticides in the hydrogeo-environment: A review of contaminant prevalence, source and mobilisation in India. *Environ. Geochem. Health* **2023**, *45*, 5481–5513. [CrossRef] [PubMed]
7. Kapka-Skrzypczak, L.; Cyranka, M.; Skrzypczak, M.; Kruszewski, M. Biomonitoring and biomarkers of organophosphate pesticides exposure—State of the art. *Ann. Agric. Environ. Med.* **2011**, *18*, 294–303.
8. Rodrigues, M.B.; da Silva, C.A.M.; Chong-Silva, D.C.; Chong-Neto, H.J. Pesticides and human health. *J. Pediatr.* **2025**, *101*, S70–S76. [CrossRef] [PubMed]
9. Baciu, T.; Borrull, F.; Aguilar, C.; Calull, M. Recent trends in analytical methods and separation techniques for drugs of abuse in hair. *Anal. Chim. Acta* **2015**, *856*, 1–26. [CrossRef]
10. LaKind, J.S.; Amina Wilkins, A.; Berlin, C.M. Environmental chemicals in human milk: A review of levels, infant exposures and health, and guidance for future research. *Toxicol. Appl. Pharmacol.* **2004**, *198*, 184–208. [CrossRef] [PubMed]
11. Ostrea, E.M.; Bielawski, D.M.; Posecion, N.C.; Corrion, M.; Villanueva-Uy, E.; Bernardo, R.C.; Jin, Y.; Janisse, J.J.; Ager, J.W. Combined analysis of prenatal (maternal hair and blood) and neonatal (infant hair, cord blood and meconium) matrices to detect fetal exposure to environmental pesticides. *Environ. Res.* **2009**, *109*, 116–122. [CrossRef]
12. Sales, C.; Cervera, M.I.; Gil, R.; Portolés, T.; Pitarch, E.; Beltran, J. Quality classification of Spanish olive oils by untargeted gas chromatography coupled to hybrid quadrupole-time of flight mass spectrometry with atmospheric pressure chemical ionization and metabolomics-based statistical approach. *Food Chem.* **2017**, *216*, 365–373. [CrossRef]
13. Salihovic, S.; Nilsson, H.; Hagberg, J.; Lindström, G. Trends in the analysis of persistent organic pollutants (POPs) in human blood. *TrAC Trends Anal. Chem.* **2013**, *46*, 129–138. [CrossRef]
14. Bruzzoniti, M.C.; Checchini, L.; De Carlo, R.M.; Orlandini, S.; Rivoira, L.; Del Bubba, M. QuEChERS sample preparation for the determination of pesticides and other organic residues in environmental matrices: A critical review. *Anal. Bioanal. Chem.* **2014**, *406*, 4089–4116. [CrossRef]
15. Europe Union. Commission Regulation (EC) No 1881/2006 of 19 December 2006 Setting Maximum Levels for Certain Contaminants in Foodstuffs (Text with EEA Relevance) 2006. Available online: https://eur-lex.europa.eu/eli/reg/2006/1881/oj/eng (accessed on 14 September 2025).
16. European Commission. EU Pesticides Database—European Commission. Available online: https://food.ec.europa.eu/plants/pesticides/eu-pesticides-database_en (accessed on 14 September 2025).
17. García-Vara, M.; Postigo, C.; Palma, P.; López de Alda, M. Development of QuEChERS-based multiresidue analytical methods to determine pesticides in corn, grapes and alfalfa. *Food Chem.* **2023**, *405*, 134870. [CrossRef] [PubMed]
18. Valverde, S.; Bernal, J.L.; Martín, M.T.; Nozal, M.J.; Bernal, J. Fast determination of neonicotinoid insecticides in bee pollen using QuEChERS and ultra-high performance liquid chromatography coupled to quadrupole time-of-flight mass spectrometry. *Electrophoresis* **2016**, *37*, 2470–2477. [CrossRef] [PubMed]
19. Jones, K.; Patel, K.; Cocker, J.; Bevan, R.; Levy, L. Determination of ethylenethiourea in urine by liquid chromatography–atmospheric pressure chemical ionisation–mass spectrometry for monitoring background levels in the general population. *J. Chromatogr. B* **2010**, *878*, 2563–2566. [CrossRef]
20. Lindh, C.H.; Littorin, M.; Johannesson, G.; Jönsson, B.A.G. Analysis of chlormequat in human urine as a biomarker of exposure using liquid chromatography triple quadrupole mass spectrometry. *J. Chromatogr. B* **2011**, *879*, 1551–1556. [CrossRef]
21. Wang, D.; Atkinson, S.; Hoover-Miller, A.; Shelver, W.L.; Li, Q.X. Simultaneous use of gas chromatography/ion trap mass spectrometry—Electron capture detection to improve the analysis of bromodiphenyl ethers in biological and environmental samples. *Rapid Commun. Mass Spectrom.* **2008**, *22*, 647–656. [CrossRef]
22. Grimalt, J.O.; Howsam, M.; Carrizo, D.; Otero, R.; de Marchi, M.R.R.; Vizcaino, E. Integrated analysis of halogenated organic pollutants in sub-millilitre volumes of venous and umbilical cord blood sera. *Anal. Bioanal. Chem.* **2010**, *396*, 2265–2272. [CrossRef]
23. Barchanska, H.; Danek, M.; Sajdak, M.; Turek, M. Review of Sample Preparation Techniques for the Analysis of Selected Classes of Pesticides in Plant Matrices. *Crit. Rev. Anal. Chem.* **2018**, *48*, 467–491. [CrossRef]
24. Sapozhnikova, Y.; Lehotay, S.J. Multi-class, multi-residue analysis of pesticides, polychlorinated biphenyls, polycyclic aromatic hydrocarbons, polybrominated diphenyl ethers and novel flame retardants in fish using fast, low-pressure gas chromatography–tandem mass spectrometry. *Anal. Chim. Acta* **2013**, *758*, 80–92. [CrossRef]
25. Wang, N.; Kong, D.; Shan, Z.; Shi, L.; Cai, D.; Cao, Y.; Liu, Y.; Pang, G. Simultaneous determination of pesticides, polycyclic aromatic hydrocarbons, polychlorinated biphenyls and phthalate esters in human adipose tissue by gas chromatography–tandem mass spectrometry. *J. Chromatogr. B* **2012**, *898*, 38–52. [CrossRef]
26. Tang, H.P. Recent development in analysis of persistent organic pollutants under the Stockholm Convention. *TrAC Trends Anal. Chem.* **2013**, *45*, 48–66. [CrossRef]

27. Feo, M.L.; Eljarrat, E.; Manaca, M.N.; Dobaño, C.; Barcelo, D.; Sunyer, J.; Alonso, P.L.; Menendez, C.; Grimalt, J.O. Pyrethroid use-malaria control and individual applications by households for other pests and home garden use. *Environ. Int.* **2012**, *38*, 67–72. [CrossRef] [PubMed]
28. Costa, C.; García-Lestón, J.; Costa, S.; Coelho, P.; Silva, S.; Pingarilho, M.; Valdiglesias, V.; Mattei, F.; Dall'aRmi, V.; Bonassi, S.; et al. Is organic farming safer to farmers' health? A comparison between organic and traditional farming. *Toxicol. Lett.* **2014**, *230*, 166–176. [CrossRef]
29. Sharma, A.; Kumar, V.; Shahzad, B.; Tanveer, M.; Sidhu, G.P.S.; Handa, N.; Kohli, S.K.; Yadav, P.; Bali, A.S.; Parihar, R.D.; et al. Worldwide pesticide usage and its impacts on ecosystem. *SN Appl. Sci.* **2019**, *1*, 1446. [CrossRef]
30. Shattuck, A.; Werner, M.; Mempel, F.; Dunivin, Z.; Galt, R. Global pesticide use and trade database (GloPUT): New estimates show pesticide use trends in low-income countries substantially underestimated. *Glob. Environ. Chang.* **2023**, *81*, 102693. [CrossRef]
31. Li, Z.; Jennings, A. Worldwide Regulations of Standard Values of Pesticides for Human Health Risk Control: A Review. *Int. J. Environ. Res. Public Health* **2017**, *14*, 826. [CrossRef] [PubMed]
32. Bayoumi, A.E. Deleterious Effects of Banned Chemical Pesticides on Human Health in Developing Countries. In *Pesticides Updates on Toxicity, Efficacy and Risk Assessment*; IntechOpen: London, UK, 2022. [CrossRef]
33. Maggi, F.; Tang, F.H.M.; Black, A.J.; Marks, G.B.; McBratney, A. The pesticide health risk index—An application to the world's countries. *Sci. Total Environ.* **2021**, *801*, 149731. [CrossRef]
34. Pathak, V.M.; Verma, V.K.; Rawat, B.S.; Kaur, B.; Babu, N.; Sharma, A.; Dewali, S.; Yadav, M.; Kumari, R.; Singh, S.; et al. Current status of pesticide effects on environment, human health and it's eco-friendly management as bioremediation: A comprehensive review. *Front. Microbiol.* **2022**, *13*, 962619. [CrossRef] [PubMed]
35. PubChem. Glyphosate. Available online: https://pubchem.ncbi.nlm.nih.gov/compound/3496 (accessed on 16 May 2025).
36. Bergwerf, H. MolView. Available online: https://molview.org/ (accessed on 2 August 2025).
37. Benbrook, C.M. Trends in glyphosate herbicide use in the United States and globally. *Environ. Sci. Eur.* **2016**, *28*, 3. [CrossRef]
38. Van Bruggen, A.H.C.; He, M.M.; Shin, K.; Mai, V.; Jeong, K.C.; Finckh, M.R.; Morris, J.G., Jr. Environmental and health effects of the herbicide glyphosate. *Sci. Total Environ.* **2018**, *616–617*, 255–268. [CrossRef]
39. Cederlund, H. Effects of spray drift of glyphosate on nontarget terrestrial plants—A critical review. *Environ. Toxicol. Chem.* **2017**, *36*, 2879–2886. [CrossRef]
40. Singh, S.; Kumar, V.; Gill, J.P.K.; Datta, S.; Singh, S.; Dhaka, V.; Kapoor, D.; Wani, A.B.; Dhanjal, D.S.; Kumar, M.; et al. Herbicide Glyphosate: Toxicity and Microbial Degradation. *Int. J. Environ. Res. Public Health* **2020**, *17*, 7519. [CrossRef]
41. Zaller, J.G.; Heigl, F.; Ruess, L.; Grabmaier, A. Glyphosate herbicide affects belowground interactions between earthworms and symbiotic mycorrhizal fungi in a model ecosystem. *Sci. Rep.* **2014**, *4*, 5634. [CrossRef]
42. Relyea, R.A. The Lethal Impact of Roundup on Aquatic and Terrestrial Amphibians. *Ecol. Appl.* **2005**, *15*, 1118–1124. [CrossRef]
43. de Morais Valentim, J.M.B.; Coradi, C.; Viana, N.P.; Fagundes, T.R.; Micheletti, P.L.; Gaboardi, S.C.; Fadel, B.; Pizzatti, L.; Candiotto, L.Z.P.; Panis, C. Glyphosate as a Food Contaminant: Main Sources, Detection Levels, and Implications for Human and Public Health. *Foods* **2024**, *13*, 1697. [CrossRef] [PubMed]
44. IARC Monograph on Glyphosate. Available online: https://www.iarc.who.int/featured-news/media-centre-iarc-news-glyphosate (accessed on 16 May 2025).
45. Panis, C.; Candiotto, L.Z.P.; Gaboardi, S.C.; Teixeira, G.T.; Alves, F.M.; da Silva, J.C.; Scandolara, T.B.; Rech, D.; Gurzenda, S.; Ponmattam, J.; et al. Exposure to Pesticides and Breast Cancer in an Agricultural Region in Brazil. *Environ. Sci. Technol.* **2024**, *58*, 10470–10481. [CrossRef]
46. Peillex, C.; and Pelletier, M. The impact and toxicity of glyphosate and glyphosate-based herbicides on health and immunity. *J. Immunotoxicol.* **2020**, *17*, 163–174. [CrossRef] [PubMed]
47. Meloni, F.; Satta, G.; Padoan, M.; Montagna, A.; Pilia, I.; Argiolas, A.; Piro, S.; Magnani, C.; Gambelunghe, A.; Muzi, G.; et al. Occupational exposure to glyphosate and risk of lymphoma:results of an Italian multicenter case-control study. *Environ. Health* **2021**, *20*, 49. [CrossRef]
48. Hsiao, C.C.; Yang, A.-M.; Wang, C.; Lin, C.-Y. Association between glyphosate exposure and cognitive function, depression, and neurological diseases in a representative sample of US adults: NHANES 2013–2014 analysis. *Environ. Res.* **2023**, *237*, 116860. [CrossRef] [PubMed]
49. Vasseur, C.; Serra, L.; El Balkhi, S.; Lefort, G.; Ramé, C.; Froment, P.; Dupont, J. Glyphosate presence in human sperm: First report and positive correlation with oxidative stress in an infertile French population. *Ecotoxicol. Environ. Saf.* **2024**, *278*, 116410. [CrossRef]
50. European Union. *Council Directive 98/83/EC of 3 November 1998 on the Quality of Water Intended for Human Consumption*; European Union: Luxembourg, 1998; Volume 330.

51. US EPA. National Primary Drinking Water Regulations 2015. Available online: https://www.epa.gov/ground-water-and-drinking-water/national-primary-drinking-water-regulations (accessed on 16 May 2025).
52. Ministério da Saúde. PORTARIA GM/MS N° 888, DE 4 DE MAIO DE 2021. Available online: https://bvsms.saude.gov.br/bvs/saudelegis/gm/2021/prt0888_07_05_2021.html (accessed on 16 May 2025).
53. Canada, H. Canadian Drinking Water Guidelines 2004. Available online: https://www.canada.ca/en/health-canada/services/environmental-workplace-health/water-quality/drinking-water/canadian-drinking-water-guidelines.html (accessed on 16 May 2025).
54. Australian Drinking Water Guidelines | NHMRC. Available online: https://www.nhmrc.gov.au/about-us/publications/australian-drinking-water-guidelines (accessed on 16 May 2025).
55. Guidelines for Drinking-Water Quality, 4th Edition, Incorporating the 1st Addendum. Available online: https://www.who.int/publications/i/item/9789241549950 (accessed on 16 May 2025).
56. Schoeters, G.; Verheyen, V.J.; Colles, A.; Remy, S.; Martin, L.R.; Govarts, E.; Nelen, V.; Hond, E.D.; De Decker, A.; Franken, C.; et al. Internal exposure of Flemish teenagers to environmental pollutants: Results of the Flemish Environment and Health Study 2016–2020 (FLEHS IV). *Int. J. Hyg. Environ. Health* **2022**, *242*, 113972. [CrossRef]
57. Ashley-Martin, J.; Huang, R.; MacPherson, S.; Brion, O.; Owen, J.; Gaudreau, E.; Bienvenu, J.-F.; Fisher, M.; Borghese, M.M.; Bouchard, M.F.; et al. Urinary concentrations and determinants of glyphosate and glufosinate in pregnant Canadian participants in the MIREC study. *Environ. Res.* **2023**, *217*, 114842. [CrossRef]
58. Ashley-Martin, J.; Marro, L.; Owen, J.; Borghese, M.M.; Arbuckle, T.; Bouchard, M.F.; Lanphear, B.; Walker, M.; Foster, W.; Fisher, M. Gestational urinary concentrations of glyphosate and aminomethylphosphonic acid in relation to preterm birth: The MIREC study. *J. Expo. Sci. Environ. Epidemiol.* **2024**, *35*, 876–881. [CrossRef]
59. Zhang, F.; Xu, Y.; Liu, X.; Pan, L.; Ding, E.; Dou, J.; Zhu, B. Concentration Distribution and Analysis of Urinary Glyphosate and Its Metabolites in Occupationally Exposed Workers in Eastern China. *Int. J. Environ. Res. Public Health* **2020**, *17*, 2943. [CrossRef]
60. Makris, K.C.; Efthymiou, N.; Konstantinou, C.; Anastasi, E.; Schoeters, G.; Kolossa-Gehring, M.; Katsonouri, A. Oxidative stress of glyphosate, AMPA and metabolites of pyrethroids and chlorpyrifos pesticides among primary school children in Cyprus. *Environ. Res.* **2022**, *212*, 113316. [CrossRef]
61. Bayebila Menanzambi, T.; Dufour, P.; Pirard, C.; Nsangu, J.; Mufusama, J.-P.; Mbinze Kindenge, J.; Djang'eIng'a, R.M.; Charlier, C. Bio-surveillance of environmental pollutants in the population of Kinshasa, Democratic Republic of Congo (DRC): A small pilot study. *Arch. Public Health* **2021**, *79*, 197. [CrossRef] [PubMed]
62. Tagne-Fotso, R.; Zeghnoun, A.; Saoudi, A.; Balestier, A.; Pecheux, M.; Chaperon, L.; Oleko, A.; Marchand, P.; Le Bizec, B.; Vattier, L.; et al. Exposure of the general French population to herbicides, pyrethroids, organophosphates, organochlorines, and carbamate pesticides in 2014–2016, Results from the Esteban study. *Int. J. Hyg. Environ. Health* **2023**, *254*, 114265. [CrossRef]
63. Conrad, A.; Schröter-Kermani, C.; Hoppe, H.-W.; Rüther, M.; Pieper, S.; Kolossa-Gehring, M. Glyphosate in German adults—Time trend (2001 to 2015) of human exposure to a widely used herbicide. *Int. J. Hyg. Environ. Health* **2017**, *220*, 8–16. [CrossRef] [PubMed]
64. Lemke, N.; Murawski, A.; Schmied-Tobies, M.I.H.; Rucic, E.; Hoppe, H.-W.; Conrad, A.; Kolossa-Gehring, M. Glyphosate and aminomethylphosphonic acid (AMPA) in urine of children and adolescents in Germany—Human biomonitoring results of the German Environmental Survey 2014–2017 (GerES V). *Environ. Int.* **2021**, *156*, 106769. [CrossRef]
65. Soukup, S.T.; Merz, B.; Bub, A.; Hoffmann, I.; Watzl, B.; Steinberg, P.; Kulling, S.E. Glyphosate and AMPA levels in human urine samples and their correlation with food consumption: Results of the cross-sectional KarMeN study in Germany. *Arch. Toxicol.* **2020**, *94*, 1575–1584. [CrossRef] [PubMed]
66. Connolly, A.; Leahy, M.; Jones, K.; Kenny, L.; Coggins, M.A. Glyphosate in Irish adults—A pilot study in 2017. *Environ. Res.* **2018**, *165*, 235–236. [CrossRef] [PubMed]
67. Porru, S.; Ferrian, M.; Mastrangelo, G.; Capovilla, D.; Corsini, E.; Fustinoni, S.; Peruzzi, M.; Colosio, C. Short occupational exposure to glyphosate and its biomonitoring via urinary levels of glyphosate and metabolite AMPA (Amino-MethylPhosphonic acid), in Italian vineyard workers. *Heliyon* **2024**, *10*, e36407. [CrossRef]
68. Mueller, W.; Jones, K.; Fuhrimann, S.; Ahmad, Z.N.B.S.; Sams, C.; Harding, A.-H.; Povey, A.; Atuhaire, A.; Basinas, I.; van Tongeren, M.; et al. Factors influencing occupational exposure to pyrethroids and glyphosate: An analysis of urinary biomarkers in Malaysia, Uganda and the United Kingdom. *Environ. Res.* **2024**, *242*, 117651. [CrossRef]
69. Balderrama-Carmona, A.P.; Valenzuela-Rincón, M.; Zamora-Álvarez, L.A.; Adan-Bante, N.P.; Leyva-Soto, L.A.; Silva-Beltrán, N.P.; Morán-Palacio, E.F. Herbicide biomonitoring in agricultural workers in Valle del Mayo, Sonora Mexico. *Environ. Sci. Pollut. Res. Int.* **2020**, *27*, 28480–28489. [CrossRef]

70. Silver, M.K.; Fernandez, J.; Tang, J.; McDade, A.; Sabino, J.; Rosario, Z.; Vega, C.V.; Alshawabkeh, A.; Cordero, J.F.; Meeker, J.D. Prenatal Exposure to Glyphosate and Its Environmental Degradate, Aminomethylphosphonic Acid (AMPA), and Preterm Birth: A Nested Case–Control Study in the PROTECT Cohort (Puerto Rico). *Environ. Health Perspect.* **2021**, *129*, 057011. [CrossRef]
71. Nova, P.; Calheiros, C.S.C.; Silva, M. Glyphosate in Portuguese Adults—A Pilot Study. *Environ. Toxicol. Pharmacol.* **2020**, *80*, 103462. [CrossRef] [PubMed]
72. Stajnko, A.; Snoj Tratnik, J.; Kosjek, T.; Mazej, D.; Jagodic, M.; Eržen, I.; Horvat, M. Seasonal glyphosate and AMPA levels in urine of children and adolescents living in rural regions of Northeastern Slovenia. *Environ. Int.* **2020**, *143*, 105985. [CrossRef]
73. Ruiz, P.; Dualde, P.; Coscollà, C.; Fernández, S.F.; Carbonell, E.; Yusà, V. Biomonitoring of glyphosate and AMPA in the urine of Spanish lactating mothers. *Sci. Total Environ.* **2021**, *801*, 149688. [CrossRef]
74. Wongta, A.; Sawarng, N.; Tongchai, P.; Sutan, K.; Kerdnoi, T.; Prapamontol, T.; Hongsibsong, S. The Pesticide Exposure of People Living in Agricultural Community, Northern Thailand. *J. Toxicol.* **2018**, *2018*, 4168034. [CrossRef]
75. Sidthilaw, S.; Sapbamrer, R.; Pothirat, C.; Wunnapuk, K.; Khacha-ananda, S. Effects of exposure to glyphosate on oxidative stress, inflammation, and lung function in maize farmers, Northern Thailand. *BMC Public Health* **2022**, *22*, 1343. [CrossRef]
76. Chaiklieng, S.; Uengchuen, K.; Gissawong, N.; Srijaranai, S.; Autrup, H. Biological Monitoring of Glyphosate Exposure among Knapsack Sprayers in Khon Kaen, Thailand. *Toxics* **2024**, *12*, 337. [CrossRef]
77. Acquavella, J.F.; Alexander, B.H.; Mandel, J.S.; Gustin, C.; Baker, B.; Chapman, P.; Bleeke, M. Glyphosate biomonitoring for farmers and their families: Results from the Farm Family Exposure Study. *Environ. Health Perspect.* **2004**, *112*, 321–326. [CrossRef]
78. Parvez, S.; Gerona, R.R.; Proctor, C.; Friesen, M.; Ashby, J.L.; Reiter, J.L.; Lui, Z.; Winchester, P.D. Glyphosate exposure in pregnancy and shortened gestational length: A prospective Indiana birth cohort study. *Environ. Health* **2018**, *17*, 23. [CrossRef] [PubMed]
79. Gerona, R.R.; Reiter, J.L.; Zakharevich, I.; Proctor, C.; Ying, J.; Mesnage, R.; Antoniou, M.; Winchester, P.D. Glyphosate exposure in early pregnancy and reduced fetal growth: A prospective observational study of high-risk pregnancies. *Environ. Health* **2022**, *21*, 95. [CrossRef] [PubMed]
80. Chang, V.C.; Andreotti, G.; Ospina, M.; Parks, C.G.; Liu, D.; Shearer, J.J.; Rothman, N.; Silverman, D.T.; Sandler, D.P.; Calafat, A.M.; et al. Glyphosate exposure and urinary oxidative stress biomarkers in the Agricultural Health Study. *JNCI J. Natl. Cancer Inst.* **2023**, *115*, 394–404. [CrossRef] [PubMed]
81. Varde, M.; Gerona, R.R.; Newman, R.B.; Reckers, A.; Braak, D.C.; Vena, J.E.; Bloom, M.S. Glyphosate exposure and preterm birth: A nested case-control pilot study. *Reprod. Toxicol.* **2023**, *117*, 108350. [CrossRef]
82. Eskenazi, B.; Gunier, R.B.; Rauch, S.; Kogut, K.; Perito, E.R.; Mendez, X.; Limbach, C.; Holland, N.; Bradman, A.; Harley, K.G.; et al. Association of Lifetime Exposure to Glyphosate and Aminomethylphosphonic Acid (AMPA) with Liver Inflammation and Metabolic Syndrome at Young Adulthood: Findings from the CHAMACOS Study. *Environ. Health Perspect.* **2023**, *131*, 037001. [CrossRef]
83. Lucia, R.M.; Liao, X.; Huang, W.-L.; Forman, D.; Kim, A.; Ziogas, A.; Norden-Krichmar, T.M.; Goodman, D.; Alvarez, A.; Masunaka, I.; et al. Urinary glyphosate and AMPA levels in a cross-sectional study of postmenopausal women: Associations with organic eating behavior and dietary intake. *Int. J. Hyg. Environ. Health* **2023**, *252*, 114211. [CrossRef]
84. Hyland, C.; Hernandez, A.; Gaudreau, É.; Larose, J.; Bienvenu, J.-F.; Meierotto, L.; Castellano, R.L.S.; Curl, C.L. Examination of urinary pesticide concentrations, protective behaviors, and risk perceptions among Latino and Latina farmworkers in Southwestern Idaho. *Int. J. Hyg. Environ. Health* **2024**, *255*, 114275. [CrossRef]
85. Curl, C.L.; Hyland, C.; Spivak, M.; Sheppard, L.; Lanphear, B.; Antoniou, M.N.; Ospina, M.; Calafat, A.M. The Effect of Pesticide Spray Season and Residential Proximity to Agriculture on Glyphosate Exposure among Pregnant People in Southern Idaho, 2021. *Environ. Health Perspect.* **2023**, *131*, 127001. [CrossRef] [PubMed]
86. Jauregui-Zunzunegui, S.; Rodríguez-Artalejo, F.; Tellez-Plaza, M.; García-Esquinas, E. Glyphosate exposure, muscular health and functional limitations in middle-aged and older adults. *Environ. Res.* **2024**, *251*, 118547. [CrossRef]
87. Chang, V.C.; Ospina, M.; Xie, S.; Andreotti, G.; Parks, C.G.; Liu, D.; Madrigal, J.M.; Ward, M.H.; Rothman, N.; Silverman, D.T.; et al. Urinary biomonitoring of glyphosate exposure among male farmers and nonfarmers in the Biomarkers of Exposure and Effect in Agriculture (BEEA) study. *Environ. Int.* **2024**, *187*, 108644. [CrossRef] [PubMed]
88. Tang, P.; Wang, Y.; Liao, Q.; Zhou, Y.; Huang, H.; Liang, J.; Zeng, X.; Qiu, X. Relationship of urinary glyphosate concentrations with glycosylated hemoglobin and diabetes in US adults: A cross-sectional study. *BMC Public Health* **2024**, *24*, 1644. [CrossRef]
89. Otaru, S.; Jones, L.E.; Carpenter, D.O. Associations between urine glyphosate levels and metabolic health risks: Insights from a large cross-sectional population-based study. *Environ. Health* **2024**, *23*, 58. [CrossRef] [PubMed]
90. Zhu, Y.; Yang, Z.; Chen, Y.; Qian, L.; Hao, C.; Hao, L.; Yang, B.; Duan, J. Association between glyphosate exposure and cardiovascular health using "Life's Essential 8" metrics in US adults. *Ecotoxicol. Environ. Saf.* **2025**, *290*, 117559. [CrossRef]

91. Feng, X.; Wang, M.; Wang, Y.; Liang, R.; Yan, C. Associations between environmental glyphosate exposure and glucose homeostasis indices in US general adults: A national population-based cross-sectional study. *Sci. Rep.* **2025**, *15*, 1627. [CrossRef]
92. He, X.; Mi, L.; Zhao, M.; Ji, Y.; Hu, Y.; Gao, Y.; Qiu, L.; Xu, K. Associations between urinary glyphosate and arthritis: An US NHANES analysis. *Front. Public Health* **2025**, *13*, 1450479. [CrossRef]
93. Martínez, M.-A.; Rodríguez, J.-L.; Lopez-Torres, B.; Martínez, M.; Martínez-Larrañaga, M.-R.; Maximiliano, J.-E.; Anadón, A.; Ares, I. Use of human neuroblastoma SH-SY5Y cells to evaluate glyphosate-induced effects on oxidative stress, neuronal development and cell death signaling pathways. *Environ. Int.* **2020**, *135*, 105414. [CrossRef]
94. Finkler, M.; Rodrigues, G.Z.P.; Kayser, J.M.; Ziulkoski, A.L.; Gehlen, G. Cytotoxic and genotoxic effects induced by associated commercial glyphosate and 2,4-D formulations using the Allium cepa bioassay. *J. Environ. Sci. Health Part B* **2022**, *57*, 133–141. [CrossRef] [PubMed]
95. Mencalha, A.; Victorino, V.J.; Cecchini, R.; Panis, C. Mapping Oxidative Changes in Breast Cancer: Understanding the Basic to Reach the Clinics. *Anticancer Res.* **2014**, *34*, 1127–1140. [PubMed]
96. Zhang, L.; Rana, I.; Shaffer, R.M.; Taioli, E.; Sheppard, L. Exposure to glyphosate-based herbicides and risk for non-Hodgkin lymphoma: A meta-analysis and supporting evidence. *Mutat. Res./Rev. Mutat. Res.* **2019**, *781*, 186–206. [CrossRef] [PubMed]
97. Authority (EFSA) EFS. Conclusion on the peer review of the pesticide risk assessment of the active substance glyphosate. *EFSA J.* **2015**, *13*, 4302. [CrossRef]
98. Glyphosate Technical Fact Sheet. Available online: https://npic.orst.edu/factsheets/archive/glyphotech.html#references (accessed on 19 May 2025).
99. Lesseur, C.; Pathak, K.V.; Pirrotte, P.; Martinez, M.N.; Ferguson, K.K.; Barrett, E.S.; Nguyen, R.H.; Sathyanarayana, S.; Mandrioli, D.; Swan, S.H.; et al. Urinary glyphosate concentration in pregnant women in relation to length of gestation. *Environ. Res.* **2022**, *203*, 111811. [CrossRef]
100. Thongprakaisang, S.; Thiantanawat, A.; Rangkadilok, N.; Suriyo, T.; Satayavivad, J. Glyphosate induces human breast cancer cells growth via estrogen receptors. *Food Chem. Toxicol.* **2013**, *59*, 129–136. [CrossRef]
101. Yahyah, H.; Kameri-Mbote, P.; Kibugi, R. Implications of pesticide use regulation on soil sustainability in Uganda. *Soil Secur.* **2024**, *16*, 100133. [CrossRef]
102. IBAMA. Pesticide Marketing Reports 2022. Available online: https://www.gov.br/ibama/pt-br/assuntos/quimicos-e-biologicos/agrotoxicos/relatorios-de-comercializacao-de-agrotoxicos/relatorios-de-comercializacao-de-agrotoxicos (accessed on 17 April 2024).
103. Wasser, J.; Berman, T.; Lerner-Geva, L.; Grotto, I.; Rubin, L. Biological monitoring of Persistent Organic Pollutants in human milk in Israel. *Chemosphere* **2015**, *137*, 185–191. [CrossRef]
104. Tanabe, S.; Kunisue, T. Persistent organic pollutants in human breast milk from Asian countries. *Environ. Pollut.* **2007**, *146*, 400–413. [CrossRef]
105. Jakszyn, P.; Goñi, F.; Etxeandia, A.; Vives, A.; Millán, E.; López, R.; Amiano, P.; Ardanaz, E.; Barricarte, A.; Chirlaque, M.D.; et al. Serum levels of organochlorine pesticides in healthy adults from five regions of Spain. *Chemosphere* **2009**, *76*, 1518–1524. [CrossRef]
106. Europe WHORO. *Health Risks of Persistent Organic Pollutants from Long-Range Transboundary Air Pollution*; World Health Organization: Geneva, Switzerland, 2003.
107. Covaci, A.; Manirakiza, P.; Schepens, P. Persistent Organochlorine Pollutants in Soils from Belgium, Italy, Greece, and Romania. *Bull. Environ. Contam. Toxicol.* **2002**, *68*, 97–103. [CrossRef]
108. Safe, S.H. Endocrine disruptors and human health—Is there a problem? An update. *Environ. Health Perspect.* **2000**, *108*, 487–493. [CrossRef]
109. WHO. Cancer Today. Available online: https://gco.iarc.who.int/today/ (accessed on 3 August 2025).
110. Guimarães, R.M.; Asmus, C.I.R.F.; Meyer, A. DDT reintroduction for malaria control: The cost-benefit debate for public health. *Cad. Saúde Pública* **2007**, *23*, 2835–2844. [CrossRef]
111. Li, J.; Wang, P.; Shi, S.; Xue, J. Background biomonitoring of residue levels of 137 pesticides in the blood plasma of the general population in Beijing. *Environ. Monit. Assess.* **2018**, *190*, 315. [CrossRef]
112. Azeredo, A.; Torres, J.P.M.; de Freitas Fonseca, M.; Britto, J.L.; Bastos, W.R.; Azevedo e Silva, C.E.; Cavalcanti, G.; Meire, R.O.; Sarcinelli, P.N.; Claudio, L.; et al. DDT and its metabolites in breast milk from the Madeira River basin in the Amazon, Brazil. *Chemosphere* **2008**, *73*, S246–S251. [CrossRef]
113. Kunisue, T.; Someya, M.; Kayama, F.; Jin, Y.; Tanabe, S. Persistent organochlorines in human breast milk collected from primiparae in Dalian and Shenyang, China. *Environ. Pollut.* **2004**, *131*, 381–392. [CrossRef]

114. Torres-Moreno, A.C.; Mejia-Grau, K.; Puente-DelaCruz, L.; Codling, G.; Villa, A.L.; Ríos-Marquez, O.; Patequiva-Chauta, L.; Cobo, M.; Johnson-Restrepo, B. Polybrominated diphenyl ethers (PBDEs), polychlorinated biphenyls (PCBs), organochlorine pesticides (OCPs) in human breast milk from Colombia: A probabilistic risk assessment approach. *Chemosphere* **2023**, *339*, 139597. [CrossRef]
115. Tomar, L.R.; Agarwal, M.P.; Avasthi, R.; Tyagi, V.; Mustafa, M.; Banerjee, B.D. Serum organochlorine pesticide levels in patients with metabolic syndrome. *Indian J. Endocrinol. Metab.* **2013**, *17*, S342. [CrossRef] [PubMed]
116. Saoudi, A.; Fréry, N.; Zeghnoun, A.; Bidondo, M.-L.; Deschamps, V.; Göen, T.; Garnier, R.; Guldner, L. Serum levels of organochlorine pesticides in the French adult population: The French National Nutrition and Health Study (ENNS), 2006–2007. *Sci. Total Environ.* **2014**, *472*, 1089–1099. [CrossRef] [PubMed]
117. Bhatnagar, V.K.; Kashyap, R.; Zaidi, S.S.A.; Kulkarni, P.K.; Saiyed, H.N. Levels of DDT, HCH and HCB Residues in Human Blood in Ahmedabad, India. *Bull. Environ. Contam. Toxicol.* **2004**, *72*, 261–265. [CrossRef] [PubMed]
118. Jaacks, L.M.; Yadav, S.; Panuwet, P.; Kumar, S.; Rajacharya, G.H.; Johnson, C.; Rawal, I.; Mohan, D.; Mohan, V.; Tandon, N.; et al. Metabolite of the pesticide DDT and incident type 2 diabetes in urban India. *Environ. Int.* **2019**, *133*, 105089. [CrossRef] [PubMed]
119. Jeon, H.L.; Hong, S.; Choi, K.; Lee, C.; Yoo, J. First nationwide exposure profile of major persistent organic pollutants among Korean adults and their determinants: Korean National Environmental Health Survey Cycle 3 (2015–2017). *Int. J. Hyg. Environ. Health* **2021**, *236*, 113779. [CrossRef] [PubMed]
120. Peng, F.-J.; Emond, C.; Hardy, E.M.; Sauvageot, N.; Alkerwi, A.; Lair, M.-L.; Appenzeller, B.M. Population-based biomonitoring of exposure to persistent and non-persistent organic pollutants in the Grand Duchy of Luxembourg: Results from hair analysis. *Environ. Int.* **2021**, *153*, 106526. [CrossRef] [PubMed]
121. lhaj, F.A.; Elhamri, H.; lhaj, Z.A.; Malisch, R.; Kypke, K.; Kabriti, M.; El Hajjaji, S.; Bellaouchou, A. First WHO/UNEP survey of the current concentrations of persistent organic pollutants in human milk in Morocco. *Food Addit. Contam. Part A* **2023**, *40*, 282–293. [CrossRef]
122. Cruz, S.; Lino, C.; Silveira, M.I. Evaluation of organochlorine pesticide residues in human serum from an urban and two rural populations in Portugal. *Sci. Total Environ.* **2003**, *317*, 23–35. [CrossRef]
123. Al-Saleh, I.; Al-Doush, I.; Alsabbaheen, A.; Mohamed, G.E.D.; Rabbah, A. Levels of DDT and its metabolites in placenta, maternal and cord blood and their potential influence on neonatal anthropometric measures. *Sci. Total Environ.* **2012**, *416*, 62–74. [CrossRef]
124. Müller, M.H.B.; Polder, A.; Brynildsrud, O.B.; Karimi, M.; Lie, E.; Manyilizu, W.B.; Mdegela, R.; Mokiti, F.; Murtadha, M.; Nonga, H.; et al. Organochlorine pesticides (OCPs) and polychlorinated biphenyls (PCBs) in human breast milk and associated health risks to nursing infants in Northern Tanzania. *Environ. Res.* **2017**, *154*, 425–434. [CrossRef]
125. Asawasinsopon, R.; Prapamontol, T.; Prakobvitayakit, O.; Vaneesorn, Y.; Mangklabruks, A.; Hock, B. Plasma levels of DDT and their association with reproductive hormones in adult men from northern Thailand. *Sci. Total Environ.* **2006**, *355*, 98–105. [CrossRef]
126. Hooper, K.; Hopper, K.; Petreas, M.X.; She, J.; Visita, P.; Winkler, J.; McKinney, M.; Mok, M.; Sy, F.; Garcha, J.; et al. Analysis of breast milk to assess exposure to chlorinated contaminants in Kazakstan: PCBs and organochlorine pesticides in southern Kazakstan. *Environ. Health Perspect.* **1997**, *105*, 1250–1254. [CrossRef]
127. US EPA. DDT—A Brief History and Status 2014. Available online: https://www.epa.gov/ingredients-used-pesticide-products/ddt-brief-history-and-status (accessed on 1 August 2025).
128. van den Berg, H.; Manuweera, G.; Konradsen, F. Global trends in the production and use of DDT for control of malaria and other vector-borne diseases. *Malar. J.* **2017**, *16*, 401. [CrossRef]
129. De Jager, C.; Farias, P.; Barraza-Villarreal, A.; Avila, M.H.; Ayotte, P.; Dewailly, E.; Dombrowski, C.; Rousseau, F.; Sanchez, V.D.; Bailey, J.L. Reduced Seminal Parameters Associated with Environmental DDT Exposure and p,p′-DDE Concentrations in Men in Chiapas, Mexico: A Cross-Sectional Study. *J. Androl.* **2006**, *27*, 16–27. [CrossRef]
130. Venners, S.A.; Korrick, S.; Xu, X.; Chen, C.; Guang, W.; Huang, A.; Altshul, L.; Perry, M.; Fu, L.; Wang, X. Preconception Serum DDT and Pregnancy Loss: A Prospective Study Using a Biomarker of Pregnancy. *Am. J. Epidemiol.* **2005**, *162*, 709–716. [CrossRef]
131. Garabrant, D.H.; Held, J.; Langholz, B.; Peters, J.M.; Mack, T.M. DDT and Related Compounds and Risk of Pancreatic Cancer. *J. Natl. Cancer Inst.* **1992**, *84*, 764–771. [CrossRef]
132. Cox, S.; Niskar, A.S.; Narayan, K.M.V.; Marcus, M. Prevalence of Self-Reported Diabetes and Exposure to Organochlorine Pesticides among Mexican Americans: Hispanic Health and Nutrition Examination Survey, 1982–1984. *Environ. Health Perspect.* **2007**, *115*, 1747–1752. [CrossRef]
133. Brody, J.G.; Moysich, K.B.; Humblet, O.; Attfield, K.R.; Beehler, G.P.; Rudel, R.A. Environmental pollutants and breast cancer. *Cancer* **2007**, *109*, 2667–2711. [CrossRef]
134. Cohn, B.A.; Wolff, M.S.; Cirillo, P.M.; Sholtz, R.I. DDT and Breast Cancer in Young Women: New Data on the Significance of Age at Exposure. *Environ. Health Perspect.* **2007**, *115*, 1406–1414. [CrossRef] [PubMed]

135. van den Berg, H. Global Status of DDT and Its Alternatives for Use in Vector Control to Prevent Disease. *Environ. Health Perspect.* **2009**, *117*, 1656–1663. [CrossRef] [PubMed]
136. CDC. Biomonitoring: Population Exposures. Environ. Public Health Track. 2024. Available online: https://www.cdc.gov/environmental-health-tracking/php/data-research/biomonitoring-population-exposures.html (accessed on 23 September 2025).
137. European Union. *Directive (EU) 2020/2184 of the European Parliament and of the Council of 16 December 2020 on the Quality of Water intended for Human Consumption (Recast) (Text with EEA Relevance)*; European Union: Luxembourg, 2020; Volume 435.
138. Overview—POPs Free. Available online: https://chm.pops.int/Implementation/POPsinarticles/Overview/tabid/2437/Default.aspx (accessed on 23 September 2025).
139. Maximum Residue Levels—Food Safety—European Commission. Available online: https://food.ec.europa.eu/plants/pesticides/maximum-residue-levels_en (accessed on 23 September 2025).
140. US EPA. Food Pestic. 2013. Available online: https://www.epa.gov/safepestcontrol/food-and-pesticides (accessed on 23 September 2025).
141. Ghosh, A.K.; Brindisi, M. Urea Derivatives in Modern Drug Discovery and Medicinal Chemistry. *J. Med. Chem.* **2020**, *63*, 2751–2788. [CrossRef] [PubMed]
142. Chaturvedi, D. Perspectives on the synthesis of organic carbamates. *Tetrahedron* **2012**, *68*, 15–45. [CrossRef]
143. Brahmachari, G. (Ed.) *Chemistry and Pharmacology of Naturally Occurring Bioactive Compounds*; CRC Press: Boca Raton, FL, USA, 2013. [CrossRef]
144. Voris, D.G.R.; Cavalcante, S.F.A.; Borges, C.V.N.; Lima, A.L.S. Carbamates: Are they "Good" or "Bad Guys"? *J. Braz. Chem. Soc.* **2024**, *35*, e-20240058. [CrossRef]
145. Hemingway, J.; Ranson, H. Insecticide Resistance in Insect Vectors of Human Disease. *Annu. Rev. Entomol.* **2000**, *45*, 371–391. [CrossRef]
146. IARC. *Occupational Exposures in Insecticide Application, and Some Pesticides*; IARC: Lyon, France, 1991.
147. IARC. *Some Thyrotropic Agents*; IARC: Lyon, France, 2001.
148. Berman, T.; Göen, T.; Novack, L.; Beacher, L.; Grinshpan, L.; Segev, D.; Tordjman, K. Urinary concentrations of organophosphate and carbamate pesticides in residents of a vegetarian community. *Environ. Int.* **2016**, *96*, 34–40. [CrossRef] [PubMed]
149. Carbamate Toxicosis in Animals—Toxicology. MSD Veterinary Manual. Available online: https://www.msdvetmanual.com/toxicology/insecticide-and-acaricide-organic-toxicity/carbamate-toxicosis-in-animals (accessed on 3 August 2025).
150. Australian Pesticides and Veterinary Medicines Authority. Aldicarb Chemical Review. Available online: https://www.apvma.gov.au/chemicals-and-products/chemical-review/listing/aldicarb (accessed on 23 September 2025).
151. World Health Organization. *The WHO Recommended Classification of Pesticides by Hazard and Guidelines to Classification 2019*; World Health Organization: Geneva, Switzerland, 2020.
152. US EPA. Label Review Manual—Chapter 7. 2013. Available online: https://www.epa.gov/pesticide-registration/label-review-manual (accessed on 3 August 2025).
153. ECHA. Guidance on Labelling and Packaging in Accordance with Regulation (EC) No 1272/2008. Available online: https://echa.europa.eu/guidance-documents/guidance-on-clp (accessed on 3 August 2025).
154. EFSA. Conclusion Regarding the Peer Review of the Pesticide Risk Assessment of the Active Substance Carbaryl | EFSA 2006. Available online: https://www.efsa.europa.eu/en/efsajournal/pub/rn-80 (accessed on 23 September 2025).
155. Conclusion on Pesticide Peer Review Regarding the Risk Assessment of the Active Substance Carbofuran | EFSA 2009. Available online: https://www.efsa.europa.eu/en/efsajournal/pub/rn-310 (accessed on 23 September 2025).
156. European Food Safety Authority (EFSA); Charistou, A.; Coja, T.; Craig, P.; Hamey, P.; Martin, S.; Sanvido, O.; Chiusolo, A.; Colas, M.; Istace, F. Guidance on the assessment of exposure of operators, workers, residents and bystanders in risk assessment of plant protection products. *EFSA J.* **2022**, *20*, e07032. [CrossRef]
157. Matošević, A.; Bosak, A. Carbamate group as structural motif in drugs: A review of carbamate derivatives used as therapeutic agents. *Arch. Ind. Hyg. Toxicol.* **2020**, *71*, 285–299. [CrossRef] [PubMed]
158. King, A.M.; Aaron, C.K. Organophosphate and Carbamate Poisoning. *Emerg. Med. Clin. N. Am.* **2015**, *33*, 133–151. [CrossRef]
159. Eyer, P. The Role of Oximes in the Management of Organophosphorus Pesticide Poisoning. *Toxicol. Rev.* **2003**, *22*, 165–190. [CrossRef]
160. Taylor, P. Anticholinesterase Agents. In *Goodman & Gilman's: The Pharmacological Basis of Therapeutics*, 13th ed.; Brunton, L.L., Hilal-Dandan, R., Knollmann, B.C., Eds.; McGraw-Hill Education: New York, NY, USA, 2017.
161. Rosman, Y.; Makarovsky, I.; Bentur, Y.; Shrot, S.; Dushnistky, T.; Krivoy, A. Carbamate poisoning: Treatment recommendations in the setting of a mass casualties event. *Am. J. Emerg. Med.* **2009**, *27*, 1117–1124. [CrossRef] [PubMed]
162. Dart, R.C. *Medical Toxicology*; Lippincott Williams & Wilkins: Philadelphia, PA, USA, 2004.

163. USA. Code of Federal Regulations 156.62—Toxicity Category. Available online: https://www.ecfr.gov/current/title-40/part-156/section-156.62 (accessed on 14 September 2025).
164. Silberman, J.; Taylor, A. Carbamate Toxicity. In *StatPearls*; StatPearls Publishing: Treasure Island, FL, USA, 2025.
165. Lee, D.H.; Jung, K.Y.; Choi, Y.H.; Cheon, Y.J. Body mass index as a prognostic factor in organophosphate-poisoned patients. *Am. J. Emerg. Med.* **2014**, *32*, 693–696. [CrossRef]
166. Saadeh, A.M.; Al-Ali, M.K.; Farsakh, N.A.; Ghani, M.A. Clinical and Sociodemographic Features of Acute Carbamate and Organophosphate Poisoning: A Study of 70 Adult Patients in North Jordan. *J. Toxicol. Clin. Toxicol.* **1996**, *34*, 45–51. [CrossRef]
167. Kumar, S.; Baggi, T.R.; Zughaibi, T. Forensic toxicological and analytical aspects of carbamate poisoning—A review. *J. Forensic Leg. Med.* **2022**, *92*, 102450. [CrossRef]
168. Zhang, J.; Guo, J.; Lu, D.; Qi, X.; Chang, X.; Wu, C.; Zhang, Y.; Liang, W.; Fang, X.; Cao, Y.; et al. Maternal urinary carbofuranphenol levels before delivery and birth outcomes in Sheyang Birth Cohort. *Sci. Total Environ.* **2018**, *625*, 1667–1672. [CrossRef]
169. Zhang, J.; Guo, J.; Wu, C.; Qi, X.; Jiang, S.; Lu, D.; Feng, C.; Liang, W.; Chang, X.; Zhang, Y.; et al. Exposure to carbamate and neurodevelopment in children: Evidence from the SMBCS cohort in China. *Environ. Res.* **2019**, *177*, 108590. [CrossRef]
170. Zhang, J.; Guo, J.; Wu, C.; Qi, X.; Jiang, S.; Zhou, T.; Xiao, H.; Li, W.; Lu, D.; Feng, C.; et al. Early-life carbamate exposure and intelligence quotient of seven-year-old children. *Environ. Int.* **2020**, *145*, 106105. [CrossRef]
171. Zhang, J.; Guo, J.; Wu, C.; Qi, X.; Jiang, S.; Lv, S.; Lu, D.; Liang, W.; Chang, X.; Zhang, Y.; et al. Carbamate pesticides exposure and delayed physical development at the age of seven: Evidence from the SMBCS study. *Environ. Int.* **2022**, *160*, 107076. [CrossRef]
172. Burgess, J.L.; Bernstein, J.N.; Hurlbut, K. Aldicarb Poisoning: A Case Report with Prolonged Cholinesterase Inhibition and Improvement After Pralidoxime Therapy. *Arch. Intern. Med.* **1994**, *154*, 221–224. [CrossRef]
173. Jaga, K.; Dharmani, C. Sources of exposure to and public health implications of organophosphate pesticides. *Rev. Panam. Salud Publica* **2003**, *14*, 171–185. [CrossRef] [PubMed]
174. Rathnayake, L.K.; Northrup, S.H. Structure and mode of action of organophosphate pesticides: A computational study. *Comput. Theor. Chem.* **2016**, *1088*, 9–23. [CrossRef]
175. Alvarez-Alvarez, I.; Andrade, R.J. Organophosphate pesticides: Another silent liver hazard? *Liver Int.* **2023**, *43*, 268–270. [CrossRef]
176. Sidhu, G.K.; Singh, S.; Kumar, V.; Dhanjal, D.S.; Datta, S.; Singh, J. Toxicity, monitoring and biodegradation of organophosphate pesticides: A review. *Crit. Rev. Environ. Sci. Technol.* **2019**, *49*, 1135–1187. [CrossRef]
177. Coman, G.; Farcas, A.; Matei, A.V.; Florian, C. Pesticides Mechanisms of Action in Living Organisms. In *Environmental Security Assessment and Management of Obsolete Pesticides in Southeast Europe*; Simeonov, L.I., Macaev, F.Z., Simeonova, B.G., Eds.; Springer: Dordrecht, The Netherlands, 2013; pp. 173–184. [CrossRef]
178. Mdeni, N.L.; Adeniji, A.O.; Okoh, A.I.; Okoh, O.O. Analytical Evaluation of Carbamate and Organophosphate Pesticides in Human and Environmental Matrices: A Review. *Molecules* **2022**, *27*, 618. [CrossRef]
179. Karalliedde, L.D.; Edwards, P.; Marrs, T.C. Variables influencing the toxic response to organophosphates in humans. *Food Chem. Toxicol.* **2003**, *41*, 1–13. [CrossRef]
180. Mileson, B.E.; Chambers, J.E.; Chen, W.L.; Dettbarn, W.; Ehrich, M.; Eldefrawi, A.T.; Gaylor, D.W.; Hamernik, K.; Hodgson, E.; Karczmar, A.G.; et al. Common Mechanism of Toxicity: A Case Study of Organophosphorus Pesticides. *Toxicol. Sci.* **1998**, *41*, 8–20. [CrossRef] [PubMed]
181. Costa, L.G. Current issues in organophosphate toxicology. *Clin. Chim. Acta* **2006**, *366*, 1–13. [CrossRef] [PubMed]
182. Androutsopoulos, V.P.; Kanavouras, K.; Tsatsakis, A.M. Role of paraoxonase 1 (PON1) in organophosphate metabolism: Implications in neurodegenerative diseases. *Toxicol. Appl. Pharmacol.* **2011**, *256*, 418–424. [CrossRef]
183. Morahan, J.M.; Yu, B.; Trent, R.J.; Pamphlett, R. A gene–environment study of the paraoxonase 1 gene and pesticides in amyotrophic lateral sclerosis. *NeuroToxicology* **2007**, *28*, 532–540. [CrossRef] [PubMed]
184. Lacasaña, M.; López-Flores, I.; Rodríguez-Barranco, M.; Aguilar-Garduño, C.; Blanco-Muñoz, J.; Pérez-Méndez, O.; Gamboa, R.; Gonzalez-Alzaga, B.; Bassol, S.; Cebrian, M.E. Interaction between organophosphate pesticide exposure and PON1 activity on thyroid function. *Toxicol. Appl. Pharmacol.* **2010**, *249*, 16–24. [CrossRef]
185. Wang, Y.; Chen, L.; Wang, C.; Hum, Y.; Gao, Y.; Zhou, Y.; Shi, R.; Zhang, Y.; Kamijima, M.; Ueyama, J.; et al. Association Between Organophosphate Pesticide Exposure and Thyroid Hormones in Pregnant Women. *Epidemiology* **2017**, *28*, S35. [CrossRef]
186. Silva, A.B.P.; Carreiró, F.; Ramos, F.; Sanches-Silva, A. The role of endocrine disruptors in female infertility. *Mol. Biol. Rep.* **2023**, *50*, 7069–7088. [CrossRef]
187. Diamanti-Kandarakis, E.; Bourguignon, J.-P.; Giudice, L.C.; Hauser, R.; Prins, G.S.; Soto, A.M.; Zoeller, R.T.; Gore, A.C. Endocrine-Disrupting Chemicals: An Endocrine Society Scientific Statement. *Endocr. Rev.* **2009**, *30*, 293–342. [CrossRef]
188. Ranjan, A.; Chauhan, A.; Jindal, T. In-Silico and In-Vitro evaluation of human acetylcholinesterase inhibition by organophosphates. *Environ. Toxicol. Pharmacol.* **2018**, *57*, 131–140. [CrossRef]

189. Uwaifo, F.; John-Ohimai, F. Dangers of Organophosphate Pesticide Exposure to Human Health. *Matrix Sci. Medica* **2020**, *4*, 27. [CrossRef]
190. Gupta, R.C.; Malik, J.K.; Milatovic, D. Chapter 37—Organophosphate and carbamate pesticides. In *Reproductive and Developmental Toxicology*; Gupta, R.C., Ed.; Academic Press: San Diego, CA, USA, 2011; pp. 471–486. [CrossRef]
191. Silvério, A.C.P.; Machado, S.C.; Azevedo, L.; Nogueira, D.A.; de Castro Graciano, M.M.; Simões, J.S.; Viana, A.L.M.; Martins, I. Assessment of exposure to pesticides in rural workers in southern of Minas Gerais, Brazil. *Environ. Toxicol. Pharmacol.* **2017**, *55*, 99–106. [CrossRef] [PubMed]
192. Kanavouras, K.; Tzatzarakis, M.N.; Mastorodemos, V.; Plaitakis, A.; Tsatsakis, A.M. A case report of motor neuron disease in a patient showing significant level of DDTs, HCHs and organophosphate metabolites in hair as well as levels of hexane and toluene in blood. *Toxicol. Appl. Pharmacol.* **2011**, *256*, 399–404. [CrossRef]
193. Hongsibsong, S.; Sittitoon, N.; Sapbamrer, R. Association of health symptoms with low-level exposure to organophosphates, DNA damage, AChE activity, and occupational knowledge and practice among rice, corn, and double-crop farmers. *J. Occup. Health* **2017**, *59*, 165–176. [CrossRef]
194. Ruckart, P.Z.; Kakolewski, K.; Bove, F.J.; Kaye, W.E. Long-term neurobehavioral health effects of methyl parathion exposure in children in Mississippi and Ohio. *Environ. Health Perspect.* **2004**, *112*, 46–51. [CrossRef]
195. Luo, F.; Sandhu, A.F.; Rungratanawanich, W.; Williams, G.E.; Akbar, M.; Zhou, S.; Song, B.-J.; Wang, X. Melatonin and Autophagy in Aging-Related Neurodegenerative Diseases. *Int. J. Mol. Sci.* **2020**, *21*, 7174. [CrossRef]
196. Chang, P.-A.; Wu, Y.-J. Motor neuron diseases and neurotoxic substances: A possible link? *Chem.-Biol. Interact.* **2009**, *180*, 127–130. [CrossRef] [PubMed]
197. IARC. *Some Drugs and Herbal Products*; IARC: Lyon, France, 2016.
198. Lukaszewicz-Hussain, A. Role of oxidative stress in organophosphate insecticide toxicity—Short review. *Pestic. Biochem. Physiol.* **2010**, *98*, 145–150. [CrossRef]
199. Wang, W.; Zhang, H.; Duan, X.; Feng, X.; Wang, T.; Wang, P.; Ding, M.; Zhou, X.; Liu, S.; Li, L.; et al. Association of genetic polymorphisms of miR-145 gene with telomere length in omethoate-exposed workers. *Ecotoxicol. Environ. Saf.* **2019**, *172*, 82–88. [CrossRef]
200. Ledda, C.; Cannizzaro, E.; Cinà, D.; Filetti, V.; Vitale, E.; Paravizzini, G.; Di Naso, C.; Iavicoli, I.; Rapisarda, V. Oxidative stress and DNA damage in agricultural workers after exposure to pesticides. *J. Occup. Med. Toxicol.* **2021**, *16*, 1. [CrossRef]
201. Frazier, L.M. Reproductive Disorders Associated with Pesticide Exposure. *J. Agromedicine* **2007**, *12*, 27–37. [CrossRef]
202. Padungtod, C.; Savitz, D.A.; Overstreet, J.W.; Christiani, D.C.; Ryan, L.M.; Xu, X. Occupational Pesticide Exposure and Semen Quality Among Chinese Workers. *J. Occup. Environ. Med.* **2000**, *42*, 982–992. [CrossRef]
203. Venkidasamy, B.; Subramanian, U.; Samynathan, R.; Rajakumar, G.; Shariati, M.A.; Chung, I.-M.; Thiruvengadam, M. Organopesticides and fertility: Where does the link lead to? *Environ. Sci. Pollut. Res.* **2021**, *28*, 6289–6301. [CrossRef] [PubMed]
204. Peiris-John, R.J.; Wickremasinghe, R. Impact of low-level exposure to organophosphates on human reproduction and survival. *Trans. R. Soc. Trop. Med. Hyg.* **2008**, *102*, 239–245. [CrossRef] [PubMed]
205. AL-Hussaini, T.K.; Abdelaleem, A.A.; Elnashar, I.; Shabaan, O.M.; Mostafa, R.; El-Baz, M.A.H.; El-Deek, S.E.; Farghaly, T.A. The effect of follicullar fluid pesticides and polychlorinated biphenyls concentrations on intracytoplasmic sperm injection (ICSI) embryological and clinical outcome. *Eur. J. Obstet. Gynecol. Reprod. Biol.* **2018**, *220*, 39–43. [CrossRef]
206. Melgarejo, M.; Mendiola, J.; Koch, H.M.; Moñino-García, M.; Noguera-Velasco, J.A.; Torres-Cantero, A.M. Associations between urinary organophosphate pesticide metabolite levels and reproductive parameters in men from an infertility clinic. *Environ. Res.* **2015**, *137*, 292–298. [CrossRef]
207. Rothlein, J.; Rohlman, D.; Lasarev, M.; Phillips, J.; Muniz, J.; McCauley, L. Organophosphate Pesticide Exposure and Neurobehavioral Performance in Agricultural and Nonagricultural Hispanic Workers. *Environ. Health Perspect.* **2006**, *114*, 691–696. [CrossRef]
208. Kraus, J.F.; Richards, D.M.; Borhani, N.O.; Mull, R.; Kilgore, W.W.; Winterlin, W. Physiological response to organophosphate residues in field workers. *Arch. Environ. Contam. Toxicol.* **1977**, *5*, 471–485. [CrossRef]
209. Pesticides Use, Pesticides Trade and Pesticides Indicators. Global, Regional and Country Trends, 1990–2020. Statistics. Available online: https://www.fao.org/statistics/highlights-archive/highlights-detail/New-FAOSTAT-data-release-Pesticides-use-pesticides-trade-and-pesticides-indicators-Global-regional-and-country-trends-1990-2020/ (accessed on 1 August 2025).
210. Ranjan, A.; Jindal, T. Overview of Organophosphate Compounds. In *Toxicology of Organophosphate Poisoning: New Insights*; Ranjan, A., Jindal, T., Eds.; Springer International Publishing: Cham, Switzerland, 2022; pp. 1–25. [CrossRef]
211. Cocker, J.; Mason, H.J.; Garfitt, S.J.; Jones, K. Biological monitoring of exposure to organophosphate pesticides. *Toxicol. Lett.* **2002**, *134*, 97–103. [CrossRef] [PubMed]

212. Karthick Rajan, D.; Mohan, K.; Rajarajeswaran, J.; Divya, D.; Thanigaivel, S.; Zhang, S. Toxic effects of organophosphate pesticide monocrotophos in aquatic organisms: A review of challenges, regulations and future perspectives. *Environ. Res.* **2024**, *244*, 117947. [CrossRef]
213. Li, A.J.; Kannan, K. Urinary concentrations and profiles of organophosphate and pyrethroid pesticide metabolites and phenoxyacid herbicides in populations in eight countries. *Environ. Int.* **2018**, *121*, 1148–1154. [CrossRef]
214. Oates, L.; Cohen, M.; Braun, L.; Schembri, A.; Taskova, R. Reduction in urinary organophosphate pesticide metabolites in adults after a week-long organic diet. *Environ. Res.* **2014**, *132*, 105–111. [CrossRef] [PubMed]
215. Colnot, T.; Dekant, W. Approaches for grouping of pesticides into cumulative assessment groups for risk assessment of pesticide residues in food. *Regul. Toxicol. Pharmacol.* **2017**, *83*, 89–99. [CrossRef]
216. Katsikantami, I.; Colosio, C.; Alegakis, A.; Tzatzarakis, M.N.; Vakonaki, E.; Rizos, A.K.; Sarigiannis, D.A.; Tsatsakis, A.M. Estimation of daily intake and risk assessment of organophosphorus pesticides based on biomonitoring data—The internal exposure approach. *Food Chem. Toxicol.* **2019**, *123*, 57–71. [CrossRef]
217. FAO; WHO. Pesticides Index—CODEX ALIMENTARIUS FAO-WHO. Available online: https://www.fao.org/fao-who-codexalimentarius/codex-texts/dbs/pestres/pesticides/en/ (accessed on 14 September 2025).
218. Authority (EFSA) EFS; Bellisai, G.; Bernasconi, G.; Binaglia, M.; Carrasco Cabrera, L.; Castellan, I.; Castoldi, A.F.; Chiusolo, A.; Chukwubike, K.; Crivellente, F.; et al. Targeted review of maximum residue levels (MRLs) for diazinon. *EFSA J.* **2023**, *21*, e08426. [CrossRef] [PubMed]
219. Zarei-Choghan, M.; Jorfi, S.; Saki, A.; Jaafarzadeh, N. Spatial distribution, ecological and health risk assessment of organophosphorus pesticides identified in the water of Naseri artificial wetland, Iran. *Mar. Pollut. Bull.* **2022**, *179*, 113643. [CrossRef]
220. López-Benítez, A.; Guevara-Lara, A.; Domínguez-Crespo, M.A.; Andraca-Adame, J.A.; Torres-Huerta, A.M. Concentrations of Organochlorine, Organophosphorus, and Pyrethroid Pesticides in Rivers Worldwide (2014–2024): A Review. *Sustainability* **2024**, *16*, 8066. [CrossRef]
221. Ferreira, V.B.; Silva, T.T.C.; Garcia, S.R.M.C.; Srur, A.U.O.S. Estimativa de ingestão de agrotóxicos organofosforados pelo consumo de frutas e hortaliças. *Cad. Saúde Colet.* **2018**, *26*, 216–221. [CrossRef]
222. Df, B. *Diretrizes para o Minitoramento de Agrotóxicos em Água para Consumo Humano*; Ministério da Saúde: Brasília, DF, Brazil, 2024.
223. Ding, G.; Wang, P.; Tian, Y.; Zhang, J.; Gao, Y.; Wang, X.; Shi, R.; Wang, G.; Shen, X. Organophosphate Pesticide Exposure and Neurodevelopment in Young Shanghai Children. *Environ. Sci. Technol.* **2012**, *46*, 2911–2917. [CrossRef]
224. Yu, R.; Liu, Q.; Liu, J.; Wang, Q.; Wang, Y. Concentrations of organophosphorus pesticides in fresh vegetables and related human health risk assessment in Changchun, Northeast China. *Food Control* **2016**, *60*, 353–360. [CrossRef]
225. Chen, Z.; Xu, Y.; Li, N.; Qian, Y.; Wang, Z.; Liu, Y. A national-scale cumulative exposure assessment of organophosphorus pesticides through dietary vegetable consumption in China. *Food Control* **2019**, *104*, 34–41. [CrossRef]
226. U.S. Department of Agriculture. Pesticide Data Program | Agricultural Marketing Service. Available online: https://www.ams.usda.gov/datasets/pdp (accessed on 3 August 2025).
227. Neff, R.A.; Hartle, J.C.; Laestadius, L.I.; Dolan, K.; Rosenthal, A.C.; Nachman, K.E. A comparative study of allowable pesticide residue levels on produce in the United States. *Glob. Health* **2012**, *8*, 2. [CrossRef]
228. Liang, C.-P.; Sack, C.; McGrath, S.; Cao, Y.; Thompson, C.J.; Robin, L.P. US Food and Drug Administration regulatory pesticide residue monitoring of human foods: 2009–2017. *Food Addit. Contam. Part A* **2021**, *38*, 1520–1538. [CrossRef]
229. CDC. Biomonitoring Data Tables for Environmental Chemicals | CDC 2024. Available online: https://www.cdc.gov/exposurereport/data_tables.html (accessed on 14 September 2025).
230. Witczak, A.; Pohoryło, A.; Abdel-Gawad, H.; Cybulski, J. Residues of some organophosphorus pesticides on and in fruits and vegetables available in Poland, an assessment based on the European union regulations and health assessment for human populations. *Phosphorus Sulfur Silicon Relat. Elem.* **2018**, *193*, 711–720. [CrossRef]
231. Wongta, A.; Sawang, N.; Tongjai, P.; Jatiket, M.; Hongsibsong, S. The Assessment of Organophosphate Pesticide Exposure among School Children in Four Regions of Thailand: Analysis of Dialkyl Phosphate Metabolites in Students' Urine and Organophosphate Pesticide Residues in Vegetables for School Lunch. *Toxics* **2022**, *10*, 434. [CrossRef] [PubMed]
232. Herrera-Moreno, J.F.; Medina-Díaz, I.M.; Bernal-Hernández, Y.Y.; Barrón-Vivanco, B.S.; González-Arias, C.A.; Moreno-Godínez, M.E.; Verdín-Betancourt, F.A.; Sierra-Santoyo, A.; Rojas-García, A.E. Organophosphorus pesticide exposure biomarkers in a Mexican population. *Environ. Sci. Pollut. Res.* **2021**, *28*, 50825–50834. [CrossRef] [PubMed]
233. Magdaleno-Magniales, U.; Salas-Espinoza, E.A.; Saldaña-Villanueva, K.; Núñez-Mojica, G.; García-Díaz, J.M.; Gaspar-Ramírez, O. Determination of highly hazardous pesticides in fruits and vegetables in the Maya region of southeast of Mexico. *J. Environ. Sci. Health Part B* **2025**, *60*, 103–110. [CrossRef]
234. Shaker, E.M.; Elsharkawy, E.E. Organochlorine and organophosphorus pesticide residues in raw buffalo milk from agroindustrial areas in Assiut, Egypt. *Environ. Toxicol. Pharmacol.* **2015**, *39*, 433–440. [CrossRef]

235. Lerro, C.C.; Koutros, S.; Andreotti, G.; Friesen, M.C.; Alavanja, M.C.; Blair, A.; Hoppin, J.A.; Sandler, D.P.; Lubin, J.H.; Ma, X.; et al. Organophosphate insecticide use and cancer incidence among spouses of pesticide applicators in the Agricultural Health Study. *Occup. Environ. Med.* **2015**, *72*, 736–744. [CrossRef] [PubMed]
236. Tang, F.H.M.; Lenzen, M.; McBratney, A.; Maggi, F. Risk of pesticide pollution at the global scale. *Nat. Geosci.* **2021**, *14*, 206–210. [CrossRef]
237. Balali-Mood, M.; Balali-Mood, K.; Moodi, M.; Balali-Mood, B. Health Aspects of Organophosphorous Pesticides in Asian Countries. *Iran. J. Public Health* **2012**, *41*, 1–14.
238. Bureau, D. Global Pesticides Market to Reach $181.2 Billion by 2031, Report. Global Agriculture 2025. Available online: https://www.global-agriculture.com/crop-protection/global-pesticides-market-to-reach-181-2-billion-by-2031-report/ (accessed on 23 September 2025).
239. Consulting FBI. Glyphosate Market Size & Share Forecasts: America, Europe, & APAC 2025–2034. Fundamental Business Insights and Consulting 2025. Available online: https://www.fundamentalbusinessinsights.com/industry-report/glyphosate-market-12216 (accessed on 23 September 2025).
240. FAO. Pesticides Use—FAOSTAT. Available online: https://www.fao.org/faostat/en/#data/RP/visualize (accessed on 23 September 2025).
241. Schütte, G.; Eckerstorfer, M.; Rastelli, V.; Reichenbecher, W.; Restrepo-Vassalli, S.; Ruohonen-Lehto, M.; Saucy, A.-G.W.; Mertens, M. Herbicide resistance and biodiversity: Agronomic and environmental aspects of genetically modified herbicide-resistant plants. *Environ. Sci. Eur.* **2017**, *29*, 5. [CrossRef] [PubMed]
242. Dowler, C. UK and EU Sent Thousands of Tonnes of Banned Pesticides Poor Countries. Unearthed 2020. Available online: https://unearthed.greenpeace.org/2020/09/10/banned-pesticides-eu-export-poor-countries/ (accessed on 23 September 2025).
243. DDT, DDE, DDD | Toxicological Profile | ATSDR. Available online: https://wwwn.cdc.gov/TSP/ToxProfiles/ToxProfiles.aspx?id=81&tid=20 (accessed on 23 September 2025).
244. US EPA. EPA Works to Ensure that Pesticides Will Not Harm Human Health or the Environment 2015. Available online: https://www.epa.gov/pesticides (accessed on 23 September 2025).
245. Mie, A.; Rudén, C. Non-disclosure of developmental neurotoxicity studies obstructs the safety assessment of pesticides in the European Union. *Environ. Health* **2023**, *22*, 44. [CrossRef] [PubMed]
246. Tudi, M.; Yang, L.; Wang, L.; Lv, J.; Gu, L.; Li, H.; Peng, W.; Yu, Q.; Ruan, H.; Li, Q.; et al. Environmental and Human Health Hazards from Chlorpyrifos, Pymetrozine and Avermectin Application in China under a Climate Change Scenario: A Comprehensive Review. *Agriculture* **2023**, *13*, 1683. [CrossRef]
247. Gaboardi, S.C.; Candiotto, L.Z.P.; Panis, C. Agribusiness in Brazil and its dependence on the use of pesticides. *Hyg. Environ. Health Adv.* **2023**, *8*, 100080. [CrossRef]
248. Li, B.A.; Li, B.M.; Bao, Z.; Li, Q.; Xing, M.; Li, B. Dichlorodiphenyltrichloroethane for Malaria and Agricultural Uses and Its Impacts on Human Health. *Bull. Environ. Contam. Toxicol.* **2023**, *111*, 45. [CrossRef] [PubMed]
249. European Human Biomonitoring Initiative | HBM4EU | Project | Fact Sheet | H2020. CORDIS | European Commission. Available online: https://cordis.europa.eu/project/id/733032 (accessed on 23 September 2025).
250. Huang, Y.; Li, Z. Global mapping of freshwater contamination by pesticides and implications for agriculture and water resource protection. *iScience* **2025**, *28*, 112861. [CrossRef]
251. Muir, D.; Sverko, E. Analytical methods for PCBs and organochlorine pesticides in environmental monitoring and surveillance: A critical appraisal. *Anal. Bioanal. Chem.* **2006**, *386*, 769–789. [CrossRef]
252. Henríquez-Hernández, L.A.; Macías-Montes, A.; Acosta-Dacal, A.; Rial-Berriel, C.; Duarte-Lopes, E.; Lopes-Ribeiro, A.L.; Alfama, P.M.; Livramento, M.; Zumbado, M.; Díaz-Díaz, R.; et al. Human biomonitoring of persistent and non-persistent pollutants in a representative sample of the general population from Cape Verde: Results from the PERVEMAC-II study. *Environ. Pollut.* **2022**, *306*, 119331. [CrossRef]

Journal of *Xenobiotics*

MDPI

Systematic Review

A Systematic Review on the Toxicology of European Union-Approved Triazole Fungicides in Cell Lines and Mammalian Models

Constantina-Bianca Vulpe [1,2], Adina-Daniela Iachimov-Datcu [2,3,*], Andrijana Pujicic [2,3] and Bianca-Vanesa Agachi [2,3]

1 Department of Scientific Research in Biology, Advanced Environmental Research Institute, West University of Timisoara, Oituz 4, 300086 Timisoara, Romania; constantina.vulpe@e-uvt.ro

2 The Academy of Romanian Scientists, Ilfov 3, 050044 Bucharest, Romania; andrijana.pujicic@e-uvt.ro (A.P.); bianca.agachi@e-uvt.ro (B.-V.A.)

3 Department of Biology, Faculty of Chemistry, Biology, Geography, West University of Timisoara, Pestalozzi 16, 300115 Timisoara, Romania

* Correspondence: daniela.iachimov@e-uvt.ro

Abstract: Triazole fungicides are widely used in agriculture but may pose risks to human health through occupational, accidental, or environmental exposure. This systematic review aimed to evaluate the toxicity of ten European Union-approved triazole fungicides in rodent models and cell lines. A total of 70 studies were included, reporting quantitative in vivo oral, dermal, or inhalation toxicity in mammals or quantitative in vitro cytotoxicity in human or mammalian cell lines; the exclusion criteria comprised publications not in English or not accessible. Literature searches were conducted in Web of Science, Google Scholar, and the Pesticide Properties DataBase (PPDB), and risk of bias in included studies was assessed using ToxRTool. Due to heterogeneity in study designs, reporting formats, and endpoints, data were synthesized descriptively. Quantitative endpoints included LD_{50}/LC_{50} values for in vivo studies and LOEC, IC_{50}, LC_{50}, and EC_{50} values for in vitro studies, while mechanistic endpoints highlighted apoptosis, oxidative stress, genotoxicity, and endoplasmic reticulum stress. Difenoconazole and tebuconazole were the most extensively studied compounds, whereas several triazoles had limited data. The limitations included heterogeneity of data and incomplete reporting, which restrict cross-study comparisons. Overall, the findings provide a comprehensive overview of potential human health hazards associated with EU-approved triazole fungicides and highlight critical knowledge gaps. The review was registered in Open Science Framework.

Keywords: fungicidal triazole class pesticides; human health toxicity; in vitro cytotoxicity; in vivo mammal studies; molecular mechanisms; ADME; network analysis

1. Introduction

Pesticides are nowadays considered essential compounds in sustainable agriculture, ensuring food security by supporting both the quality and quantity of agricultural yields. They comprise a broad range of substances with specific properties designed to protect crops from various pests, including unwanted plants (herbicides), animals (insecticides, rodenticides, and acaricides), and fungi (fungicides). Nevertheless, their extensive use also leads to continuous exposure of both the environment and humans to these compounds. Due to their diverse chemical structures and mechanisms of action, pesticides may exert harmful effects on human health and the environment, thereby impacting the holistic One Health concept [1,2].

Chemical food safety is a critical principle for public health at the global level. Various organizations, such as the World Health Organization, have developed principles and methods for assessing the risks posed by pesticides [3]. Furthermore, the status of chemical contamination in food—such as pesticide residues—is monitored periodically to maintain control over exposure [4].

At the European level, the European Union (EU) has developed one of the most detailed and comprehensive frameworks for ensuring chemical safety in food [5]. In this framework, pesticide use is regulated by EU Pesticide Regulation (EC) No. 1107/2009 [6] concerning the placing of plant protection products on the market, as well as EU Directive 2009/128/EC [7], regarding community action to achieve the sustainable use of pesticides. Other plans encompass regulations for organic farming and the adoption of integrated pest management strategies [8,9]. Maximum residue limits for pesticides in food and feed are established under Regulation (EC) No. 396/2005 [10], providing legally defined thresholds to ensure food safety. Furthermore, the European Union publishes an annual report on pesticide residues in food [11].

Fungicides are a class of pesticides widely used to protect agriculturally important plants from fungal diseases that can compromise both crop yield and product quality. Triazole fungicides can be classified as a group of compounds that contain a triazole ring in their chemical structures. Their mechanism of action is based on the inhibition of lanosterol 14α-demethylase, an essential enzyme in the ergosterol-biosynthesis pathway. By blocking this step, triazole fungicides disrupt fungal cell-membrane formation and impair growth. Evidence from in silico and in vitro studies shows strong binding of triazole derivatives to this enzyme, marked reductions in ergosterol content, and ERG11 gene downregulation in *Candida albicans*, confirming their sterol-biosynthesis-inhibitory activity [12–15].

Triazole fungicides are frequently utilized in agriculture to protect a broad range of economically important crops from fungal diseases. However, their extensive application also raises concerns regarding potential environmental contamination and human exposure. Considering that we are in the One Health Era, a thorough analysis of both environmental and human health effects is essential. Since triazole fungicides can have diverse impacts on the environment and human health, this study focuses on their potential effects on humans as part of the One Health framework. It is important to note that not all triazole fungicides are approved for use within the European Union. Therefore, this study focuses specifically on the triazole fungicides authorized for use in the EU.

There is an extensive body of review literature addressing various aspects of pesticides, which include topics such as their utility and importance in agriculture and human society [16,17], environmental and food contamination [18–22], analytical methods for their detection in environmental and food matrices [23–25], methods of remediation [26,27], human exposure [1,28–30], toxicological effects on both the environment [22,31–35] and human health (including mammalian models) [30,33–38], as well as regulatory perspectives [2,39]. However, although some of these reviews briefly mention triazole fungicides or provide examples from this class [2,16,17,20,21,25–27,30,33,39], only a few include a detailed analysis of fungicides [22,29,31,32,34,35], and, to the best of our knowledge, openly accessible papers with a detailed analysis specifically of triazole fungicides appear to be limited.

In terms of the literature on fungicides, the number of review studies is smaller than that for pesticides in general. These works cover various aspects of fungicides, including their mechanisms of action [40,41], effects on target human fungal pathogens [42], development of resistance [40], metabolism and environmental fate [43], methods of removal or degradation [43], environmental contamination and ecotoxicology [40,41,43–47], and impacts on human health [48,49]. Among these, only a limited number include or address triazole fungicides in detail [40,42,44], rather than focusing solely on a few examples from this class, while some reviews may not address them at all.

Review studies focusing primarily on triazole fungicides are limited in number. The few existing studies examine aspects such as their toxicity to target fungal species [50,51], environmental toxicity [12,52–54], and effects on human health and laboratory mammal models [55–58].

The limited number of review studies specifically addressing triazole fungicides is also reflected in the results obtained from the Web of Science database (as of 12 November 2025) [59]. Using the Advanced Search and Query Builder functions with the key search phrases (TI = (pesticide)) AND TI = (review), (TI = (fungicide)) AND TI = (review), and (TI = (triazole fungicide)) AND TI = (review), it was determined that articles containing the terms fungicide and review in their title represent only 3.6% of those containing pesticide and review. Moreover, articles including triazole fungicide and review account for merely 8.4% of those with fungicide and review, and only 0.3% of those with pesticide and review. In total, only six such articles were identified, of which only one report effects on human health or mammalian models [55].

Given the limited number of existing review studies specifically addressing triazole fungicides and the methodological constraints observed in those available—including the restricted number of compounds analyzed, the lack of quantitative and comparative in vivo–in vitro assessments, and the predominantly narrative approach without adherence to systematic frameworks such as PRISMA—the present review is both timely and necessary. It aims to systematically integrate and critically evaluate current data on triazole fungicides, encompassing different exposure types, toxicological endpoints, and experimental models. This comprehensive synthesis seeks to provide a more robust understanding of their potential effects on human health and the environment, contributing to the broader One Health perspective and supporting evidence-based regulatory and scientific advancements.

The scope of this study is to systematically evaluate the human health toxicity of European Union-approved triazole fungicides. This review integrates evidence from in vitro studies in mammalian cell lines and in vivo studies in rodent models, encompassing oral, dermal, and inhalation exposure routes. Both quantitative data, such as dose–response relationship values, and qualitative data, including mechanistic insights into toxicity pathways, are systematically evaluated in this review. This study stands out by providing a thorough and comparative assessment of the toxicological effects of EU-approved triazole fungicides, addressing a significant knowledge gap given that limited studies have comparatively assessed these compounds, and even fewer focus specifically on human health. The objectives of this study comprise:

1. Identification and characterization of the EU-approved triazole fungicides;
2. Identification of exposure data and toxicokinetic profile;
3. Evaluation of in vitro toxicity in mammalian cell lines;
4. Examination of potential molecular mechanisms of toxicity derived from in vitro data;
5. Review and analysis of in vivo toxicity data in rodent models, including oral, dermal, and inhalation exposures;
6. Assessment of qualitative evidence on specific and general human health issues linked to triazole fungicide exposure according to the Pesticide Properties DataBase records.

2. European Union-Approved Triazole Fungicides

For a fungicide to be approved in the European Union, the applicant must submit a complete dossier, and the active substance must meet the approval criteria laid out in Annex II of Regulation (EC) No 1107/2009. These criteria encompass assessments of human health impacts, environmental fate and behavior, ecotoxicology, and groundwater safety, each supported by specific threshold values or conditions that must be satisfied for approval to be granted. Upon expiry of the approval period, renewal is possible only if the active substance continues to comply with the current approval criteria in force at the time of re-evaluation [6].

A total of 35 triazole fungicides were analyzed with respect to their regulatory status within the European Union (Table 1). In this research, only triazole fungicides that remain approved for use according to current EU regulations were selected as compounds of interest [60].

Table 1. Triazole fungicides and their regulatory status within the European Union (N/A-not available).

Triazole Fungicide	Status of Approval	Approval Expiration Date
Azaconazole	Not approved	N/A
Bitertanol	Not approved	29 August 2013
Bromuconazole	Approved	30 April 2027
Cyproconazole	Not approved	31 May 2021
Difenoconazole	Approved	15 March 2026
Diniconazole	Not approved	N/A
Epoxiconazole	Not approved	30 April 2020
Etaconazole	Not approved	N/A
Fenbuconazole	Not approved	30 April 2021
Fluconazole	N/A	N/A
Fluotrimazole	N/A	N/A
Fluoxytioconazole	N/A	N/A
Fluquinconazole	Not approved	31 December 2021
Flusilazole	Not approved	N/A
Flutriafol	Not approved	31 May 2021
Furconazole	Not approved	N/A
Hexaconazole	Not approved	N/A
Imibenconazole	Not approved	N/A
Ipconazole	Not approved	31 May 2023
Ipfentrifluconazole	N/A	N/A
Mefentrifluconazole	Approved	20 March 2029
Metconazole	Approved	31 August 2031
Myclobutanil	Not approved	31 May 2021
Paclobutrazol	Approved	31 August 2026
Penconazole	Approved	15 October 2026
Propiconazole	Not approved	19 December 2018
Prothioconazole	Approved	31 March 2027
Quinconazole	N/A	N/A
Simeconazole	Not approved	N/A
Tebuconazole	Approved	15 August 2026
Tetraconazole	Approved	31 March 2027
Triadimefon	Not approved	N/A
Triadimenol	Not approved	31 August 2019
Tricyclazole	Not approved	N/A
Triticonazole	Approved	31 January 2027

Of the 35 triazole fungicides analyzed, 10 were found to have an approved status within the EU. These compounds represent the triazole fungicides of interest included in this review (Table 2).

Table 2. Molecular weight and 2D chemical structures of the EU-approved triazole fungicides (MW = molecular weight; data retrieved from PubChem [61]).

Bromuconazole (BRO)	Difenoconazole (DIF)
MW: 377.1 g/mol	MW: 406.3 g/mol
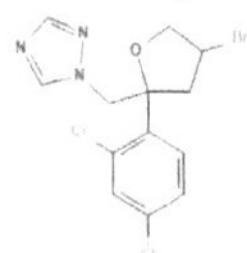	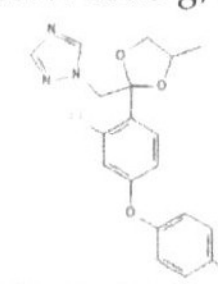
Mefentrifluconazole (MEF)	**Metconazole (MET)**
MW: 397.8 g/mol	MW: 319.8 g/mol
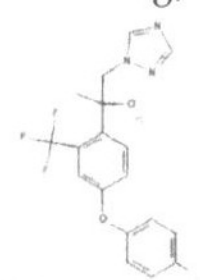	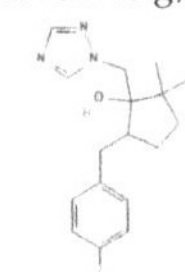
Paclobutrazol (PAC)	**Penconazole (PEN)**
MW: 293.79 g/mol	MW: 284.18 g/mol
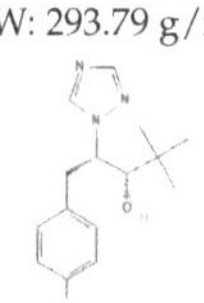	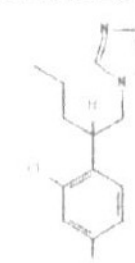
Prothioconazole (PRO)	**Tebuconazole (TEB)**
MW: 344.3 g/mol	MW: 307.82 g/mol
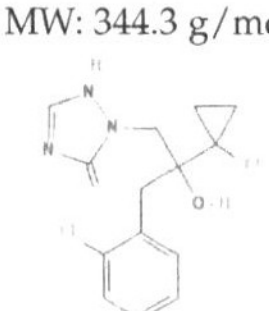	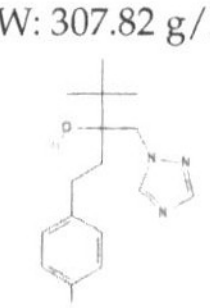
Tetraconazole (TET)	**Triticonazole (TRI)**
MW: 372.14 g/mol	MW: 317.8 g/mol
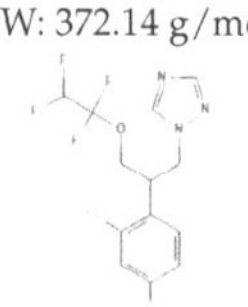	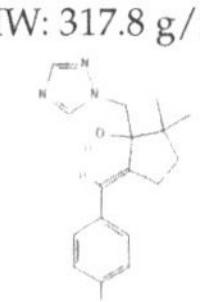

Although these ten triazole fungicides are approved for use within the European Union, they may still pose potential risks to human health through occupational exposure during their production or application, as well as through accidental or environmental exposure. An additional aggravating factor is the potential for improper handling by non-specialized individuals, which may result in accidental direct exposure during application or indirect exposure through environmental contamination. Therefore, understanding the potential toxicological effects of these fungicides on human health is essential.

3. Exposure Data and Toxicokinetic Profile of EU-Approved Triazole Fungicides

The human body may come into contact with triazole fungicides through several potential exposure routes, including oral exposure via the ingestion of contaminated food, dermal exposure through direct contact with concentrated formulations during handling, and inhalation of vapors released when these are applied in agriculture. However, the available scientific literature provides relatively limited information addressing and detailing these exposure pathways.

The Pesticide Properties DataBase includes information on parameters such as the Acceptable Daily Intake (ADI), Acute Acceptable Operator Exposure Level (AAOEL), and Acceptable Operator Exposure Level—Systemic (AOEL) [62]. Nevertheless, these data are primarily derived from studies conducted on mammalian models and are not available for all triazole fungicides currently approved for use within the European Union.

Data on ADI and AOEL are reported for all ten triazole fungicides of interest, ranging from 0.004 mg/kg bw/day for TET to 0.036 mg/kg bw/day for PRO in the case of ADI, and from 0.01 mg/kg bw/day for TRI to 0.16 mg/kg bw/day for DIF for AOEL. Data regarding AAOEL are not available for BRO, DIF, PAC, PEN, TEB, and TET; for the remaining compounds, AAOEL values range from 0.01 mg/kg bw/day for MET to 0.2 mg/kg bw/day for PRO [63–72].

Information on the toxicokinetic profile, encompassing absorption, distribution, metabolism, and excretion (ADME) of EU-approved triazole fungicides is quite limited, both in terms of experimental data and in silico predictions. Such information could not be identified for BRO, MEF and PEN.

Computational predictions using ADMETlab 2.0 and admetSAR 2.0 indicated that DIF has good bioavailability, high intestinal absorption, good oral bioavailability, high plasma protein binding, moderate clearance, and the ability to penetrate the blood–brain barrier [73]. Another in silico study, using FAFDrugs4 and SwissADME, indicated that MET, PAC, TEB, TET, and TRI exhibit good oral bioavailability, can be readily absorbed in the gastrointestinal tract, and are able to reach systemic circulation. These compounds are also predicted to have a moderate ability to penetrate the skin and to cross the blood–brain barrier, and may be metabolized by cytochrome P450 enzymes (CYPs). Additionally, TRI is predicted to be a P-glycoprotein (P-gp) substrate, suggesting potential active efflux both from the gastrointestinal tract back into the lumen and from the brain [74]. TRI was further evaluated, in another study, using ADMETlab 2.0, which predicted high intestinal absorption, moderate clearance, and high plasma protein binding [75].

For PRO, ADME data were derived from in vitro CYP inhibition assays and a physiologically-based pharmacokinetic (PBPK) model. These studies indicated no significant inhibition of CYP2C9, CYP2C19, or CYP3A substrates at dietary exposure levels corresponding to the ADI [76]. Another study indicated that the phase I metabolism of PRO is primarily mediated by CYP2C19 in the liver, based on in vitro experiments using human liver microsomes and recombinant human CYP450 enzymes to convert prothioconazole to its metabolite prothioconazole-desthio [77]. Although in silico data were identified, it should be taken into account that these predictions may differ from experimental results, as has been observed for organophosphates [78].

4. Literature Data-Mining and Analysis Workflow

Data mining from the scientific literature regarding the toxicity of triazole fungicides of interest on human health, as well as other relevant information, and the subsequent selection and analysis of these data, were conducted in accordance with PRISMA guidelines (Figure 1). The protocol for this systematic review was registered in Open Science Framework [79], registration DOI:10.17605/OSF.IO/NVZ5X.

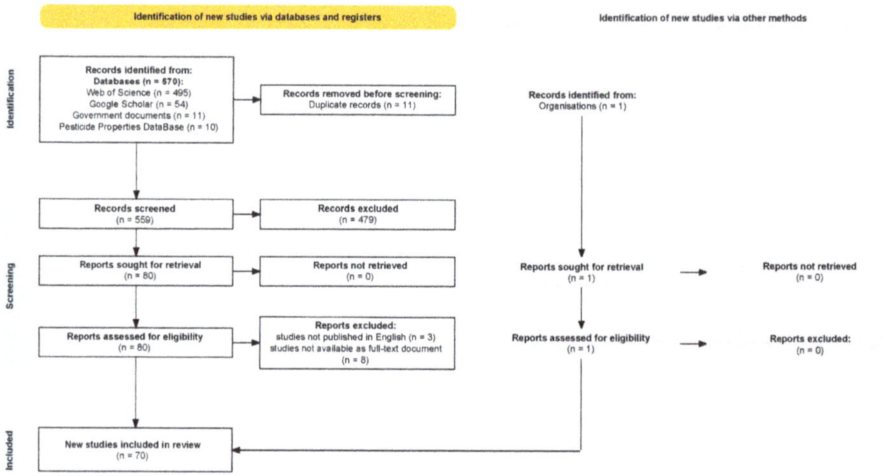

Figure 1. PRISMA diagram illustrating the systematic workflow for data acquisition, screening, eligibility assessment, and final inclusion in the analysis (diagram created using the PRISMA 2020 Flow Diagram Shiny App [80]).

The PICO (Population, Intervention/Exposure, Comparator, Outcomes) research question for this systematic review was: What are the toxicological effects of triazole fungicides in in vitro cell lines and in vivo mammalian models, including in vivo endpoints (LD_{50}, LC_{50}), in vitro endpoints (IC_{50}, EC_{50}, LC_{50}, LOEC), molecular mechanisms of cytotoxicity, and human health issues derived from the Pesticide Properties DataBase?

Following the PICO framework, studies were identified through extensive literature searches across multiple databases, including Web of Science (last search conducted on 12 November 2025), Google Scholar (last consulted on 5 November 2025), and Pesticide Properties DataBase (last searched on 7 November 2025).

The Web of Science (WoS) database [59] was specifically used to retrieve relevant data on the fungicides of interest. This search targeted both in vivo and in vitro studies related to human health, reporting quantitative toxicity measures such as the median lethal dose (LD_{50}), median lethal concentration (LC_{50}), half-maximal inhibitory concentration (IC_{50}), half-maximal effective concentration (EC_{50}), or lowest observed effect concentration (LOEC). The following query structure was used, where the generic term "fungicide" was replaced sequentially with the name of each of the ten fungicides of interest in the search strings: TS = (fungicide) AND ((TS = (in vivo) OR TS = (in vitro) OR TS = (human health*)) AND (TS = (LC_{50}) OR TS = (IC_{50}) OR TS = (EC_{50}) OR TS = (LD_{50}) OR TS = (LOEC))).

An additional search was performed in Google Scholar (GS) [81], where an exhaustive screening of relevant literature was undertaken. Keywords used in the search string included the names of the ten fungicides of interest combined with terms such as "in vivo", "rat", "mouse", "rabbit", "*Rattus norvegicus*", "*Mus musculus*", "*Oryctolagus cuniculus*", "LD_{50}", and "LC_{50}" for in vivo studies. For in vitro studies, the search terms included each fungicide name together with "in vitro", "cell", "cytotoxicity", "viability", and "IC_{50}". Depending on the relevance of the search results, articles were screened from the first pages displayed, prioritizing those that reported experimental toxicity data. In addition to peer-reviewed scientific literature, supplementary information sources such as governmental toxicological reports, Safety Data Sheets (SDS) provided by fungicide manufacturers, and records from Pesticide Properties DataBase (PPDB) [62] were also examined.

The screening and selection of studies were performed collaboratively by the review authors, with specific responsibilities assigned to each database. In the Web of Science (WoS) database, one reviewer analyzed the total number of records retrieved using the predefined search strategy and identified those meeting the established inclusion criteria. In Google Scholar (GS), in vitro studies were searched and screened by one reviewer, whereas in vivo studies were independently identified by another reviewer, both following the same eligibility criteria described above. Data retrieved from PPDB were collected by a different author, who extracted both quantitative and qualitative toxicological information relevant to the fungicides of interest. The screening was performed stepwise: titles were first analyzed for the presence of relevant keywords or related content, abstracts of potentially relevant records were then screened, and full texts were examined if they were deemed relevant to the field of interest. No automation tools were used in any stage of the screening or selection process.

Data extraction was performed by a single reviewer per study. Predefined spreadsheets were used to standardize the process. For in vivo studies, the spreadsheet included fields for the fungicide analyzed, toxicological endpoints (LD_{50} or LC_{50}), type of exposure, and the organism used. For in vitro studies, the spreadsheet included fields for the fungicide, tested concentrations, cell line, toxicological parameters (IC_{50}, LC_{50}, EC_{50}, or LOEC), and qualitative information on molecular mechanisms of cytotoxicity. For data extracted from PPDB, quantitative fields included Mammals—Acute oral LD_{50}, Mammals—Dermal LD_{50}, and Mammals—Inhalation LC_{50}, while qualitative fields included Specific human health issues and General human health issues. No automation tools were used at any stage of screening or data collection.

The outcomes of interest included quantitative toxicity measures and qualitative information relevant to human health. For in vivo studies, quantitative endpoints were LD_{50} (oral, dermal) and LC_{50} (inhalation), type of exposure, and the organism used. For in vitro studies, endpoints were IC_{50}, LC_{50}, EC_{50}, LOEC, tested concentrations, and cell line. Qualitative outcomes included molecular mechanisms of cytotoxicity (in vitro) and specific or general human health issues (PPDB). For each included study, all data corresponding to the predefined inclusion criteria were extracted and considered, with only clearly reported values used or values that could be standardized or normalized, such as through unit conversion or calculation from reported ratios. No other data were sought.

In total, 570 sources were identified, including scientific articles, 11 governmental reports (GRs), 1 Safety Data Sheet (SDS) from manufacturing companies and 10 reports from PPDB. Of these, 495 scientific articles were retrieved from WoS, and 54 from GS. The sources included 11 duplicates, resulting in 559 studies. Following this process, 479 studies were excluded because, although they were retrieved from WoS using the search phrases, they were not relevant to human health or did not report the quantitative data of interest.

Following the screening stage, 81 studies were selected for detailed analysis. Of these, 11 studies were excluded based on two criteria: studies were not published in English and some studies were not available as full-text documents. No study meeting the inclusion criteria was excluded unless it also met one or more of the exclusion criteria. Consequently, a total of 70 studies were included in this systematic review.

Of the 70 studies analyzed, 36 provided quantitative data from in vitro experiments on cell lines [82–117], while 34 reported quantitative data from in vivo studies in laboratory rodents (this number also includes governmental and company reports, as well as PPDB entries) [63–72,118–141] (Figure 2). Qualitative data were extracted exclusively from these 70 selected studies.

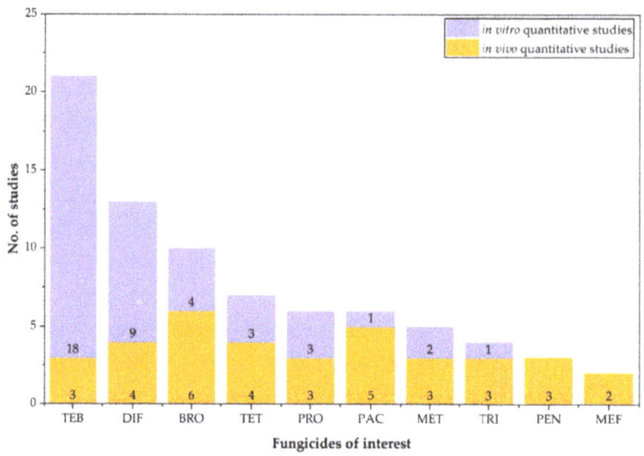

Figure 2. Number of studies analyzed for in vitro (light blue bar) and in vivo (light orange bar) quantitative data of the EU approved triazole fungicides toxicological studies (total counts include duplicate studies shown in both categories).

Data analysis involved normalization of measurement units where applicable. For this review, oral and dermal LD_{50} values from mammalian exposure studies were expressed in mg/kg body weight (BW), LC_{50} values from inhalation exposure studies in mg/L, and IC_{50}, LC_{50}, EC_{50}, and LOEC values from in vitro tests in micromolar (μM). Some toxicity endpoints reported as ratios were standardized by converting to their corresponding absolute values to allow comparison across studies. Studies were grouped into in vitro and in vivo categories for synthesis based on the experimental model used. This separation was made to account for differences in biological relevance and data comparability between test systems.

Studies were assigned to syntheses based on endpoint type, exposure route, and experimental model to ensure that only comparable data were grouped for analysis. For each outcome, the effect measures reported in the original studies were used. Values were extracted as presented in the articles, with minimal normalization or standardization applied where necessary. Very few studies reported measures of variability, such as standard deviations; therefore, the extracted data do not include this information.

Quantitative data for both in vivo and in vitro studies were tabulated and provided in Supplementary Material (Tables S1 and S3). Statistical analysis (nonparametric Kruskal-Wallis and Dunn tests), as well as data visualization were conducted using multiple tools, including OriginPro 2025 (OriginLab Corporation, Northampton, MA, USA) for the plotting and analysis of quantitative data, and Cytoscape (version 3.10.4) for the construction and visualization of network relationships between the analyzed fungicides and qualitative toxicity endpoints.

Quantitative data were analyzed and compared descriptively, as most studies did not report standard deviation values, preventing the conduct of group statistics with effect estimate or a meta-analysis. Heterogeneity was explored descriptively through comparative analysis of subgroups, as standard deviation values were largely missing and substantial differences existed in how outcomes were reported across studies. The risk of bias for each of the results was further examined using ToxRTool [142], by one reviewer. Risk of bias due to missing results was considered descriptively, as many studies reported only threshold values or lacked sufficient statistical information, preventing formal quantitative assessment. Formal certainty assessment (e.g., GRADE) was not performed due to high heterogeneity, different in vitro cell lines with few studies per fungicide, and sparse or threshold-based data in in vivo studies per exposure type, preventing reliable grading of confidence. Sensitivity analyses were not performed due to the limited number of comparable studies and the heterogeneity of the available data.

5. In Vitro Toxicity of EU-Approved Triazole Fungicides in Mammalian Cell Lines

The analysis of the selected articles allowed the identification of quantitative data for all fungicides of interest, with the exception of MEF and PEN, for which data gaps were observed. The in vitro data were obtained across a wide range of cell lines, including cancerous, transformed, and immortalized lines, derived from human and other mammalian species. The cell lines with the most studies and data across multiple fungicides were HepG2, KGN, and hGCs. In contrast, CHO, H295R, HepG2, and MCF-7 were the cell lines with the most data available for the same fungicide, specifically TEB (Figure 3).

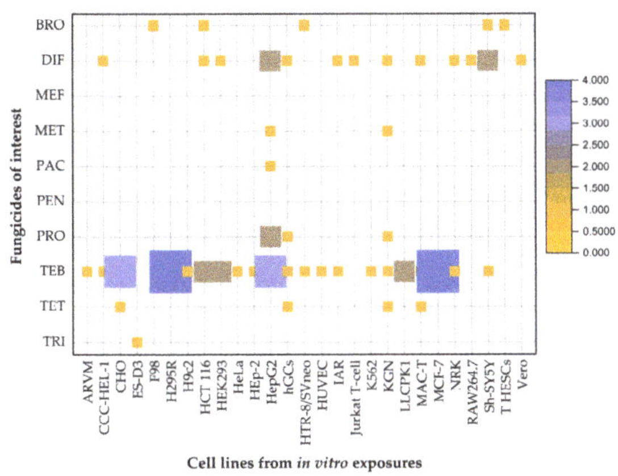

Figure 3. Matrix bubble plot depicting the number and categories of cell lines reported for each fungicide of interest. Larger and more blue bubbles represent a higher number of studies for the corresponding cell line and fungicide, while orange indicates fewer studies. The scale reflects study count both by color and bubble size. The fungicides included in the plot are: Bromuconazole (BRO), Difenoconazole (DIF), Mefentrifluconazole (MEF), Metconazole (MET), Paclobutrazol (PAC), Penconazole (PEN), Prothioconazole (PRO), Tebuconazole (TEB), Tetraconazole (TET), and Triticonazole (TRI). Cell lines represented in the plot are: adult rat ventricular myocytes (ARVM), human hepatocytes (CCC-HEL-1), Chinese hamster ovary cells (CHO), mouse embryonic multipotent stem cell line (ES-D3), rat glial-like cells (F98), human adrenocortical cell line (H295R), rat heart myoblast cells (H9c2), human colon cancer cell line (HCT 116), human renal cells (HEK293), human cervical carcinoma derived immortalized cell line (HeLa), human HeLa-contaminated laryngeal carcinoma-derived epithelial cells (HEp-2), human hepatocellular carcinoma-derived epithelial cells (HepG2), human primary granulosa cells (hGCs), human trophoblast immortalized cell line (HTR-8/SVneo), human endothelial cell line (HUVEC), rat liver epithelial cells (IAR), human T lymphocyte cell line (Jurkat T-cell), human leukemia-derived lymphoblast cells (K562), human granulosa line (KGN), pig kidney-derived epithelial cells (LLCPK1), cattle mammary alveolar epithelial cell (MAC-T), human breast cancer cell line (MCF-7), rat kidney cell line (NRK), mouse macrophage cell line (RAW264.7), human neuroblastoma cell line (SH-SY5Y), human uterine immortalized fibroblasts (T HESCs), African green monkey kidney-derived epithelial cells (Vero).

The in vitro data included IC_{50}, LC_{50}, EC_{50}, and LOEC values, which were used to generate forest-like plots, with the error bars representing the tested concentration ranges. DIF and TEB had the largest number of data points, whereas PAC and TRI each had only a single reported value (Figure 4).

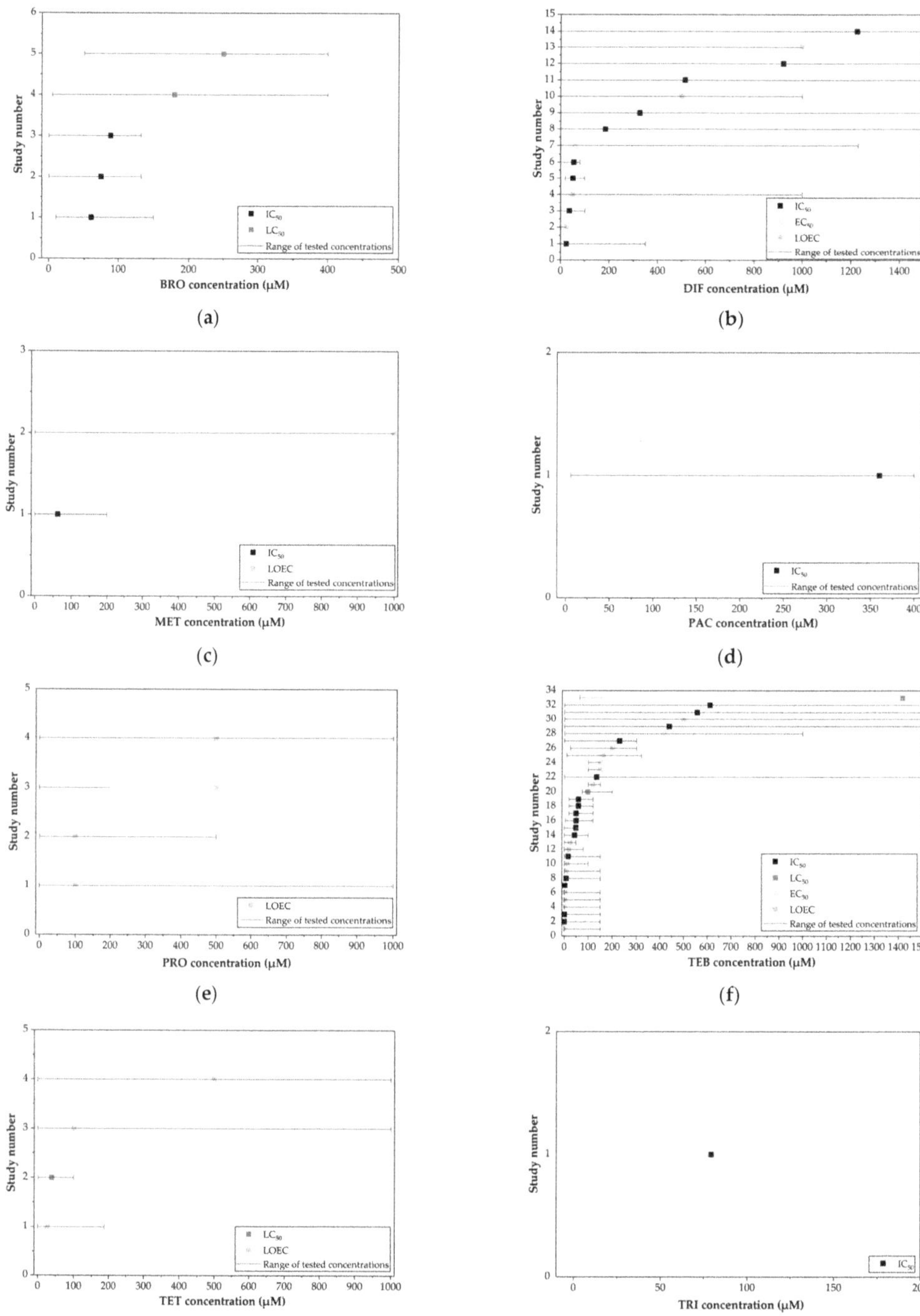

Figure 4. Forest-like plot of in vitro toxicity values (IC_{50}, LC_{50}, EC_{50}, LOEC) for EU-Approved Triazole Fungicides with Tested Concentration Ranges: (**a**) BRO, (**b**) DIF, (**c**) MET, (**d**) PAC, (**e**) PRO, (**f**) TEB, (**g**) TET and (**h**) TRI. The values shown in this figure correspond to data presented in Table S1 of the Supplementary Material.

The combined median values of the analyzed toxicological indices indicated that TET was the most cytotoxic, exhibiting the lowest value, whereas DIF was the least cytotoxic, with the highest value. The overall toxicity ranking was TET > MET > TRI > BRO > TEB > DIF. No comparable results could be identified for MEF, PEN, and PRO. In the in vitro studies, heterogeneity was primarily driven by the animal strain used, followed by inconsistencies in reporting formats (LOEC values versus toxicological indices).

Risk of bias analysis using the ToxRTool software (2009 version) [142] showed that only a limited number of studies met the criteria for classification into Class 1 (Supplementary Table S1). These included LOEC data for PRO in HepG2 cells, TEB in HepG2 cells, and TEB in SH-SY5Y cells, as well as toxicological indices for TEB in HeLa and HepG2 cells and TRI in ES-D3 cells. This indicates that only a small subset of the available evidence met high-quality methodological standards, suggesting that overall confidence in the body of evidence is constrained by variable study quality and potential sources of bias in the remaining studies. Potential risk of bias from missing results in in vitro studies was evaluated descriptively, but the diversity of endpoints and cell lines precluded statistical assessment.

6. Molecular Mechanisms Underlying In Vitro Toxicity of EU-Approved Triazole Fungicides

Most of the analyzed in vitro studies also assessed the mechanisms of toxicity by evaluating specific molecular markers. Among these molecular mechanisms, cellular apoptosis was the most frequently observed across the largest number of fungicides and studies, followed by oxidative stress. The fungicides for which the molecular mechanisms of cytotoxicity were most extensively investigated were DIF and TEB, whereas MEF, PEN, and TRI had no data regarding these aspects (Figure 5).

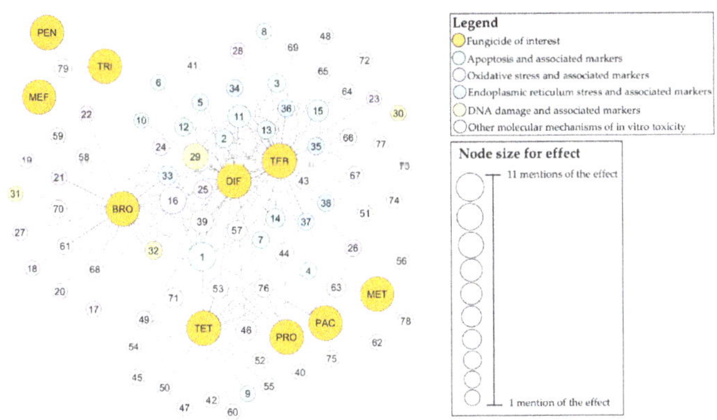

Figure 5. Cytoscape-generated network of EU-approved triazole fungicides and associated molecular mechanisms of in vitro cytotoxicity. Numbers shown on network nodes correspond to specific molecular-level effects, detailed in Table S2 of the Supplementary Material.

Regarding apoptosis, the most frequently assessed markers were the Bax/Bcl-2 ratio, reduction of mitochondrial membrane potential, and induction of poly (ADP-ribose) polymerase (PARP) cleavage. In addition, the expression levels of caspases 3, 7, 8, and 9 were also evaluated. For oxidative stress, reactive oxygen species (ROS) level was the most commonly measured parameter, alongside malondialdehyde (MDA), catalase (CAT), superoxid dismutase (SOD), glutathione peroxidase (GPX), and glutathione-S-transferase (GST). For endoplasmic reticulum (ER) stress, parameters included the expression of activating transcription factor 4 (ATF4), binding-immunoglobulin protein/glucose regulated protein78 (Bip/GRP78), C/EBP homologous protein (CHOP), and endoplasmic oxidoreductin-1-like (ERO1-Lα).

7. In Vivo Toxicity of EU-Approved Triazole Fungicides in Rodent Models

The analysis of the selected articles enabled the identification of quantitative in vivo toxicity data for all fungicides of interest, with the exception of MEF regarding oral exposure, for which a data gap was observed. The available data included LD_{50} values for oral and dermal exposures and LC_{50} values for inhalation studies. Most of the results were obtained from experiments conducted on rats, while a smaller number of studies involved mice (oral exposure) and rabbits (inhalation exposure) (Figure 6). For the in vivo data, only a limited number of exact categorical values were identified; therefore, the analysis and visualization also included data reported as being greater than a specific threshold value.

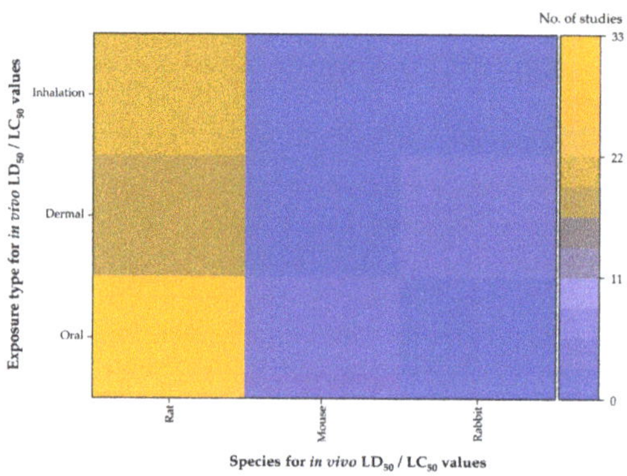

Figure 6. Heatmap visualization of in vivo LD_{50}/LC_{50} values for the analyzed triazole fungicides, organized by exposure type (oral, dermal, inhalation) and test species (rat, mouse, rabbit).

The oral LD_{50} values indicated that the most toxic fungicides for rodents were BRO, followed by TET, whereas the least toxic were PRO and TRI. BRO also had the highest number of studies reporting quantitative oral toxicity data. For fungicides with available data in both rats and mice, some difference between the LD_{50} values of the two species were observed (Figure 7).

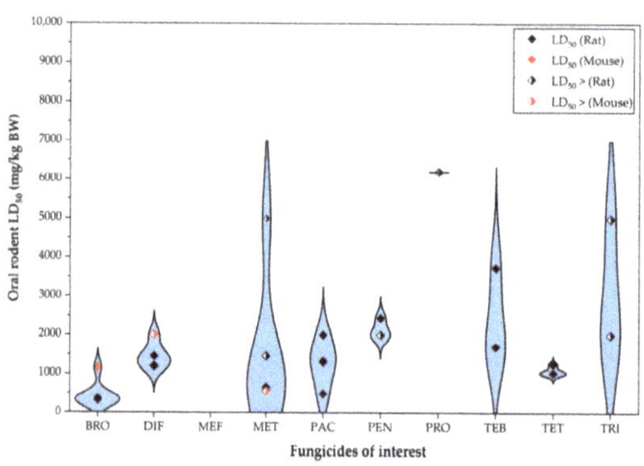

Figure 7. Violin plot illustrating the distribution of oral LD_{50} values for the triazole fungicides of interest. The values shown in this figure correspond to data presented in Table S3 of the Supplementary Material.

The dermal LD_{50} values were relatively similar among the analyzed fungicides, except for MEF and PEN, which showed higher LD_{50} values, indicating lower dermal toxicity compared to the others (Figure 8).

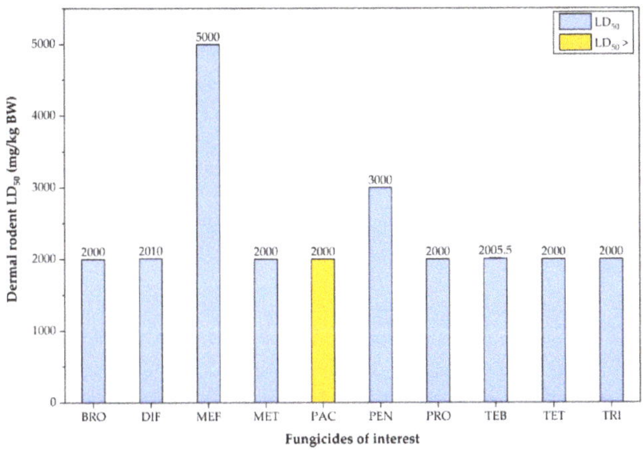

Figure 8. Column plot illustrating the distribution of dermal LD_{50} values for the triazole fungicides of interest. The values shown in this figure correspond to data presented in Table S3 of the Supplementary Material.

The LC_{50} values for inhalation exposure indicated higher inhalation toxicity for PAC, DIF, and TET, with recorded values of 3.13, 3.30, and 3.66 mg/L, respectively (Figure 9).

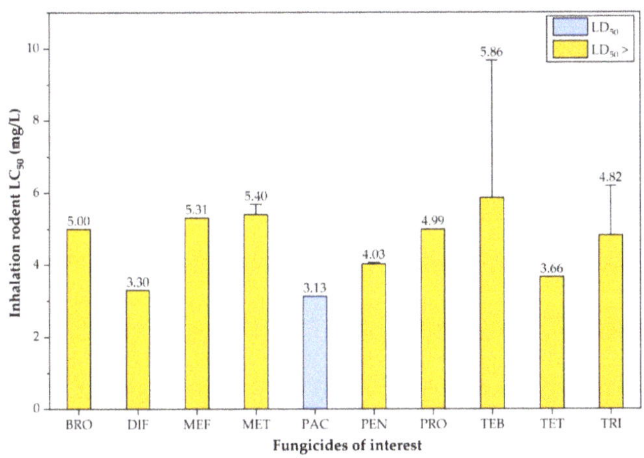

Figure 9. Column plot illustrating the distribution of inhalation LC_{50} values for the triazole fungicides of interest. The values shown in this figure correspond to data presented in Table S3 of the Supplementary Material.

The in vivo data indicated that BRO, DIF, PAC, and TET exhibit higher toxicity in rodents, whereas MEF, PEN, PRO, and TRI were associated with lower toxicity levels. The heterogeneity of the in vivo data was driven by differences in the reporting of toxicological indices—specifically the use of exact values versus threshold (greater than) values—as well as variation in the species used across studies.

Risk of bias assessment showed that all in vivo studies were classified as Class 3, indicating limited methodological detail and uncertainty regarding study quality. Risk of bias from missing results in in vivo studies was assessed descriptively; sparse and threshold-based data prevented statistical analysis. Statistical analysis further showed that significant differences were identified only within the in vivo dataset, specifically for oral toxicity values in rats. The analysis included only exact reported values, excluding threshold-type data. A Kruskal–Wallis test revealed statistically significant differences between fungicides, and post-hoc Dunn tests indicated significant differences between BRO and DIF, and between BRO and TEB, with BRO demonstrating higher toxicity in both comparisons.

8. Qualitative Human Health Effects Associated with EU-Approved Triazole Fungicide Exposure

Network analysis of the PPDB data regarding general and specific human health effects associated with the EU-approved triazole fungicides revealed marked differences in data availability. DIF and TEB were the most extensively characterized, with DIF linked to endocrine disruption, thyroid toxicity, liver toxicity, cardiotoxicity, and nephrotoxicity, and TEB associated with hematotoxicity, liver toxicity, eye irritation, and endocrine disruption. MET was reported only as a respiratory tract irritant, while PEN was noted solely for endocrine disruption. No specific human health data were available for BRO, PAC, PRO, or TRI (Figure 10).

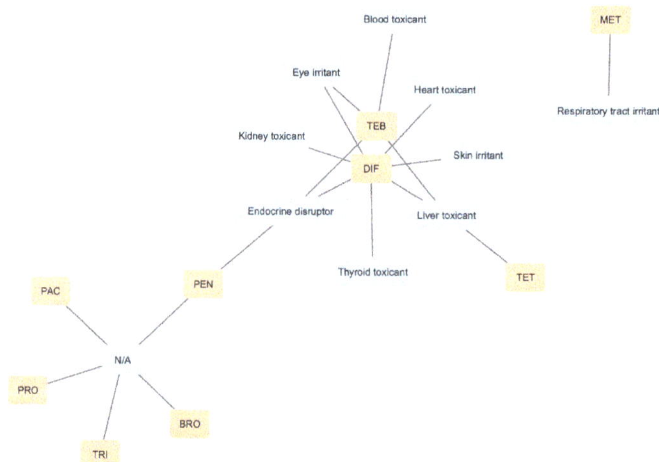

Figure 10. Cytoscape-generated network of EU-approved triazole fungicides and associated human health effects identified from PPDB database. Nodes representing the fungicides of interest are colored orange, while those corresponding to specific human health issues are shown in blue (N/A – not available).

This analysis highlights substantial gaps in human health hazard data for several EU-approved triazole fungicides and underscores the need for targeted toxicological investigation, particularly for those compounds with little to no recorded information.

9. Discussion

A previous review on the genotoxic effects of triazole fungicides [55] reported the potential of these compounds to induce DNA damage and oxidative stress in vivo, consistent with the patterns observed in our analysis of in vitro data. In addition to these findings, our review addresses EU-approved triazole fungicides and integrates evidence from in vitro cytotoxicity, mechanistic studies, and toxicokinetic data, alongside in vivo results. This complementary approach provides a broader perspective on multiple toxicity domains and potential human health effects, extending the insights reported in the previous review, which focused specifically on in vivo genotoxicity in mammals.

Another study investigating the impact of triazole fungicides on female fertility and embryonic development [57] reported that these compounds may inhibit or interfere with enzymes such as CYP19A1, induce oxidative stress, and trigger apoptosis, while a second study on chiral triazole fungicides in mammalian models [58] highlighted oxidative stress and endocrine disruption, which further supports the patterns observed in our analysis of in vitro studies.

The triazole fungicides analyzed in this paper were observed to induce oxidative stress in cells, consistent with previous findings for propiconazole [143]. Another effect observed for triazole fungicides in our review was the induction of apoptosis in in vitro studies, consistent with previous findings for other fungicides [144]. A study investigating the neurotoxicity of certain fungicides in primary cultured mouse cortical neurons [145] reported depolarization of the mitochondrial membrane potential, a mechanism of toxicity also observed for triazole fungicides in our review.

A study identified an LD_{50} value for acute oral exposure to phenylpyrrole in rats [146], reporting a value greater than 5000 mg/kg BW, which is consistent with some of the results observed for triazole fungicides in our review.

These aspects discussed above, including key outcomes reported for triazole and other fungicides in the literature, are further summarized in Table S4 of the Supplementary Materials, providing a visual overview of how these findings align with or differ from the results of the current study on EU-approved triazoles.

The available literature indicates that this review aligns with observed patterns; however, heterogeneity in data, reporting formats, and methodologies—both within this review and in other studies—prevents a more direct and detailed comparison.

The toxicological and mechanistic data compiled in this review are potentially relevant for regulatory evaluation and risk assessment of EU-approved triazole fungicides, as they provide additional evidence on human health effects and toxicity pathways. According to Regulation (EC) No 1107/2009 [6], approval dossiers must include, among other relevant data, scientific peer-reviewed literature published within the last ten years addressing effects on human health, the environment, and non-target species. In this context, the present review compiles in vitro, in vivo, mechanistic and toxicokinetic data, providing complementary evidence to support a comprehensive assessment of triazole fungicides in the EU.

The review reveals notable gaps in the available data on EU-approved triazole fungicides, including limited in vitro, in vivo, and toxicokinetic information for several compounds. Mechanistic and human health-related evidence is incomplete, restricting comprehensive risk assessment and comparison across all fungicides. These limitations, combined with the widespread use of triazoles, underscore the need for further research and monitoring to support safe and sustainable application. Addressing these gaps will inform risk management and regulatory practices to better protect human health within the One Health framework. Moreover, future research on environmental impacts and ecotoxicity of triazole fungicides needs to be further explored to provide a more comprehensive One Health perspective.

10. Conclusions

This systematic review provides an integrated evaluation of the toxicological evidence related to ten triazole fungicides currently approved for use within the European Union. Both in vitro and in vivo data indicate that these compounds exhibit varying degrees of toxicity, with bromuconazole, difenoconazole, paclobutrazol, and tetraconazole showing comparatively higher toxic potential in rodent models, while mefentrifluconazole, penconazole, prothioconazole, and triticonazole demonstrated lower acute toxicity.

The in vitro findings revealed consistent cytotoxic effects across a wide range of mammalian cell lines, with tetraconazole emerging as the most cytotoxic compound based on combined quantitative indices. The most frequently investigated mechanisms of toxicity included cellular apoptosis and oxidative stress, followed by endoplasmic reticulum stress and genotoxicity, suggesting that mitochondrial dysfunction and redox imbalance play central roles in triazole-induced cytotoxicity.

Toxicokinetic information, encompassing absorption, distribution, metabolism, and excretion (ADME), remains limited and often derived from in silico predictions rather than experimental data. Available evidence points to good oral absorption, high plasma protein binding, and potential interactions with cytochrome P450 enzymes for several fungicides, underlining the need for more comprehensive pharmacokinetic and metabolic studies.

Qualitative assessments based on the PPDB database highlighted associations between some triazole fungicides and specific human health concerns, such as developmental toxicity, as well as general systemic effects. However, data gaps persist, particularly regarding chronic exposure and cumulative effects.

Overall, this review emphasizes the importance of continued toxicological and mechanistic investigation of EU-approved triazole fungicides to refine human health risk assessments, ensure regulatory safety, and guide the development of less hazardous alternatives.

This study provides an important contribution to the One Health framework regarding triazole fungicides, with a focus on their effects on human health, and highlights the need to continue the research to include environmental and ecotoxicological aspects.

Supplementary Materials: The following supporting information can be downloaded at https://www.mdpi.com/article/10.3390/jox15060208/s1. Table S1: In vitro toxicity data for the EU-approved triazole fungicides of interest. Data highlighted in bold are classified as Category 1 according to the ToxRTool evaluation from the risk-of-bias analysis; Table S2: Assigned numbers for molecular mechanisms of in vitro cytotoxicity for Cytoscape network diagram; Table S3: In vivo toxicity data for the EU-approved triazole fungicides of interest; Table S4: Comparative summary of key outcomes from literature-reported fungicide studies, including triazoles and other fungicides, for contextualization with findings from the current study on selected EU-approved triazole in vitro and in vivo toxicity. References [55–58], [63–72], [82–141], and [143–146] are cited in the Supplementary Materials.

Author Contributions: C.-B.V., A.-D.I.-D., A.P. and B.-V.A. contributed to the study's conceptualization, data mining and analysis, data curation, writing and visualization. All authors have read and agreed to the published version of the manuscript.

Funding: This article was supported by the UVT 1000 Develop Fund of the West University of Timisoara.

Institutional Review Board Statement: Not applicable.

Informed Consent Statement: Not applicable.

Data Availability Statement: No new data were created or analyzed in this study.

Conflicts of Interest: The authors declare no conflicts of interest.

References

1. Fagundes, T.R.; Coradi, C.; Vacario, B.G.L.; de Morais Valentim, J.M.B.; Panis, C. Global Evidence on Monitoring Human Pesticide Exposure. *J. Xenobiot.* **2025**, *15*, 187. [CrossRef]
2. Buonsenso, F. Scientific and Regulatory Perspectives on Chemical Risk Assessment of Pesticides in the European Union. *J. Xenobiot.* **2025**, *15*, 173. [CrossRef]
3. World Health Organization; Food and Agriculture Organization of the United Nations. Principles and Methods for the Risk Assessment of Chemicals in Food. Available online: https://www.who.int/publications/i/item/9789241572408 (accessed on 15 November 2025).
4. World Health Organization; Food and Agriculture Organization of the United Nations. Pesticide Residues in Food: Report 2024: Joint FAO/WHO Meeting on Pesticide Residues. Available online: https://www.who.int/publications/i/item/9789240113954 (accessed on 15 November 2025).
5. D'Amore, T.; Smaoui, S.; Varzakas, T. Chemical Food Safety in Europe Under the Spotlight: Principles, Regulatory Framework and Roadmap for Future Directions. *Foods* **2025**, *14*, 1628. [CrossRef] [PubMed]
6. European Parliament & Council. Regulation (EC) No 1107/2009 of 21 October 2009 on the Placing of Plant Protection Products on the Market (Repealing Directives 79/117/EEC and 91/414/EEC). 2009. Available online: https://eur-lex.europa.eu/legal-content/EN/TXT/?uri=celex:32009R1107 (accessed on 14 November 2025).
7. European Parliament & Council. Directive 2009/128/EC of the European Parliament and of the Council of 21 October 2009 Establishing a Framework for Community Action to Achieve the Sustainable Use of Pesticides (Text with EEA Relevance). 2009. Available online: https://eur-lex.europa.eu/eli/dir/2009/128/oj/eng (accessed on 14 November 2025).
8. Zubrod, J.P.; Feckler, A.; Englert, D.; Koksharova, N.; Rosenfeldt, R.R.; Seitz, F.; Schulz, R.; Bundschuh, M. Inorganic fungicides as routinely applied in organic and conventional agriculture can increase palatability but reduce microbial decomposition of leaf litter. *J. Appl. Ecol.* **2015**, *52*, 310–322. [CrossRef]
9. Triantafyllidis, V.; Kosma, C.; Karabagias, I.K.; Zotos, A.; Pittaras, A.; Kehayias, G. Fungicides in Europe During the Twenty-first Century: A Comparative Assessment Using Agri-environmental Indices of EU27. *Water Air Soil Pollut.* **2022**, *233*, 52. [CrossRef]
10. European Food Safety Authority. How Pesticides are Regulated in the EU—EFSA and the Assessment of Active Substances. 2002. Available online: https://www.efsa.europa.eu/sites/default/files/corporate_publications/files/Pesticides-ebook-180424.pdf (accessed on 14 November 2025).
11. European Food Safety Authority; Carrasco Cabrera, L.; Di Piazza, G.; Dujardin, B.; Marchese, E.; Medina Pastor, P. The 2023 European Union report on pesticide residues in food. *EFSA J.* **2025**, *23*, e9398. [CrossRef]
12. Roman, D.L.; Voiculescu, D.I.; Filip, M.; Ostafe, V.; Isvoran, A. Effects of Triazole Fungicides on Soil Microbiota and on the Activities of Enzymes Found in Soil: A Review. *Agriculture* **2021**, *11*, 893. [CrossRef]
13. Thabit, T.M.A.; Abdelkareem, E.M.; Bouqellah, N.A.; Shokr, S.A. Triazole Fungicide Residues and Their Inhibitory Effect on Some Trichothecenes Mycotoxin Excretion in Wheat Grains. *Molecules* **2021**, *26*, 1784. [CrossRef]
14. Rather, I.A.; Sabir, J.S.M.; Asseri, A.H.; Wani, M.Y.; Ahmad, A. Triazole Derivatives Target 14α-Demethylase (LDM) Enzyme in *Candida albicans* Causing Ergosterol Biosynthesis Inhibition. *J. Fungi* **2022**, *8*, 688. [CrossRef]
15. Gutierrez-Perez, C.; Cramer, R.A. Targeting fungal lipid synthesis for antifungal drug development and potentiation of contemporary antifungals. *npj Antimicrob. Resist.* **2025**, *3*, 27. [CrossRef]
16. Cooper, J.; Dobson, H. The benefits of pesticides to mankind and the environment. *Crop Prot.* **2007**, *26*, 1337–1348. [CrossRef]
17. Popp, J.; Pető, K.; Nagy, J. Pesticide productivity and food security. A review. *Agron. Sustain. Dev.* **2013**, *33*, 243–255. [CrossRef]
18. Gevao, B.; Semple, K.T.; Jones, K.C. Bound pesticide residues in soils: A review. *Environ. Pollut.* **2000**, *108*, 3–14. [CrossRef]
19. Ritter, W.F. Pesticide contamination of ground water in the United States—A review. *J. Environ. Sci. Health Part B* **1990**, *25*, 1–29. [CrossRef]
20. Donkor, A.; Osei-Fosu, P.; Dubey, B.; Kingsford-Adaboh, R.; Ziwu, C.; Asante, I. Pesticide residues in fruits and vegetables in Ghana: A review. *Environ. Sci. Pollut. Res.* **2016**, *23*, 18966–18987. [CrossRef]
21. Rad, S.M.; Ray, A.K.; Barghi, S. Water Pollution and Agriculture Pesticide. *Clean Technol.* **2022**, *4*, 1088–1102. [CrossRef]
22. de Souza, R.M.; Seibert, D.; Quesada, H.B.; de Jesus Bassetti, F.; Fagundes-Klen, M.R.; Bergamasco, R. Occurrence, impacts and general aspects of pesticides in surface water: A review. *Process Saf. Environ. Prot.* **2020**, *135*, 22–37. [CrossRef]
23. Botnaru, A.A.; Lupu, A.; Morariu, P.C.; Nedelcu, A.H.; Morariu, B.A.; Di Gioia, M.L.; Lupu, V.V.; Dragostin, O.M.; Caba, I.-C.; Anton, E.; et al. Innovative Analytical Approaches for Food Pesticide Residue Detection: Towards One Health-Oriented Risk Monitoring. *J. Xenobiot.* **2025**, *15*, 151. [CrossRef]

24. Liu, M.; Khan, A.; Wang, Z.; Liu, Y.; Yang, G.; Deng, Y.; He, N. Aptasensors for pesticide detection. *Biosens. Bioelectron.* **2019**, *130*, 174–184. [CrossRef]
25. Grimalt, S.; Dehouck, P. Review of analytical methods for the determination of pesticide residues in grapes. *J. Chromatogr. A* **2016**, *1433*, 1–23. [CrossRef] [PubMed]
26. Satish, G.P.; Ashokrao, D.M.; Arun, S.K. Microbial degradation of pesticide: A review. *Afr. J. Microbiol. Res.* **2017**, *11*, 992–1012. [CrossRef]
27. Marican, A.; Durán-Lara, E.F. A review on pesticide removal through different processes. *Environ. Sci. Pollut. Res.* **2018**, *25*, 2051–2064. [CrossRef]
28. Maroni, M.; Colosio, C.; Ferioli, A.; Fait, A. Biological Monitoring of Pesticide Exposure: A review. Introduction. *Toxicology* **2000**, *143*, 5–118. [CrossRef] [PubMed]
29. Aprea, C.; Colosio, C.; Mammone, T.; Minoia, C.; Maroni, M. Biological monitoring of pesticide exposure: A review of analytical methods. *J. Chromatogr. B* **2002**, *769*, 191–219. [CrossRef]
30. Kim, K.-H.; Kabir, E.; Jahan, S.A. Exposure to pesticides and the associated human health effects. *Sci. Total Environ.* **2017**, *575*, 525–535. [CrossRef]
31. Johnsen, K.; Jacobsen, C.S.; Torsvik, V.; Sørensen, J. Pesticide effects on bacterial diversity in agricultural soils—A review. *Biol. Fertil. Soils* **2001**, *33*, 443–453. [CrossRef]
32. Pekár, S. Spiders (Araneae) in the pesticide world: An ecotoxicological review. *Pest Manag. Sci.* **2012**, *68*, 1438–1446. [CrossRef]
33. Zhou, W.; Li, M.; Achal, V. A comprehensive review on environmental and human health impacts of chemical pesticide usage. *Emerg. Contam.* **2025**, *11*, 100410. [CrossRef]
34. Lushchak, V.I.; Matviishyn, T.M.; Husak, V.V.; Storey, J.M.; Storey, K.B. Pesticide toxicity: A mechanistic approach. *EXCLI J.* **2018**, *17*, 1101–1136. [CrossRef] [PubMed]
35. Rani, L.; Thapa, K.; Kanojia, N.; Sharma, N.; Singh, S.; Grewal, A.S.; Srivastav, A.L.; Kaushal, J. An extensive review on the consequences of chemical pesticides on human health and environment. *J. Clean. Prod.* **2021**, *283*, 124657. [CrossRef]
36. Yan, D.; Zhang, Y.; Liu, L.; Yan, H. Pesticide exposure and risk of Alzheimer's disease: A systematic review and meta-analysis. *Sci. Rep.* **2016**, *6*, 32222. [CrossRef]
37. Kalyabina, V.P.; Esimbekova, E.N.; Kopylova, K.V.; Kratasyuk, V.A. Pesticides: Formulants, distribution pathways and effects on human health—A review. *Toxicol. Rep.* **2021**, *8*, 1179–1192. [CrossRef]
38. Kaur, R.; Choudhary, D.; Bali, S.; Bandral, S.S.; Singh, V.; Ahmad, M.A.; Rani, N.; Singh, T.G.; Chandrasekaran, B. Pesticides: An alarming detrimental to health and environment. *Sci. Total Environ.* **2024**, *915*, 170113. [CrossRef] [PubMed]
39. Skevas, T.; Oude Lansink, A.G.J.M.; Stefanou, S.E. Designing the emerging EU pesticide policy: A literature review. *NJAS-Wagening. J. Life Sci.* **2013**, *64*, 95–103. [CrossRef]
40. Baibakova, E.V.; Nefedjeva, E.E.; Suska-Malawska, M.; Wilk, M.; Sevriukova, G.A.; Zheltobriukhov, V.F. Modern Fungicides: Mechanisms of Action, Fungal Resistance and Phytotoxic Effects. *Annu. Res. Rev. Biol.* **2019**, *32*, 1–16. [CrossRef]
41. Yang, C.; Hamel, C.; Vujanovic, V.; Gan, Y. Fungicide: Modes of Action and Possible Impact on Nontarget Microorganisms. *Int. Sch. Res. Not.* **2011**, *2011*, 130289. [CrossRef]
42. Bastos, R.W.; Rossato, L.; Goldman, G.H.; Santos, D.A. Fungicide effects on human fungal pathogens: Cross-resistance to medical drugs and beyond. *PLoS Pathog.* **2021**, *17*, e1010073. [CrossRef] [PubMed]
43. Gikas, G.D.; Parlakidis, P.; Mavropoulos, T.; Vryzas, Z. Particularities of Fungicides and Factors Affecting Their Fate and Removal Efficacy: A Review. *Sustainability* **2022**, *14*, 4056. [CrossRef]
44. Zubrod, J.P.; Bundschuh, M.; Arts, G.; Brühl, C.A.; Imfeld, G.; Knäbel, A.; Payraudeau, S.; Rasmussen, J.J.; Rohr, J.; Scharmüller, A.; et al. Fungicides: An Overlooked Pesticide Class? *Environ. Sci. Technol.* **2019**, *53*, 3347–3365. [CrossRef]
45. Cullen, M.G.; Thompson, L.J.; Carolan, J.C.; Stout, J.C.; Stanley, D.A. Fungicides, herbicides and bees: A systematic review of existing research and methods. *PLoS ONE* **2019**, *14*, e0225743. [CrossRef]
46. Rondeau, S.; Raine, N.E. Fungicides and bees: A review of exposure and risk. *Environ. Int.* **2022**, *165*, 107311. [CrossRef]
47. Komárek, M.; Čadková, E.; Chrastný, V.; Bordas, F.; Bollinger, J.-C. Contamination of vineyard soils with fungicides: A review of environmental and toxicological aspects. *Environ. Int.* **2010**, *36*, 138–151. [CrossRef] [PubMed]
48. Lini, R.S.; Scanferla, D.T.P.; de Oliveira, N.G.; Aguera, R.G.; Santos, T.d.S.; Teixeira, J.J.V.; Kaneshima, A.M.d.S.; Mossini, S.A.G. Fungicides as a risk factor for the development of neurological diseases and disorders in humans: A systematic review. *Crit. Rev. Toxicol.* **2024**, *54*, 35–54. [CrossRef] [PubMed]
49. Thomas Cusick, H. Fungicides. In *Farm Toxicology: A Primer for Rural Healthcare Practitioners*; Meggs, W.J., Langley, R.L., Eds.; Springer Nature: Cham, Switzerland, 2025; pp. 85–91.
50. Machado, F.J.; Santana, F.M.; Lau, D.; Del Ponte, E.M. Quantitative Review of the Effects of Triazole and Benzimidazole Fungicides on Fusarium Head Blight and Wheat Yield in Brazil. *Plant Dis.* **2017**, *101*, 1633–1641. [CrossRef] [PubMed]

51. Beyer, M.; Klix, M.B.; Klink, H.; Verreet, J.A. Quantifying the effects of previous crop, tillage, cultivar and triazole fungicides on the deoxynivalenol content of wheat grain—A review. *J. Plant Dis. Prot.* **2006**, *113*, 241–246. [CrossRef]
52. Roman, D.L.; Voiculescu, D.I.; Ostafe, V.; Ciorsac, A.; Isvoran, A. A review of the toxicity of triazole fungicides approved to be used in European Union to the soil and aqueous environment. *Ovidius Univ. Ann. Chem.* **2022**, *33*, 113–120. [CrossRef]
53. Huang, T.; Jiang, H.; Zhao, Y.; He, J.; Cheng, H.; Martyniuk, C.J. A comprehensive review of 1,2,4-triazole fungicide toxicity in zebrafish (*Danio rerio*): A mitochondrial and metabolic perspective. *Sci. Total Environ.* **2022**, *809*, 151177. [CrossRef]
54. Pobezhimova, T.P.; Korsukova, A.V.; Dorofeev, N.V.; Grabelnych, O.I. Physiological effects of triazole fungicides in plants. *Proc. Universities. Appl. Chem. Biotechnol.* **2019**, *9*, 461–476. [CrossRef]
55. Correia, L.V.B.; de Aguiar, G.C.; Pereira, A.M.R.d.S.; Thomaz, L.d.S.C.; de Oliveira, I.C.C.d.S.; Mari, R.d.B.; Perobelli, J.E.; Ribeiro, D.A.; da Silva, R.C.B. Triazole fungicides induce genotoxicity via oxidative stress in mammals in vivo: A comprehensive review. *Rev. Environ. Health* **2025**. [CrossRef]
56. Paul Friedman, K.; Papineni, S.; Marty, M.S.; Yi, K.D.; Goetz, A.K.; Rasoulpour, R.J.; Kwiatkowski, P.; Wolf, D.C.; Blacker, A.M.; Peffer, R.C. A predictive data-driven framework for endocrine prioritization: A triazole fungicide case study. *Crit. Rev. Toxicol.* **2016**, *46*, 785–833. [CrossRef]
57. Sharma, S.; Pandey, G. Understanding the impact of triazoles on female fertility and embryo development: Mechanisms and implications. *Toxicol. Rep.* **2025**, *14*, 101948. [CrossRef] [PubMed]
58. Wang, N.; Liu, Y.; Li, J.; Xu, L.; Liang, X. Advances in Selective Bioactivity and Toxicity of Chiral Triazole Fungicides and Their Selective Behavior in Mammals. *Chirality* **2025**, *37*, e70055. [CrossRef] [PubMed]
59. Web of Science. Web of Science Advanced Search Query Builder. Available online: https://www.webofscience.com/wos/woscc/advanced-search (accessed on 12 November 2025).
60. European Commission. EU Pesticides Database—Active Substances. Available online: https://ec.europa.eu/food/plant/pesticides/eu-pesticides-database/start/screen/active-substances (accessed on 5 November 2025).
61. Kim, S.; Chen, J.; Cheng, T.; Gindulyte, A.; He, J.; He, S.; Li, Q.; Shoemaker, B.A.; Thiessen, P.A.; Yu, B.; et al. PubChem 2025 update. *Nucleic Acids Res.* **2024**, *53*, D1516–D1525. [CrossRef]
62. Lewis, K.A.; Tzilivakis, J.; Warner, D.J.; Green, A. An international database for pesticide risk assessments and management. *Hum. Ecol. Risk Assess. Int. J.* **2016**, *22*, 1050–1064. [CrossRef]
63. Pesticide Properties DataBase. Bromuconazole. Available online: https://sitem.herts.ac.uk/aeru/ppdb/en/Reports/97.htm#3 (accessed on 30 October 2025).
64. Pesticide Properties DataBase. Difenoconazole. Available online: https://sitem.herts.ac.uk/aeru/ppdb/en/Reports/230.htm#3 (accessed on 30 October 2025).
65. Pesticide Properties DataBase. Mefentrifluconazole. Available online: https://sitem.herts.ac.uk/aeru/ppdb/en/Reports/3098.htm#3 (accessed on 30 October 2025).
66. Pesticide Properties DataBase. Metconazole. Available online: https://sitem.herts.ac.uk/aeru/ppdb/en/Reports/451.htm#3 (accessed on 30 October 2025).
67. Pesticide Properties DataBase. Paclobutrazol. Available online: https://sitem.herts.ac.uk/aeru/ppdb/en/Reports/504.htm#3 (accessed on 30 October 2025).
68. Pesticide Properties DataBase. Penconazole. Available online: https://sitem.herts.ac.uk/aeru/ppdb/en/Reports/509.htm#3 (accessed on 30 October 2025).
69. Pesticide Properties DataBase. Prothioconazole. Available online: https://sitem.herts.ac.uk/aeru/ppdb/en/Reports/559.htm#3 (accessed on 30 October 2025).
70. Pesticide Properties DataBase. Tebuconazole. Available online: https://sitem.herts.ac.uk/aeru/ppdb/en/Reports/610.htm#3 (accessed on 30 October 2025).
71. Pesticide Properties DataBase. Tetraconazole. Available online: https://sitem.herts.ac.uk/aeru/ppdb/en/Reports/626.htm#3 (accessed on 30 October 2025).
72. Pesticide Properties DataBase. Triticonazole. Available online: https://sitem.herts.ac.uk/aeru/ppdb/en/Reports/673.htm#3 (accessed on 30 October 2025).
73. Voiculescu, D.I.; Roman, D.L.; Ostafe, V.; Isvoran, A. A Cheminformatics Study Regarding the Human Health Risks Assessment of the Stereoisomers of Difenoconazole. *Molecules* **2022**, *27*, 4682. [CrossRef]
74. Gridan, I.M.; Alecu, A.C.; Isvoran, A. Prediction of ADME-Tox properties and toxicological endpoints of triazole fungicides used for cereals protection. *ADMET DMPK* **2019**, *7*, 161–173. [CrossRef]
75. Roman, D.L.; Voiculescu, D.I.; Matica, M.A.; Baerle, V.; Filimon, M.N.; Ostafe, V.; Isvoran, A. Assessment of the Effects of Triticonazole on Soil and Human Health. *Molecules* **2022**, *27*, 6554. [CrossRef]

76. Perovani, I.S.; Yamamoto, P.A.; da Silva, R.M.; Lopes, N.P.; de Moraes, N.V.; de Oliveira, A.R.M. Unveiling CYP450 inhibition by the pesticide prothioconazole through integrated in vitro studies and PBPK modeling. *Arch. Toxicol.* **2025**, *99*, 2845–2854. [CrossRef]
77. Perovani, I.S.; Santos Barbetta, M.F.; Moreira da Silva, R.; Lopes, N.P.; Moraes de Oliveira, A.R. In vitro-in vivo correlation of the chiral pesticide prothioconazole after interaction with human CYP450 enzymes. *Food Chem. Toxicol.* **2022**, *163*, 112947. [CrossRef]
78. Pampalakis, G. Underestimations in the In Silico-Predicted Toxicities of V-Agents. *J. Xenobiot.* **2023**, *13*, 615–624. [CrossRef] [PubMed]
79. Open Science Framework. View Only Link of Open Science Framwork Registration That Is Under Embargo. Available online: https://osf.io/nvz5x/overview?view_only=96796d5395344b10b3755d3e79dde0a8 (accessed on 16 November 2025).
80. Haddaway, N.R.; Page, M.J.; Pritchard, C.C.; McGuinness, L.A. PRISMA2020: An R package and Shiny app for producing PRISMA 2020-compliant flow diagrams, with interactivity for optimised digital transparency and Open Synthesis. *Campbell. Syst. Rev.* **2022**, *18*, e1230. [CrossRef]
81. Google. Google Scholar. Available online: https://scholar.google.com (accessed on 5 November 2025).
82. Almedeny, S.A.; Sahib, Z.H.; Al Mukhtar, E.J.; Alkelaby, K.K. Evaluation of the Cytotoxic Effect of the Epoxyconazole and Difenoconazole on Human Colorectal Cancer HCT116 Cell Line. *Int. J. Drug Deliv. Technol.* **2022**, *12*, 805–808. [CrossRef]
83. Andrioli, N.B.; Nieves, M.; Poltronieri, M.; Bonzon, C.; Chaufan, G. Genotoxic effects induced for sub-cytotoxic concentrations of tebuconazole fungicide in HEp-2 cell line. *Chem. -Biol. Interact.* **2023**, *373*, 110385. [CrossRef] [PubMed]
84. Barrón Cuenca, J.; de Oliveira Galvão, M.F.; Ünlü Endirlik, B.; Tirado, N.; Dreij, K. In vitro cytotoxicity and genotoxicity of single and combined pesticides used by Bolivian farmers. *Environ. Mol. Mutagen.* **2022**, *63*, 4–17. [CrossRef]
85. Bülbül, E.; Özhan, G.; Bülbül, E.; Bülbül, G.Ö.E.; Özhan, G. Cytotoxic effects of triazole fungucides. *J. Fac. Pharm. Istanb. Univ.* **2012**, *42*, 23–31.
86. Daniel, S.L.; Hartman, G.L.; Wagner, E.D.; Plewa, M.J. Mammalian Cell Cytotoxicity Analysis of Soybean Rust Fungicides. *Bull. Environ. Contam. Toxicol.* **2007**, *78*, 474–478. [CrossRef]
87. de Jong, E.; Barenys, M.; Hermsen, S.A.B.; Verhoef, A.; Ossendorp, B.C.; Bessems, J.G.M.; Piersma, A.H. Comparison of the mouse Embryonic Stem cell Test, the rat Whole Embryo Culture and the Zebrafish Embryotoxicity Test as alternative methods for developmental toxicity testing of six 1,2,4-triazoles. *Toxicol. Appl. Pharm.* **2011**, *253*, 103–111. [CrossRef] [PubMed]
88. Hu, L.; Wang, X.; Qian, M.; Zhang, H.; Jin, Y. Impacts of prothioconazole and prothioconazole-desthio on bile acid and glucolipid metabolism: Upregulation of CYP7A1 expression in HepG2 cells. *Pestic. Biochem. Phys.* **2024**, *198*, 105702. [CrossRef]
89. Jeong, S.A.; Song, J.; Ham, J.; An, G.; Song, G.; Lim, W. Tetraconazole interrupts mitochondrial function and intracellular calcium levels leading to apoptosis of bovine mammary epithelial cells. *Pestic. Biochem. Phys.* **2023**, *191*, 105366. [CrossRef]
90. Kim, M.; Park, W.; Lim, W.; Song, G.; Park, S. Bromuconazole impairs implantation process through cellular stress response in human trophoblast and endometrial cells. *Pestic. Biochem. Phys.* **2025**, *214*, 106632. [CrossRef]
91. Kjærstad, M.B.; Taxvig, C.; Nellemann, C.; Vinggaard, A.M.; Andersen, H.R. Endocrine disrupting effects in vitro of conazole antifungals used as pesticides and pharmaceuticals. *Reprod. Toxicol.* **2010**, *30*, 573–582. [CrossRef]
92. Kwon, H.-C.; Kim, D.-H.; Jeong, C.-H.; Kim, Y.-J.; Han, J.-H.; Lim, S.-J.; Shin, D.-M.; Kim, D.-W.; Han, S.-G. Tebuconazole Fungicide Induces Lipid Accumulation and Oxidative Stress in HepG2 Cells. *Foods* **2021**, *10*, 2242. [CrossRef]
93. Lee, W.-Y.; Lee, R.; Park, H.-J. Tebuconazole Induces ER-Stress-Mediated Cell Death in Bovine Mammary Epithelial Cell Lines. *Toxics* **2023**, *11*, 397. [CrossRef]
94. Leite, F.G.; Silva, C.d.P.M.; Miranda, R.G.; Dorta, D.J. Comparison of in vitro toxicity in HepG2 cells: Toxicological role of Tebuconazole-tert-butyl-hydroxy in exposure to the fungicide Tebuconazole. *Pestic. Biochem. Phys.* **2024**, *202*, 105954. [CrossRef]
95. Li, R.; Liu, B.; Xu, W.; Yu, L.; Zhang, C.; Cheng, J.; Tao, L.; Li, Z.; Zhang, Y. DNA damage and cell apoptosis induced by fungicide difenoconazole in mouse mononuclear macrophage RAW264.7. *Environ. Toxicol.* **2022**, *37*, 650–659. [CrossRef]
96. Li, Y.-C.; Liu, S.-Y.; Li, H.-R.; Meng, F.-B.; Qiu, J.; Qian, Y.-Z.; Xu, Y.-Y. Use of Transcriptomics to Reveal the Joint Immunotoxicity Mechanism Initiated by Difenoconazole and Chlorothalonil in the Human Jurkat T-Cell Line. *Foods* **2024**, *13*, 34. [CrossRef]
97. Liu, Y.; Xu, N.; Song, X.; Deng, M.; Sun, R.; Wang, P.; Cao, L. Specific Hepatorenal Toxicity and Cross-Species Susceptibility of Eight Representative Pesticides. *Toxics* **2025**, *13*, 911. [CrossRef]
98. Luo, Y.; Lu, S.; Sun, X.; Gao, Y.; Sun, G.; Yang, M.; Sun, X. Paclobutrazol exposure induces apoptosis and impairs autophagy in hepatocytes via the AMPK/mTOR signaling pathway. *J. Biochem. Mol. Toxicol.* **2021**, *35*, e22874. [CrossRef] [PubMed]
99. Na, M.-J.; Lee, W.-Y.; Park, H.-J. Difenoconazole Induced Damage of Bovine Mammary Epithelial Cells via ER Stress and Inflammatory Response. *Cells* **2024**, *13*, 1715. [CrossRef] [PubMed]
100. Nie, J.; Zhou, J.; Shen, Y.; Lin, R.; Hu, H.; Zeng, K.; Bi, H.; Huang, M.; Yu, L.; Zeng, S.; et al. Studies on the interaction of five triazole fungicides with human renal transporters in cells. *Toxicol. In Vitr.* **2023**, *88*, 105555. [CrossRef] [PubMed]

101. Othmène, Y.B.; Kevin, M.; Ahmed, K.; Salem, I.B.; Anissa, B.; Salwa, A.-E.; Christophe, L. Tebuconazole induces ROS-dependent cardiac cell toxicity by activating DNA damage and mitochondrial apoptotic pathway. *Ecotoxicol. Environ. Saf.* **2020**, *204*, 111040. [CrossRef]
102. Othmène, Y.B.; Kevin, M.; Anissa, B.; Ahmed, K.; Intidhar, B.S.; Manel, B.; Salwa, A.-E.; Christophe, L. Triazole fungicide tebuconazole induces apoptosis through ROS-mediated endoplasmic reticulum stress pathway. *Environ. Toxicol. Pharmacol.* **2022**, *94*, 103919. [CrossRef]
103. Othmène, Y.B.; Salem, I.B.; Hamdi, H.; Annabi, E.; Abid-Essefi, S. Tebuconazole induced cytotoxic and genotoxic effects in HCT116 cells through ROS generation. *Pestic. Biochem. Phys.* **2021**, *174*, 104797. [CrossRef]
104. Pan, L.; Lu, L.; Wang, J.; Zheng, C.; Fu, Y.; Xiao, S.; Jin, Y.; Zhuang, S. The fungicide difenoconazole alters mRNA expression levels of human CYP3A4 in HepG2 cells. *Environ. Chem. Lett.* **2017**, *15*, 673–678. [CrossRef]
105. Rjiba-Touati, K.; Ayed-Boussema, I.; Hamdi, H.; Abid, S. Genotoxic damage and apoptosis in rat glioma (F98) cell line following exposure to bromuconazole. *NeuroToxicology* **2023**, *94*, 108–116. [CrossRef]
106. Rjiba-Touati, K.; Ayed-Boussema, I.; Hamdi, H.; Azzebi, A.; Abid, S. Bromuconazole fungicide induces cell cycle arrest and apoptotic cell death in cultured human colon carcinoma cells (HCT116) via oxidative stress process. *Biomarkers* **2022**, *27*, 659–670. [CrossRef]
107. Rjiba-Touati, K.; Hamdi, H.; M'nassri, A.; Rich, S.; Mokni, M.; Abid, S. Brain injury, genotoxic damage and oxidative stress induced by Bromuconazole in male Wistar rats and in SH-SY5Y cell line. *Biomarkers* **2022**, *27*, 599–607. [CrossRef]
108. Sanchez, C.L.; Souders, C.L.; Pena-Delgado, C.J.; Nguyen, K.T.; Kroyter, N.; Ahmadie, N.E.; Aristizabal-Henao, J.J.; Bowden, J.A.; Martyniuk, C.J. Neurotoxicity assessment of triazole fungicides on mitochondrial oxidative respiration and lipids in differentiated human SH-SY5Y neuroblastoma cells. *NeuroToxicology* **2020**, *80*, 76–86. [CrossRef]
109. Serra, L.; Estienne, A.; Bongrani, A.; Ramé, C.; Caria, G.; Froger, C.; Jolivet, C.; Henriot, A.; Amalric, L.; Corbin, E.; et al. The epoxiconazole and tebuconazole fungicides impair granulosa cells functions partly through the aryl hydrocarbon receptor (AHR) signalling with contrasted effects in obese, normo-weight and polycystic ovarian syndrome (PCOS) patients. *Toxicol. Rep.* **2024**, *12*, 65–81. [CrossRef] [PubMed]
110. Serra, L.; Estienne, A.; Caria, G.; Ramé, C.; Jolivet, C.; Froger, C.; Henriot, A.; Amalric, L.; Guérif, F.; Froment, P.; et al. In vitro exposure to triazoles used as fungicides impairs human granulosa cells steroidogenesis. *Environ. Toxicol. Pharmacol.* **2023**, *104*, 104295. [CrossRef] [PubMed]
111. Sevim, Ç.; Taghizadehghalehjoughi, A.; Kara, M. Effects of chlorpyrifos-methyl, chlormequat, deltamethrin, glyphosate, pirimiphos-methyl, tebuconazole and their mixture on oxidative stress and toxicity in HUVEC cell line. *İstanb. J. Pharm.* **2021**, *51*, 183–190. [CrossRef]
112. Tian, S.; Yan, S.; Meng, Z.; Sun, W.; Yan, J.; Huang, S.; Wang, Y.; Zhou, Z.; Diao, J.; Li, L.; et al. Widening the Lens on Prothioconazole and Its Metabolite Prothioconazole-Desthio: Aryl Hydrocarbon Receptor-Mediated Reproductive Disorders through in Vivo, in Vitro, and in Silico Studies. *Environ. Sci. Technol.* **2022**, *56*, 17890–17901. [CrossRef] [PubMed]
113. Wang, T.; Ma, M.; Chen, C.; Yang, X.; Qian, Y. Three widely used pesticides and their mixtures induced cytotoxicity and apoptosis through the ROS-related caspase pathway in HepG2 cells. *Food Chem. Toxicol.* **2021**, *152*, 112162. [CrossRef]
114. Wang, X.; Ni, H.; Xu, W.; Wu, B.; Xie, T.; Zhang, C.; Cheng, J.; Li, Z.; Tao, L.; Zhang, Y. Difenoconazole induces oxidative DNA damage and mitochondria mediated apoptosis in SH-SY5Y cells. *Chemosphere* **2021**, *283*, 131160. [CrossRef] [PubMed]
115. Xu, J.; Xiong, H.; Zhang, X.; Muhayimana, S.; Liu, X.; Xue, Y.; Huang, Q. Comparative cytotoxic effects of five commonly used triazole alcohol fungicides on human cells of different tissue types. *J. Environ. Sci. Health Part B* **2020**, *55*, 438–446. [CrossRef]
116. Xu, X.-B. Chiral analysis and semi-preparative separation of metconazole stereoisomers by supercritical fluid chromatography and cytotoxicity assessment in vitro. *J. Sep. Sci.* **2024**, *47*, 2300655. [CrossRef]
117. Zhou, J.; Zhang, J.; Li, F.; Liu, J. Triazole fungicide tebuconazole disrupts human placental trophoblast cell functions. *J. Hazard. Mater.* **2016**, *308*, 294–302. [CrossRef]
118. Abbassy, M.A.; Marzouk, M.A.; Nasr, H.M.; Mansy, A. Effect of imidacloprid and tetraconazole on various hematological and biochemical parameters in male albino rats (*Rattus norvegious*). *J. Political Sci. Public Aff.* **2014**, *2*, 1000122. [CrossRef]
119. Australian Pesticides and Veterinary Medicines Authority. Evaluation of the New Active Prothioconazole in the Product Redigo Fungicidal Seed Treatment. 2007. Available online: https://www.apvma.gov.au/sites/default/files/publication/13941-prs-prothioconazole.pdf (accessed on 10 September 2025).
120. El-Shershaby, A.E.-F.M.; Lashein, F.E.-D.M.; Seleem, A.A.; Ahmed, A.A. Toxicological potential of penconazole on early embryogenesis of white mice *Mus musculus* in either pre- or post-implantation exposure. *Environ. Sci. Pollut. Res.* **2020**, *27*, 9943–9956. [CrossRef]
121. European Food Safety Authority. Conclusion regarding the peer review of the pesticide risk assessment of the active substance triticonazole. *EFSA J.* **2005**, *3*, 33ar. [CrossRef]

122. European Food Safety Authority. Conclusion regarding the peer review of the pesticide risk assessment of the active substance bromuconazole. *EFSA J.* **2008**, *6*, 136r. [CrossRef]
123. European Food Safety Authority. Conclusion regarding the peer review of the pesticide risk assessment of the active substance tetraconazole. *EFSA J.* **2008**, *152*, 1–86. [CrossRef]
124. European Food Safety Authority. Conclusion on the peer review of the pesticide risk assessment of the active substance bromuconazole. *EFSA J.* **2010**, *8*, 1704. [CrossRef]
125. European Food Safety Authority. Conclusion on the peer review of the pesticide risk assessment of the active substance paclobutrazol. *EFSA J.* **2010**, *8*, 1876. [CrossRef]
126. European Food Safety Authority. Conclusion on the peer review of the pesticide risk assessment of the active substance difenoconazole. *EFSA J.* **2011**, *9*, 1967. [CrossRef]
127. Food and Agriculture Organization of the United Nations. *Toxicological Evaluation of Penconazole*; Food and Agriculture Organization of the United Nations: Rome, Italy, 2015.
128. Hariyadi, H.R. The presence of bromuconazole fungicide pollutant in organic waste anaerobic fermentation. *IOP Conf. Ser. Earth Environ. Sci.* **2017**, *60*, 012022. [CrossRef]
129. Khwanes, S.A.; Mohamed, R.A.; Ibrahim, K.A.; Abd El-Rahman, H.A. Ginger reserves testicular spermatogenesis and steroidogenesis in difenoconazole-intoxicated rats by conducting oxidative stress, apoptosis and proliferation. *Andrologia* **2022**, *54*, e14241. [CrossRef]
130. Klix, M.B.; Verreet, J.-A.; Beyer, M. Comparison of the declining triazole sensitivity of Gibberella zeae and increased sensitivity achieved by advances in triazole fungicide development. *Crop Prot.* **2007**, *26*, 683–690. [CrossRef]
131. Li, X.; Lian, T.; Su, B.; Liu, H.; Wang, Y.; Wu, X.; He, J.; Wang, Y.; Xu, Y.; Yang, S.; et al. Construction of a physiologically based pharmacokinetic model of paclobutrazol and exposure estimation in the human body. *Toxicology* **2024**, *505*, 153841. [CrossRef]
132. Mohamed, R.E.s.; El-Sayed, M.M.; Arief, M.M.; Mahmoud, A.A. Evaluation of the ameliorative effect of *Cinnamon cassia* against metabolic disorder and thyroid hormonal disruption following treatment with difenoconazole fungicide in the male albino rats. *Egypt. J. Chem.* **2022**, *65*, 55–67. [CrossRef]
133. Osman, A.H.; Elshama, S.S.; Osman, A.S.; El-Hameed, A.K.A. Toxicological and pathological evaluation of prolonged bromuconazole fungicide exposure in male rats. *Med. J. Cairo Univ.* **2011**, *79*, 555–564. Available online: https://scholar.cu.edu.eg/?q=yomnakhaled/files/cairo_univeristy_journal.pdf (accessed on 10 September 2025).
134. Rjiba-Touati, K.; Hamdi, H.; M'nassri, A.; Guedri, Y.; Mokni, M.; Abid, S. Bromuconazole caused genotoxicity and hepatic and renal damage via oxidative stress process in Wistar rats. *Environ. Sci. Pollut. Res.* **2022**, *29*, 14111–14120. [CrossRef]
135. Solera ATO, L. Safety Data Sheet - Tebuconazole. 2015. Available online: https://www.greenbook.net/solera-source-dynamics/tebuconazole-36f (accessed on 6 November 2025).
136. Tesh, S.A.; Tesh, J.M.; Fegert, I.; Buesen, R.; Schneider, S.; Mentzel, T.; van Ravenzwaay, B.; Stinchcombe, S. Innovative selection approach for a new antifungal agent mefentrifluconazole (Revysol®) and the impact upon its toxicity profile. *Regul. Toxicol. Pharmacol.* **2019**, *106*, 152–168. [CrossRef]
137. United States Environmental Protection Agency. Pesticide Fact Sheet—Tetraconazole. 2005. Available online: https://www3.epa.gov/pesticides/chem_search/reg_actions/registration/fs_PC-120603_01-Apr-05.pdf (accessed on 10 September 2025).
138. United States Environmental Protection Agency. Pesticide Fact Sheet—Triticonazole. 2005. Available online: https://www3.epa.gov/pesticides/chem_search/cleared_reviews/csr_PC-125620_1-Nov-00_a.pdf (accessed on 10 September 2025).
139. United States Environmental Protection Agency. Pesticide Fact Sheet—Metconazole. 2007. Available online: https://www3.epa.gov/pesticides/chem_search/reg_actions/registration/fs_PC-125619_01-Sep-07.pdf (accessed on 10 September 2025).
140. Xu, M.; Yang, F. Integrated gender-related effects of profenofos and paclobutrazol on neurotransmitters in mouse. *Ecotoxicol. Environ. Saf.* **2020**, *190*, 110085. [CrossRef]
141. Yue, K.; Liu, Z.; Pi, Z.; Li, H.; Wang, Y.; Song, F.; Liu, Z. Network Pharmacology Combined with Metabolomics Approach to Investigate the Toxicity Mechanism of Paclobutrazol. *Chem. Res. Toxicol.* **2022**, *35*, 626–635. [CrossRef]
142. Schneider, K.; Schwarz, M.; Burkholder, I.; Kopp-Schneider, A.; Edler, L.; Kinsner-Ovaskainen, A.; Hartung, T.; Hoffmann, S. "ToxRTool", a new tool to assess the reliability of toxicological data. *Toxicol. Lett.* **2009**, *189*, 138–144. [CrossRef]
143. Kwon, H.C.; Sohn, H.; Kim, D.H.; Shin, D.M.; Jeong, C.H.; Chang, Y.H.; Yune, J.H.; Kim, Y.J.; Kim, D.-W.; Kim, S.H.; et al. In Vitro and In Vivo Study on the Toxic Effects of Propiconazole Fungicide in the Pathogenesis of Liver Fibrosis. *J. Agric. Food Chem.* **2021**, *69*, 7399–7408. [CrossRef] [PubMed]

144. Lin, N.; Garry, V.F. In vitro studies of cellular and molecular developmental toxicity of adjuvants, herbicides, and fungicides commonly used in Red River Valley, Minnesota. *J. Toxicol. Environ. Health Part A* **2000**, *60*, 423–439. [CrossRef]
145. Regueiro, J.; Olguín, N.; Simal-Gándara, J.; Suñol, C. Toxicity evaluation of new agricultural fungicides in primary cultured cortical neurons. *Environ. Res.* **2015**, *140*, 37–44. [CrossRef] [PubMed]
146. Safandeev, V.V.; Beloedova, N.S.; Poroșin, M.A.; Bogdanova, A.V.; Sinitskaya, T.A. Characterization of a phenylpyrrole derivative fungicide in an acute oral toxicology study in rats. *Health Care Kyrg. Sci. Pract. J.* **2023**, *1*, 54–58. [CrossRef]

Journal of *Xenobiotics*

MDPI

Article

From Farmworkers to Urban Residents: Mapping Multi-Class Pesticide Exposure Gradients in Morocco via Urinary Biomonitoring

Zineb Ben Khadda [1], Andrei-Flavius Radu [2,3,*], Souleiman El Balkhi [4,5,6], Fagroud Mustapha [7], Yahya El Karmoudi [8], Gabriela Bungau [2,*], Pierre Marquet [4,5], Tarik Sqalli Houssaini [1,9] and Sanae Achour [10,11]

1 Laboratory of Epidemiology and Research in Health Sciences, Faculty of Medicine and Pharmacy, Sidi Mohammed Ben Abdellah University, Fez 30070, Morocco; benkhadda.zaynab@gmail.com (Z.B.K.); tarik.sqalli@usmba.ac.ma (T.S.H.)

2 Doctoral School of Biomedical Sciences, University of Oradea, 410087 Oradea, Romania

3 Department of Psycho-Neurosciences and Recovery, Faculty of Medicine and Pharmacy, University of Oradea, 410073 Oradea, Romania

4 Department of Pharmacology, Toxicology and Pharmacovigilance, CHU Limoges, 87000 Limoges, France; souleiman.elbalkhi@chu-limoges.fr (S.E.B.); pierre.marquet@unilim.fr (P.M.)

5 P&T, UMR1248, Inserm, University Limoges, 87032 Limoges, France

6 Pharmacology-Toxicology and Pharmacovigilance Department, Centre de Biologie Et de Recherche en Santé (CBRS), 2, Av. Martin Luther King, 87042 Limoges, France

7 Department of Agronomy, National School of Agriculture, Meknes 50001, Morocco; mfagroud@gmail.com

8 Laboratory of Ecology, Systematics, and Conservation of Biodiversity, LESCB URL-CNRST N° 18, FS, Abdelmalek Essaadi University, Tetouan 93002, Morocco; yahyaelkarmoudi@gmail.com

9 Department of Nephrology, University of Hospital Hassan II, Fez 30050, Morocco

10 Laboratory of Pharmacology and Toxicology, University Hospital Hassan II, Fez 30050, Morocco; sanae.achour@usmba.ac.ma

11 Biomedical and Translational Research Laboratory, Faculty of Medicine and Pharmacy of the Fez, University of Sidi Mohamed Ben Abdellah, Fez 30070, Morocco

* Correspondence: andreiflavius.radu@uoradea.ro (A.-F.R.); gbungau@uoradea.ro (G.B.)

Abstract: Pesticide exposure gradients between occupational, para-occupational, and general populations remain poorly characterized in North African agricultural contexts. This study evaluates urinary pesticide levels among farmers, indirectly exposed individuals, and a control group in Morocco's Fez-Meknes region. A cross-sectional survey measured pesticide concentrations using LC-MS/MS in urine samples collected from 154 adults residing in both rural and urban areas. A questionnaire was used to gather information from participants regarding factors that may elevate the risk of pesticide exposure. The results revealed that farmers exhibited the highest concentrations of pesticides in their urine, including compounds classified as Ia/Ib by the World Health Organization. Indirectly exposed individuals showed moderate levels of contamination, with notable detections such as dichlofluanid (22.13 μg/L), while the control group had residual traces of neonicotinoids, notably imidacloprid (2.05 μg/L). Multivariate analyses revealed several sociodemographic factors significantly associated with increased pesticide exposure. The main risk factors identified included low education, residence in an agricultural area, and the consumption of untreated water (wells/rivers). Conversely, wearing personal protective equipment was associated with reduced urinary concentrations. This study highlights intense occupational exposure among farmers, secondary environmental contamination among residents living near treated areas, and the widespread dispersion of pesticide residues into urban areas.

Keywords: pesticide biomonitoring; LC-MS/MS; farmers; toxicology; exposure risk factors; urinary biomonitoring

1. Introduction

The intensification of environmental degradation, driven by accelerated urbanization and an expanding range of human activities, has resulted in the widespread dissemination of hazardous substances into natural ecosystems. As the use of synthetic chemicals has grown rapidly across industrial, agricultural, domestic, and technical domains, human populations are increasingly at risk of both direct and indirect exposure. This growing burden of chemical pollutants poses significant challenges to public and environmental health, especially in regions where surveillance, regulation, and risk mitigation strategies remain limited or inconsistently applied [1,2].

In contemporary agricultural systems, the use of pesticides presents a complex paradox. While contributing significantly to crop yield optimization and food availability, it simultaneously introduces substantial threats to both human health and environment [3]. From a chemical standpoint, pesticides are broadly grouped into several principal categories, including organochlorines, organophosphates, carbamates, pyrethroids, and pyrethrins. These compounds are primarily organic in nature and may be either synthetically produced or derived from botanical sources [4].

Global regulatory frameworks for pesticide management vary considerably. High-income countries typically enforce more rigorous legislation and oversight mechanisms. In contrast, many low- and middle-income nations face challenges related to insufficient institutional capacity, technical resources, and enforcement infrastructure, which limits the effective implementation of pesticide regulations [5].

Morocco's extensive agricultural sector, spanning diverse agroecological zones, relies heavily on pesticide applications to enhance crop productivity and mitigate biotic stressors [6]. However, the escalating use of agrochemicals has raised critical public health concerns, particularly regarding occupational exposure among farming communities. Epidemiological evidence indicates inadequate adherence to safety protocols in rural areas, exacerbating exposure risks to pesticide residues [7].

Exposure to these chemicals can occur through inhalation, ingestion, or skin contact and is associated with various adverse effects [8]. Epidemiological studies have highlighted links between chronic exposure to pesticides and an increased risk of chronic kidney disease and neurodegenerative diseases, such as Parkinson's disease and Alzheimer's disease [9,10]. Additionally, some pesticides are classified as endocrine disruptors, interfering with the hormonal system and potentially causing reproductive problems, birth defects, and hormone-dependent cancers [11,12]. Children and pregnant women are particularly vulnerable to environmental and chemical exposures due to their developmental stages. Prenatal or early exposure to harmful substances can significantly impact neurological and cognitive development, leading to long-term consequences [13,14].

These impacts highlight the importance of biomonitoring as an essential tool, which not only allows for the precise quantification of individual and collective exposure by measuring residues of these substances in biological matrices such as blood, urine, or tissues [15,16] but also correlates this exposure to specific pathologies [17,18]. Biomonitoring plays a central role in identifying vulnerable populations, including children, pregnant women, and agricultural workers, due to their increased sensitivity or occupational exposure [19–21]. Furthermore, the data it provides constitute an essential scientific basis for the following: (i) strengthening pesticide regulations; (ii) promoting sustainable agricultural practices; (iii) developing prevention strategies adapted to at-risk groups [22].

The present article aims to investigate human exposure to a broad spectrum of pesticides through urinary biomonitoring in distinct population groups within the Fez-Meknes region of Morocco. By evaluating pesticide residue levels in different groups, this study offers novel insights into exposure gradients in a North African agricultural context. This work addresses a significant gap in the literature by providing relevant data from a region where biomonitoring studies remain scarce. The findings contribute to a better understanding of the risks of current agricultural practices and identify critical exposure items.

2. Materials and Methods

2.1. Chemicals and Reagents

All analytical standards (i.e., pesticides and metabolites), internal isotopic standards (IS), and solvents and reagents required for chromatographic analyses were purchased from certified suppliers (LGC standards, Teddington, United Kingdom, VWR International, Pennsylvania, the United States, Sigma-Aldrich, Darmstadt, Germany and Cluzeau Info Labo, Brive-la-Gaillarde, France). Purified water was obtained using a Millipore system, Darmstadt, Germany, and QuEChERS salts were supplied by Macherey Nagel, Düren, North Rhine-Westphalia, Germany. The complete list of analytical standards is provided in Supplementary File S1 for reference.

2.2. Study Area and Sampling Design

Two separate municipalities were selected as study sites in the Fez-Meknes region of Morocco: (1) the rural municipality of Aïn Aïcha, an agricultural area characterized by intensive pesticide use in cereal and olive crops; (2) the urban district of Fez, a primarily residential area with public green spaces.

These sites, among others in the Fez-Meknes region, had been previously and distinctively assessed for agricultural practices and environmental exposure levels by our research team [6]. We recruited 154 adult participants divided into three groups: 56 male farmers from Aïn Aïcha, 48 rural residents with indirect exposure (mainly women), and 50 urban controls (medical personnel from Fez). The observed sex distribution across the study groups reflects the socio-cultural and occupational roles specific to the rural Moroccan context. In this setting, farming activities involving direct handling and the application of pesticides are traditionally performed by men, which explains why the farmer group consists exclusively of males. Conversely, women in rural areas are more likely to be indirectly exposed through domestic tasks, occasional farm assistance, or environmental proximity to treated fields.

This study was conducted in a rural municipality with intensive agricultural activity and significant herbicide use, thus providing a relevant setting for assessing the health effects related to this exposure. Sample size calculation was performed using statistical power analysis ($\alpha = 0.05$), taking into account the anticipated effect size, expected data variability, and the power required to detect significant associations between herbicide exposure and health parameters. Practical considerations related to participant accessibility, logistical constraints of recruitment and biological sample collection, and available resources were also incorporated in the final sample size determination. By ethical principles, all participants provided written informed consent after receiving full information on the objectives and conduct of the study. Data collection combined the following: (1) a standardized questionnaire administered face-to-face by trained interviewers, allowing the collection of sociodemographic characteristics, occupational history, and herbicide exposure patterns; (2) biological samples collected according to standardized protocols to ensure the quality of the analyses.

The farmers involved in the study were asked to report on the plant protection products they used. In our initial publication, we focused on glyphosate and its metabolite amino-methyl phosphonic acid (AMPA), given their widespread use in the region [13]. In this article, we extend our analysis to other identified pesticides. This complementary approach allows for a more comprehensive assessment of multiple exposures and their potential impacts on health and the environment in this agricultural area.

This study received ethical approval from the Regional Health Directorate of Fez-Meknes (ref. #91#/16/06/2023) and was conducted by the principles of the Declaration of Helsinki. Participants collected a first-morning urine sample in sterile 100 mL polypropylene containers. A minimum volume of 20 mL was required for a sample to be considered valid and included in the analysis.

A strict storage protocol was applied: samples were immediately refrigerated at 4 °C during transport to the ERES laboratory of the Faculty of Medicine, Pharmacy, and Odontology of Fez, and then stored at −80 °C for subsequent analysis within an average of 3 h post-collection. The assays were performed by the Pharmacology and Toxicology Laboratory of the Limoges University Hospital (France). After being stored at −80 °C for several weeks upon reception, the samples were thawed, and aliquots were prepared for further analysis. Aliquots intended for general screening were analyzed directly, while those designated for analysis of pyrethroid metabolites and dialkylphostates were stored at −80 °C until further testing implying that these specific aliquots underwent two freeze–thaw cycles.

To ensure data confidentiality, all personal information was anonymized before processing, by current regulations on the protection of personal data. This methodology ensured both the quality of the analyses and compliance with ethical standards throughout the study.

2.3. *Pesticides Preparation and Extraction*

2.3.1. General Pesticide Screening

A liquid–solid extraction protocol was applied to 1 mL of a urine or serum sample, into which 15 µL of a 1:10 diluted working solution (Sol El LC) containing analytical standards (e.g., albendazole, carbendazim, phosalone) and an internal standard (IS) was spiked. The sample was then treated with 5 mL of acetonitrile (ACN) and 1 spatula of CHROMABOND® QuEChERS Mix I (Macherey-Nagel, REF 730970.1, Sorbent-LOT 0125/30). This salt mixture consists of 4 g of anhydrous magnesium sulfate ($MgSO_4$, ≥97%), 1 g of sodium chloride (NaCl), 0.5 g of disodium hydrogen citrate sesquihydrate ($Na_2H\cdot citrate\cdot 1.5\ H_2O$), and 1 g of trisodium citrate dihydrate ($Na_3\cdot citrate\cdot 2\ H_2O$). These salts facilitate phase separation and buffer the pH during extraction. After vortexing, the samples were centrifuged to induce phase separation. The upper organic layer (ACN) was carefully transferred to a new glass tube, followed by the addition of one extra spatula of $MgSO_4$ to ensure the dehydration of residual aqueous content. A second vortex and centrifugation step were performed. The resulting supernatant was evaporated to dryness under a gentle nitrogen stream. The dried extract was reconstituted in 100 µL of a 50:50 mixture of methanol and aqueous buffer (MeOH/Tampon 8060), and 2 µL of the reconstituted sample were injected into a liquid chromatography system coupled to tandem mass spectrometry (LC-MS/MS). This protocol was validated across a range of concentrations (0.01 to 10 µg/L), with appropriate controls and fortified samples processed in parallel to ensure analytical performance.

2.3.2. Pyrethroid Metabolites Biomonitoring

Quantification of urinary pyrethroid metabolites was performed after enzymatic hydrolysis with β-glucuronidase (from *Helix pomatia*), followed by liquid–liquid extraction with n-hexane and analysis by LC-MS/MS. The stable isotopes 3-PBA-^{13}C and trans-Cl_2CA-6D were used as IS.

A volume of 5 mL of urine was mixed with 1.25 mL of 1 M sodium acetate buffer (pH 4.8) and 25 μL of IS, and then 20 μL of β-glucuronidase was added. After incubation at 37 °C for 16 h, 6 mL of n-hexane and 1 mL of concentrated HCl (37%) were incorporated. The mixture was stirred for 10 min, and then centrifuged at 2000× *g* for 5 min. The organic phase was collected in a 15 mL glass tube, while the aqueous phase was extracted a second time with 6 mL of n-hexane under the same conditions. The two organic phases were combined, and then 3 mL of 0.1 M NaOH was added. After stirring (10 min) and centrifugation (2000× *g*, 5 min), the upper phase was removed. Then, 6 mL of n-hexane and 200 μL of HCl (37%) were added to the residue. After further stirring and centrifugation, the organic phase was collected and evaporated to dryness under a nitrogen stream.

The dry residue was reconstituted in 80 μL of a water/0.1% formic acid mixture (70:30, *v/v*), and 10 μL of this solution was injected into the LC-MS/MS system. Calibration curves were established by applying the same procedure to standards spiked with the appropriate working solutions.

2.3.3. Organophosphate Metabolite Biomonitoring

An analysis of urinary metabolites of organophosphorus pesticides was performed by LC-MS/MS after an organic solvent extraction step. Specific stable isotopes were used as ISs for determination.

The assay was conducted as follows: 20 μL of IS was added to 2 mL of urine sample. Then, 1 mL of 6 M HCl, 5 mL of diethyl ether, and 4 g of sodium chloride were added. The mixture was stirred for 15 min and centrifuged for 5 min at 2000× *g*. The organic phase was recovered into a 15 mL glass tube, while the remaining aqueous phase was re-extracted with 5 mL of ethyl acetate, stirred, and centrifuged under the same conditions as previously described. The two organic phases were combined, and then evaporated to dryness under nitrogen flow. The resuspension of the dried samples was performed in 1 mL of (50/50) 2 mM ammonium formate (pH 3) and methanol, and finally, 5 μL of the solution was injected into the LC-MS/MS system. The calibration standards were subjected to the same treatment after they were spiked with the appropriate volume of the working solutions.

2.4. *Instrumental Analysis*

2.4.1. General Pesticide Screening

The LC-MS/MS system used in this assay is composed of an LC-MS 8060 triple quadrupole mass spectrometer and a Shimadzu NEXERA X2 series, Kyoto, Japan. The column used for the separation is a Raptor Biphenyl column (100 × 2.10 mm, 2.7 μm particles), Pennsylvania, United States.

The mobile phases for the pumps were prepared as follows: for pump A, a combination of 0.002% formic acid and ammonium formate (2 mmol/L) in water was used, while for pump B, the same combination was used but this time in methanol instead of water. The flow rate was set at 0.3 mL/min and phase B was set at 10% from 0 to 0.5 min, 60% from 13 to 28 min, and 100% until the end of the run at 35 min.

The positive and negative ionization modes were used in the identification and quantification of pesticides. Pesticide identification and quantification were carried out in both positive and negative ionization modes using multiple reaction monitoring (MRM), targeting one quantifier and one qualifier ion for each of the 320 compounds included in this targeted screening [23]. For a positive identification, the ratio between the quantifier and qualifier ion transitions had to be within ±20% of the ratio determined from the calibration standards.

2.4.2. Pyrethroid Metabolites

The LC-MS/MS system used in this assay is composed of an AB SCIEX API 5500 QTrap triple quadrupole mass spectrometer and a Shimadzu LC-20AD. The column used for the separation is an Atlantis T3 column (150 × 2.10 mm, 5 µm particles) (Waters, Milford, MA, USA). The mobile phases for the pumps were prepared as follows: phase A contained 0.1% formic acid and phase B was a mixture of (95/5) methanol acidified with 0.1% formic acid. For phase A, the flow rate was set at 0.2 mL/min with a gradient of phase B that increases from 50% to 80% between 0.5 and 10 min and then decreases to 50% at 12 min until the end of run at 18 min. The identification and quantification of 3-PBA, 4-FPBA, Cl_2CA (cis and trans), and Br_2CA were conducted in negative ionization mode using MRM. For each compound, a quantifier ion transition was used (213.0/92.9 for 3-PBA, 231.0/93.1 for 4-FPBA, 208.9/36.9 for Cl_2CA, and 342.9/80.8 for Br_2CA), along with a corresponding qualifier ion transition (213.0/65.1, 231.0/65.1, 207.0/35.0, and 296.8/80.9, respectively). To confirm identification, the ratio between the quantifier and qualifier transitions, originating from the same precursor ion, was required to fall within ±20% of the ratio established by calibration standards.

2.4.3. Organophosphate Metabolite

Analyses were performed using an LC-MS/MS system integrating a triple quadrupole mass spectrometer (Shimadzu 8060) coupled with the NEXERA X2 chromatographic system (Shimadzu, Japan). Chromatographic separation was performed on an INERTSIL ODS-3 column (100 × 2.1 mm, 5 µm, GL Sciences Inc., Tokyo, Japan). Mobile phase A consisted of a 2 mM ammonium formate solution (pH 3), while phase B consisted of a 2 mM acetonitrile/ammonium formate mixture (90:10, *v*/*v*; pH 3). The gradient program included an isocratic elution at 90% B from 0 to 3 min, followed by a reduction to 20% B between 3 and 7 min, and then a return to 90% B for up to 9 min. The mobile phase flow rate was maintained at 0.2 mL/min. The identification and quantification of the metabolites DMP, DMTP, DMDTP, DEP, DETP, and DEDTP were performed in negative MRM mode, following specific transitions for each compound (quantifier and qualifier ions).

The LC-MS/MS method was validated according to standard analytical criteria, including linearity, correlation coefficient, limits of detection (LOD) and quantification, precision, accuracy, and extraction yield. This method has been accredited by COFRAC.

2.5. Statistical Analysis

Descriptive statistics, including frequency tables, were performed using IBM SPSS Statistics 26.0 software (IBM Corp., Armonk, NY, USA, 2019) to present demographic characteristics and pesticide exposure levels. All statistical tests were performed with a significance level of $\alpha = 0.05$. Pesticide concentrations are expressed in µg/L. Samples with concentrations below the LOD were assigned a zero value for statistical analysis. This conservative approach minimizes exposure overestimation in population-level comparisons. Principal component analysis, partial least squares-discriminant analysis (PLS-DA), and multiple regression were performed using the R exploratory multivariate data analysis ropls package version 1.26.428. A map showing the location of the study area was prepared using ArcGIS 10.3.1 software. Figures were generated using GraphPad Prism (version 9.0) and Adobe Illustrator 2020 for graphical refinement.

3. Results

3.1. Survey Data

Table 1 presents the sociodemographic characteristics of the 154 participants from the Fez-Meknes region of Morocco, divided into three groups: farmers (56 individuals), indirectly exposed individuals (48 individuals), and the control group (50 individuals). (Figure 1).

Table 1. Demographic characteristics of the total study population.

Feature	Overall	Farmers	Exposed Indirectly	Control	p Value
Cohort	154	56	48	50	
Age (years)					
Mean ± SD	47.45 ± 14.48	53.53 ± 13.88	49.83 ± 14.38	38.36 ± 10.39	
Median	47.00	53.00	50.50	34.50	0.232 [a]
Range	[22–87]	[29–80]	[22–87]	[26–70]	
Interquartile range	23.50	21.00	21.50	14.75	
Gender					
Man	73 (47.4%)	56 (100%)	6 (12.5%)	11 (22%)	<0.001 [b]
Female	81 (52.6%)	0 (0%)	42 (87.5%)	39 (78%)	
Education					
None	60 (38.96%)	29 (51.7%)	31 (64.6%)	0 (0%)	
Primary school	40 (25.97%)	25 (44.6%)	15 (31.25%)	0 (0%)	<0.001 [c]
Secondary school	18 (11.68%)	2 (3.7%)	2 (4.15%)	14 (28%)	
University	36 (23.39%)	0 (0%)	0 (0%)	36 (72%)	
Living on farm					
Yes	98 (63.64%)	56 (100%)	42 (87.5%)	0 (0%)	<0.001 [b]
No	56 (36.36%)	0 (0%)	6 (12.5%)	50 (100%)	
Drinking water source					
Tap	64 (41.56%)	7 (12.5%)	3 (6.25%)	50 (100%)	<0.001 [b]
Wells and rivers	90 (58.44)	49 (87.5%)	45 (93.75%)	0 (0%)	
Origin of the food consumed					
Market	86 (55.9)	14 (25)	22 (45.8)	50 (100)	<0.001 [b]
Farm	68 (44.1)	42 (75)	26 (54.2)	0 (0)	
Pesticide use					
Yes	80 (51.94%)	56 (100%)	24 (50%)	0 (0%)	0.005 [b]
No	74 (48.06%)	0 (0%)	24 (50%)	50 (100%)	
Use of PPE					
Yes	13 (8.5)	13 (23.2)	0 (0)	0 (0)	<0.001 [b]
No	141 (91.5)	43 (76.8)	48 (100)	50 (100)	

[a] Student *t*-test; [b] Chi square test; [c] Fisher's exact test. Percentages in parentheses represent proportions within each category for each group (farmers, exposed indirectly, controls). PPE, personal protective equipment.

The mean age of participants was 47.45 ± 14.48 years, with significant differences between groups ($p < 0.05$): farmers were on average older (53.53 ± 13.88 years) than indirectly exposed individuals (49.83 ± 14.38 years) and the control group (38.36 ± 10.39 years).

Farmers were exclusively male (100%), while the indirectly exposed and control groups were 87.5% and 78% female, respectively ($p < 0.001$). In terms of education, most farmers had no education (51.7%) or had only attended primary school (44.6%), whereas the control group had a higher level of education, with 28% having reached secondary school and 72% having reached university level ($p < 0.001$). Regarding place of residence, all farmers lived on farms, compared to 87.5% of indirectly exposed individuals and no individual in the control group ($p < 0.001$). The source of drinking water also varied significantly between groups: 87.5% of farmers and 93.7% of indirectly exposed individuals used wells or rivers, while 100% of the control group used tap water ($p < 0.001$) (Table 1). Among the participants, 75% of farmers consumed mainly farm products while all controls reported consuming mainly market products ($p < 0.001$). While all farmers use pesticides, only 23.2% of them (n = 13) adopt personal protective equipment (PPE).

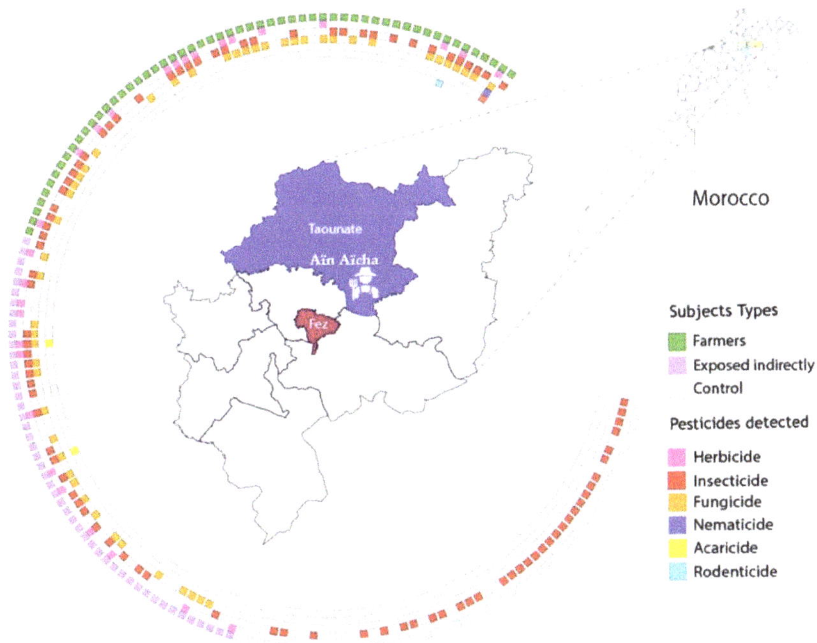

Figure 1. Classification of participants (N = 154) according to the categories of pesticides detected: representation by concentric rings showing (1) the exposure groups (farmers, indirectly exposed, controls) in the outer ring, and (2) the types of contaminants (herbicides, insecticides, fungicides, nematicides, acaricides, rodenticides) in the inner rings.

3.2. Pesticide Contamination in Urine Samples

A total of 30 unique pesticide compounds were detected (above the LOD) in samples collected from 56 farmers, 48 indirectly exposed individuals, and 50 control individuals. Table 2 shows the results of the detection of these 30 pesticides in the three groups studied: farmers, indirectly exposed individuals, and the control group.

These pesticides are distributed as follows: 10 insecticide ingredients, 11 fungicide ingredients, 5 herbicide ingredients, 2 acaricides, 1 nematicide, and 1 rodenticide. They were classified according to their hazard by the World Health Organization (WHO). Farmers have the highest exposure, with frequent detections of thiamethoxam (24 detections), dichlofluanid (10 detections), and trichlorophenol 2,4,6 (12 detections). Maximum concentrations were observed for pthalimide (30.90 µg/L), endosulfan-lactone (8.24 µg/L), and tetrachlorvinphos (7.36 µg/L). This was particularly the case for dodine (15 detections) and thiamethoxam (15 detections). Indirectly exposed individuals also showed significant levels of exposure, characterized by high levels of dodine (31.3% of samples) and thiamethoxam, as well as notable maximum concentrations of dichlofluanid (22.13 µg/L) and dinotefuran

(18.62 µg/L). In addition, the control group revealed significant contamination, particularly by imidacloprid (28 detections) and thiamethoxam (14 detections), with concentrations reaching 2.05 µg/L and 0.34 µg/L, respectively. It is important to note that the number of unique pesticide types (n = 30) differs from the total number of individual detections, as several pesticides were found in multiple participants. The diagram shows the distribution and mean concentrations of pesticides (in µg/L) detected in the three groups (Figure 2).

Table 2. Urinary pesticide contaminant profiles in 154 adults.

Pesticide			>LOD			Maximum Concentration		
Substance	Type	Hazardous Class	Farmers	Exposed Indirectly	C	Farmers	Exposed Indirectly	C
Benfluralin	Herbicide	II	0	1	0	0.00	0.05	0
Buprofezin	Insecticide	U	1	0	0	0.66	0	0
Cadusafos	Nematicide	Ia	1	0	0	0.14	0	0
Chlorobenzilate	Acaricide	II	1	0	0	0.02	0	0
Clomazone	Herbicide	II	2	2	0	0.18	0.11	0
DEET	Insecticide	U	1	0	0	1.54	0	0
Dichlofluanid	Fungicide	U	10	7	0	7.10	22.13	0
Dicloran	Fungicide	U	0	1	0	0.00	0.09	0
Diphenylamine	Fungicide	U	8	0	0	0.23	0	0
Endosulfan Lactone	Insecticide	Ib	1	0	0	8.24	0	0
Fenpropimorph	Fungicide	II	5	4	0	0.18	0.16	0
Malaoxon	Insecticide	Ib	0	2	0	0.00	3.9	0
Metribuzin	Herbicide	II	1	1	0	0.05	0.052	0
Oxadiazon	Herbicide	II	1	0	0	0.02	0	0
Oxamyl	Insecticide	Ib	3	1	0	0.50	0.15	0
Penthiopyrad	Fungicide	U	1	0	0	0.06	0	0
Phenyl 2, phenol	Fungicide	U	10	0	0	1.21	0	0
Phthalimide	Fungicide	U	9	4	0	30.90	20.40	0
Pyrimethanil	Fungicide	U	1	2	0	0.06	0.10	0
Trichlorophenol 2,4,6	Herbicide	II	12	12	0	0.29	0.34	0
Bifenazate	Acaricide	U	0	1	0	0.00	0.73	0
Carbendazim	Fungicide	II	4	2	0	0.92	0.15	0
Clofentezine	Acaricide	U	0	1	0	0.00	0.11	0
Dinotefuran	Insecticide	U	16	13	9	4.82	18.62	4.44
Dodine	Fungicide	II	1	15	0	2.20	2.97	0
Imidacloprid	Insecticide	II	0	2	28	0.00	1.29	2.05
Tetrachlorvinphos	Insecticide	II	1	0	0	7.36	0	0
Thiabendazole	Fungicide	II	0	1	0	0.00	1.14	0
Thiamethoxam	Insecticide	II	24	15	14	9.26	9.31	0.34
Warfarin	Rodenticide	Ib	1	0	0	0.11	0	0

Concentration distribution by WHO Hazard is defined according to the WHO recommended classification of pesticides by hazard, including the following: class Ia, extremely hazardous; class Ib, highly hazardous; class II, moderately hazardous; U, unlikely to present acute hazard. LOD, limit of detection; C, control.

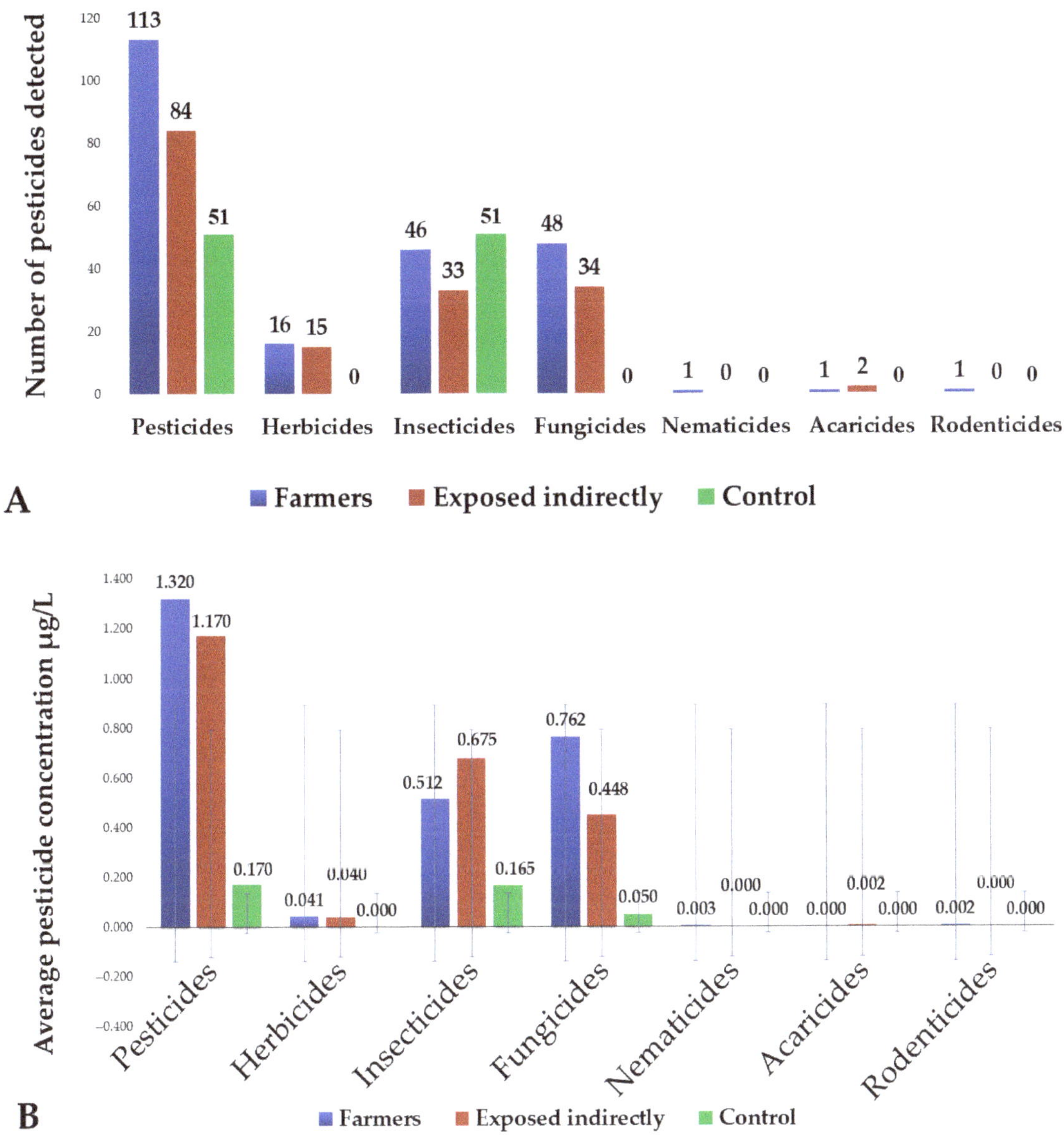

Figure 2. Histogram illustrating the number of pesticides detected (**A**) and the average pesticide concentration µg/L (**B**) in the three groups studied: farmers (blue), indirectly exposed individuals (red), and control group (green).

Farmers had the highest number of pesticides detected (113), with a predominance of insecticides (46) and fungicides (48). The highest mean concentrations were observed for insecticides (1.24 ± 0.18 µg/L) and fungicides (0.83 ± 0.12 µg/L) in this group. Indirectly exposed individuals showed intermediate levels, with a 40–60% reduction in concentrations compared to farmers. The control group mainly showed residual traces. Other pesticide categories, including herbicides, acaricides, nematicides, and rodenticides, were detected at frequencies below 15% and at concentrations not exceeding 0.2 μg/L in all groups studied. In farmers, the highest concentrations were observed for several compounds, with particularly notable values for phthalimide (10.36 μg/L), endosulfan lactone (8.24 μg/L), and tetrachlorvinphos (7.36 μg/L). Indirectly exposed individuals had significant levels of phthalimide (7.5 μg/L), dichlofluanid (6.13 μg/L), and dinotefuran (4 μg/L), while the control group mainly had residual contamination with neonicotinoids (dinotefuran at 2.1 μg/L and imidacloprid at 0.4 μg/L) (Figure 3).

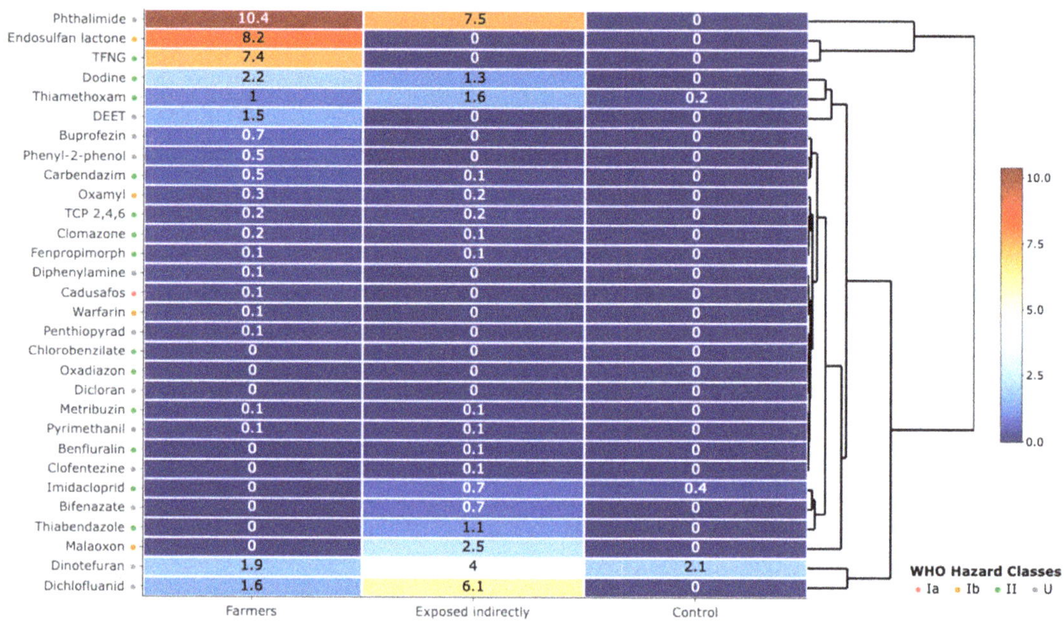

Figure 3. Supervised heat map representing the average concentrations of the 30 detected contaminants (rows) in farmers (N = 56), indirectly exposed individuals (N = 48) and the control group (N = 50), divided into three classes (columns). The colored dots indicate the pesticide hazard classes according to the WHO classification: class Ia (red), class Ib (orange), class II (green), and class U (gray).

Pesticides classified Ia/Ib by the WHO (cadusafos, endosulfan lactone, oxamyl) were exclusively detected in farmers, with concentrations reaching 8.24 μg/L for endosulfan lactone. In addition, certain compounds such as dichlofluanid and phthalimide, although classified as U, present particularly high concentrations (>5 μg/L).

PLS-DA score analysis revealed a clear separation between the three study groups (farmers, indirectly exposed individuals, and control group) based on urinary pesticide concentrations (Figure 4).

Farmers clustered in a distinct region of the factorial design, reflecting specific pesticide concentration profiles. The control group clustered in an opposite area, indicating significantly lower exposure levels. Indirectly exposed individuals occupied an intermediate position, suggesting moderate exposure. VIP analysis identified eight pesticides as main discriminant markers (VIP > 1) between exposure groups: diphenylamine, 2-phenylphenol, imidacloprid, TCP 2,4,6, phthalimide, carbendazim, fenpropimorph, and oxamyl (Figure 5).

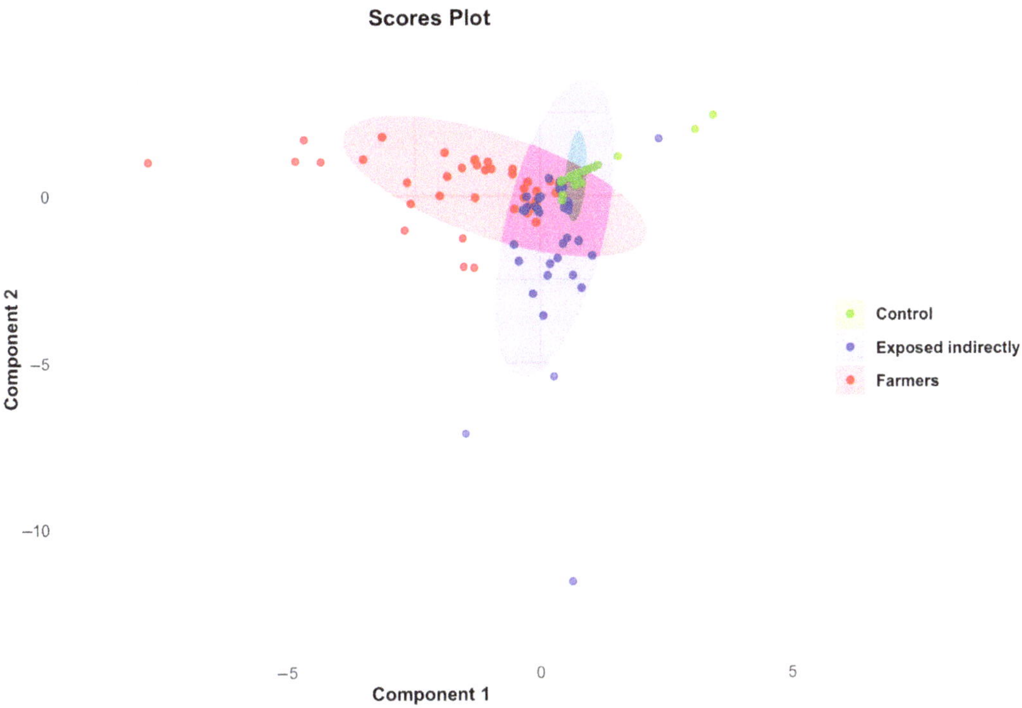

Figure 4. PLS-DA score plot based on the 30 detected contaminants, showing the separation between the three groups: farmers (N = 56; red), indirectly exposed individuals (N = 48; blue), and control group (N = 50; light green).

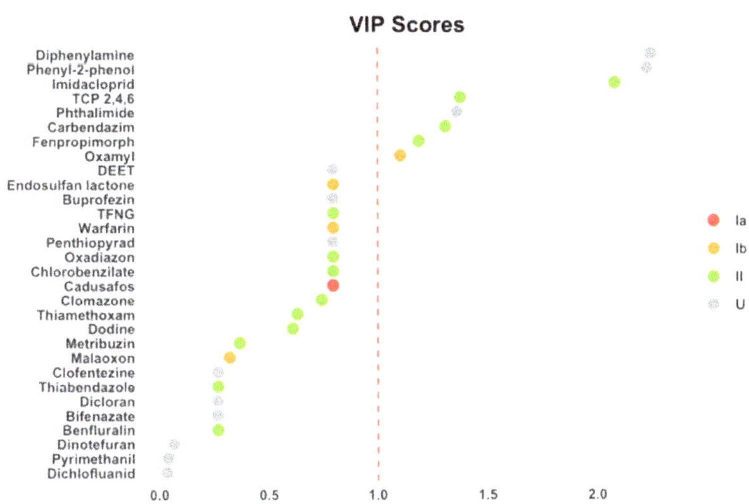

Figure 5. Dot plot displaying VIP scores for the most important pesticides identified by the PLS-DA model. The colored dots indicate the pesticide hazard classes according to the WHO classification: class Ia (red), class Ib (orange), class II (green), and class U (gray).

Correlation matrix analysis revealed distinct patterns of pesticide co-exposure. Strong correlations ($\tau > 0.6$) were observed between WHO class Ia/Ib insecticides (cadusafos, endosulfan lactone) and commonly used neonicotinoids (imidacloprid, thiamethoxam). Triazole fungicides, such as dodine and carbendazim, showed intermediate-to-strong correlations ($\tau = 0.58$–0.65) and formed a distinct cluster (Figure 6).

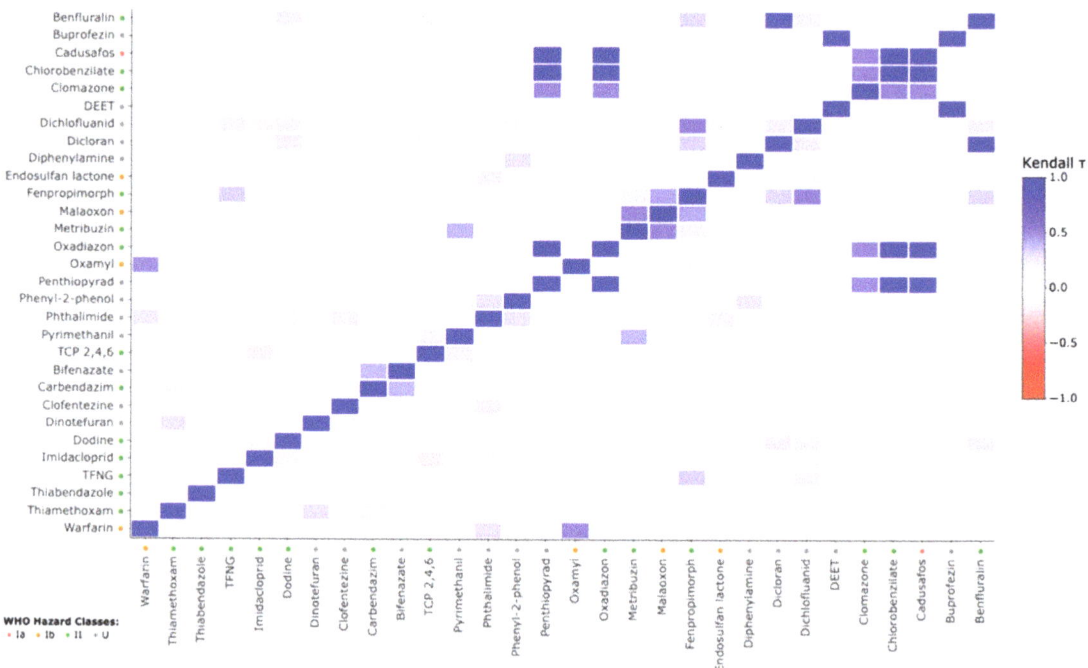

Figure 6. Sorted correlation matrix of the 30 contaminants detected in urine samples. Kendall's τ coefficients are used to represent the strength and significance of the correlations, according to the scale shown in the legend to the right. The colored dots indicate the pesticide hazard classes according to the WHO classification: class Ia (red), class Ib (orange), class II (green), and class U (gray).

3.3. *Correlation of Urinary Pesticide Concentrations with Sociodemographic Factors*

Multivariate regression analysis, the detailed results of which are presented in Table 3, assessed the associations between urinary concentrations of different types of pesticides and various sociodemographic and behavioral variables. It was found that there was no statistically significant association between the age or sex of the participants and the urinary concentrations of the pesticides studied. However, this analysis revealed several significant associations. More specifically, individuals with a university education level had significantly lower urinary concentrations of acaricides ($p = 0.014$) compared to those without formal education. Furthermore, residence in an agricultural environment was associated with higher urinary concentrations of acaricides ($p = 0.001$). Similarly, the consumption of water from wells or rivers was linked to higher urinary concentrations of rodenticides. Finally, the use of Personal Protective Equipment (PPE) was associated with significantly lower urinary concentrations of herbicides ($p = 0.019$) and rodenticides ($p = 0.008$). These four variables retained after the multivariate analysis were then included in a supervised multivariate approach to better elucidate their individual contributions to pesticide contamination.

Table 3. Summary of the multiple regression model: concentrations by pesticide type vs. explanatory variables.

Var	Herbicide (µg/L)			Insecticide (µg/L)			Fungicide (µg/L)			Acaricide (µg/L)			Nematicide (µg/L)			Rodenticide (µg/L)		
	M	Est Coeff.	p	M	Est Coeff.	p	M	Est Coeff.	p	M	Est Coeff.	p	M	Est Coeff.	p	M	Est Coeff.	p
Age	-	0.001 (−0.001–0.001)	0.645	-	−0.008 (−0.022–0.007)	0.297	-	−0.088 (−0.031–0.012)	0.390	-	−0.101 (-)	0.313	-	−0.076 (-)	0.443	-	−0.151 (-)	0.150
Gender		−0.007 (−0.032–0.017)	0.562		−0.125 (−0.542–0.293)	0.556		−0.439 (−1.048–0.171)	0.157		0.029 (−0.003–0.004)	0.765		0.009 (−0.004–0.004)	0.926		−0.099 (−0.005–0.002)	0.337
Man	0.034	-		0.444	-		0.675	-		0.000	-		0.001	-		0.001	-	
Female	0.021			0.457			0.183			0.001			0.000			0.000		
Education		−0.014 (−0.031–0.004)	0.134		−0.255 (−0.558–0.047)	0.097		−0.353 (−0.794–0.089)	0.117		−0.419 (−0.006–−0.001)	0.014		−0.042 (−0.003–0.003)	00.805		−0.223 (−0.004–0.001)	0.209
None	0.044	-		0.637	-		0.821	-		0.001	-		0.000	-		0.001	-	
Primary school	0.035			0.510			0.324			0.000			0.003			0.000		
Secondary school	0.007			0.271			0.078			0.000			0.000			0.000		
University	0.000			0.161			0			0.000			0.000			0.000		
Living on farm	-	−0.018 (−0.062–0.026)	0.419	-	0.085 (−0.564–0.938)	0.623	-	−0.133 (−1.229–0.962)	0.810	-	0.566 (0.004–0.017)	0.001	-	−0.214 (−0.013–0.003)	0.208	-	0.083 (−0.005–0.008)	0.639
Yes	0.040	-		0.530	-		0.654	-		0.000	-		0.001	-		0.001	-	
No	0.004			0.313			0.000			0.002			0.000			0.000		
Drinking water	-	−0.009 (−0.046–0.028)	0.644	-	0.247 (−0.093–1.168)	0.094	-	−0.298 (−1.219–0.623)	0.524	-	0.110 (−0.003–0.007)	0.445	-	−0.369 (−0.015–−0.002)	0.011	-	−0.072 (−0.007–0.004)	0.634
Tap	0.008	-		0.186	-		0.782	-		0.000	-		0.002	-		0.000	-	
Wells and rivers	0.039			0.621			0.001			0.001			0.000			0.001		
Origin of the food consumed	-	0.009 (−0.017–0.034)	0.494	-	0.066 (−0.291–0.571)	0.521	-	−0.048 (−0.677–0.582)	0.881	-	−0.195 (−0.007–0.000)	0.054	-	−0.075 (−0.006–0.003)	0.452	-	−0.064 (−0.005–0.003)	0.543
Farm	0.036	-		0.500	-		0.670	-		0.002	-		0.002	-		0.001	-	
Market	0.020			0.413			0.215			0.000			0.000			0.000		
Pesticide use	-	−0.039 (−0.039–0.029)	0.777	-	0.521 (−0.085–1.080)	0.093	-	−0.096 (−0.946–0.754)	0.824	-	0.029 (−0.004–0.005)	0.833	-	−0.054 (−0.007–0.005)	0.005	-	−0.038 (−0.006–0.004)	0.792
Yes	0.040	-		0.456	-		0.692	-		0.000	-		0.001	-		0.001	-	
No	0.013			0.840			0.118			0.001			0.000			0.000		
Use of PPE	-	0.204 (0.008–0.086)	0.019	-	0.000 (−0.664–0.662)	0.998	-	0.748 (−0.220–1.715)	0.129	-	−0.044 (−0.007–0.004)	0.612	-	−0.231 (−0.016–0.002)	0.008	-	0.074 (−0.003–0.008)	0.415

Table 3. *Cont.*

Var	Herbicide (μg/L)			Insecticide (μg/L)			Fungicide (μg/L)			Acaricide (μg/L)			Nematicide (μg/L)			Rodenticide (μg/L)		
	M	Est Coeff.	*p*	M	Est Coeff.	*p*	M	Est Coeff.	*p*	M	Est Coeff.	*p*	M	Est Coeff.	*p*	M	Est Coeff.	*p*
Yes	0.007	-		0.392	-		0.256	-		0.010	-		0.106	-		0.000	-	
No	0.002			0.457			0.000			0.000			0.000			0.000		
R2		0.127		-	0.089			0.079			0.119			0.122			0.034	

Var, variables; M, mean; Est coeff, estimation coefficient (95% CI).

4. Discussion

In this research, an exploratory observational survey was conducted to assess pesticide exposure among residents of the Fez-Meknes region. This work is a continuation of an initial investigation on urinary biomonitoring of glyphosate and its main metabolite, AMPA, among farmers, indirectly exposed individuals, and a control group [13]. Although this initial study demonstrated exposure to these substances, available data on other pesticide classes remain limited. Therefore, the present study aims to expand knowledge by examining exposure to a broader range of pesticides from various chemical families to obtain a more comprehensive assessment of toxicological risk for this population. We measured pesticide concentrations in urine samples from 154 adults living in rural and urban areas, using validated analytical methods [23]. A control was evaluated at an interval of 10 samples at varying levels, varying between the LOQ and a median concentration level. All aspects related to analytical validation, including instrumental calibration and quality assurance procedures, quality control using replicates, and the assessment of matrix effects and sensitivity, are comprehensively detailed in Supplementary File S2.

Urine constitutes a primary biological matrix for biomonitoring due to its ability to reflect recent exposure levels, making it a valuable tool for assessing occupational and environmental risks [24]. Urine collection is non-invasive and can be easily conducted in large population studies, facilitating extensive biomonitoring efforts [25]. We therefore assessed pesticide exposure among farmers, their indirectly exposed wives, and the control group.

Our results indicate heterogeneous but widespread pesticide contamination across all groups studied, with a marked prevalence among farmers, reflecting acute occupational exposure. A total of 30 active substances were detected above the LOD, including a majority being insecticides (10 compounds) and fungicides (11 compounds), reflecting their dominant use in regional phytosanitary practices [6]. The exposure burden is significantly higher among farmers, both in terms of frequency and concentration, characterized by the detection of 113 pesticides including 46 insecticides and 48 fungicides, directly reflecting intensive agricultural practices. The presence at particularly high concentrations of phthalimide (30.90 μg/L) and endosulfan-lactone (8.24 μg/L), two compounds recognized as endocrine disruptors (EDCs), raises major public health concerns due to their demonstrated ability to interfere with hormonal regulatory mechanisms. These substances exert their harmful effects mainly by binding to hormone receptors, disrupting the synthesis, metabolism and activity of endogenous hormones, including estrogens and androgens, essential for the maintenance of many functions [26–28]. This biochemical interaction can lead to significant hormonal imbalance, which can compromise human health systemically. Indeed, chronic exposure to these EDCs is increasingly correlated with an increased incidence of serious pathologies such as hormone-dependent cancers (especially breast and prostate cancers), metabolic disorders including obesity and type 2 diabetes, as well as various alterations of reproductive function, such as infertility and abnormalities of embryonic and fetal development [29,30].

The exclusive detection of WHO Ia/Ib pesticides, including cadusafos and oxamyl, in the agricultural population represents a critical public health problem. Chronic exposure to these substances can lead to various neurological disorders, hence the need to raise awareness among agricultural populations and take preventive measures [31]. The classification of dichlofluanid and phthalimide as "low hazard" (U) despite their high concentrations (>5 μg/L) raises serious concerns regarding traditional toxicological assessments. These classifications often neglect chronic effects at low doses and the possibility of combined effects, which may lead to an underestimation of the risks associated with these substances.

A recent ecotoxicological study demonstrates that dichlofluanid has significant chronic toxicity in marine organisms, with adverse effects observed at environmentally relevant concentrations as low as 1 μg/L in sea urchin embryos and 100 μg/L in copepods [32]. Chronic exposure to low doses can lead to cumulative health effects, as evidenced by studies linking pesticide exposure to chronic respiratory diseases [33].

These findings call for a rethinking of conventional toxicological frameworks, including incorporating the potential effects of repeated exposures. Our correlation matrix (Figure 6) highlights co-detections in pairwise combinations of neonicotinoids, triazole fungicides, and organophosphate insecticides belonging to different chemical classes. This pattern indicates potential cumulative or synergistic effects via overlapping biological mechanisms such as endocrine disruption and neurotoxicity. A recent study has proven that co-exposure to pyrethroids and phenylpyrazoles results in synergistic neurotoxicity via shared mitochondrial signaling and proteolytic chimneys [34].

In this context, approaches such as cumulative risk assessment (e.g., EPA Cumulative Risk Index) or Adverse Outcome Pathways offer promising integrative frameworks that are better suited to the complex exposures observed in agricultural and peri-agricultural populations [35].

In indirectly exposed individuals, the analysis reveals an intermediate exposure profile, with an estimated 40–60% decrease in concentrations compared to farmers. However, the frequent detection of pesticides such as dodine (31.3% of samples) and thiamethoxam, as well as the high concentrations measured for dichlofluanid (22.13 μg/L) and dinotefuran (18.62 μg/L), indicate diffuse environmental contamination, probably due to air drift, dust transport or soil leaching [36,37].

On the other hand, the detection of imidacloprid residues (28 samples, up to 2.05 μg/L) and thiamethoxam (14 samples) in the control group confirms the environmental dispersion of neonicotinoids in Morocco. Although these substances are subject to restrictions in the European Union, their use remains widespread in Moroccan agriculture, particularly in intensive crops (market gardening, citrus fruits, and cereals) [38]. These compounds persist in the environment and exhibit high mobility, with a demonstrated ability to diffuse into aquatic ecosystems and accumulate in food chains [39]. Their detection, even at low doses, remains a concern due to their potential neurotoxic effects, particularly on brain development [40,41].

Multivariate regression analysis (Table 3) highlighted key determinants of pesticide exposure in our study population, highlighting the complexity of exposure pathways. We observed that individuals with a higher level of education, particularly university education, had significantly lower urinary acaricide concentrations (p = 0.014). Similar observations were reported in an integrated analysis of three epidemiological studies: the PIPAH cohort in the UK, which focused on professional pesticide users, a study of farmers in Malaysia, and the PESTROP project in Uganda [42]. Higher education levels are generally associated with a better understanding of the risks associated with pesticide use, an increased awareness of potential health effects, and more frequent adoption of personal protective measures [43]. In Morocco, it has been shown that the majority of farmers in the Fez-Meknes region have a low level of education, which constitutes an obstacle to the adoption of safe agricultural practices [6].

Moreover, residence in an agricultural setting is frequently correlated with increased pesticide exposure, due to proximity to application sources [44,45]. A statistically significant association was observed between place of residence and urinary concentrations of acaricides (p = 0.001), confirming persistent passive environmental exposure in individuals living in these areas. This exposure can occur through inhalation of contaminated particles in the air, skin contact with household surfaces or ingestion of contaminated dust, especially in homes close to treated fields [46,47]. In this context, the quality of water used for domestic purposes could represent an additional source of exposure. Indeed, the consumption of water from untreated wells or rivers has been correlated with significantly higher urinary concentrations of rodenticides. This result suggests possible contamination of water resources by agricultural residues, via runoff, infiltration, or drift of products applied to crops [48]. Thus, the combination of several environmental vectors, such as air, dust, surfaces, and water, could explain the increased levels of biologically detectable pesticides in human biological matrices in rural areas, even in the absence of direct occupational exposure [49,50].

Along with passive exposure experienced by rural residents, active or occupational exposure represents another major route of pesticide absorption, particularly among farmers [51]. In this context, the use of PPE such as gloves, masks, suits, or goggles appears to be an effective preventive measure [52]. Our results show that regular PPE use was associated with significantly lower urinary concentrations of herbicides and rodenticides, consistent with recent research findings showing that behavioral practices, including PPE use, can alter biological exposure profiles [53,54]. For example, a longitudinal study embedded in the ECHO program in the United States found that changes in behavior were associated with measurable changes in urinary concentrations of various contaminants, including pesticides [55]. However, in many agricultural areas, the use of PPE remains insufficient or inappropriate, often due to economic barriers, a lack of awareness of the risks or the perception that such equipment hinders comfort and efficiency at work [56–58].

This study provides a comprehensive assessment of pesticide exposure in Morocco using a biomonitoring approach, presenting several major strengths. First, the inclusion of three distinct groups (farmers, indirectly exposed individuals, and urban controls) allows for robust comparisons of occupational and environmental exposure profiles. The use of COFRAC-accredited LC-MS/MS methods ensures high analytical reliability, with the detection of 30 pesticide compounds belonging to different classes. The study design integrates both quantitative exposure data and detailed sociodemographic information, allowing for a nuanced analysis of risk factors. Moreover, the application of PLS-DA multivariate analysis effectively differentiates exposure profiles between groups.

However, several methodological limitations must be acknowledged regarding the assessment of long-term exposure trends or the establishment of direct causal relationships between pesticide exposure and potential health outcomes. Although urinary biomonitoring is a valuable and validated tool for occupational and environmental risk assessment, offering a non-invasive and easily implementable method for population studies, it primarily reflects recent exposure. Therefore, the urine analyses conducted here cannot reflect chronic accumulation or seasonal variations in pesticide use.

In our study, we performed a one-time spot urine sampling, which provides a snapshot of recent exposure rather than a long-term exposure estimate. The majority of the pesticides detected in our cohort are characterized by short biological half-lives, typically ranging from a few hours to a few days. For example, thiamethoxam $t_1/_2$ is 1–3 h in human plasma, with rapid urinary excretion of its metabolite clothianidin [59]; imidacloprid, $t_1/_2 \approx 5$ h in humans, with urinary elimination of hydroxylated and nitroso metabolites; carbendazim, $t_1/_2 \approx 6$–12 h in humans; rapid first-pass metabolism [60]; phthalimide (a metabolite of folpet and captan), rapidly hydrolyzed and cleared within 24 h in rodent models [61]. Given these short half-lives, the detection of parent compounds or metabolites in urine indicates recent exposure, typically within the preceding 24–72 h. However, without longitudinal sampling or dietary/exposure diaries, it is not possible to extrapolate this to average or chronic exposure levels.

Furthermore, the targeted selection of the municipalities of Aïn Aïcha and Fez, while relevant for our local objective, limits the generalizability of the results to all Moroccan agricultural areas. Regarding the demographic characteristics of the sample, the marked gender imbalance (100% male farmers, 87.5% female indirectly exposed individuals, and 78% female controls) reflects the socio-professional reality on the ground in this region of Morocco. Although this may a priori raise questions about the interpretation of sex-specific exposure patterns, our multiple regression model included sex as an explanatory variable, and the results indicated a lack of statistically significant association between participant sex and urinary concentrations of the studied pesticides for the compounds considered. Other methodological limitations include the failure to account for potential pesticide residues in consumed food, as well as in air, soil, and surface water. Thus, although urinary biomonitoring reflects overall exposure, precise identification of exposure pathways, particularly dietary and environmental, remains limited. Future research should include these measurements to better characterize exposure sources.

5. Conclusions

Our research highlights significantly different pesticide exposure levels according to the proximity to agricultural activities in the Fez-Meknes region. Farmers exhibit the highest concentrations, including pesticides classified as highly hazardous by the WHO. Intermediate levels of contamination are observed among those indirectly exposed, and residual traces are detected even in urban areas.

The study reveals significant associations between elevated pesticide levels and various sociodemographic factors, including lower educational attainment, residence in agricultural areas, and the use of untreated water sources. In contrast, wearing PPE is associated with a significant reduction in exposure risk. These findings underscore the urgent need for strengthened pesticide management policies and targeted educational interventions, particularly in rural settings, to reduce health risks linked to pesticide exposure and promote safer agricultural practices.

Supplementary Materials: The following supporting information can be downloaded at: https://www.mdpi.com/article/10.3390/jox15040120/s1, Supplementary File S1, list of internal standards; and jox-3727112, Supplementary File S2, general screening method and quality control.

Author Contributions: Conceptualization, Z.B.K., S.E.B., G.B., T.S.H. and S.A.; Data curation, Z.B.K., S.E.B., F.M., Y.E.K., G.B., P.M., T.S.H. and S.A.; Formal analysis, S.E.B., F.M., Y.E.K., G.B., P.M., T.S.H. and S.A.; Investigation, Z.B.K., A.-F.R., S.E.B. and F.M.; Methodology, Z.B.K., A.-F.R., F.M., Y.E.K. and P.M.; Resources, S.E.B., F.M. and Y.E.K.; Software, Z.B.K.; Supervision, A.-F.R., T.S.H. and S.A.; Validation, A.-F.R., G.B., P.M., T.S.H. and S.A.; Visualization, All authors; Writing—original draft, All authors; Writing—review and editing, Z.B.K., A.-F.R., G.B. and T.S.H. All authors have read and agreed to the published version of the manuscript.

Funding: This research was supported by the project 'Evaluation of Health and Carcinogenic Risk Related to Occupational Pesticide Exposure' (No. 920/AMP 2019) for the Cancer Research Institute (IRC) in Fez, Morocco, and the University of Oradea, Oradea, Romania supported the APC.

Institutional Review Board Statement: The study was conducted in accordance with the Declaration of Helsinki and approved by the Health Regional Direction of the Fez Meknes Region (#91#/16/06/2023).

Informed Consent Statement: Informed consent was obtained from all subjects involved in the study.

Data Availability Statement: The raw data supporting the conclusions of this article will be made available by the authors on request.

Acknowledgments: The authors would like to express their gratitude to the University of Oradea, Romania for its valuable support and active collaboration, which were instrumental in carrying out the present research and for supporting the APC. In addition, the authors wish to express their sincere gratitude to researchers Haitam Lahmamsi, Imane Berni, Said Ezrari, and Mohamed Afghani for their insightful input during the study design and interpretation phases.

Conflicts of Interest: The authors declare no conflicts of interest.

Abbreviations

The following abbreviations are used in this manuscript:

LC-MS/MS	Liquid chromatography-tandem mass spectrometry
IS	Internal standard
MRM	Multiple reaction monitoring
LOD	Limit of detection
PLS-DA	Partial least squares-discriminant analysis
PPE	Personal protective equipment
WHO	World Health Organization
EDCs	Endocrine disruptors

References

1. Filipoiu, D.C.; Bungau, S.G.; Endres, L.; Negru, P.A.; Bungau, A.F.; Pasca, B.; Radu, A.F.; Tarce, A.G.; Bogdan, M.A.; Behl, T.; et al. Characterization of the Toxicological Impact of Heavy Metals on Human Health in Conjunction with Modern Analytical Methods. *Toxics* **2022**, *10*, 716. [CrossRef] [PubMed]
2. Nicolopoulou-Stamati, P.; Maipas, S.; Kotampasi, C.; Stamatis, P.; Hens, L. Chemical Pesticides and Human Health: The Urgent Need for a New Concept in Agriculture. *Front. Public Health* **2016**, *4*, 148. [CrossRef] [PubMed]
3. Zhou, W.; Li, M.; Achal, V. A Comprehensive Review on Environmental and Human Health Impacts of Chemical Pesticide Usage. *Emerg. Contam.* **2025**, *11*, 100410. [CrossRef]
4. Shekhar, C.; Khosya, R.; Thakur, K.; Mahajan, D.; Kumar, R.; Kumar, S.; Sharma, A.K.A. Systematic Review of Pesticide Exposure, Associated Risks, and Long-Term Human Health Impacts. *Toxicol. Rep.* **2024**, *13*, 101840. [CrossRef] [PubMed]
5. Ahmad, M.F.; Ahmad, F.A.; Alsayegh, A.A.; Zeyaullah, M.; AlShahrani, A.M.; Muzammil, K.; Saati, A.A.; Wahab, S.; Elbendary, E.Y.; Kambal, N.; et al. Pesticides Impacts on Human Health and the Environment with Their Mechanisms of Action and Possible Countermeasures. *Heliyon* **2024**, *10*, e29128. [CrossRef] [PubMed]
6. Ben Khadda, Z.; Fagroud, M.; El Karmoudi, Y.; Ezrari, S.; Berni, I.; De Broe, M.; Behl, T.; Bungau, S.G.; Sqalli Houssaini, T. Farmers' Knowledge, Attitudes, and Perceptions Regarding Carcinogenic Pesticides in Fez Meknes Region (Morocco). *Int. J. Environ. Res. Public Health* **2021**, *18*, 10879. [CrossRef] [PubMed]
7. Bahouq, M.; Bahouq, H.; Soulaymani, A. Bibliographic Review of Phytopharmacovigilance Actions and Measures on Plant Protection Products in Morocco. *E3S Web Conf.* **2021**, *319*, 1034. [CrossRef]
8. Quindroit, P.; Crépet, A.; Brochot, C. Estimating human exposure to pyrethroids' mixtures from biomonitoring data using physiologically based pharmacokinetic modeling. *Environ. Res.* **2021**, *192*, 110281. [CrossRef] [PubMed]
9. Chin-Chan, M.; Navarro-Yepes, J.; Quintanilla-Vega, B. Environmental Pollutants as Risk Factors for Neurodegenerative Disorders: Alzheimer and Parkinson Diseases. *Front. Cell. Neurosci.* **2015**, *9*, 124. [CrossRef] [PubMed]
10. ben Khadda, Z.; Fagroud, M.; El Karmoudi, Y.; Ezrari, S.; Elhanafi, L.; Radu, A.F.; Bungau, S.G.; Houssaini, T.S. Association between Pesticide Exposure and End-Stage Renal Disease: A Case-Control Study from Morocco Based on the STROBE Guidelines. *Ecotoxicol. Environ. Saf.* **2024**, *288*, 117360. [CrossRef] [PubMed]
11. Marx-Stoelting, P.; Niemann, L.; Ritz, V.; Ulbrich, B.; Gall, A.; Hirsch-Ernst, K.I.; Pfeil, R.; Solecki, R. Assessment of Three Approaches for Regulatory Decision Making on Pesticides with Endocrine Disrupting Properties. *Regul. Toxicol. Pharmacol.* **2014**, *70*, 590–604. [CrossRef] [PubMed]
12. Ewence, A.; Brescia, S.; Johnson, I.; Rumsby, P.C. An Approach to the Identification and Regulation of Endocrine Disrupting Pesticides. *Food Chem. Toxicol.* **2015**, *78*, 214–220. [CrossRef] [PubMed]

13. ben Khadda, Z.; Bungau, S.G.; El Balkhi, S.; Ezzari, S.; Radu, A.-F.; Houssaini, T.S.; Achour, S. Urinary Biomonitoring of Exposure to Glyphosate and Its Metabolite Amino-Methyl Phosphonic Acid among Farmers and Non-Farmers in Morocco. *Environ. Toxicol. Pharmacol.* **2025**, *113*, 104620. [CrossRef] [PubMed]
14. Palma-Gudiel, H.; Fañanás, L. Prenatal Exposures and Behavioral Epigenetics in Human Infants and Children. In *Developmental Human Behavioral Epigenetics*; Elsevier: Amsterdam, The Netherlands, 2021; pp. 83–90.
15. Cao, L.; Kang, Q.; Tian, Y. Pesticide Residues: Bridging the Gap between Environmental Exposure and Chronic Disease through Omics. *Ecotoxicol. Environ. Saf.* **2024**, *287*, 117335. [CrossRef] [PubMed]
16. Hopf, N.B.; Rousselle, C.; Poddalgoda, D.; Lamkarkach, F.; Bessems, J.; Schmid, K.; Jones, K.; Takaki, K.; Casteleyn, L.; Jeddi, M.Z.A. Harmonized Occupational Biomonitoring Approach. *Environ. Int.* **2024**, *191*, 108990. [CrossRef] [PubMed]
17. Manno, M. Biological Monitoring for Environmental Health Risk Assessment. In *Environmental Geochemistry*; Elsevier: Amsterdam, The Netherlands, 2024; pp. 101–119.
18. Santonen, T.; Aitio, A.; Fowler, B.A.; Nordberg, M. Biological Monitoring and Biomarkers. In *Handbook on the Toxicology of Metals*; Elsevier: Amsterdam, The Netherlands, 2015; pp. 155–171.
19. Guimarães, M.T.; Cunha, M.G.; Carvalho, D.P.; Ribeiro, T.S.; Martins, L.C.; Braga, A.L.F.; Pereira, L.A.A. Influence of Environmental Contamination on Pregnancy Outcomes. *Environ. Sci. Pollut. Res.* **2015**, *22*, 14950–14959. [CrossRef] [PubMed]
20. Di Ciaula, A.; Gentilini, P.; Diella, G.; Lopuzzo, M.; Ridolfi, R. Biomonitoring of Metals in Children Living in an Urban Area and Close to Waste Incinerators. *Int. J. Environ. Res. Public Health* **2020**, *17*, 1919. [CrossRef] [PubMed]
21. Balderrama-Carmona, A.P.; Valenzuela-Rincón, M.; Zamora-Álvarez, L.A.; Adan-Bante, N.P.; Leyva-Soto, L.A.; Silva-Beltrán, N.P.; Morán-Palacio, E.F. Herbicide Biomonitoring in Agricultural Workers in Valle Del Mayo, Sonora Mexico. *Environ. Sci. Pollut. Res.* **2020**, *27*, 28480–28489. [CrossRef] [PubMed]
22. Ogidi, O.I.; Onwuagba, C.G.; Richard-Nwachukwu, N. Biomonitoring Tools, Techniques and Approaches for Environmental Assessments. In *Biomonitoring of Pollutants in the Global South*; Springer: Berlin/Heidelberg, Germany, 2024; pp. 243–273.
23. Zarrouk, E.; El Balkhi, S.; Saint-Marcoux, F. Critical Evaluation of High-Resolution and Low-Resolution Mass Spectrometry for Biomonitoring of Human Enviromental Exposure to Pesticides. *Environ. Technol. Innov.* **2024**, *2*, 549. [CrossRef]
24. Nijssen, R.; Lommen, A.; van den Top, H.; van Dam, R.; Meuleman-Bot, C.; Tienstra, M.; Zomer, P.; Sunarto, S.; van Tricht, F.; Blokland, M. Assessment of Exposure to Pesticides: Residues in 24 h Duplicate Diets versus Their Metabolites in 24 h Urine Using Suspect Screening and Target Analysis. *Anal. Bioanal. Chem.* **2024**, *416*, 635–650. [CrossRef] [PubMed]
25. Filippi, I.; Fernández, P.; Grimalt, J.O.; Butinof, M.; Amé, M.V.; Muñoz, S.E. Glyphosate and AMPA in Saliva and Other Traditional Human Matrices. New Findings for Less Invasive Biomonitoring to the Exposure to Pesticides. *Environ. Adv.* **2024**, *15*, 100474. [CrossRef]
26. Ahn, C.; Jeung, E.-B. Endocrine-Disrupting Chemicals and Disease Endpoints. *Int. J. Mol. Sci.* **2023**, *24*, 5342. [CrossRef] [PubMed]
27. Fenichel, P.; Brucker-Davis, F.; Chevalier, N. Perturbateurs Endocriniens–Reproduction et Cancers Hormono-Dépendants. *Presse Med.* **2016**, *45*, 63–72. [CrossRef] [PubMed]
28. Yaglova, N.V.; Yaglov, V. V Endocrine Disruptors Are a Novel Direction of Endocrinologic Scientific Investigation. *Ann. Russ. Acad. Med. Sci.* **2012**, *67*, 56–61. [CrossRef]
29. Calaf, G.M.; Ponce-Cusi, R.; Aguayo, F.; Muñoz, J.P.; Bleak, T.C. Endocrine Disruptors from the Environment Affecting Breast Cancer. *Oncol. Lett.* **2020**, *20*, 19–32. [CrossRef] [PubMed]
30. Asteria, C.; Morpurgo, P.S.; Cerutti, N.; Hall, J.M. Endocrine Disruptors in Obesity. *Front. Endocrinol.* **2024**, *15*, 1526898. [CrossRef] [PubMed]
31. Arab, A.; Mostafalou, S. Neurotoxicity of Pesticides in the Context of CNS Chronic Diseases. *Int. J. Environ. Health Res.* **2022**, *32*, 2718–2755. [CrossRef] [PubMed]
32. Cruz, A.C.F.; Kamikawachi, R.C.; de Figueiredo Eufrasio Pauly, G.; Campos, B.G.; Castro, Í.B.; de Souza Abessa, D.M. Adverse Effects of Dichlofluanid on Neotropical Marine Organisms. *Discov. Ocean.* **2025**, *2*, 3. [CrossRef]
33. Jestin-Guyon, N.; Raherison-Semjen, C. Pesticide Exposure and Chronic Respiratory Diseases. *Rev. Mal. Respir.* **2024**, *41*, 343–371. [CrossRef] [PubMed]
34. Hirano, T.; Ikenaka, Y.; Nomiyama, K.; Honda, M.; Suzuki, N.; Hoshi, N.; Tabuchi, Y. An Adverse Outcome Pathway-Based Approach to Assess the Neurotoxicity by Combined Exposure to Current-Use Pesticides. *Toxicology* **2023**, *500*, 153687. [CrossRef] [PubMed]
35. Jornod, F.; Rugard, M.; Tamisier, L.; Coumoul, X.; Andersen, H.R.; Barouki, R.; Audouze, K. AOP4EUpest: Mapping of Pesticides in Adverse Outcome Pathways Using a Text Mining Tool. *Bioinformatics* **2020**, *36*, 4379–4381. [CrossRef] [PubMed]
36. Pérez-Lucas, G.; Vela, N.; El Aatik, A.; Navarro, S. Environmental Risk of Groundwater Pollution by Pesticide Leaching through the Soil Profile. In *Pesticides-Use and Misuse and Their Impact in the Environment*; IntechOpen: London, UK, 2018; ISBN 1838800476.

37. Bucheli, T.D.; Barmettler, E.; Bartolomé, N.; Hilber, I.; Hornak, K.; Meuli, R.G.; Reininger, V.; Riedo, J.; Rösch, A.; Sutter, P. Pesticides in Agricultural Soils: Major Findings from Various Monitoring Campaigns in Switzerland. *Chimia* **2023**, *77*, 750–757. [CrossRef] [PubMed]
38. Tampieri, F.; Durighello, A.; Biondo, O.; Gąsior, M.; Knyś, A.; Marotta, E.; Paradisi, C. Kinetics and Products of Air Plasma Induced Oxidation in Water of Imidacloprid and Thiamethoxam Treated Individually and in Mixture. *Plasma Chem. Plasma Process.* **2019**, *39*, 545–559. [CrossRef]
39. Kurwadkar, S.; Evans, A.; DeWinne, D.; White, P.; Mitchell, F. Modeling Photodegradation Kinetics of Three Systemic Neonicotinoids—Dinotefuran, Imidacloprid, and Thiamethoxam—In Aqueous and Soil Environment. *Environ. Toxicol. Chem.* **2016**, *35*, 1718–1726. [CrossRef] [PubMed]
40. Sheets, L.P.; Li, A.A.; Minnema, D.J.; Collier, R.H.; Creek, M.R.; Peffer, R.C. A Critical Review of Neonicotinoid Insecticides for Developmental Neurotoxicity. *Crit. Rev. Toxicol.* **2016**, *46*, 153–190. [CrossRef] [PubMed]
41. Sass, J.B.; Donley, N.; Freese, W. Neonicotinoid Pesticides: Evidence of Developmental Neurotoxicity from Regulatory Rodent Studies. *Front. Toxicol.* **2024**, *6*, 1438890. [CrossRef] [PubMed]
42. Mueller, W.; Jones, K.; Fuhrimann, S.; Naim, Z.; Sidek, B.; Sams, C.; Harding, A.; Povey, A.; Atuhaire, A.; Basinas, I.; et al. Factors Influencing Occupational Exposure to Pyrethroids and Glyphosate: An Analysis of Urinary Biomarkers in Malaysia, Uganda and the United Kingdom. *Environ. Res.* **2024**, *242*, 117651. [CrossRef] [PubMed]
43. Saeed, M.F.; Shaheen, M.; Ahmad, I.; Zakir, A.; Nadeem, M.; Chishti, A.A.; Shahid, M.; Bakhsh, K.; Damalas, C.A. Pesticide Exposure in the Local Community of Vehari District in Pakistan: An Assessment of Knowledge and Residues in Human Blood. *Sci. Total Environ.* **2017**, *587*, 137–144. [CrossRef] [PubMed]
44. Brouwer, M.; Kromhout, H.; Vermeulen, R.; Duyzer, J.; Kramer, H.; Hazeu, G.; De Snoo, G.; Huss, A. Assessment of Residential Environmental Exposure to Pesticides from Agricultural Fields in the Netherlands. *J. Expo. Sci. Environ. Epidemiol.* **2018**, *28*, 173–181. [CrossRef] [PubMed]
45. Faure, E.; Béranger, R.; Fervers, B.; Schüz, J.; Blain, J. A GIS-Based Method to Define Geographical Determinants of Environmental Exposure to Agricultural Pesticides in France. *Rev. D'épidémiologie Santé Publique* **2018**, *66*, S333. [CrossRef]
46. Connolly, A.; Koch, H.M.; Bury, D.; Koslitz, S.; Kolossa-Gehring, M.; Conrad, A.; Murawski, A.; McGrath, J.A.; Leahy, M.; Brüning, T. A Human Biomonitoring Study Assessing Glyphosate and Aminomethylphosphonic Acid (AMPA) Exposures among Farm and Non-Farm Families. *Toxics* **2022**, *10*, 690. [CrossRef] [PubMed]
47. Sugeng, A.J. *From Field to Home: Assessing Air Infiltration and Soil Track-In Transport Pathways of Agricultural Pesticides into Farmworkers' Home and Identifying Risk Factors for Increased in-Home Pesticide Levels*; The University of Arizona: Tucson, AZ, USA, 2016; ISBN 1339748061.
48. Nogueira, W.V.; da Silva, J.L.; Braga, E.P.P.; Kupski, L.; Garda-Buffon, J. The Do Pesticides and Mycotoxins in Water Pose an Exposure Risk to Humans? *Agua Y Territ. Landsc.* **2025**, *25*, 283–296. [CrossRef]
49. Silva, V.; Gai, L.; Harkes, P.; Tan, G.; Ritsema, C.J.; Alcon, F.; Contreras, J.; Abrantes, N.; Campos, I.; Baldi, I. Pesticide Residues with Hazard Classifications Relevant to Non-Target Species Including Humans Are Omnipresent in the Environment and Farmer Residences. *Environ. Int.* **2023**, *181*, 108280. [CrossRef] [PubMed]
50. Rajput, S.; Sharma, R.; Kumari, A.; Kaur, R.; Sharma, G.; Arora, S.; Kaur, R. Pesticide Residues in Various Environmental and Biological Matrices: Distribution, Extraction, and Analytical Procedures. *Environ. Dev. Sustain.* **2022**, *24*, 6032–6052. [CrossRef]
51. Tudi, M.; Li, H.; Li, H.; Wang, L.; Lyu, J.; Yang, L.; Tong, S.; Yu, Q.J.; Ruan, H.D.; Atabila, A. Exposure Routes and Health Risks Associated with Pesticide Application. *Toxics* **2022**, *10*, 335. [CrossRef] [PubMed]
52. Garrigou, A.; Laurent, C.; Berthet, A.; Colosio, C.; Jas, N.; Daubas-Letourneux, V.; Jackson Filho, J.-M.; Jouzel, J.-N.; Samuel, O.; Baldi, I. Critical Review of the Role of PPE in the Prevention of Risks Related to Agricultural Pesticide Use. *Saf. Sci.* **2020**, *123*, 104527. [CrossRef]
53. Verbeek, J.H.; Rajamaki, B.; Ijaz, S.; Sauni, R.; Toomey, E.; Blackwood, B.; Tikka, C.; Ruotsalainen, J.H.; Balci, F.S.K. Personal Protective Equipment for Preventing Highly Infectious Diseases Due to Exposure to Contaminated Body Fluids in Healthcare Staff. *Cochrane Database Syst. Rev.* **2020**, *4*, CD011621. [PubMed]
54. Sapbamrer, R.; Thammachai, A. Factors Affecting Use of Personal Protective Equipment and Pesticide Safety Practices: A Systematic Review. *Environ. Res.* **2020**, *185*, 109444. [CrossRef] [PubMed]
55. Romano, M.E.; Buckley, J.P.; Li, X.; Herbstman, J.B.; Kannan, K.; Lee, S.; Schantz, S.L.; Trasande, L.; Karagas, M.R.; Perera, F. Changes in Urinary Concentrations of Contemporary and Emerging Chemicals in Commerce during the COVID-19 Pandemic: Insights from the Environmental Influences on Child Health Outcomes (ECHO) Program. *PLoS ONE* **2025**, *20*, e0317358. [CrossRef] [PubMed]
56. James, M.; Chauncey, A.; Ogle-Mustafa, C.; Chichester, M. Determinants of the Use of Personal Protective Equipment: A Literature Review. *J. Educ. Soc. Behav. Sci.* **2023**, *36*, 1–14. [CrossRef]

57. Trubetskov, A.D.; Makhonko, M.N.; Shkrobova, N.V.; Shelekhova, T. V Problems of Using Personal Protective Equipment in Modern Conditions. *Russ. J. Occup. Health Ind. Ecol.* **2023**, *63*, 336–343. [CrossRef]
58. Yu*, M.; Griffin, L. A Pilot Study of Healthcare Workers' Experience With Personal Protective Equipment. In Proceedings of the Human Factors and Ergonomics Society Annual Meeting, Atlanta, GA, USA, 10–14 October 2022; SAGE Publications: Los Angeles, CA, USA, 2022; Volume 66, pp. 898–902. [CrossRef]
59. Yi, L.; Zhang, S.; Chen, X.; Wang, T.; Yi, X.; Yeerkenbieke, G.; Shi, S.; Lu, X. Evaluation of the Risk of Human Exposure to Thiamethoxam by Extrapolation from a Toxicokinetic Experiment in Rats and Literature Data. *Environ. Int.* **2023**, *173*, 107823. [CrossRef] [PubMed]
60. Tao, Y.; Phung, D.; Dong, F.; Xu, J.; Liu, X.; Wu, X.; Liu, Q.; He, M.; Pan, X.; Li, R.; et al. Urinary Monitoring of Neonicotinoid Imidacloprid Exposure to Pesticide Applicators. *Sci. Total Environ.* **2019**, *669*, 721–728. [CrossRef] [PubMed]
61. European Food Safety Authority (EFSA); Álvarez, F.; Arena, M.; Auteri, D.; Leite, S.B.; Binaglia, M.; Castoldi, A.F.; Chiusolo, A.; Cioca, A.; Colagiorgi, A.; et al. Peer Review of the Pesticide Risk Assessment of the Active Substance Folpet. *EFSA J.* **2023**, *21*, e08139. [CrossRef] [PubMed]

Journal of *Xenobiotics*

MDPI

Article

Four-Year Monitoring Survey of Pesticide Residues in Tomato Samples: Human Health and Environmental Risk Assessment

Alessandro Atzei [1,*], Hamza Bouakline [2], Francesco Corrias [1] and Alberto Angioni [1,*]

1 Department of Life and Environmental Sciences, University of Cagliari, Monserrato, 09042 Cagliari, Italy; francesco.corrias@unica.it

2 Environment and Applied Chemistry Laboratory (LCAE), Faculty of Sciences, Mohammed First University, Oujda 60000, Morocco; h.bouakline@ump.ac.ma

* Correspondence: alessandro.atzei@unica.it (A.A.); aangioni@unica.it (A.A.); Tel.: +39-070-6758615 (Alberto Angioni)

Abstract: A four-year survey was conducted to monitor the presence of multiple pesticide residues contaminating tomatoes, with the aim of evaluating the potential health and environmental risks. A multiresidue liquid chromatography–triple mass spectrometry with a multiple reaction monitoring (LC-MS/MS-MRM) method was fully validated and used to test 252 pesticides in 360 samples analysed. According to SANTE guidelines, the proposed method was considered suitable for the purpose. Dietary risk assessment was conducted using the Hazard Quotient (HQ) approach and the European Food Safety Authority (EFSA) Pesticide Residue Intake Model; meanwhile, the cumulative environmental risk assessment was conducted using the Concentration Addition (CA) and Independent Action (IA) methods. Data obtained revealed multiple contaminations in most fields examined over the years. Twenty-two pesticide residues were identified, comprising 68.2% fungicides, 27.3% insecticides, and the remaining 4.5% acaricides. Higher levels were detected for Boscalid in 2022 in three fields, with an average value of 0.42 mg/kg. Multi-residue contamination occurred each year; the lowest abundance was detected in 2023 (3.9%), and the highest in 2022 (12.3%), with 5 pesticide residues as the maximum number of compounds detected in one sample in 2022. The consumer risk assessment identified no potential health concerns for adults or toddlers, and the combined risk was considered acceptable. The environmental assessment showed maximum cumulative ratio (MCR) values that were always ≥ 1, indicating a contribution to the toxicity of the mixture, only slightly higher than that of the single compound with the highest toxicity. The results of this study highlight the critical need to include cumulative dietary exposure assessments in pesticide risk evaluations, especially for food products that are susceptible to contamination by multiple residues.

Keywords: monitoring survey; pesticide residues; multiresidues methods; risk assessment; consumer health

1. Introduction

The agricultural food production system depends on the use of plant protection products (PPPs) to preserve crops from pests and decay worldwide. The use of PPPs has an impact on human health and the environment [1]. Even when pesticide treatments are carried out following good agricultural practices, there is a possibility that these compounds may leave quantifiable residues on the commodities. Moreover, multiple residues could be encountered when multiple pesticides are applied, and the toxicological synergistic or additive effect cannot be predicted [2]. The European Green Deal with the farm-to-fork strategy highlights the necessity to decrease and progressively eliminate the use of pesticides. The European Union's (EU) key actions are represented by the zero-pollution action plan and the biodiversity strategy, as well as the regulation on the sustainable use of pesticides [3]. These measures implement integrated pest management (IPM) methods, using products with lower impact on human health and the environment, prioritising non-chemical alternatives and setting up appropriately sized buffer zones [4]. Regulation (EC) No. 396/2005 and its amendments establish maximum residue levels (MRLs) for food products and animal feed. MRLs represent the legal limit and are obtained by correlating a complete dataset on toxicology and residue behaviour, together with methods of analysis [5]. However, zero-residue commodities reflect the current practices of food producers, indicating levels of pesticide residue below the limit of quantitative determination (LOQ) of official analytical methods [6].

With this perspective, the analysis should search for authorised and unauthorised pesticides for a specific crop. However, pesticide applications in crops often involve a mixture of PPPs, leading to complex pesticide mixtures; therefore, multiresidue methods capable of analysing a considerable number of pesticides in one spot represent the current challenge for analytical laboratories and food factories in monitoring purposes. These methods are mainly based on LC-MS/MS and GC-MS/MS (gas chromatography/triple mass spectrometry) techniques [7]. EU monitoring programmes are set annually according to Implementing Regulation (EU) 2024/989 and comprise 10 foodstuffs of plant origin and 208 parameters, 2 of animal origin and 33 parameters. In 2026, the foodstuffs selected will be oranges, pears, kiwis, cauliflower, onions, carrots, potatoes, beans (dried), rye (grains), brown rice (=de-husked rice), poultry fat, and beef liver [8].

Among the most cultivated vegetables in the EU, tomatoes (*Solanum lycopersicum* L.) account for almost 20%, with nearly a quarter used for processing (juices, canning, purees, sauces, and ketchup) [9].

Italy accounts for 15% of world production and 56% of the European crop; 60% of Italian processed tomatoes are exported worldwide. In Sardinia, 500 hectares are cultivated on average, producing nearly 500,000 quintals of industrial tomatoes, with 60% grown in the Medio Campidano and 40% in the Oristano area (data not published, supplied by the processing industry).

Tomatoes' primary pathogens include fungal infections such as late blight (*Phytophthora infestans*) and *Alternaria* spp., as well as bacterial diseases and insects such as *Tuta absoluta*, *Helicoverpa armigera*, *Trialeurodes vaporarium*, and *Agrotis ipsilon* [10]. In Italy, the sustainable use of pesticides is regulated by the National Action Plan, allowing the use of 159 pesticides on tomatoes. The authorised compounds consist of fungicides (68), insecticides (54), nematicides (13), acaricides (12), herbicides (16), plant growth regulators (4), molluscicides (2), and pheromones (2) [11,12]. In 2020, the Italian Ministry of Health published a report on the control of residues of plant protection products in food, highlighting that out of a total of 4759 samples, half of the samples (54.6%) were found to be free from detectable residues, and 43.9% reported residue values below the legal limit, while the remaining 1.5% reported residues above the legal limit [13]. Ouakhssase et al. monitored 432 pesticides in tomato samples from a local market in Morocco. The results showed 56% of samples were contaminated with one or more pesticide residues [14]. El Sherif et al. reported the recovery of 19 pesticides on 18 tomato samples from an Egyptian local market, mostly with multiresidue contamination [15]. Monitoring plans are usually carried out on samples collected in the market or during imports from foreign countries and are intended to verify compliance with the legal limits for authorised pesticides, not considering a risk assessment for cumulative contamination. These plans provide information on the contamination of ready-to-eat plant products. However, data on products intended for industrial processing are accessible only to manufacturing companies. Therefore, there is a lack of cumulative risk assessment studies focusing on pre-processing industrial commodities both from a specific region and over multiple consecutive years.

This work presents the results of a four-year monitoring survey conducted in Sardinia, which aimed to evaluate the contamination of tomatoes with pesticide residues after field treatment before harvest for processing into by-products for industrial use. Analyses of pesticide residues were performed by UHPLC-MS/MS. The paper reports on the evaluation of pesticide multiple sample contamination risk assessment vs. human dietary and the environmental compart.

2. Materials and Methods

2.1. Standards and Reagents

Acetonitrile (ACN) and methanol were LC/MS-grade solvents purchased from Sigma Aldrich (Milan, Italy). Ammonium formate 5 M (product code: G1946-85021, Agilent, Milan, Italy), formic acid (reagent-grade purity > 95%) and certified analytical standards (98.0% purity, Table S1) of pesticides at 100 mg/L in ACN (LC/MS Pesticide Comprehensive Test Mix Kit, part number: 5190-0551) were from Agilent Technologies (Milan, Italy). An intermediate solution of the pesticide mixture was prepared at 1 mg/L in ACN. Calibration curves consisting of five diluting points were prepared daily from the intermediate solution in tomato extract, free of pesticides. QuEChERS salt1 (part number: 5982-6650, Agilent, Milan, Italy) and salt2 (part number: 5982-5056, Agilent, Milan, Italy) were used for extraction and purification. A MilliQ Millipore purification system produced water with a conductivity below 18.2 MΩ (MilliQ Integral, Merck, Milan, Italy).

2.2. Sample Collection and Processing

Tomatoes at industrial maturity were collected according to the Commission Directive 2002/63/EC [16] over four consecutive years from 2021 to 2024, from all cultivated fields in the provinces of Medio Campidano and Oristano for a total of 360 samples (126-2021, 81-2022, 76-2023, and 79-2024) and 1375.4 ha sampled. After harvest, tomato samples were brought to the laboratory and stored in a refrigerator at 4 °C until processing (within 24 h after collection). The entire sample collected (5 kg) was chopped and homogenised using a semi-industrial blender; therefore, an aliquot was withdrawn and subjected to the extraction procedure. Three replicates for each sample were performed.

2.3. Extraction Procedure

Tomato samples were extracted according to Corrias et al. [17]. Briefly, ten grams of the homogenised sample were weighed into a 50 mL tube, mixed with 10 mL of ACN, and vortexed (Reax Top, Heidolph, Schwabach, Germany) for 1 min. Subsequently, 6.5 g of QuEChERS salt1 was added, and the tube was vortexed for 2 min and agitated in a rotary shaker for 15 min. The sample was centrifuged for 5 min at 1.957 relative centrifugal force (RCF) and 10 °C (Centrifuge 5810 R, Eppendorf AG, Hamburg, Germany). The supernatant (5 mL) was collected and transferred into a 15 mL tube containing 1 g of QuEChERS salt2. The tube was vortexed for 2 min, followed by agitation on a rotary shaker for 15 min. The solution was centrifuged for 5 min at 1.957 relative centrifugal force (RCF) and 10 °C. The organic fraction was filtered through a 0.45 µm PTFE filter (Thermo Scientific, Milan,

Italy) and transferred into a 1.8 mL vial for UHPLC-MS/MS analysis immediately after the preparation step.

2.4. UHPLC-MS/MS Analysis and Method Validation

The analytical method was validated following the SANTE/11312/2021 guidelines [18]. The multiresidue UHPLC-MS/MS-MRM method updates a previously validated method by adding 136 pesticides for a final multiresidue analysis of 252 compounds [17]. Briefly, a Triple Quad 6470-1290 Infinity LC-MS/MS system (Agilent Technologies, Milan, Italy) was used with a ZORBAX Eclipse Plus C18 (2.1 × 150 mm, 1–8 μm) column and a binary gradient, H_2O 5 mM in ammonium formate + 0.1% formic acid (A), and methanol 5 mM in ammonium formate + 0.1% formic acid (B). The analytical parameters evaluated included linearity, matrix effect, mean recovery (RE%), repeatability intra-day precision (RSDr), intermediate precision inter-day precision (RSDWR), and limit of quantification (LOQs). Pesticide-free tomato samples from organic growing fields were used as a blank matrix for calibration standards, recoveries, and matrix effect. Method linearity was assessed based on the coefficient of determination (r^2) of five-point calibration curves constructed for each compound with concentrations ranging from 0.001 to 0.1 mg/L.

Expanded measurement uncertainty (U, with a coverage factor of k = 2, used to obtain a confidence interval of 95%) (Equation (1)) was calculated within the lab validation approach by combining component uncertainties of reproducibility and trueness (u_c, with a coverage factor of k = 2 to achieve a confidence level of approximately 95%) (Equation (2)) [19]. Expanded uncertainty was calculated according to the following equations for all compounds:

$$U = 2 \times u_c \tag{1}$$

$$u_c = \sqrt{(u(Rw)^2 + u(bias)^2)} \tag{2}$$

U = expanded measurement uncertainty; u_c combined measurement uncertainty.

u(bias) = uncertainty component for the bias (trueness).

u(Rw) = uncertainty component for the precision (within lab reproducibility of the recovery).

2.5. Risk Assessment

The dietary risk assessment was carried out using two approaches: the Hazard Quotient (HQ) method for assessing long-term and acute dietary exposure to chemical contaminants, and the standard tool EFSA PRIMo (Pesticide Residue Intake Model, revision 3.1). This tool consists of an Excel-based calculation spreadsheet that is applied at the EU level in the framework of setting and reviewing maximum residue levels for pesticides under Regulation (EC) No 396/2005 and in the peer review of pesticides under Regulation (EU) No 1107/2009 [5,20,21] (Figure 1). This tool allows for the assessment of both long-term (chronic) and short-term risks (acute). Chronic risk assessment was performed comparing the TMDI (Theoretical Maximum Daily Intake) to the ADI (Acceptable Daily Intake). If the TMDI is lower than the ADI, consumer risk is highly unlikely, and no further dietary risk assessment is needed. For the acute risk assessment, IESTI (International Estimated Short-term Intakes) was compared to the ARfD (Acute Reference Dose). If the IESTI is lower than the ARfD, the acute dietary risk is considered acceptable.

The HQ was determined using the following Equation (3) [22]:

$$\text{Hazard Quotient (HQ)} = \boldsymbol{HQ} = \frac{\boldsymbol{EDI}}{\boldsymbol{ADI\ or\ ARfD}} \tag{3}$$

EDI: Estimated Daily Intake; (mean residue level (mg/kg) × Fi)/Bw.

Fi: median European food consumption data, adult: 0.072 kg/d, toddler (2–3 years): 0.020 kg/d.

Bw: median European body weight (Adult 71.4 kg, toddler 20 kg).

ADI and ARfD: mg/kg bw/d.

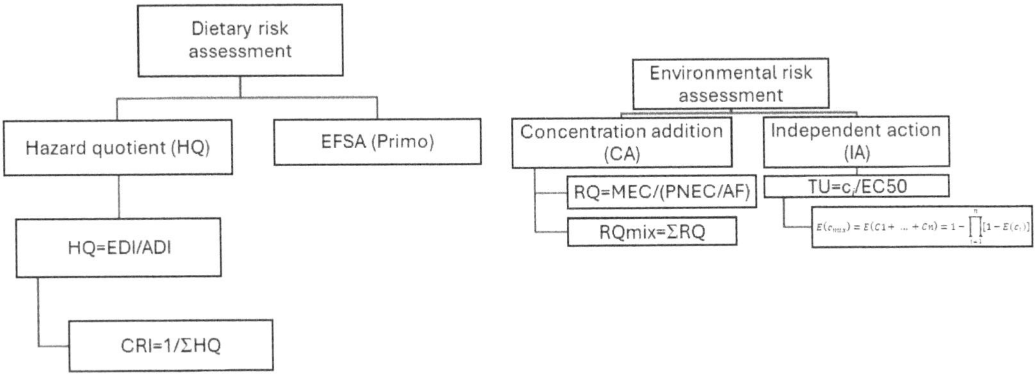

Figure 1. Flowchart of dietary and environmental risk assessment for cumulative contaminations.

TDMI was implemented in EFSA PRIMO using the following equation:

$$\boldsymbol{TMDI} = \frac{\sum_{i=1}^{n}(STM\ R_i \times C_i)}{bw}$$

STM R_i: mean residue concentration, calculated from the individual results;

MC_i: mean consumption for a given RAC;

bw: mean body weight for the subgroup of the population related to the LP or mean consumption (in kg).

IESTI was implemented in EFSA PRIMO using the following equation:

$$\boldsymbol{IESTI} = \frac{(HR \times v \times LP)}{bw}$$

$$\%ARfD = \frac{IESTI}{ARfD} \times 100$$

LP: Large portion reported (in kg/day) (97.5th percentile of eaters (or alternative percentile, depending on the number of reported eating occasions);

HR: Highest residue according to residue definition for enforcement in composite sample (in mg/kg);

v: variability factor, depending on the unit weight of the whole product (URAC), for tomatoes is 3;

Adults and toddlers (2–3 years) were considered for HQ risk assessment and the Cumulative risk assessment (CRA). An HQ value > 1 indicates a potential health risk.

Human and mammalian Cumulative Risk Assessment (CRA) was conducted, assessing the Cumulative Risk Index (CRI) as the reciprocal of the sum of the HQ values of each pesticide in the mixture of interest (Hazard Index, HI) [23,24]. The CRI can be calculated considering the NOAEL or ADI values [24]. In this paper, to be consistent with the single pesticide residues' long-term dietary risk assessment for chemical contaminants evaluation, the ADI was used. When the CRI is greater than 1, the combined risk is considered acceptable.

The cumulative environmental risk assessment was executed using the Concentration Addition (CA) and Independent Action (IA) methods (Figure 1). The Concentration Addition model assumes similar modes of action among the mixture components and is calculated by the sum (RQmix) (5) of the single risk quotient (RQ) (4) calculated by dividing the measured environmental concentration (MEC or c_i) by the predicted no-effect concentration (PNEC) selected as the lowest acute environmental concentration (worst case) and dividing it by a nominative safety assessment factor (AF), or considering the toxic unit (TU) as the environmental concentration for the EC50 (TU = c_i/EC50), respectively. For the calculation made in this paper, the EC50 values for *Daphnia magna* were selected (Table S2), as they are the most sensitive. The risk associated with the mixture was evaluated as high risk (RQmix $\geq$ 1), moderate risk (0.1 $\leq$ RQmix < 1), low risk (0.01 $\leq$ RQmix < 0.1), or rare risk (RQmix < 0.01) [25].

$$\mathbf{RQ} = \frac{MEC}{PNEC/AF} \tag{4}$$

$$\text{RQmix} = \Sigma\text{RQ} \tag{5}$$

Considering the alternative approach to independent Action (IA), $E(c_{mix})$ (6) is the sum of the independent events, considered as a toxic unit (TU).

$$E(c_{mix}) = E(C1 + \ldots + Cn) = 1 - \prod_{i=1}^{n}[1 - E(c_i)] \tag{6}$$

$E(c_{mix})$ is the IA-expected overall effect (scaled to the range 0–1) of a mixture composed of n chemicals, and $E(c_i)$ gives the effect of chemical *i* if applied singly in a concentration c_i.

The Independent Action model assumes dissimilar modes of action and estimates the combined effect as the probability that at least one component causes an effect, assuming independent action.

The maximum cumulative ratio (MCR) was calculated by comparing the toxicity of the mixture to the highest toxicity of a single component (7) [26].

MCR = Toxicity of the mixture/Highest toxicity of a single mixture component. (7)

3. Results and Discussion

3.1. Method Validation

No interfering peaks were observed within the retention time window of the pesticides for both quantification and confirmation ions, showing acceptable specificity, and no additional cleanup step was necessary (Figure 2). The instrumental response of pesticides was affected (10.3%) by coextracted compounds from the tomato matrix. Among pesticide residues, 64.3% were suppressed, and 35.7% increased, with a range of −52.5% (invermectin B1) to 173.6% (fuberidazol). According to SANTE criteria, the matrix effect in the range of −20 < ME% < 20 was considered insignificant [27].

To compensate for the matrix effect (ME), five-point calibration curves were prepared in blank matrix extracts. Linearity was above the threshold required by the method validation guidelines [18], with correlation coefficients (r^2) ranging from 0.9959 to 1.0000, and a maximum RSD% of 1.96 (Table S1). The instrument limits of quantification (LOQs) were far below the default MRL set by the European Commission (0.01 mg/kg) [28], with values ranging from 0.002 mg/kg for beflubutamid to 0.007 mg/kg for ivermectin B1a (Table S1). Recoveries ranged from 65.10 to 119.53% at the LOQ level, and from

65.28 to 119.56% at 10xLOQ. However, only the 9% and the 3%, respectively, were below the 70% recovery in agreement with SANTE principles (Table S1) [18]. Repeatability (RSDr) and reproducibility (RSDwR) yielded average results below 12.82%, with maximum and minimum RSD% values of 19.27% and 0.41% in RSDr, and 13.17% and 0.45% in RSDwR. The expanded measure of uncertainty showed values ranging from 3.1% (thiocarb) to 35.8% (flubendiamide), always below 50% of the default values for both spiking levels for all pesticides.

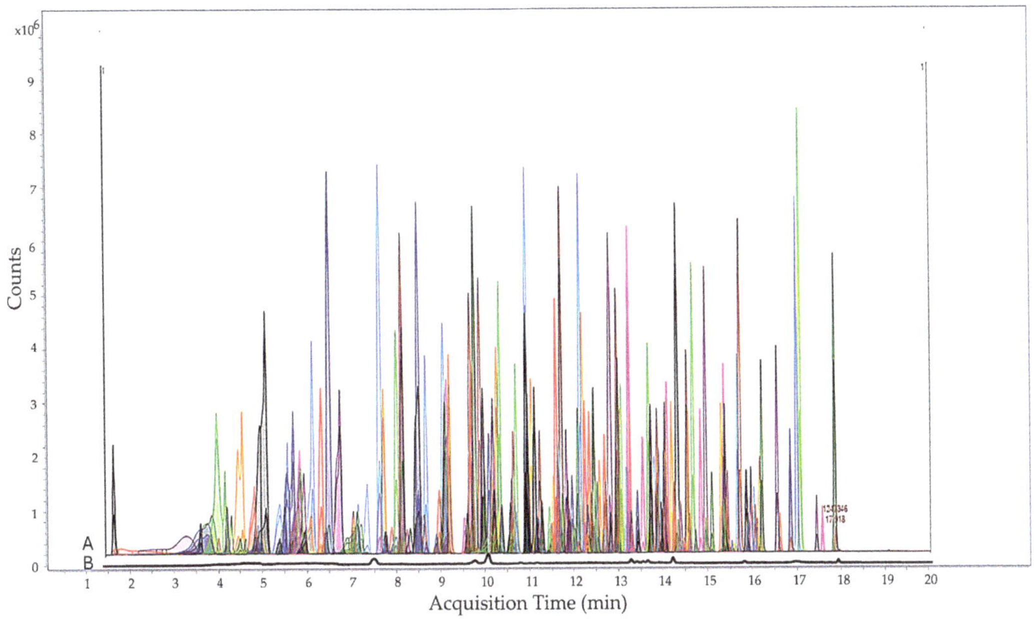

Figure 2. MRM chromatogram of tomato matrix fortified at LOQ with 252 pesticides (A) and of a blank tomato matrix (B).

3.2. Monitoring Assessment of Pesticide Residues in Tomato Samples

The main tomato adversities requiring the use of plant protection products in Sardinia are represented by the yellow noctule (*Heliothis armigera*), moth (*Tuta absoluta*), downy mildew (*Phytophtora infestans*), and powdery mildew (*Leveillula taurica*). Therefore, fungicides and insecticides are used by farmers to prevent plant damage and fruit losses. Despite the high number of compounds available in the market, in Sardinia, farmers use only a limited number of pesticides, following integrated production regulations and applying techniques for preventing and monitoring pest attacks. Agronomy experts assist farmers in using biological pest control methods, appropriate cultivation techniques, and plant protection products that pose less risk to human health and the environment. Among the pesticides used, on average in all monitored years, 47% were insecticides, 38% fungicides, and 15% herbicides. The most used were chlorantraniliprole, spinosad, fenpyroximate, hexythiazox, emamectin, lambda-cyalothrin, and etofenprox, whereas among fungicides were cymoxanil, penconazole, metalaxyl-M, propamocarb, pyraclostrobin, and mandipropamid. Quizalofop-ethyl and pendimethalin were the most used herbicides. The analysis revealed the presence of 22 pesticide residues in the four years, comprising 68.2% fungicides, 27.3% insecticides, and the remaining 4.5% acaricides; no herbicide was recorded during the analysis. Sample contamination varied over the years, with a positive trend in the last two years, and with an increasing number of samples free from residues (Figure 3). All samples showed residue values below the MRL set on tomatoes; higher levels were detected for Boscalid in 2022 with an average value of 0.42 mg/kg. Overall, the high number of fields sampled and the distinct locations with different phytosanitary problems at different periods of maturity resulted in a very varied and uneven distribution of residues. Most of the contaminated samples had a single pesticide residue (n = 1). The number of single pesticide residues detected was similar in the four years, ranging from 12 (2021) to 19 (2022). The most significant difference was that the maximum levels of pesticide residues showed a downward trend from 2022 to 2024.

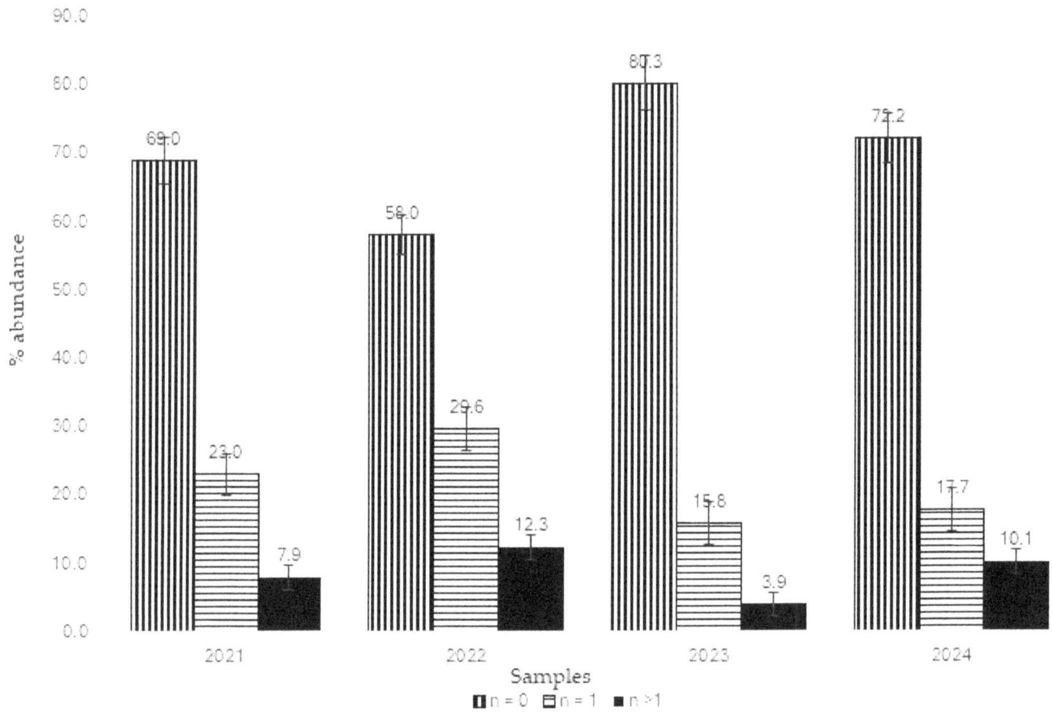

Figure 3. Samples (%) contaminated with pesticide residues during the four-year monitoring period. n = number of pesticide residues detected.

Multi-residue contamination occurred each year for a total of 42 different mixtures in the four years; the lower abundance was detected in 2023 (3.9%), and the higher in 2022 (12.3%) (Figure 3).

The maximum number of residues detected in one sample was 5 in 2022. In 2021 and 2022, the mixtures with 3 pesticide residues showed a similar abundance (50%) and decreased till 2024 (25%), whereas those with 2 residues were higher in 2023 (66.7%) and 2024 (75%) (Figure S1). Tomatoes for industrial processing are harvested during the summer from the last week of July to the first week of September. Throughout the period under review, the average temperature was similar in all years, ranging within 20–35 °C, whereas rains were registered at the beginning of August 2021 and in the middle of August 2023. This fact can explain the higher number of fungicide treatments in these years, due to the favourable climatic conditions that allowed for the development of fungi and moulds. A progressive improvement in farming practices and activities related to the selection of less persistent and less toxic pesticides had a positive impact on crop management, increasing tomato safety.

The presence of multiresidue contamination raises concerns about potential cumulative and synergistic effects, requiring the evaluation of a comprehensive risk assessment of the interaction among pesticide residues below the MRL [5,29].

Previous monitoring studies in Morocco on tomatoes analysed 39 samples from the market, searching 432 pesticide residues. The analysis showed residues mainly below the MRL, and multiresidue contamination was detected with a maximum of 7 pesticides in a single sample, but only in one case above the MRL. The compounds detected at a higher frequency were azoxystrobin, spinosad and difenconazole [14]. A different trend was highlighted in a monitoring study in Egypt, showing the presence of 15 pesticide residues with concentrations above the MRL. However, the analytical method applied did not allow the determination of nonvolatile pesticides, the number of pesticide residues investigated was limited to 19 compounds, and the samples analysed were only 18 tomatoes and 18 cucumbers [15]. Another survey made in Morocco in the years 2018–2019 analysed 202 pesticides in 51 samples and showed the presence of 69% of compounds contaminated with residues, always below the MRL [30]. A monitoring study performed in Italy in the years 2020–2021 on fruits and vegetables from the market showed that the 25.8% of the samples had detectable residues, always below the MRL. The most present were tebuconazole and boscalid [31]. From 2013 to 2015, a survey was performed in China on 132 samples of peaches collected from the field. A total of 105 pesticides were searched in the fruits, leading to 39 residues detected. The major number of residues were insecticides, followed by fungicides and acaricides, around 0.6% of contaminated fruits had residues exceeding the MRL. The 76.3% showed multiple residues, with the main mixture composed of two pesticides. The highest number of pesticides detected in a single sample was 13, and the most found were carbendazim, chlorpyriphos, acetamiprid, cypermethrin, and imidacloprid [32]. The official control system of the Italian Ministry of Health in 2020 reported that samples of tomato with detectable pesticide residues were around 50%, with only 0.8% with residues above the MRL. The irregular samples had residues of deltamethrin and chlorfenapyr. In the period between 2015 and 2020, tomatoes were the most sampled crops, and the principal fruits were irregular samples [13]. Data from monitoring surveys on tomatoes for industrial processing are not easily available in the literature, as they are primarily collected by companies for internal control purposes.

In the present survey, most of the residues found in different years were fungicides; in 2021 they were spinosad, mandipropamid, tebuconazole, metalaxyl, zoxamide, and ametotractin; in 2022, and almost all pesticide residues were detected with comparable abundance, even if pyraclostrobin, and zoxamide had the highest frequency. In 2023, the ten pesticides detected in several mixtures showed residues in a maximum of two mixtures; however, in 2024, acetamiprid, azoxystrobin, and spinosad were the most represented. Overall, the total number of residues decreased each year from 2021 to 2024 (Figure S1).

The pesticides found act through different modes of action (MoA); they are all characterised by a low acute toxicity in mammals and no significant carcinogenic, genotoxic, or teratogenic potential at environmentally relevant exposure levels.

Comparing the data with available monitoring studies, the analysis showed fewer samples contaminated with pesticide residues, a reduced number of mixtures, and a lower number of residues for each single sample.

3.3. Risk Assessment

The HQ approach is a first-tier empirical risk assessment method used to assess acute or chronic risk [24]. The EDI_{adult} values ranged from 1.01×10^{-5} to 3.49×10^{-4} mg/kg bw/d, whereas $EDI_{toddlers}$ ranged from 1.01×10^{-5} to 3.44×10^{-4} mg/kg bw/d. All HQ values were <1, indicating that the single pesticide residues detected in tomato samples did not pose any health risk for consumers. Emamectin exhibited the highest HQ value, accounting for 4.51×10^{-2} and 4.45×10^{-2} for adults and toddlers, respectively (Table 1).

Chronic risk assessment carried out with the EFSA's PRIMo tool showed ADI% ranging from 0% to 36% for adults and the general population (Table S2) and from 0% to 10% for children, toddlers, and infants (Table S3). As none of the residues exceeded 100% of the ADI, no chronic risk for consumers is expected.

Table 1. Consumer chronic risk assessment based on the HQ approach.

		Adults		Toddlers	
Pesticide	**ADI (mg/kg/d)**	**EDI (mg/kg bw/d)**	**HQ**	**EDI (mg/kg bw/d)**	**HQ**
Acetamiprid	0.005 *	1.26×10^{-4}	2.53×10^{-2}	1.25×10^{-4}	2.49×10^{-2}
Acetamiprid	0.025	1.26×10^{-4}	5.05×10^{-3}	1.25×10^{-4}	4.98×10^{-3}
Cymoxanil	0.013	5.39×10^{-5}	4.15×10^{-3}	5.32×10^{-5}	4.09×10^{-3}
Metalaxyl	0.08	6.48×10^{-5}	8.11×10^{-4}	6.39×10^{-5}	7.99×10^{-4}
Azoxystrobin	0.2	1.18×10^{-4}	5.92×10^{-4}	1.17×10^{-4}	5.84×10^{-4}
Boscalid	0.04	3.49×10^{-4}	8.73×10^{-3}	3.44×10^{-4}	8.61×10^{-3}
Mandipropamid	0.15	3.77×10^{-5}	2.51×10^{-4}	3.71×10^{-5}	2.48×10^{-4}
Dimethomorph	0.05	1.02×10^{-4}	2.04×10^{-3}	1.01×10^{-4}	2.02×10^{-3}
Myclobutanil	0.025	4.06×10^{-5}	1.62×10^{-3}	4.00×10^{-5}	1.60×10^{-3}
Tetraconazole	0.004	5.58×10^{-5}	1.39×10^{-2}	5.50×10^{-5}	1.38×10^{-2}
Penconazole	0.03	2.49×10^{-2}	9.37×10^{-4}	2.77×10^{-5}	9.24×10^{-4}
Tebuconazole	0.03	4.09×10^{-3}	3.94×10^{-3}	1.17×10^{-4}	3.89×10^{-3}
Zoxamide	0.5	7.99×10^{-4}	2.15×10^{-4}	1.06×10^{-4}	2.12×10^{-4}
Spinosad (sum A + D)	0.024	5.84×10^{-4}	3.10×10^{-3}	7.34×10^{-5}	3.06×10^{-3}
Pyraclostrobin	0.03	8.61×10^{-3}	2.11×10^{-3}	6.25×10^{-5}	2.08×10^{-3}
Clofentezin	0.017	2.48×10^{-4}	2.00×10^{-3}	3.35×10^{-5}	1.97×10^{-3}
Difenoconazole	0.01	2.02×10^{-3}	2.33×10^{-2}	2.30×10^{-4}	2.30×10^{-2}
Ametoctradin	10	1.60×10^{-3}	1.07×10^{-5}	1.06×10^{-4}	1.06×10^{-5}
Metaflumizone	0.01	1.38×10^{-2}	1.01×10^{-3}	1.00×10^{-5}	1.00×10^{-3}
Emamectin	0.0005	2.49×10^{-2}	4.51×10^{-2}	2.23×10^{-5}	4.45×10^{-2}
Etofenprox	0.03	4.04×10^{-5}	1.35×10^{-3}	3.98×10^{-5}	1.33×10^{-3}

* ADI changed in 2024.

Acute risk assessment performed with the HQ showed values below 1, indicating that the single pesticide residues detected in tomato samples did not pose any acute health risk to consumers. Azoxystrobin, mandipropamid, zoxamide, clofentezin, and ametotractin reported an insignificant ARfD in the EFSA conclusions, and the acute risk was not assessed (Table S4).

Acute risk assessment carried out with the EFSA's PRIMo tool showed ARfD% ranging from 0.1% to 45.9% for adults and from 0.2% to 55.8% for toddlers, not posing any acute health risk to consumers. However, when the new ARfD value for acetamiprid (0.005 mg/kg bw/d) was entered into the system, the ARfD% increased to 125.4%, exceeding the safety value of 100% (Table S5) for BE children with a daily intake of tomatoes of 180 g, whereas data from HQ calculations for an edible portion of 20 g led to HQ < 1. New considerations should therefore be made for the single compound acetamiprid, regarding the food daily intake (FDI) for a selected group of consumers.

IESTI values indicate that, based on the detected residual levels, there is no short-term risk to adults or toddlers from the pesticides evaluated in the present study. Although all IESTI remains below the established ARfDs (Table S6), some substances contribute significantly to the overall exposure, particularly in toddlers (Table S4).

Researchers and legislators mainly assess the safety of chemicals one at a time. When the potential toxicological risk is below the reference value for acute and chronic toxicity, the use of the pesticide is considered acceptable. However, the new trend in risk assessment is to address chemical mixtures of toxic compounds. Consumers are exposed daily to a high number of toxic compounds at the same time or over a restricted period. Among these, there is a high probability of being exposed to more than one pesticide within the same food or food product. The EFSA's latest report, published in 2023, reported the data from the European Union, with 58.1% of samples free from detectable pesticides, but 39.8% containing one or more pesticide residues below the MRL [33]. Combined exposure can occur with pesticides present in the same composite samples containing multiple residues or from combined foods containing one or more pesticide residues. Furthermore, exposure can occur through different pathways, such as oral, inhalation, or dermal, and each of these is considered independently, even if it is the totality of the exposure that determines the real risk. In this paper, we considered the pesticide residue in tomato samples, and therefore, the only route of contamination is oral. Moreover, it is a single ingestion, and chronic assumption of the mixtures detected is unrealistic.

Whatever the possibility of encountering multiple contaminations, the interaction among PPPs can follow different patterns: dose-addition, response-addition, synergism, or antagonism. Therefore, understanding the effective toxicity of a mixture is a challenging task. The assessment of the cumulative dietary exposure to multiple pesticide residues is in line with the EFSA and EPA Commission guidelines for cumulative risk assessment (CRA), which deal with the correct approach to define a risk assessment when more than one toxic is found in food commodities. The EFSA recommends that cumulative risk assessments for pesticides consider all the substances that belong to the common mechanism group (CMG), with a similar MoA when exposure occurs through the consumption of the same food commodity. Regarding the interaction among specific pesticides at low, non-effective doses, a case-by-case approach should be put in place [23].

The correct criteria to be applied for cumulative risk assessment should consider several issues and could be adapted to different situations, due also to the availability of toxicological data of different natures. When performing human risk assessment, the mechanism of mammalian toxicity (concerning human and animal health), exposure is assessed by combining occurrence data on chemicals with consumption data, whereas concentration data are commonly used for environmental and ecological evaluations. The occurrence and representative concentrations of PPPs residues in food commodities should be used for risk assessment; however, these data are few, and even if not perfectly suitable, data from supervised or monitoring trials can be used. When the components of the mixture and their exposure levels are chemically defined, it is possible to apply the component-based approach with exposure and hazard data of the individual components.

Since the pesticide concentrations found were highly variable in the different mixtures and in many cases close to the LOQ, biotesting is not a viable path, and offline modelling approaches should be considered for risk assessment purposes of the mixtures.

First, the PPP's chemical and toxicity activities were defined to set the appropriate assessment group of the mixture. The pesticide detected in the mixture belonged to various categories: fungicides (68.2%), insecticides (27.3%), and acaricides (4.5%). The toxicities of the pesticides included in the mixtures for humans and other mammals were classified between II and IV according to the Environmental Protection Agency (EPA), with 80% Class III and IV having a moderate to low toxicity (Table 2).

The toxicity of the compounds detected in the mixtures acts at distinct districts, including the mitochondria, the endocrine system, and energy production, through the inhibition of the cytochrome bc1 complex in the mitochondria. However, most of them have specific activity on insect and fungal biochemical mechanisms, which do not overlap with human ones.

Therefore, grouping the pesticides in mixtures for their MoA is not an easy task. In this paper, we decided to consider all pesticides in the mixtures without considering the MoA, even if this could lead to an overestimation of the cumulative effect.

The analysis of the pesticide mixture showed no critical concern for human safety. All mixtures had CRI values well above the baseline value of 1, ranging from 16.4 to 578.0 in adults and from 26.1 to 581.4 in toddlers (Table 3).

Table 2. Toxicological and chemical characteristics, together with the main uses of the pesticides detected in the mixture.

Plant Protection Product	Toxicity Class (EPA)	NOAEL mg·Kg bw·d	Toxicity Activity	Chemical Class	Main Uses
Acetamiprid	II/III	10	Acetylcholine, nicotinic receptors	neonicotinoid	insecticide
Cymoxanil	II	0.01	Inhibits cytochrome c oxidase in the fourth complex of the electron transport chain	organic nitriles	fungicide
Metalaxyl	III	9.4	Disrupts fungal nucleic acid synthesis by inhibiting RNA polymerase I	phenylamide	fungicide
Azoxystrobin	III/IV	20	Quinone outside inhibitor (QoI)-type	strobilurin	fungicide
Boscalid	III/IV	22	Mitochondrial damage	aromatic anilides	fungicide
Mandipropamid	IV	247.6	Inhibitor of phospholipid biosynthesis	mandelamide	fungicide
Dimethomorph	III	16	Endocrine disruptor	morpholine group	fungicide
Myclobutanil	II	2.5	Induce liver microsomal enzymes	triazole	fungicide
Tetraconazole	III/IV	2.95	Interact with nuclear receptors (constitutive androstane receptor, CAR; pregnane X receptor, PXR), liver toxicity.	triazole	fungicide
Penconazole	III	69	Increase oxidative damage and modify enzyme activity	triazole	fungicide
Tebuconazole	III/IV	3	Activate nuclear receptors (aryl hydrocarbon receptor, AHR), trigger oxidative stress and apoptosis.	triazole	fungicide
Zoxamide	III/IV	50	Mitotic arrest by disrupting cell division, binding to β-tubulin and inhibiting tubulin polymerisation	benzamide	fungicide
Spinosad (Sum A + D)	III	4.89	Nicotinic acetylcholine receptor	spynosin	insecticide
Pyraclostrobin	IV	3.4	Quinone outside inhibitor (QoI)-type	strobilurin	fungicide
Clofentezin	III	1.95	Effect on the liver (enzyme induction) Affects thyroid function and impairs steroid hormone regulation	tetrazine	acaricide
Difenoconazole	III/IV	3.7	Disruptions to the endocrine system and nervous system	dioxolanes	fungicide
Ametoctradin	IV	1000	Disrupting mitochondrial energy production	triazolopyrimidine	fungicide
Metaflumizone	IV	40	Sodium channel blocker	semicarbazone	insecticide
Emamectin	II	0.6	Chloride channel activator by binding gamma aminobutyric acid (GABA) receptor and glutamate-gated chloride channels disrupting nerve signals within arthropods	second-generation avermectin	insecticide
Etofenprox	IV	23	Disturbs insect nervous systems following direct contact or ingestion	pyrethroid derivative	insecticide

Table 3. Cumulative risk assessment for adults and toddlers for the pesticide mixtures detected in the survey.

	2021	2022	2023	2024
		Residues (mg/kg ± RSD%) *		
min	0.058 ± 3.61	0.068 ± 5.59	0.067 ± 5.27	0.104 ± 9.17
max	0.228 ± 3.80	1.061 ± 1.27	0.584 ± 9.77	0.403 ± 5.73
median	0.133 ± 11.46	0.204 ± 2.46	0.188 ± 1.96	0.260 ± 1.53
		CRI adult		
min	22.64 ± 16.03	21.90 ± 1.26	81.12 ± 9.03	54.14 ± 13.84
max	3096.03 ± 20.10	879.30 ± 9.15	1250.86 ± 5.07	679.97 ± 10.10
median	446.49 ± 4.14	237.70 ± 4.98	177.53 ± 2.63	115.61 ± 4.49
		CRI toddler		
min	22.96 ± 16.03	22.21 ± 1.26	82.26 ± 9.03	54.91 ± 13.84
max	3139.64 ± 20.10	891.68 ± 9.15	1268.48 ± 5.07	689.54 ± 10.10
median	452.78 ± 4.14	241.03 ± 4.97	180.03 ± 2.63	117.24 ± 4.49

* Sum of residues of the single mixtures.

The pesticides that contributed most to the HQ values were spinosad, boscalid, and pyraclostrobin in 2021, spinosad and pyraclostrobin in 2022, acetamiprid, tebuconazole, and pyraclostrobin in 2023, and acetamiprid, metalaxyl, tebuconazole, and spinosad in 2024.

Currently, the European environmental risk assessment of pesticides (ERA) [20] considers almost exclusively single applications of single PPP on a single crop. This means that ERA considers only a mixture of single pesticides present in the same PPP, which could be a formulation of one or more pesticides and different additives. Therefore, no assumptions are reported by ERA regarding uncontrolled pesticide mixtures present in the environment. The literature data confirmed the absence of a regulation in any part of the world regarding this last issue [34]. For environmental risk assessment of a single compound, the Predicted Environmental Concentration to the Predicted No-Effect Concentration (PEC/PNEC) ratio, the Toxicity Exposure Ratio (TER), or a specified Margin of Safety factor could be applied. It is hypothesised that by safeguarding the most sensitive trophic level, all other organism groups are concomitantly protected, and that protecting the structure of an ecosystem also protects ecosystem functions. Further data on environmental exposure or hazard may be needed to clarify the environmental risk, depending on the initial risk assessment's results. These data are generated from a series of studies known as higher-tier studies, which are becoming more comprehensive. A relevant comparison must be made at each tier between the estimated exposure and the estimated hazard. There are, therefore, separate tiers for both exposure and hazard estimation. However, it should be considered that when environmental toxicity is considered, different targets should be evaluated during ecotoxicological assessments, such as birds and mammals, terrestrial vertebrates from wildlife, bees and other arthropods, and aquatic organisms. The EFSA approach to aquatic organisms' claims for a 10-step decision tree comparing measured vs. calculated product toxicity to check for a toxicity driver, synergism, and/or antagonism [35]. However, the aquatic guidance assumes that "observed effects are, in many cases, related to the effects of one or two pesticides". Aquatic organisms' calculations start using the FOCUS stepwise procedure, which allows for defining a risk profile and mitigation measures. For each pesticide in the mixture, the most sensitive aquatic taxa can be different; all these variables require hundreds of calculations. Therefore, alternative strategies based on worst-case risk quotients (RQs) can be more easily applicable, providing sufficiently reliable information on toxicity.

In this study, the measured pesticide concentrations in tomato matrices were assumed to be equal to those released in surface water. However, this is a worst-case assumption that does not consider the dilution effect or degradation in the water medium. This strategy was adopted due to the varying environmental conditions in the field, with the capacity of rivers and ponds changing rapidly with elevated summer temperatures and reduced rainfall.

The idea behind a mixture risk assessment is to use evaluation variables from earlier studies so that unnecessary experiments can be avoided. Therefore, toxicity calculations can be based on established reference models: Concentration Addition (CA) (also known as dose addition) and Independent Action (IA) (also known as response addition) [25,36–38].

Concentration Addition by RQmix showed mainly moderate risk (50.0%), and 25% had a low or a high-risk level (Table 4). Risk quotients provide an immediate understanding of (un)acceptability of a risk without additional information (AF or acceptability criteria), giving a worst-case aquatic risk assessment; the sum can be directly used for a risk assessment of the mixture. The theory behind IA is that the chemicals in a mixture do not physically, chemically, or biologically interact; therefore, some doubt exists considering the use on living organisms. Nevertheless, it has been discovered to offer reliable predictions of chemical mixtures with varying sites or modes of action in test systems comprising algae and bacteria [39].

IA and CA predictions are similar, with some pesticides having a higher weight in driving the toxicity value. CA was the most conservative and usually predicts a higher mixture toxicity [40].

Table 4. CA and IA environmental risk assessment for the pesticide mixtures detected in the survey.

	2021	2022	2023	2024
		CA		
min	0.10	0.01	0.52	0.00
max	14.10	15.36	235.0	0.95
median	0.46	10.93	1.86	0.93
		IA		
min	0.01	0.01	0.22	0.00
max	1.37	15.25	5.64	00.095
median	0.51	2.62	1.56	0.093

Pyraclostrobin and etofenprox were the main pesticides responsible for the RQ and IA values ≥ 1 in all years except in 2024, in which the moderate risk in CA was attributed to azoxystrobin.

Pyraclostrobin is a synthetic fungicide that acts as a quinone outside inhibitor (QoI) that targets fungal respiration by inhibiting the cytochrome bc1 complex. The NOAEL for short-term oral toxicity is 6 mg/kg BW, the acceptable daily intake (ADI) is established at 0.03 mg/kg bw/d, and the acute AOEL (AAOEL) is established at 0.015 mg/kg bw, on the same basis as the ARfD, applying the AF of 100 and a correction for oral absorption value of 50%; pyraclostrobin is also classified as STOT RE 212 based on effects in liver, gastrointestinal tract and the nasal cavity through prolonged or repeated exposure.

Pyraclostrobin has been associated with adverse effects on neuronal proliferation and potential endocrine-disrupting activity, particularly anti-androgenic effects [41,42]. These effects may contribute to a common outcome of neurodevelopmental impairment, justifying their inclusion in cumulative assessment groups (CAGs) targeting the nervous system.

Etofenprox is an insecticide pyrethroid derivative, and it disturbs insect nervous systems following direct contact or ingestion. The acceptable daily intake (ADI) of etofenprox is 0.03 mg/kg bw/d based on the long-term mouse study and applying an AF of 100, which is supported by the long-term rat study. The acceptable operator exposure level (AOEL) is

0.06 mg/kg bw/d based on the oral 90-day rat study, applying an AF of 100 and a correction factor of 30% for low oral absorption. The acute reference dose (ARfD) was set at 1.0 mg/kg bw, based on the NOAEL of 100 mg/kg bw/d from the rabbit developmental toxicity study and an AF of 100. The available acute toxicity data for etofenprox suggested a classification as very toxic to aquatic organisms, based on the acute toxicity to fish (LC50 = 2.7 μg a.s./L), daphnia (LD_{50} = 1.2 μg a.s./L), and algae (EC_{50} > 150 μg a.s./L).

The environmental impact of the mixture is calculated based on the residues found in tomato samples and leads to values that would suggest a negative impact on aquatic organisms. However, considering the average tomato amount per hectare (75 tons), multiplying by the amount of pyraclostrobin or etofenprox for kg of tomato, the residues left in the field will be almost ¼ of the application dose reported on the label.

The IA evaluation seems to be more efficient in describing the environmental toxicity with respect to CA for the pesticides in the mixtures. This fact was probably related to the dissimilarity of the MoA, which represents a valid criterion for selecting the most appropriate concept.

The MCR values were always $\geq$1, indicating a contribution to the toxicity of the mixture, only slightly higher than the single compound with the highest toxicity.

4. Conclusions

Numerous risks are faced by both humans and wild animals, including environmental pollutants via multiple exposure pathways. The negative impact of a single chemical is sometimes disregarded because its concentration in the environment is so low. The issue of mixture toxicity has become a hot topic, especially when it comes to the additive and synergistic effects of environmental mixtures. The present monitoring survey allowed the assessment of tomato contamination by pesticide residues over four years, from 2021 to 2024. This study remarks that when PPP treatments are made following the label, with field observation for treatment decisions, and using PPPs with low toxicity, the residues are not only below the MRL but are near the instrumental limit of determination.

The samples showed a higher presence of contamination from a single compound rather than contamination from multiple residues. The high number of fields sampled in different areas with different pest problems has led to dissimilar pesticide mixture contamination. The consumer risk assessment carried out with the HQ method and EFSA-PRIMO identified no potential health concerns for adults or toddlers, and the combined risk was considered acceptable, except for acetamiprid in 2024, which changed the ADI from 0.025 to 0.005, leading to an IESTI value of 115% of the ARfD.

The MoA of the pesticides detected was different among the mixtures. Nevertheless, the environmental risk assessment was carried out both by CA and IA modes. CA showed values in some cases too high, whereas IA may provide a more realistic estimate given the likely dissimilar modes of action. The EFSA suggests using CA when pesticides can be included in the same cumulative assessment group; however, this strategy exhibits many deficiencies when pesticide mixtures are constituted by many different pesticides with different MoA. In contrast, IA does not have a solid foundation of scientific research to validate its use in these cases.

The study has some limitations related to several factors, including the actual deposition of residues and the extent of their dilution in surface water. The number of pesticides analysed is consistent with the agricultural production in the areas considered (tomatoes, artichokes, melons, and fennel) but does not account for any unauthorised use of pesticides on other horticultural crops or the metabolites formed after treatments. Furthermore, it does not take into account soil contamination and possible leaching into groundwater.

Regarding surface water, the primary critical issue is the inability to make accurate estimates of dilution due to the significant variability of watercourses and ponds near crops during the summer season. However, it should be noted that the assessment of the risks of pesticides to aquatic invertebrates in the assessment dossiers does not take into account dilution or degradation. The definition of correct methods for assessing the dietary risk and the environmental risk of pesticide mixtures at concentrations below the MRL is still in its infancy, due to the difficulty of verifying an additive or synergistic effect, or an unpredictable effect of the mixture. Furthermore, the variability in the composition of mixtures represents an additional gap that is difficult to resolve. The implementation of bioassays to validate IA predictions and computational systems capable of simultaneously assessing different pesticides with different MOAs could be the most effective solution. Moreover, a deeper study should be performed on the fate and risk assessment of the degradation products.

Supplementary Materials: The following supporting information can be downloaded at https://www.mdpi.com/article/10.3390/jox15050171/s1, Figure S1: Pesticide Frequency; Table S1: Linear regression equations of the new validated pesticides and matrix effects; Table S2: Chronic risk assessment (JMPR methodology—IEDI/TMDI) for EU populations (Adult and General Population) using EFSA's PRIMo tool revision 3.1; Table S3: Chronic risk assessment for EU populations (JMPR methodology—IEDI/TMDI) (Children, toddler and infant) using EFSA's PRIMo tool revision 3.1; Table S4: Consumer acute risk assessment based on the HQ approach; Table S5: Acute risk assessment for EU populations (toddler and adults) using EFSA's PRIMo tool revision 3.1; Table S6: Table of all available data on the acute toxicity to aquatic invertebrates [43–62].

Author Contributions: Conceptualization, A.A. (Alessandro Atzei); methodology, A.A. (Alessandro Atzei) and A.A. (Alberto Angioni); software, A.A. (Alessandro Atzei); validation, A.A. (Alessandro Atzei) and H.B.; formal analysis, A.A. (Alessandro Atzei) and F.C.; resources, A.A. (Alberto Angioni); data curation, A.A. (Alessandro Atzei); writing—original draft preparation, A.A. (Alberto Angioni) and A.A. (Alessandro Atzei); writing—review and editing, A.A. (Alberto Angioni) and F.C.; supervision, A.A. (Alberto Angioni); All authors have read and agreed to the published version of the manuscript.

Funding: This research received no external funding.

Institutional Review Board Statement: Not applicable.

Informed Consent Statement: Not applicable.

Data Availability Statement: The original contributions presented in this study are included in the article/Supplementary Materials. Further inquiries can be directed to the corresponding author(s).

Acknowledgments: We acknowledge Project Title eINS Ecosystem of Innovation for Next Generation Sardinia—CUP F53C22000430001—Grant Assignment Decree No. 1056 adopted on 23 June 2022 by the Italian Ministry of Ministry of University and Research (MUR), under the National Recovery and Resilience Plan (NRRP), Mission 4 Component 2 Investment 1.5—Call for tender No. 3277 published on 30 December 2021 by the Italian Ministry of University and Research (MUR) funded by the European Union—NextGenerationEU. Project Code ECS0000038.

Conflicts of Interest: The authors declare no conflicts of interest.

References

1. Tudi, M.; Ruan, D.H.; Wang, L.; Lyu, J.; Sadler, R.; Connell, D.; Chu, C.; Phung, D.T. Agriculture Development, Pesticide Application and Its Impact on the Environment. *Int. J. Environ. Res. Public Health* **2021**, 18, 1112. [CrossRef]
2. Yang, M.; Wang, Y.; Yang, G.; Wang, Y.; Liu, F.; Chen, C. A review of cumulative risk assessment of multiple pesticide residues in food: Current status, approaches and future perspectives. *Trends Food Sci. Technol.* **2024**, 144, 104340. [CrossRef]

3. Wesseler, J.H.H. The EU's farm-to-fork strategy: An assessment from the perspective of agricultural economics. *Appl. Econ. Perspect Policy* **2022**, *44*, 13239. [CrossRef]
4. Integrated Pest Management (IPM). European Commission, Food Safety. Available online: https://food.ec.europa.eu/plants/pesticides/sustainable-use-pesticides/integrated-pest-management-ipm_en (accessed on 2 June 2025).
5. European Commission. Regulation (EC) no 396/2005 of the European Parliament and of the council of 23 February 2005 on maximum residue levels of pesticides in or on food and feed of plant and animal origin and amending Council Directive 91/414/EEC. *Off. J. Eur. Union* **2005**, *L70*, 1–16.
6. Djekic, I.; Smigic, N.; Udovicki, B.; Tomic, N. "Zero Residue" Concept—Implementation and Certification Challenges. *Standards* **2023**, *3*, 177–186. [CrossRef]
7. Marín-Sáez, J.; López-Ruiz, R.; Romero-González, R.; Garrido Frenich, A. Multiresidue Methods for Determination of Pesticides and Related Contaminants in Food by Liquid Chromatography. In *Liquid Chromatography, 3rd Edition, Applications, Vol 2 Handbooks in Separation Science*; Fanali, S., Chankvetadze, B., Haddad, P.R., Poole, C.F., Riekkola, M.L., Eds.; Elsevier: Amsterdam, The Netherlands, 2023; Chapter 21; pp. 705–732. [CrossRef]
8. European Commission. *Commission Implementing Regulation (EU) 2024/989 of 2 April 2024 Concerning A Coordinated Multiannual Control Programme of The Union for 2025, 2026 and 2027 to Ensure Compliance With Maximum Residue Levels of Pesticides and to Assess The Consumer Exposure to Pesticide Residues in and on Food of Plant and Animal Origin and Repealing Implementing Regulation (EU) 2023/731*; European Union: Luxembourg, 2024. Available online: http://data.europa.eu/eli/reg_impl/2024/989/oj (accessed on 2 June 2025).
9. The Fruit and Vegetable Sector in the EU—A Statistical Overview. Eurostat Data Extracted in January and February 2024. Available online: https://ec.europa.eu/eurostat/statistics-explained/index.php?oldid=656462 (accessed on 5 March 2025).
10. Karačić, V.; Miljaković, D.; Marinković, J.; Ignjatov, M.; Milošević, D.; Tamindžić, G.; Ivanović, M. Bacillus Species: Excellent Biocontrol Agents against Tomato Diseases. *Microorganisms* **2024**, *12*, 457. [CrossRef]
11. European Commission. *European Commission Regulation (EC) No 10235/2016 of the European Parliament and of the Council of 23 February 2021 on Evaluation of Data Submitted to Confirm MRLs Following the Review of Existing MRLs*; European Union: Luxembourg, 2021.
12. PAN (Piano di Azione Nazionale). *Piano di Azione Nazionale per l'Uso Sostenibile dei Prodotti Fitosanitari. Decreto 22 Gennaio 2014—Ai SensI dell'Articolo 6 del Decreto Legislativo 14 Agosto 2012, n. 150 Recante: Attuazione della Direttiva 2009/128/CE che Istituisce un Quadro per l'Azione Comunitaria ai Fini dell'Utilizzo Sostenibile dei Pesticidi*; Ministero delle Politiche Agricole Alimentari e Forestali: Rome, Italy, 2014.
13. Ministero della Salute. *Controllo Ufficiale sui Residui dei Prodotti Fitosanitari negli Alimenti—Rapporto 2020—Ministero della Salute 0018801-24/11/2022-GAB-MDS-A*; Ministero della Salute: Rome, Italy, 2022. Available online: https://www.salute.gov.it/new/sites/default/files/imported/C_17_pubblicazioni_3284_allegato.pdf (accessed on 11 June 2025).
14. Ouakhssase, A.; Addi, E.A. Monitoring 432 potential pesticides in tomatoes produced and commercialized in Souss Massa region Morocco, using LC-MS/MS and GC-MS/MS. *Environ. Pol.* **2023**, *337*, 122611. [CrossRef]
15. El Sherif, D.F.; Elabasy, N.N.; Sayed, M.A. Monitoring and risk assessment of pesticide residues in tomatoes and cucumbers traded in Fayoum governorate markets, Egypt. *FJARD* **2024**, *38*, 616–627. [CrossRef]
16. European Commission. *Commission Directive 2002/63/EC of 11 July 2002 Establishing Community Methods of Sampling for the Official Control of Pesticide Residues in or on Products of Plant and Animal Origin Repealing Directive 79/700/EEC*; European Commission: Luxembourg, 2002.
17. Corrias, F.; Atzei, A.; Lai, C.; Dedola, F.; Ibba, E.; Zedda, G.; Canu, F.; Angioni, A. Effects of Industrial Processing on Pesticide Multiresidues Transfer from Raw Tomatoes to Processed Products. *Foods* **2020**, *9*, 1497. [CrossRef] [PubMed]
18. European Commission Directorate-General for Health and Food Safety. *Guidance Document on Analytical Quality Control and Method Validation Procedures for Pesticide Residues Analysis in Food and Feed (SANTE/11312/2021 Rev. 1)*; European Commission: Luxembourg, 2021.
19. EURACHEM/CITAC. *EURACHEM/CITAC Guide Quantifying Uncertainty in Analytical Measurement*, 3rd ed.; EURACHEM/CITAC: London, UK, 2012.
20. Regulation (EC) No 1107/2009 of the European Parliament of the Council of 21 October 2009 Concerning the Placing of Plant Protection Products on the Market Repealing Council Directives 79/117/EEC and 91/414/EEC. Available online: http://data.europa.eu/eli/reg/2009/1107/oj (accessed on 11 June 2025).
21. European Food Safety Authority (EFSA). *Pesticide Residue Intake Model—EFSA PRIMo Revision 3.1 (Update of EFSA PRIMo Revision 3)*; Technical Report, approved 22 March 2019; EFSA: Parma, Italy, 2019. [CrossRef]
22. Simon, T. *Environmental Risk Assessment. A Toxicological Approach*, 2nd ed.; available at VitalSource Bookshelf; CRC Press: Boca Raton, FL, USA; Taylor & Francis: London, UK, 2020; ISBN E-Book 9780429286001.

23. Di Piazza, G.; Dujardin, B.; Levorato, S.; Medina, P.; Mohimont, L.; Solazzo, E.; Costanzo, V. Prioritisation of pesticides and target organ systems for dietary cumulative risk assessment based on the 2019–2021 monitoring cycle. *EFSA J.* **2024**, *22*, e8554. [CrossRef]
24. Boobis, A.R.; Ossendorp, B.C.; Banasiak, U.; Hameyd, P.Y.; Sebestyene, I.; Moretto, A. Cumulative risk assessment of pesticide residues in food. *Toxicol. Lett.* **2008**, *180*, 137–150. [CrossRef] [PubMed]
25. Kar, S.; Sanderson, H.; Roy, K.; Benfenati, E.; Leszczynski, J. Ecotoxicological assessment of pharmaceuticals and personal care products using predictive toxicology approaches. *Green Chem.* **2020**, *22*, 1458. [CrossRef]
26. Price, P.S.; Han, X. Maximum cumulative ratio (MCR) as a tool for assessing the value of performing a cumulative risk assessment. *Int. J. Environ. Res. Public Health* **2011**, *8*, 2212–2225. [CrossRef] [PubMed]
27. European Commission. *Document SANTE/11813/2017. Guidance Document on Analytical Quality Control and Method Validation Procedures for Pesticide Residues and Analysis in Food and Feed*; European Commission: Luxembourg, 2017.
28. European Commission. EU Pesticides Database. 2023. Available online: https://food.ec.europa.eu/plants/pesticides/eu-pesticides-database_en (accessed on 11 June 2025).
29. Grosssteiner, I.; Mienne, A.; Lucas, L.; Yvonnet, P.L.; Trenteseaux, C.; Fontaine, K.; Sarda, X. Cumulative risk assessment with pesticides in the framework of MRL setting. *EFSA J.* **2023**, *21* (Suppl. S1), e211009. [CrossRef]
30. Choubbane, H.; Ouakhssase, A.; Chahid, A.; Taourirte, M.; Aamouche, A. Pesticides in fruits and vegetables from the Souss Massa region, Morocco. *Food Addit. Contam. B Surveill.* **2022**, *15*, 79–88. [CrossRef] [PubMed]
31. Nardelli, V.; Calitri, A.; Casamassima, F.P.; Della Rovere, I.; Ingegno, M.T.; Iammarino, M. Monitoring of Pesticides in Fruit and Vegetables Marketed in Italy during 2020/2021 and Food Safety Evaluations. In Proceedings of the 14th European Pesticide Residue Workshop (EPRW 2022), Bologna, Italy, 19–23 September 2022. Abstract PT–07.
32. Li, Z.; Nie, J.; Yan, Z.; Cheng, Y.; Lan, F.; Huang, Y.; Chen, Q.; Zhao, X.; Li, A. A monitoring survey and dietary risk assessment for pesticide residues on peaches in China. *Regul. Toxicol. Pharmacol.* **2018**, *97*, 152–162. [CrossRef]
33. EFSA Scientific Report. The 2021 European Union report on pesticide residues in food. *EFSA J.* **2023**, *21*, e07939. [CrossRef]
34. Weisner, O.; Frische, T.; Liebmann, L.; Reemtsma, T.; Roß-Nickoll, M.; Schäfer, R.B.; Schäffer, A.; Scholz-Starke, B.; Vormeier, P.; Knillmann, S.; et al. Risk from pesticide mixtures—The gap between risk assessment and reality. *Sci. Total Environ.* **2021**, *796*, 149017. [CrossRef]
35. EFSA (EFSA Panel on Plant Protection Products and their Residues). Guidance on tiered risk assessment for plant protection products for aquatic organisms in edge-of-field surface waters. *EFSA J.* **2013**, *11*, 3290. [CrossRef]
36. Backhaus, T.; Faust, M. Predictive Environmental Risk Assessment of Chemical Mixtures: A Conceptual Framework. *Environ. Sci. Technol.* **2012**, *46*, 2564–2573. [CrossRef]
37. Backhaus, T.; Altenburger, R.; Faust, M.; Frein, D.; Frische, T.; Johansson, P.; Kehrer, A.; Porsbring, T. Proposal for environmental mixture risk assessment in the context of the biocidal product authorization in the EU. *Environ. Sci. Eur.* **2013**, *25*, 1–9. [CrossRef]
38. Dietrich, C.; Wang, M.; Ebeling, M.; Gladbach, A. An efficient and pragmatic approach for regulatory aquatic mixture risk assessment of pesticides. *Environ. Sci. Eur.* **2022**, *34*, 16. [CrossRef]
39. Cedergreen, N.; Christensen, A.M.; Kamper, A.; Kudsk, P.; Mathiassen, S.K.; Streibig, J.C.; Sorensen, H. A Review of independent action compared to concentration addition as reference models for mixture of compounds with different molecular target sites. *Environ. Toxicol. Chem.* **2008**, *27*, 1621–1632. [CrossRef] [PubMed]
40. Kortenkamp, A.; Backhaus, T.; Faust, M. *State of the Art Report on Mixture Toxicity*; Report to the EU Commission, Directorate General for the Environment, No. 070307/2007/485103/ETU/D.1; EU Commission: Brussels, Belgium, 2009.
41. Laetz, C.A.; Baldwin, D.H.; Hebert, V.R.; Stark, J.D.; Scholz, N.L. The synergistic toxicity of pesticide mixtures: Implications for risk assessment and the conservation of endangered Pacific salmon. *Environ. Health Perspect.* **2009**, *117*, 348–353. [CrossRef] [PubMed]
42. Bui, T.T.T.; Vo, P.L.T.; Truong, L. Developmental neurotoxicity of pesticide mixtures in zebrafish: Synergistic effects on behavior and acetylcholinesterase activity. *Chemosphere* **2022**, *301*, 134416. [CrossRef]
43. European Food Safety (EFSA). Peer review of the pesticide risk assessment of theactive substance acetamiprid. Conclusion on Pesticide Peer review. *EFSA J.* **2016**, *14*, 4610. [CrossRef]
44. European Food Safety (EFSA). Conclusion regarding the peer review of the pesticide risk assessment of the active substance cymoxanil. Conclusion on Pesticide Peer review. *EFSA Sci. Rep.* **2008**, *167*, 1–116. [CrossRef]
45. U.S. Environmental Protection Agency. Available online: https://archive.epa.gov/pesticides/chemicalsearch/chemical/foia/web/pdf/113501/113501-068.pdf (accessed on 11 June 2025).
46. European Food Safety (EFSA). Conclusion on the peer review of the pesticide risk assessment of the active substance azoxystrobin. Conclusion on Pesticide Peer review. *EFSA J.* **2010**, *8*, 1542. [CrossRef]

47. European Food Safety (EFSA). Evaluation of the new active BOSCALID in the product FILAN FUNGICIDE. Australian Pesticides and Veterinary Medicines Authority 2004. Available online: https://www.apvma.gov.au/sites/default/files/publication/13621-prs-boscalid.pdf (accessed on 11 June 2025).
48. European Food Safety (EFSA). Conclusion on the peer review of the pesticide risk assessment of the active substance mandipropamid. Conclusion on Pesticide Peer review. *EFSA J.* **2012**, *10*, 2935. [CrossRef]
49. European Food Safety (EFSA). Conclusion regarding the peer review of the pesticide risk assessment of the active substance dimethomorph. Conclusion on Pesticide Peer review. *EFSA Sci. Rep.* **2006**, *82*, 1–69. [CrossRef]
50. European Food Safety (EFSA). Conclusion on the peer review of the pesticide risk assessment of the activesubstance myclobutanil. Conclusion on Pesticide Peer review. *EFSA J.* **2010**, *8*, 1682. [CrossRef]
51. European Food Safety (EFSA). Conclusion regarding the peer review of the pesticide risk assessment of the active substance tetraconazole. Conclusion on Pesticide Peer review. *EFSA Sci. Rep.* **2008**, *152*, 1–86. [CrossRef]
52. European Food Safety (EFSA). Conclusion regarding the peer review of the pesticide risk assessment of the active substance penconazole. Conclusion on Pesticide Peer review. *EFSA Sci. Rep.* **2008**, *175*, 1–104. [CrossRef]
53. European Food Safety (EFSA). Conclusion regarding the peer review of the pesticide risk assessment of the active substance tebuconazole. Conclusion on Pesticide Peer review. *EFSA Sci. Rep.* **2008**, *176*, 1–109. [CrossRef]
54. European Food Safety (EFSA). Conclusion regarding the peer review of the pesticide risk assessment of the active substance zoxamide. Conclusion on Pesticide Peer review. *EFSA J.* **2017**, *15*, 4980. [CrossRef]
55. European Food Safety (EFSA). Conclusion regarding the peer review of the pesticide risk assessment of the active substance spinosad. Conclusion on Pesticide Peer review. *EFSA J.* **2018**, *16*, 5252. [CrossRef]
56. European Food Safety (EFSA). Conclusion regarding the peer review of the pesticide risk assessment of the active substance pyraclostrobin. Conclusion on Pesticide Peer review. *EFSA J.* **2025**, *23*, e9257. [CrossRef]
57. European Food Safety (EFSA). Conclusion regarding the peer review of the pesticide risk assessment of the active substance clofentezin. Conclusion on Pesticide Peer review. *EFSA J.* **2021**, *19*, 6817. [CrossRef]
58. European Food Safety (EFSA). Conclusion regarding the peer review of the pesticide risk assessment of the active substance difenconazole. Conclusion on Pesticide Peer review. *EFSA J.* **2011**, *9*, 1967. [CrossRef]
59. European Food Safety (EFSA). Conclusion regarding the peer review of the pesticide risk assessment of the active substance ametotractin. Conclusion on Pesticide Peer review. *EFSA J.* **2012**, *10*, 2921. [CrossRef]
60. European Food Safety (EFSA). Conclusion regarding the peer review of the pesticide risk assessment of the active substance metaflumizone. Conclusion on Pesticide Peer review. *EFSA J.* **2013**, *11*, 3373. [CrossRef]
61. European Food Safety (EFSA). Conclusion regarding the peer review of the pesticide risk assessment of the active substance emamectin. Conclusion on Pesticide Peer review. *EFSA J.* **2012**, *10*, 2955. [CrossRef]
62. European Food Safety (EFSA). Conclusion regarding the peer review of the pesticide risk assessment of the active substance etofenprox. Conclusion on Pesticide Peer review. *EFSA Sci. Rep.* **2008**, *213*, 1–131. [CrossRef]

Journal of
Xenobiotics

MDPI

Article

Temporal Trends, Multiple Residue Incidence, and Chronic Health Risk of Pesticides in Egyptian Onions: A Four-Year Market Surveillance

Farag Malhat [1,2,*], Shokr Shokr [1,2], Sara Heikal [3] and Nour El-Hoda Zidan [4]

1 Central Agricultural Pesticide Laboratory, Pesticide Residues and Environmental Pollution Department, Agricultural Research Center, Dokki, Giza 12618, Egypt; drshokr63@yahoo.com
2 Central Agricultural Pesticide Laboratory, National Center for Monitoring Pesticide Residues and Pollutants, Agricultural Research Center, Dokki, Giza 12618, Egypt
3 Pesticide Department, Faculty of Agriculture, Menofiya University, Shebien El-Kom 32514, Egypt; saraheikal2020@gmail.com
4 Pesticides Chemistry and Toxicology Department, Faculty of Agriculture, Kafrelsheikh University, Kafr El-Sheikh 33516, Egypt; nourelhoda_az@hotmail.com
* Correspondence: farag_malhat@yahoo.com

Abstract: This study analyzed 5731 Egyptian onion samples collected from 2021 to 2024 to detect 430 pesticide residues and evaluate occurrence patterns, EU Maximum Residue Limits (MRLs) compliance, multiple residue prevalence, and consumer health implications. The analysis revealed temporal fluctuations in contamination, with residue-free samples ranging from 51.19% (2023) to 75.59% (2022) and MRL exceedances varying from 1.34% (2022) to 8.33% (2023). Throughout 2021–2024, fungicides dominated pesticide residues, declining from 70% to 50%, while insecticides increased from 30% to 40%. Pesticide detection patterns shifted significantly over the study period, with carbendazim decreasing from 20.99% to 2.35%, azoxystrobin fluctuating between 3.86% and 18.78%, and dimethomorph rising to 16.67%, reflecting evolving agricultural practices. Multiple residue occurrence fluctuated dramatically, from 9.76% (2022) to a peak of 30.82% (2023), while single residue occurrence remained more stable, ranging between 14.06% and 22.90%. Several pesticides exceeded EU MRLs, including imazalil, chlorpropham, chlorpyrifos, imidacloprid, and profenofos. Despite these violations, the EFSA PRIMo 3.1 model's chronic risk assessment showed that exposure levels for all population groups remained well below Acceptable Daily Intake (ADI) thresholds, with a maximum %ADI of 2.27% for chlorpyrifos in high-consumption regions.

Keywords: *Allium cepa*; pesticide residues prevalence; MRL violations; multiple residues; chronic risk assessment

1. Introduction

The onion (*Allium cepa* L.), a global staple cultivated worldwide as a profitable cash crop and indispensable culinary ingredient, is also recognized for its nutraceutical and medicinal properties [1]. Onions are nutrient-dense vegetables packed with essential vitamins, minerals, and bioactive compounds like quercetin and allicin that help to mitigate oxidative stress, reduce blood pressure, and offer anticancer and immune-boosting benefits [2]. In 2023, Egypt ranked third globally in onion production, with approximately

3.8 million tonnes and a gross production value of about USD 1.6 billion, while its onion exports totaled approximately 533,000 tonnes, valued at USD 207 million [3].

In Egypt, onion cultivation is significantly impacted by a range of pests and diseases that attack the crop both in the field and during storage. Key pests include onion thrips, aphids, and bulb flies, while diseases like downy mildew and purple blotch are prevalent during the growing season [4]. Additionally, post-harvest losses are often caused by various fungal and bacterial rots, such as neck rot, black mold, and bacterial soft rot, which thrive in storage conditions [5]. Consequently, the application of insecticides, fungicides, and bactericides is a necessary practice for farmers to protect yields and ensure the quality of the onion harvest.

However, pesticide residues can persist, leading to environmental contamination and health risks, highlighting the need for integrated pest management and residue monitoring. Monitoring pesticide levels in agricultural products fulfills several important purposes: ensuring regulatory compliance by quantifying residue levels on crops and verifying proper application practices; enhancing consumer safety and confidence; protecting environmental ecosystems; and facilitating international trade by meeting export standards [6,7].

Many researchers have examined the occurrence of pesticide residues in onions available in markets across the world, including those in China [8–10], the Republic of Korea [11], Iran [12], Vietnam [13], the Czech Republic [14], Spain [15], United Arab Emirates [16], Saudi Arabia [17], Bolivia [18], Ethiopia [19], Brazil [20], and the European Union [21].

Despite these valuable contributions, a comprehensive, multi-year market surveillance specifically focused on Egyptian onions, detailing temporal trends in pesticide residue levels, the incidence of multiple residues, and a thorough chronic health risk assessment tailored to this specific crop, remains underexplored, as previous Egyptian studies on pesticide residues in onions have several notable gaps. For instance, Malhat et al. [22,23] did not specify the seasonal distribution of their onion samples, which potentially limits the understanding of temporal variations in pesticide residue levels, while El-Sheikh et al. [24] restricted their analysis to a single governorate, thereby limiting external validity.

This study aimed to comprehensively monitor 430 pesticide residues in onions from Egyptian markets over a four-year period (2021–2024), evaluate compliance with EU MRLs, and assess potential health risks to consumers. The study sought to identify temporal trends in pesticide detection rates, determine the most frequently detected pesticide compounds and their concentration patterns, analyze the prevalence of single versus multiple residue contamination, and conduct chronic dietary risk assessments across diverse population groups using the EFSA PRIMo 3.1 model to establish whether current pesticide residue levels pose appreciable health risks to consumers through onion consumption.

2. Materials and Methods

2.1. Reagents, Chemicals, and Standards

The study employed a range of high-quality chemicals and tools sourced from reputable suppliers. Methanol (LC-grade) and acetonitrile (HPLC-grade) were supplied by Supelco (Merck), located in Darmstadt, Germany. Ammonium formate of analytical grade and formic acid of LC–MS grade, both possessing purities of 99% or greater and 98–100% respectively, were obtained from Sigma-Aldrich (Merck KGaA) in Darmstadt, Germany. QuEChERS kits, along with a salt extraction kit containing crucial components such as anhydrous magnesium sulfate and sodium chloride, were supplied by UCT, located in Bristol, PA, USA. Additionally, the dispersive solid-phase extraction kit from UCT contained specific quantities of primary–secondary amine (PSA) and magnesium sulfate. A collection of 430 individual high-purity pesticide standards (>99% purity) was sourced from

Dr. Ehrenstorfer GmbH (Augsburg, Germany) and Sigma-Aldrich (Steinheim, Germany). The deionized water utilized in the experiments was generated through a Milli-Q water purification system.

2.2. Standard Solution Preparation

Standard stock solutions were formulated in toluene at a concentration of 1000 μg/mL and kept at −18 °C. From each of these solutions, a pesticide mixture was created, with each compound reaching a concentration of 10 μg/mL. Next, mixed standard working solutions at different concentrations were prepared by diluting the multi-analyte intermediate standard solution. The solutions were employed to prepare matrix-matched calibration standards and to effectively carry out tests for method recovery.

2.3. Gathering of Samples

From 2021 to 2024, inspectors collected a total of 5731 onion samples from various producers at wholesale markets in five Egyptian governorates (Gharbia, Qalyubia, Giza, Beheira, and El-Minofyia) to monitor for 430 multi-class pesticide residues. Adhering to the standardized sampling protocol outlined by the CODEX Alimentarius guidelines [25], these samples were collected randomly, and each weighed at least 3 kg. The samples were homogenized with a Hobart food processor model HCM62–1 (Hobart Corp., Troy, OH, USA) and stored in sealed, polypropylene containers in a refrigerator at 4 °C until analysis within approximately 24 h.

2.4. Preparation of Samples

Sample preparation was conducted according to the EN 15662:2018 [26] standard, using module E1 for extraction and module C2 for clean-up. A 10 g homogenized sample was placed in a 50 mL tube, and 10 mL of acetonitrile was subsequently added. The mixture was vigorously stirred for three minutes before adding the QuEChERS salt blend containing both anhydrous magnesium sulfate and sodium chloride to facilitate efficient phase separation. The sample was subsequently treated in a Geno/Grinder® (SPEX SamplePrep, LLC, Metuchen, NJ, USA) for one minute and then centrifuged at 3000× *g* for five minutes.

During the clean-up, dispersive solid-phase extraction was utilized to eliminate interfering components from the sample matrix. A 1 mL portion of the crude extract was transferred into a 2 mL polypropylene minicentrifuge tube preloaded with 50 mg PSA and 300 mg anhydrous magnesium sulfate. The mixture was vortexed for one minute, followed by an additional five-minute centrifugation at 3000× *g*. The purified extract was then placed into a GC-MS/MS vial for analysis. A 1 mL sample of the unpurified acetonitrile extract was subjected to LC-MS/MS analysis without any prior cleanup steps.

2.5. Chromatographic Analysis

2.5.1. LC-MS/MS Analysis

An AB SCIEX (USA) QTRAP 6500+ mass spectrometer, interfaced with a Shimadzu (Kyoto, Japan) LC-20 series HPLC system, was utilized for the analysis. A Phenomenex (Torrance, CA, USA) Synergi C18 reverse-phase column, kept at a temperature of 40 °C, was used to separate multi-residue compounds using gradient elution. The mobile phase system included Phase A, comprising 10 mM ammonium formate (pH 4) in a 90:10 water–methanol mixture (v/v), and Phase B, consisting of pure methanol, with a flow rate of 0.4 mL/min. The gradient elution program began with 100% Phase A for the first minute, followed by a linear decrease to 0% Phase A over 14 min. The system was then held at 0% Phase A for 3 min (15–18 min) to ensure full elution of late-eluting compounds, after which the composition was returned to 100% Phase A within 0.5 min and re-equilibrated for 2 min. The total run time was 20 min, with an injection volume of 2 μL.

Electrospray ionization (ESI) was performed using polarity-switching technology, allowing detection in both positive and negative ion modes with a 5.5 kV spray voltage. An ion source temperature of 400 °C was used, with curtain gas and collision gas both pressurized at 20 psi, while the nebulizer and auxiliary gas pressures were 35 psi. Multiple reaction monitoring (MRM) mode was employed in the mass spectrometer to monitor transitions from precursor ions to the two most abundant product ions for each analyte, in accordance with SANTE guidelines. Instrument control and data acquisition were managed using Analyst software (version 1.7.1) from AB SCIEX (Framingham, MA, USA). Retention times, quantification and confirmation MRM transitions, and collision energy parameters for LC compounds are listed in Table S1, while the associated total ion chromatograms (TICs) are shown in Figure S1.

2.5.2. GC-MS/MS Analysis

Pesticides underwent thorough analysis utilizing a Thermo Scientific (Waltham, MA, USA) Trace 1300 gas chromatography system connected to a mass spectrometer equipped with an electron ionization source. An Agilent Technologies (Santa Clara, CA, USA) HP-5MS capillary column (30 m × 0.25 mm I.D., 0.25 μm film thickness) was employed for the separation. A 1.0 μL sample was injected using a split/splitless injector that was set to splitless mode. The programmed temperature for the PTV inlet began at 100 °C, held for 0.02 min, and then increased by 12 °C per second up to 320 °C, where it was maintained for 0.75 min. Helium was used as the carrier gas at a flow rate of 1.611 mL/min. The temperature program was: 80 °C, increased by 20 °C/min to 170 °C, then increased by 20 °C/min to 310 °C, with a final 3.5 min hold. The transfer line and ion source of the mass spectrometer were maintained at 250 °C and 300 °C, respectively. Compound detection was achieved using MRM mode, wherein two transitions were analyzed for each compound emerging from the GC (Table S2). All targeted pesticides were fully separated within the 15 min run time, evidenced by the TIC in Figure S2.

2.6. Method Validation

To validate the analytical method, we assessed key parameters—linearity, matrix effect, selectivity, sensitivity, accuracy, and precision—as required by the EU SANTE/11312/2021 guidelines [27]. To assess matrix effects, two sets of calibration curves were generated for both LC-MS/MS and GC-MS/MS analysis. One set was prepared in a pure solvent (acetonitrile), while the other was matrix-matched using an onion extract. For each curve, we used six concentration levels: 1, 5, 10, 25, 50, and 100 μg/L. A correlation coefficient of at least 0.99 was deemed acceptable for linearity.

The method's selectivity was assessed by comparing chromatograms from solvent blanks and control onion samples against a pesticide mix standard, allowing us to differentiate analytes from background noise and other matrix components. The method's sensitivity was assessed by determining the limit of quantification (LOQ) in spiked samples, which represents the minimum spiking concentration that satisfied the validation performance requirements.

We ensured acceptable accuracy, within the range of 70–120%, and precision, with a relative standard deviation (RSD) of no more than 20% (Table S3), by spiking blank samples with mixed standard solutions at concentrations of 0.005, 0.01, and 0.05 mg kg^{-1}. Following this spiking, the samples were analyzed according to the method protocol, utilizing matrix-matched calibration curves for quantification. We assessed repeatability and within-laboratory reproducibility through multiple repetitions to calculate intra-day ($n = 5$) and inter-day ($n = 15$, over three consecutive days) RSD%. Lastly, we calculated the expanded measurement uncertainty (MU) using the reproducibility and trueness parameters, ensuring it remained within the guideline-specified threshold of 50% or below.

2.7. Quality Control and Quality Assurance

The analytical method for detecting multiple classes of pesticides in onions was validated at the Central Agricultural Pesticide Laboratory (CAPL) in Egypt, in accordance with the requirements of its ISO/IEC 17025:2017 [28] accreditation. The laboratory maintained data reliability by implementing a thorough quality assurance program that incorporated proficiency testing, spiked testing, intra-laboratory testing, and retention and replicate testing, all adhering to ISO/IEC 17025:2017 requirements.

2.8. Evaluation of Potential Health Hazards Associated with Onions Contaminated with Pesticides

To evaluate the long-term risks from pesticide residues in onions, the European Food Safety Authority's (EFSA) Pesticide Residue Intake Model (EFSA PRIMo, revision 3.1) was employed [29]. The model integrates data from national food surveys carried out by EU Member States, encompassing information on onion intake and substantial portion sizes. The International Estimated Daily Intake (IEDI) is determined by taking into account different dietary patterns, the mean level of pesticide residues in onions, average body weight, and typical food consumption. To calculate chronic exposure as a percentage of the Acceptable Daily Intake (%ADI), the IEDI is divided by the ADI for the pesticides and then multiplied by 100. If this %ADI value surpasses 100%, it suggests that pesticide residues in onions could present a potential health risk to humans.

3. Results and Discussion

3.1. Determination of Pesticide Residues in Onion Samples and Evaluation of Adherence to EU MRLs

An examination of pesticide residue concentrations in onion samples obtained from markets in Egypt over the period from 2021 to 2024 reveals fluctuating trends in detection rates, compliance with EU MRLs, and exceedances, underscoring the dynamic interplay between agricultural practices, regulatory enforcement, and market surveillance (Figure 1). In 2021, with 262 samples analyzed, 32 distinct pesticides were detected, resulting in 54.58% of samples free of detectable residues, 42.75% complying with MRLs, and a relatively low 2.67% exceeding limits. The following year, 2022, saw a substantial increase in sample size to 2458, with 43 pesticides identified, yielding the highest proportion of residue-free samples at 75.59% and the lowest exceedance rate of 1.34%, alongside 23.07% within MRLs, which may reflect improved pest management. However, 2023 marked a concerning escalation, as 756 samples revealed 67 pesticides, the highest diversity observed, coupled with the lowest residue-free rate of 51.19%, 40.48% compliant with MRLs, and a peak exceedance of 8.33%, potentially attributable to intensified pesticide use amid pest pressures. By 2024, analysis of 2255 samples detected 54 pesticides, with 71.53% residue-free, 24.43% $\leq$ MRL, and 4.04% > MRL, indicating a partial recovery in compliance.

Residue-free rates in our study (51.19–75.59%) fall within the global range, which extends from 100% in Bolivian samples [18] to 0% in Vietnamese onions [13]. Other reported rates include 84.6% in the Republic of Korea [11], 65.8% in China [9], 88.3% in Brazil [20], and 25% in Saudi Arabia [17].

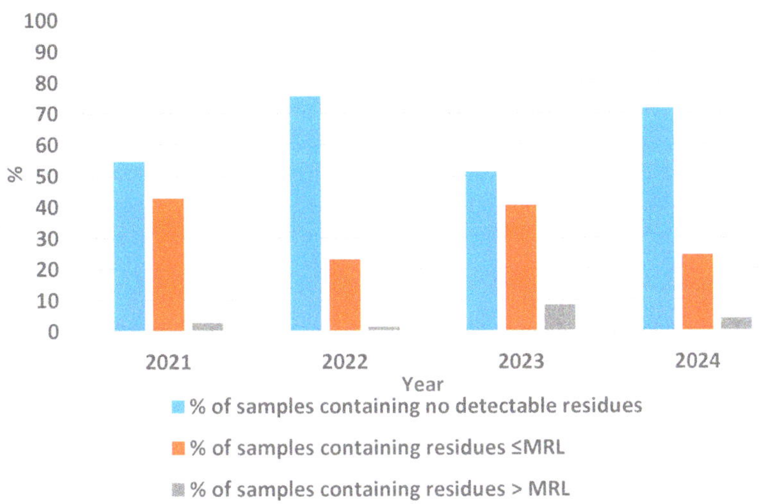

Figure 1. Temporal trends in pesticide residue detection and MRL compliance in Egyptian onions (2021–2024).

Our exceedance rates (1.34–8.33%) represent moderate levels when compared globally. While these rates are higher than the zero exceedances reported in the Republic of Korea [11], they are comparable to findings from other regions: 6% in China [9], 5.6% in United Arab Emirates (UAE) imports [16], 10% in Saudi Arabia [17], 8.2% in Brazil [20], and 11.7% in Egypt [22].

3.2. The Ten Pesticides Most Commonly Detected in Onion Samples

The frequency distribution of pesticide detections exhibited remarkable temporal shifts across the four-year study period, revealing dynamic changes in agricultural practices and regulatory responses in Egyptian onion production. Carbendazim demonstrated the most dramatic decline, plummeting from its peak detection frequency of 20.99% in 2021 to progressively lower levels of 10.09% in 2022, 5.56% in 2023, and ultimately 2.35% in 2024 (Table 1). Conversely, azoxystrobin maintained consistently high detection rates despite significant fluctuations, starting at 20.99% in 2021, and dropping to 3.86% in 2022, then resurging to become the most frequently detected pesticide at 18.78% in 2023 and 11.97% in 2024. Dimethomorph exhibited the most striking emergence pattern, transitioning from minimal detection frequencies of 2.67% in 2021 and 1.51% in 2022 to becoming the third most frequently detected pesticide at 16.67% in 2023 and 7.23% in 2024. Metalaxyl showed a persistent presence as the second most frequently detected pesticide with notable fluctuation from 10.69% in 2021 to 5.57% in 2022, peaking at 17.33% in 2023, and settling at 8.07% in 2024. Several pesticides demonstrated sporadic appearance patterns, with chlorpyrifos emerging prominently in 2022–2023 but disappearing from the top ten by 2024, while profenofos and lambda-cyhalothrin showed intermittent presence, and imidacloprid appeared only in the latter two years, collectively indicating shifting pesticide selection preferences and potentially evolving resistance management strategies among Egyptian onion producers.

Table 1. Yearly comparison of predominant pesticides detected in Egyptian onions (2021–2024).

2021 (*n* = 262)					2022 (*n* = 2458)				
Pesticide	**Frequency%**	**Mean (mg/kg)**	**Range (Min.–Max.)**	**EU MRL (mg/kg)**	**Pesticide**	**Frequency%**	**Mean (mg/kg)**	**Range (Min.–Max.)**	**EU MRL (mg/kg)**
Carbendazim	20.99	0.006	<LOQ–0.025	0.10	Carbendazim	10.09	0.011	<LOQ–0.197	0.10
Azoxystrobin	20.99	0.005	<LOQ–0.013	10.00	Metalaxyl	5.57	0.005	<LOQ–0.025	0.03
Metalaxyl	10.69	0.005	<LOQ–0.027	0.03	Azoxystrobin	3.86	0.006	<LOQ–0.030	10.00
Boscalid	4.96	0.005	<LOQ–0.009	5.00	Cypermethrin	2.97	0.007	<LOQ–0.057	0.10
Dimethomorph	2.67	0.005	<LOQ–0.006	0.60	Chlorpyrifos	2.32	0.008	<LOQ–0.064	0.01
Difenoconazole	2.29	0.005	<LOQ–0.005	0.50	Thiophanate-methyl	2.24	0.020	<LOQ–0.231	0.10
Lambda-Cyhalothrin	2.29	0.005	<LOQ–0.006	0.20	Difenoconazole	1.79	0.007	<LOQ–0.049	0.50
Imazalil	2.29	0.029	<LOQ–0.054	0.01	Dimethomorph	1.51	0.008	<LOQ–0.042	0.60
Profenofos	2.29	0.005	<LOQ–0.005	0.02	Clothianidin	1.30	0.005	<LOQ–0.026	0.01
Clothianidin	2.29	0.005	<LOQ–0.005	0.01	Chlorpropham	1.10	0.014	<LOQ–0.049	0.01
2023 (*n* = 756)					**2024 (*n* = 2255)**				
Pesticide	**Frequency%**	**Mean (mg/kg)**	**Range (Min.–Max.)**	**EU MRL (mg/kg)**	**Pesticide**	**Frequency%**	**Mean (mg/kg)**	**Range (Min.–Max.)**	**EU MRL (mg/kg)**
Azoxystrobin	18.78	0.018	<LOQ–0.213	10.00	Azoxystrobin	11.97	0.030	<LOQ–2.204	10.00
Metalaxyl	17.33	0.019	<LOQ–0.277	0.03	Metalaxyl	8.07	0.011	<LOQ–0.284	0.03
Dimethomorph	16.67	0.037	<LOQ–0.297	0.60	Dimethomorph	7.23	0.046	<LOQ–1.102	0.60
Difenoconazole	9.26	0.011	<LOQ–0.079	0.50	Difenoconazole	6.08	0.022	<LOQ–0.559	0.50
Chlorpyrifos	6.08	0.030	<LOQ–0.297	0.01	Profenofos	3.59	0.032	<LOQ–0.336	0.02
Carbendazim	5.56	0.020	<LOQ–0.281	0.10	Lambda-Cyhalothrin	2.75	0.015	<LOQ–0.225	0.20
Boscalid	5.42	0.020	<LOQ–0.102	5.00	Carbendazim	2.35	0.011	<LOQ–0.070	0.10
Cypermethrin	5.29	0.022	<LOQ–0.183	0.10	Chlorpropham	2.17	0.021	<LOQ–0.125	0.01
Lambda-Cyhalothrin	4.89	0.024	<LOQ–0.205	0.20	Cypermethrin	1.95	0.010	<LOQ–0.064	0.10
Imidacloprid	4.89	0.015	<LOQ–0.089	0.01	Imidacloprid	1.95	0.040	<LOQ–0.304	0.01

The analysis revealed that mean concentrations of several pesticides exceeded EU MRLs across multiple years. In 2021, imazalil demonstrated the most severe violation with mean concentrations of 0.029 mg/kg, nearly three times the EU MRL of 0.01 mg/kg. The pattern of violations expanded in 2022, where chlorpropham exhibited mean concentrations of 0.014 mg/kg, exceeding its 0.01 mg/kg limit by 40%. In 2023, chlorpyrifos violations intensified with mean concentrations reaching 0.030 mg/kg, three times the acceptable limit, and imidacloprid joined the violators with mean residues of 0.015 mg/kg exceeding its 0.01 mg/kg MRL. By 2024, the violation pattern persisted with profenofos showing mean concentrations of 0.032 mg/kg against a 0.02 mg/kg limit, chlorpropham continuing its violations with mean residues of 0.021 mg/kg, and imidacloprid demonstrating the most severe violation with mean concentrations of 0.040 mg/kg, quadruple the EU MRL.

A comprehensive analysis of pesticide residues from 2021 to 2024 revealed fungicides as the most predominant class, constituting 70% of the most frequently detected pesticides in 2021 with a combined detection frequency of 65.57%, followed by 60% in both 2022 and 2023 with frequencies of 25.06% and 72.72% respectively, and 50% in 2024 at 35.70%; within this class, benzimidazoles like carbendazim and thiophanate-methyl were notable early on, with carbendazim decreasing from 20.99% in 2021 to 2.35% in 2024 and thiophanate-methyl detected only in 2022 at 2.24%, while strobilurins such as azoxystrobin fluctuated at 20.99% in 2021, 3.86% in 2022, 18.78% in 2023, and 11.97% in 2024; phenylamides like metalaxyl showed volatile frequencies of 10.69% in 2021, 5.57% in 2022, 17.33% in 2023, and 8.07% in 2024; morpholines, particularly dimethomorph, increased significantly from

2.67% in 2021 to 16.67% in 2023 and 7.23% in 2024; and triazoles like difenoconazole rose from 2.29% in 2021 to 9.26% in 2023 and 6.08% in 2024. Insecticides ranked second, representing 30% of top detections in 2021 and 2022 with combined frequencies of 6.87% and 6.59% respectively, rising to 40% in 2023 and 2024 at 21.15% and 10.24%; key insecticides included organophosphates such as chlorpyrifos at 2.32% in 2022 and 6.08% in 2023, and profenofos at 2.29% in 2021 and 3.59% in 2024, alongside neonicotinoids like clothianidin and imidacloprid, with the latter detected only in 2023 and 2024 at 4.89% and 1.95%. Finally, the sprouting inhibitor chlorpropham appeared sporadically, accounting for 10% of frequent detections in 2022 at 1.10% and in 2024 at 2.17%, signaling post-harvest treatments during onion storage and distribution. The observed shift in detection frequency, with a relative decrease in fungicides and an increase in insecticides, directly corresponds with evolving local agricultural practices and pest management strategies. For example, the declining detection of the fungicide carbendazim aligns with its increasing regulatory scrutiny and its gradual replacement by newer, more targeted alternatives in crop protection schedules. Conversely, the consistent detection of a compound like dimethomorph is linked to its specific and continued use against prevalent pathogens such as downy mildew in key regional crops.

The specific pesticides exceeding EU MRLs in our study are consistent with findings from earlier Egyptian studies, including profenofos and chlorpyrifos [24], as well as carbendazim, metalaxyl, chlorpyrifos, and profenofos reported by Malhat et al. [22,23]. Similar exceedances have been reported by Tadesse et al. [19] for chlorpyrifos and malathion in Ethiopian onions, Chau et al. [13] for cypermethrin (up to 43 times MRL) and difenoconazole in Vietnamese onions, Jardim and Caldas [20] for organophosphorus compounds in Brazilian onions, Osaili et al. [16] for profenofos and pyridaben in UAE onions, Ramadan et al. [17] for methomyl violations in Saudi Arabian onions, and Veiga-del-Baño et al. [15] for chlorpropham exceedances in Spanish onions.

Additionally, Chu et al. [9] identified fungicides as prevalent in Chinese onions, though with different compounds and 6% MRL exceedances occurring mainly from insecticides such as carbofuran and cyhalothrin. Horská et al. [14] also frequently detected azoxystrobin in Czech onions.

3.3. Incidence of Single and Multiple Residue Detections in Onion Samples

The pesticide residue data reveals notable shifts in contamination patterns across the four-year period, particularly in the distribution between single and multiple residue occurrences, as shown in Figure 2. Single pesticide residues showed relative stability, fluctuating within a narrow range from 22.90% in 2021 to a low of 14.06% in 2024, with intermediate values of 14.65% (2022) and 17.99% (2023). In contrast, multiple pesticide residues exhibited more dramatic variations, starting at 22.52% in 2021, dropping sharply to 9.76% in 2022, representing a 56.6% decrease, before surging to a peak of 30.82% in 2023, the highest recorded level across all years. This 2023 spike in multiple residues coincided with the lowest proportion of residue-free samples (51.19%), suggesting a period of intensified pesticide application. The year 2024 showed improvement with multiple residues declining to 14.41%, approaching parity with single residue levels (14.06%), while residue-free samples recovered to 71.53%. The ratio of multiple to single residues varied considerably: approximately 1:1 in 2021, 0.67:1 in 2022, 1.71:1 in 2023, and 1.02:1 in 2024, indicating that 2023 was an outlier year where samples were more likely to contain multiple pesticides than single ones. This pattern suggests that contamination severity, rather than mere presence, fluctuated significantly, with 2022 and 2024 representing years of better pesticide management compared to 2021 and especially 2023.

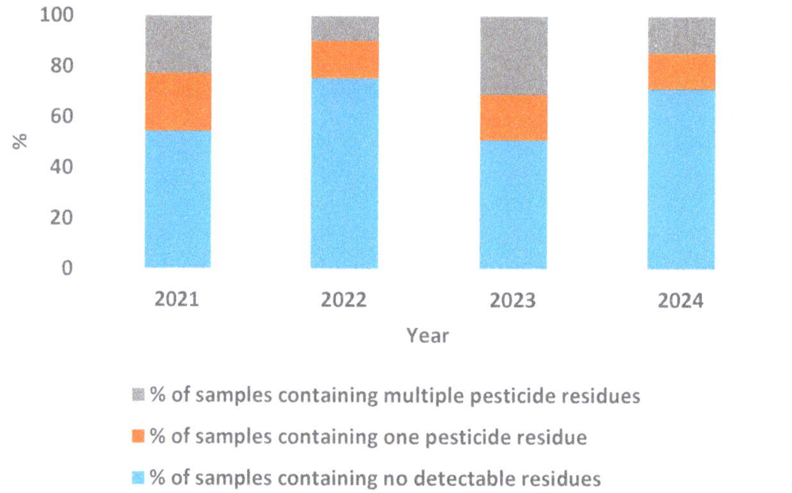

Figure 2. Distribution of single and multiple pesticide residues in Egyptian onions across four years.

While samples containing 9 pesticide residues were absent in 2021, they emerged at 0.04% in 2022, peaked at 0.40% in 2023, then declined to 0.09% in 2024, suggesting a tenfold increase from 2022 to 2023 followed by a 77.5% reduction (Table 2). Mid-range contamination (5–6 residues) showed progressive intensification, with 6-residue samples increasing from zero detection in 2021 to 0.67% by 2024, while 5-residue samples grew from 1.15% to peak at 3.84% in 2023 before moderating to 1.11% in 2024. The 3-residue category exhibited a notable decline from 6.49% in 2021 to 2.44% in 2022, then rebounded to 6.22% in 2023 before settling at 3.46% in 2024, while 4-residue samples peaked at 4.63% in 2023 from a baseline of 4.20% in 2021. The most substantial contamination category, samples with 2 pesticide residues, demonstrated relative stability around 10% (10.31% in 2021, 5.90% in 2022, 10.32% in 2023, and 6.52% in 2024), though the 2022 and 2024 values indicate approximately 40% reductions from peak years.

Table 2. Frequency of multiple pesticide residues in Egyptian market onions: Distribution by number of residues per sample (2021–2024).

Year	2021	2022	2023	2024
Multiple Pesticides	%	%	%	%
9 pesticide residues	0.00	0.04	0.40	0.09
8 pesticide residues	0.00	0.04	0.40	0.13
7 pesticide residues	0.38	0.00	1.06	0.40
6 pesticide residues	0.00	0.08	2.91	0.67
5 pesticide residues	1.15	0.33	3.84	1.11
4 pesticide residues	4.20	0.98	4.63	1.86
3 pesticide residues	6.49	2.44	6.22	3.46
2 pesticide residues	10.3	5.90	10.32	6.52
1 pesticide residue	22.9	14.7	17.9	14.1

Nevertheless, the limited knowledge regarding interactions between multiple pesticides and human health makes their occurrence a cause for concern. Research [6,30,31] suggests that these exposures could lead to synergistic or additive toxic effects that heighten health risks. Climate change exacerbates pesticide contamination in agriculture by elevating pest pressure, which necessitates more frequent pesticide applications, and by modifying environmental conditions that slow down the breakdown of these chemicals. As a result, food crops end up accumulating greater levels of pesticide residues [32].

Comparative analyses with global studies reveal that our data's temporal depth and emphasis on multi-residue complexity are relatively unique, with other regions showing varying but often less detailed instances of multiple residues. For instance, Chu et al. [9] reported multiple residues in 19.6% of Chinese onion samples, with up to 14 pesticides per sample. Similarly, Chau et al. [13] detected up to seven co-occurring residues in Vietnamese onions.

3.4. Health Risk Assessment

The chronic risk assessment of pesticide residues in onions, as evaluated using the EFSA PRIMo 3.1 model and presented in Table 3, reveals that exposure levels across diverse population groups remain well below the Acceptable Daily Intake (ADI) thresholds, with all calculated percentages (%ADI) substantially less than 100%, thereby indicating no appreciable long-term health risks from consumption of these contaminated onions. Notably, the highest %ADI values are observed for chlorpyrifos (ADI 0.001 mg/kg bw/day), ranging from 1.06% in the Portuguese general population to 2.27% in the GEMS/Food G06 cluster (encompassing populations from Armenia, Cuba, Egypt, Greece, Iran, Lebanon, and Turkey), which likely reflects higher per capita onion consumption patterns in these regions compared to others, such as the GEMS/Food G07 cluster (including Australia, Finland, France, and the United Kingdom (UK) where values are consistently lower. Other pesticides with relatively elevated %ADI include lambda-cyhalothrin (up to 0.727% in G06) and cypermethrin (up to 0.333% in G06), yet these figures underscore a broad margin of safety, as even the most conservative estimates, factoring in variations in dietary habits, body weights, and consumption data from clusters like GEMS/Food G10 and GEMS/Food G15, do not approach hazardous levels. Pesticides like clothianidin and azoxystrobin demonstrate negligible contributions (e.g., <0.004% across groups), highlighting the dominance of organophosphate and pyrethroid residues in overall exposure profiles. This comprehensive assessment across 45 distinct population groups strengthens the reliability of these findings and suggests that current agricultural practices and regulatory frameworks are effectively maintaining pesticide residues in onions at levels that ensure consumer safety across diverse dietary contexts and consumption patterns globally.

These results align with international studies confirming the safety of onion consumption despite pesticide residues. For instance, Ahn et al. [11] reported %ADI values up to 0.584% in South Korean onions and concluded that there were no significant long-term risks. Similarly, Bai et al. [8] found negligible chronic exposure to propiconazole in Chinese onions, with intakes well below ADI thresholds. Chu et al. [9] observed %ADI values below 100% in Chinese onions, while El-Sheikh et al. [24] reported acceptable levels with up to 21.13% ADI for profenofos. Additionally, Malhat et al. [22,23], Khaled et al. [33], and Gamal et al. [34] confirmed hazard indices and EDI values below ADI, indicating low risks. The European Food Safety Authority [21] assessed EU onions and found low chronic risks.

Table 3. Chronic dietary exposure to pesticides in Egyptian onions: International Estimated Daily Intake (IEDI) and % Acceptable Daily Intake (%ADI) across diverse populations.

Pesticide (ADI)	MS Diet	GEMS/Food G06	Romania General	GEMS/Food G10	GEMS/Food G15	GEMS/Food G08	Sweden General	Finland 3-Year-Old Children	GEMS/Food G07	Portugal General
Azoxystrobin (0.2)	IEDI	2.27×10^{-2}	2.25×10^{-2}	1.75×10^{-2}	1.56×10^{-2}	1.53×10^{-2}	1.35×10^{-2}	1.17×10^{-2}	1.06×10^{-2}	1.06×10^{-2}
	% of ADI	1.14×10^{-2}	1.13×10^{-2}	8.77×10^{-3}	7.78×10^{-3}	7.64×10^{-3}	6.75×10^{-3}	5.86×10^{-3}	5.31×10^{-3}	5.28×10^{-3}
Boscalid (0.04)	IEDI	1.51×10^{-2}	1.50×10^{-2}	1.17×10^{-2}	1.04×10^{-2}	1.02×10^{-2}	9.00×10^{-3}	7.82×10^{-3}	7.08×10^{-3}	7.03×10^{-3}
	% of ADI	3.78×10^{-2}	3.75×10^{-2}	2.92×10^{-2}	2.59×10^{-2}	2.55×10^{-2}	2.25×10^{-2}	1.95×10^{-2}	1.77×10^{-2}	1.76×10^{-2}
Carbendazim (0.02)	IEDI	1.51×10^{-2}	1.50×10^{-2}	1.17×10^{-2}	1.04×10^{-2}	1.02×10^{-2}	9.00×10^{-3}	7.82×10^{-3}	7.08×10^{-3}	7.03×10^{-3}
	% of ADI	7.57×10^{-2}	7.50×10^{-2}	5.85×10^{-2}	5.18×10^{-2}	5.10×10^{-2}	4.50×10^{-2}	3.91×10^{-2}	3.54×10^{-2}	3.52×10^{-2}
Chlorpropham (0.05)	IEDI	1.59×10^{-2}	1.58×10^{-2}	1.23×10^{-2}	1.09×10^{-2}	1.07×10^{-2}	9.45×10^{-3}	8.21×10^{-3}	7.43×10^{-3}	7.39×10^{-3}
	% of ADI	3.18×10^{-2}	3.15×10^{-2}	2.46×10^{-2}	2.18×10^{-2}	2.14×10^{-2}	1.89×10^{-2}	1.64×10^{-2}	1.49×10^{-2}	1.48×10^{-2}
Chlorpyrifos (0.001)	IEDI	2.27×10^{-2}	2.25×10^{-2}	1.75×10^{-2}	1.56×10^{-2}	1.53×10^{-2}	1.35×10^{-2}	1.17×10^{-2}	1.06×10^{-2}	1.06×10^{-2}
	% of ADI	2.27×10^{0}	2.25×10^{0}	1.75×10^{0}	1.56×10^{0}	1.53×10^{0}	1.35×10^{0}	1.17×10^{0}	1.06×10^{0}	1.06×10^{0}
Clothianidin (0.097)	IEDI	3.78×10^{-3}	3.75×10^{-3}	2.92×10^{-3}	2.59×10^{-3}	2.55×10^{-3}	2.25×10^{-3}	1.95×10^{-3}	1.77×10^{-3}	1.76×10^{-3}
	% of ADI	3.90×10^{-3}	3.87×10^{-3}	3.01×10^{-3}	2.67×10^{-3}	2.63×10^{-3}	2.32×10^{-3}	2.01×10^{-3}	1.82×10^{-3}	1.81×10^{-3}
Cypermethrin (0.005)	IEDI	1.67×10^{-2}	1.65×10^{-2}	1.29×10^{-2}	1.14×10^{-2}	1.12×10^{-2}	9.90×10^{-3}	8.60×10^{-3}	7.79×10^{-3}	7.74×10^{-3}
	% of ADI	3.33×10^{-1}	3.30×10^{-1}	2.57×10^{-1}	2.28×10^{-1}	2.24×10^{-1}	1.98×10^{-1}	1.72×10^{-1}	1.56×10^{-1}	1.55×10^{-1}
Difenoconazole (0.01)	IEDI	1.67×10^{-2}	1.65×10^{-2}	1.29×10^{-2}	1.14×10^{-2}	1.12×10^{-2}	9.90×10^{-3}	8.60×10^{-3}	7.79×10^{-3}	7.74×10^{-3}
	% of ADI	1.67×10^{-1}	1.65×10^{-1}	1.29×10^{-1}	1.14×10^{-1}	1.12×10^{-1}	9.90×10^{-2}	8.60×10^{-2}	7.79×10^{-2}	7.74×10^{-2}
Dimethomorph (0.05)	IEDI	3.48×10^{-2}	3.45×10^{-2}	2.69×10^{-2}	2.38×10^{-2}	2.34×10^{-2}	2.07×10^{-2}	1.80×10^{-2}	1.63×10^{-2}	1.62×10^{-2}
	% of ADI	6.96×10^{-2}	6.90×10^{-2}	5.38×10^{-2}	4.77×10^{-2}	4.69×10^{-2}	4.14×10^{-2}	3.60×10^{-2}	3.26×10^{-2}	3.24×10^{-2}
Imazalil (0.025)	IEDI	2.19×10^{-2}	2.18×10^{-2}	1.70×10^{-2}	1.50×10^{-2}	1.48×10^{-2}	1.31×10^{-2}	1.13×10^{-2}	1.03×10^{-2}	1.02×10^{-2}
	% of ADI	8.78×10^{-2}	8.70×10^{-2}	6.78×10^{-2}	6.01×10^{-2}	5.91×10^{-2}	5.22×10^{-2}	4.53×10^{-2}	4.11×10^{-2}	4.08×10^{-2}
Imidacloprid (0.06)	IEDI	3.03×10^{-2}	3.00×10^{-2}	2.34×10^{-2}	2.07×10^{-2}	2.04×10^{-2}	1.80×10^{-2}	1.56×10^{-2}	1.42×10^{-2}	1.41×10^{-2}
	% of ADI	5.05×10^{-2}	5.00×10^{-2}	3.90×10^{-2}	3.46×10^{-2}	3.40×10^{-2}	3.00×10^{-2}	2.61×10^{-2}	2.36×10^{-2}	2.34×10^{-2}
Lambda-Cyhalothrin (0.0025)	IEDI	1.82×10^{-2}	1.80×10^{-2}	1.40×10^{-2}	1.24×10^{-2}	1.22×10^{-2}	1.08×10^{-2}	9.38×10^{-3}	8.50×10^{-3}	8.44×10^{-3}
	% of ADI	7.27×10^{-1}	7.20×10^{-1}	5.61×10^{-1}	4.98×10^{-1}	4.89×10^{-1}	4.32×10^{-1}	3.75×10^{-1}	3.40×10^{-1}	3.38×10^{-1}
Metalaxyl (0.08)	IEDI	1.44×10^{-2}	1.43×10^{-2}	1.11×10^{-2}	9.85×10^{-3}	9.68×10^{-3}	8.55×10^{-3}	7.43×10^{-3}	6.73×10^{-3}	6.68×10^{-3}
	% of ADI	1.80×10^{-2}	1.78×10^{-2}	1.39×10^{-2}	1.23×10^{-2}	1.21×10^{-2}	1.07×10^{-2}	9.28×10^{-3}	8.41×10^{-3}	8.35×10^{-3}
Profenofos (0.03)	IEDI	2.42×10^{-2}	2.40×10^{-2}	1.87×10^{-2}	1.66×10^{-2}	1.63×10^{-2}	1.44×10^{-2}	1.25×10^{-2}	1.13×10^{-2}	1.13×10^{-2}
	% of ADI	8.07×10^{-2}	8.00×10^{-2}	6.24×10^{-2}	5.53×10^{-2}	5.43×10^{-2}	4.80×10^{-2}	4.17×10^{-2}	3.78×10^{-2}	3.75×10^{-2}
Thiophanate-methyl (0.02)	IEDI	1.51×10^{-2}	1.50×10^{-2}	1.17×10^{-2}	1.04×10^{-2}	1.02×10^{-2}	9.00×10^{-3}	7.82×10^{-3}	7.08×10^{-3}	7.03×10^{-3}
	% of ADI	7.57×10^{-2}	7.50×10^{-2}	5.85×10^{-2}	5.18×10^{-2}	5.10×10^{-2}	4.50×10^{-2}	3.91×10^{-2}	3.54×10^{-2}	3.52×10^{-2}

GEMS/Food G06: General population (Armenia, Cuba, Egypt, Greece, Iran, Lebanon, and Turkey). GEMS/Food G07: General population (Australia, Bermuda, Finland, France, Iceland, Luxembourg, Norway, Switzerland, the UK, and Uruguay). GEMS/Food G08: General population (Austria, Germany, Poland, and Spain). GEMS/Food G10: General population (Belarus, Bulgaria, Canada, Croatia, Cyprus, Estonia, Italy, Japan, Latvia, Malta, New Zealand, the Republic of Korea, the Russian Federation, and the USA). GEMS/Food G15: General population (the Czech Republic, Denmark, Hungary, Ireland, Lithuania, Portugal, Romania, Serbia, Slovakia, Slovenia, and Sweden).

4. Conclusions

This four-year study of 5731 Egyptian onion samples from 2021 to 2024 highlights evolving pesticide residue dynamics, with detection of up to 67 pesticides in 2023, alongside variable compliance with EU MRLs. Residue-free samples peaked at 75.59% in 2022 but dropped to 51.19% in 2023, correlating with the highest MRL exceedance rate of 8.33% and multiple residue incidence of 30.82%. By 2024, improvements were evident, with 71.53% residue-free samples and exceedances reduced to 4.04%. The dramatic shift in pesticide class distribution—fungicides declining from 70% to 50% while insecticides increased from 30% to 40%—is likely driven by a complex combination of factors, including changing pest pressures, adaptive management strategies, regulatory requirements, and economic considerations influencing farmer choices. The elimination of carbendazim usage (from 20.99% to 2.35%) alongside the emergence of dimethomorph as a primary fungicide demonstrates active resistance management and regulatory compliance efforts. The sharp fluctuations in multiple residue occurrence, marked by a jump to 30.82% in 2023 and a drop to 14.41% in 2024, highlight the critical need for continuous monitoring programs to identify periods that may require intervention. Nevertheless, EFSA PRIMo 3.1-based chronic risk assessments across global population groups confirmed exposures far below ADI limits, with chlorpyrifos reaching a maximum 2.27% ADI in high-consumption clusters. These findings underscore the need for sustained regulatory enhancements to minimize exceedances while affirming overall consumer safety.

Supplementary Materials: The following supporting information can be downloaded at https://www.mdpi.com/article/10.3390/jox15060192/s1. Table S1: LC-MS/MS optimization parameters including molecular masses to charge ratio (m/z) in the quads (q1&q3) retention time (RT), decluster potential (DP), entrance potential (EP), collision energy (CE) and collision cell exit potential (CXP) for all target MRM transitions of LC-Compounds.; Table S2: GC-MS/MS optimization parameters including retention time (RT), molecular masses to charge ratio (m/z) in the quads (q1&q3), and collision energy (CE) for all target MRM transitions of GC-Compounds.; Table S3: Limits of quantification (LOQ), linearity (R2), average recoveries (Rec.), and relative standard deviations (RSD) for 430 pesticides analyzed in onions.; Figure S1: LC-MS/MS chromatogram of the analytical standard of LC analytes at 0.1 mg/kg.; Figure S2: GC-MS/MS chromatogram of analytical standard of GC analytes at 0.1 mg/kg.

Author Contributions: Conceptualization, F.M. and S.S.; Methodology, F.M.; Formal Analysis, F.M.; Investigation, F.M.; Data Curation, F.M. and N.E.-H.Z.; Writing—Original Draft Preparation, N.E.-H.Z.; Writing—Review and Editing, F.M., S.S., S.H. and N.E.-H.Z.; Visualization, N.E.-H.Z.; Supervision, S.S.; Project Administration, S.S. All authors have read and agreed to the published version of the manuscript.

Funding: This research received no external funding.

Institutional Review Board Statement: Not applicable.

Informed Consent Statement: Not applicable.

Data Availability Statement: The original contributions presented in the study are included in the article/Supplementary Material. Further inquiries can be directed to the corresponding author.

Conflicts of Interest: The authors declare no conflicts of interest.

References

1. Stoica, F.; Rațu, R.N.; Veleșcu, I.D.; Stănciuc, N.; Râpeanu, G. A comprehensive review on bioactive compounds, health benefits, and potential food applications of onion (*Allium cepa* L.) skin waste. *Trends Food Sci. Technol.* **2023**, *141*, 104173. [CrossRef]
2. Gupta, A.J.; Kaldate, S.; Volaguthala, S.; Mahajan, V. Onion nutritional and nutraceutical composition and therapeutic potential of its phytochemicals assessed through preclinical and clinical studies. *J. Funct. Foods* **2025**, *129*, 106889. [CrossRef]
3. FAOSTAT. Available online: https://www.fao.org/faostat/en/#rankings/countries_by_commodity (accessed on 30 September 2025).
4. Ahmed, M.F.A.; Amin, M.M.; Roshdy, F.H. Impact of eco-friendly treatments on managing onion downy mildew. *J. Plant Food Sci.* **2024**, *2*, 142–148. [CrossRef]
5. Khalifa, M.; Mahmoud, N.; Abou-Zeid, N. Management of onion bulb rots during storage using pre-and post-harvest control treatments. *Egypt. J. Phytopathol.* **2016**, *44*, 1–16. [CrossRef]
6. Eissa, F.; Zidan, N.E.H.; Sebaei, A.S.; Mohamed, M.E.B. Pesticide residues in fruits and vegetables: Analysis and risk assessment of EU RASFF notifications between 1999 and 2022. *J. Food Compos. Anal.* **2024**, *134*, 106556. [CrossRef]

7. Malhat, F.; Abdel-Megeed, M.; Saber, E.S.; Shokr, S.A.S.; Eissa, F. Assessment of pesticide residue patterns and multiple residue occurrence in Egyptian medicinal and aromatic plants: A comprehensive market screening. *Int. J. Environ. Anal. Chem.* **2025**, 1–16. [CrossRef]
8. Bai, A.; Chen, A.; Chen, W.; Luo, X.; Liu, S.; Zhang, M.; Liu, Y.; Zhang, D. Study on degradation behaviour, residue distribution, and dietary risk assessment of propiconazole in celery and onion under field application. *J. Sci. Food Agric.* **2021**, *101*, 1998–2005. [CrossRef]
9. Chu, N.; Shu, X.; Meng, X.; Zhang, X.; Yang, J.; Li, B. Determination and dietary exposure assessment of 79 pesticide residues in Chinese onion (*Allium fistulosum* L.). *CyTA-J. Food* **2023**, *21*, 41–48. [CrossRef]
10. Guan, W.; Li, C.; Liu, X.; Zhou, S.; Ma, Y. Graphene as dispersive solid-phase extraction materials for pesticides LC-MS/MS multi-residue analysis in leek, onion and garlic. *Food Addit. Contam. Part A* **2014**, *31*, 250–261. [CrossRef]
11. Ahn, J.W.; Jeon, Y.H.; Hwang, J.I.; Kim, H.Y.; Kim, J.H.; Chung, D.H.; Kim, J.E. Monitoring of pesticide residues and risk assessment for fruit vegetables and root vegetables of environment-friendly certified and general agricultural products. *Korean J. Environ. Agric.* **2012**, *31*, 164–169. [CrossRef]
12. Gordan, H.; Mahdavi, V.; Behbahan, A.K. Pesticides residue analysis and associated human health risk assessment in tomato and onion. *Environ. Monit. Assess.* **2025**, *197*, 718. [CrossRef]
13. Chau, N.D.G.; Van Hop, N.; Long, H.T.; Duyen, N.T.M.; Raber, G. Multi-residue analytical method for trace detection of new-generation pesticides in vegetables using gas chromatography–tandem mass spectrometry. *J. Environ. Sci. Health Part B* **2020**, *55*, 417–428. [CrossRef] [PubMed]
14. Horská, T.; Kocourek, F.; Stará, J.; Holý, K.; Mráz, P.; Krátký, F.; Kocourek, V.; Hajšlová, J. Evaluation of pesticide residue dynamics in lettuce, onion, leek, carrot and parsley. *Foods* **2020**, *9*, 680. [CrossRef] [PubMed]
15. Veiga-del-Baño, J.M.; Oliva, J.; Cámara, M.Á.; Andreo-Martínez, P.; Motas, M. Quick Monitoring of Tomato and Onion Samples During Routine Regulatory Analysis of Pesticide Residues. *Arch. Environ. Contam. Toxicol.* **2025**, *89*, 23–33. [CrossRef]
16. Osaili, T.M.; Al Sallagi, M.S.; Dhanasekaran, D.K.; Odeh, W.A.B.; Al Ali, H.J.; Al Ali, A.A.; Radwan, H.; Obaid, R.S.; Holley, R. Pesticide residues in fresh vegetables imported into the United Arab Emirates. *Food Control* **2022**, *133*, 108663. [CrossRef]
17. Ramadan, M.F.; Abdel-Hamid, M.M.; Altorgoman, M.M.; AlGaramah, H.A.; Alawi, M.A.; Shati, A.A.; Shweeta, H.A.; Awwad, N.S. Evaluation of pesticide residues in vegetables from the Asir Region, Saudi Arabia. *Molecules* **2020**, *25*, 205. [CrossRef]
18. Skovgaard, M.; Renjel Encinas, S.; Jensen, O.C.; Andersen, J.H.; Condarco, G.; Jørs, E. Pesticide residues in commercial lettuce, onion, and potato samples from Bolivia—A threat to public health? *Environ. Health Insights* **2017**, *11*, 1178630217704194. [CrossRef]
19. Tadesse, B.; Asrat, M.; Zewge, B.; Tesfaw, A. Quantitative Analysis of Pesticides in Vegetables: A Study on Tomatoes and Onions from Ethiopian Farms. *Meas. Food* **2025**, *19*, 100242. [CrossRef]
20. Jardim, A.N.O.; Caldas, E.D. Pesticide residues in food of plant origin commercialized in Brazil from 2010 to 2020—An update from the two national monitoring programs. *Food Control* **2024**, *165*, 110674. [CrossRef]
21. European Food Safety Authority (EFSA); Carrasco Cabrera, L.; Di Piazza, G.; Dujardin, B.; Marchese, E.; Medina Pastor, P. The 2022 European Union report on pesticide residues in food. *EFSA J.* **2024**, *22*, e8753. [CrossRef]
22. Malhat, F.; Saber, A.; Saber, E.S.; Shokr, S.A.S.; Abdel-Megeed, M. Nationwide surveillance and cumulative risk assessment of pesticide residues in Egyptian vegetables: Results from 2018 to 2021. *Separations* **2024**, *11*, 318. [CrossRef]
23. Malhat, F.; Saber, E.S.; Shokr, S.A.S.; Eissa, F. Pesticide residues in Egyptian vegetables: A comprehensive analysis of compliance, co-occurrence of multiple residues, and health risk assessment. *J. Food Compos. Anal.* **2025**, *143*, 107634. [CrossRef]
24. El-Sheikh, E.S.A.; Ramadan, M.M.; El-Sobki, A.E.; Shalaby, A.A.; McCoy, M.R.; Hamed, I.A.; Ashour, M.B.; Hammock, B.D. Pesticide residues in vegetables and fruits from farmer markets and associated dietary risks. *Molecules* **2022**, *27*, 8072. [CrossRef] [PubMed]
25. CAC. Codex Alimentarius Commission Joint FAO/WHO Food Standards Program. Available online: https://www.fao.org/fao-who-codexalimentarius/codex-texts/list-standards/en/ (accessed on 30 September 2025).
26. *EN 15662: 2018*; Foods of Plant Origin. Multimethod for the Determination of Pesticide Residues Using GC- and LC-Based Analysis Following Acetonitrile Extraction/Partitioning and Clean-Up by Dispersive SPE. Modular QuEChERS-Method. BSI Standards Publication: London, UK, 2018.
27. Pihlström, T.; Fernández-Alba, A.R.; Amate, C.F.; Poulsen, M.E.; Hardebusch, B.; Anastassiades, M.; Lippold, R.; Cabrera, L.C.; de Kok, A.; ORegan, F. Analytical quality control and method validation procedures for pesticide residues analysis in food and feed SANTE 11312/2021. *Sante* **2021**, *11312*. Available online: https://food.ec.europa.eu/system/files/2023-11/pesticides_mrl_guidelines_wrkdoc_2021-11312.pdf (accessed on 30 September 2025).
28. *ISO/IEC 17025:2017*; General Requirements for the Competence of Testing and Calibration Laboratories (3rd ed.). International Organization for Standardization: Geneva, Switzerland, 2017. Available online: https://www.iso.org/files/live/sites/isoorg/files/store/en/PUB100424.pdf (accessed on 10 September 2025).

29. European Food Safety Authority (EFSA); Anastassiadou, M.; Brancato, A.; Carrasco Cabrera, L.; Ferreira, L.; Greco, L.; Jarrah, S.; Kazocina, A.; Leuschner, R.; Magrans, J.O.; et al. Pesticide Residue Intake Model-EFSA PRIMo revision 3.1: (update of EFSA PRIMo revision 3). *EFSA Supporting Publ.* **2019**, *16*, 1605E. [CrossRef]
30. Malhat, F.; Saber, E.S.; Shokr, S.A.S.; Eissa, F. Pesticide residues in Egyptian strawberries over a four-year period: Occurrence, adherence to maximum residue limits, co-occurrence, and effects on consumer health. *J. Food Compos. Anal.* **2025**, *146*, 107978. [CrossRef]
31. Malhat, F.; Saber, E.S.; Shokr, S.A.S.; Eissa, F. Pesticide residue profiles and health risk assessment of domestically grown and imported peaches in Egyptian markets. *J. Food Compos. Anal.* **2025**, *147*, 108004. [CrossRef]
32. Zidan, N.A.; Abdel-Halim, K.Y.; Eissa, F. Climate Change Impacts on Pesticide Use and Fate in the Environment. In *Climate Change Impacts on Toxins and Health Effects*; Saad-Hussein, A., El-Tawil, O., El-Sheikh, E.S.A., Eds.; Springer: Singapore, 2025; pp. 149–170. [CrossRef]
33. Khaled, O.; Ryad, L.; Eissa, F. Determination of tetracycline residues in potatoes and soil by LC-MS/MS: Method development, validation, and risk assessment. *Food Chem.* **2024**, *461*, 140841. [CrossRef]
34. Gamal, A.; Abdel-Megeed, M.; Mahmoud, H.; Soliman, M.; Al-Anany, M.; Eissa, F. Pesticide residues in green, roasted, and capsule coffee from the Egyptian market: Occurrence, processing effects, and health risk assessment. *Food Chem.* **2025**, *486*, 144671. [CrossRef]

Journal of *Xenobiotics*

Article

Prioritization and Sensitivity of Pesticide Risks from Root and Tuber Vegetables

Milica Lučić [1] and Antonije Onjia [2,*]

[1] Innovation Center of the Faculty of Technology and Metallurgy, 11120 Belgrade, Serbia; milica.lucic@tmf.bg.ac.rs
[2] Department of Analytical Chemistry and Quality Control, Faculty of Technology and Metallurgy, University of Belgrade, 11120 Belgrade, Serbia
* Correspondence: onjia@tmf.bg.ac.rs

Abstract: This study investigated pesticide residues in 580 vegetable samples collected from markets in Serbia, encompassing potatoes, carrots, celery, radishes, horseradish, ginger, onions, and leeks. In total, 33 distinct pesticides were detected using validated HPLC-MS/MS and GC-MS/MS analytical methods. Multiple residues were identified in 19 samples, while 29 samples exceeded established maximum residue levels (MRLs). Acute and chronic dietary risks were assessed for both adults and children. Although individual hazard quotients (HQs) for adults and children remained below the threshold of concern (HQ < 1), the cumulative acute risk reached up to 63.1% of the Acute Reference Dose (ARfD) for children and 51.1% ARfD for adults, with ginger and celery posing the highest risks. Similarly, cumulative chronic risks remained below the safety threshold, with the Acceptable Daily Intake (ADI) percentages reaching a maximum of 5.9% ADI for adults and increased vulnerability of 11.0% ADI among children. Monte Carlo simulations were applied to account for variability and uncertainty in chronic exposure estimates. The hazard index (HI) results showed that adverse health effects for both population groups remained within acceptable safety limits (HI < 1), although higher susceptibility was observed in children. Sensitivity analysis identified body weight and vegetable consumption rates as the most influential factors affecting chronic risk variability.

Keywords: potato; carrot; ginger; celery; onion; radish; QuEChERS; GC-MS/MS; HPLC-MS/MS; risk ranking; RASFF

1. Introduction

Pesticides are widely used in agriculture to control pests and plant diseases. These substances comprise a broad class of chemicals intentionally introduced into the environment to control weeds (herbicides), insect infestations (insecticides), fungal infections (fungicides), and rodents (rodenticides) [1,2]. In addition to their primary role in agriculture, pesticides are employed post-harvest to preserve crops and in non-agricultural settings such as residential, commercial, and public areas. However, the predominant application remains in agriculture, where the intensification of food production has inevitably led to increased pesticide use [1,3]. Due to their inherent toxicity, pesticides pose significant risks not only to human health but also to a wide range of non-target organisms. The adverse effects of pesticide exposure can be both acute and chronic, depending on the exposure dose and route [4]. Residual pesticide contamination in crops is a major concern because of their environmental persistence, potential for bioaccumulation, and toxicological impact. Upon release into the environment, pesticides may undergo degradation through abiotic processes or undergo transformation processes in organisms, yet their intermediate metabolites can also contribute to ecological and health risks [5,6].

Chronic exposure to pesticides has been associated with an increased risk of various health disorders, including cancer [7], endocrine disruption, cardiovascular diseases [8], respiratory conditions [9], and allergic reactions [10]. Moreover, prolonged or repeated pesticide exposure may negatively affect reproductive health, contributing to infertility and an elevated risk of miscarriage [11].

Despite the potential risks of pesticide contamination, vegetables remain an essential part of a healthy diet due to their high nutritional value. Vegetables are a fundamental component of a balanced diet, with regular consumption contributing positively to overall human health. Based on the edible plant part, vegetables are generally categorized into leafy, fruiting, root, tuber, and bulb vegetables [12]. Among root vegetables commonly included in the daily diet are carrot, celery, radish, horseradish, and parsley root. Potatoes are the most widely consumed tuber vegetable and rank as the third most produced food crop globally, following wheat and rice [13], and also serve as an important source of carbohydrates, dietary fiber, potassium, vitamin C, and other essential nutrients. Other frequently consumed vegetables, such as carrots, celery, onions, and ginger, are important sources of dietary fiber, vitamins, minerals, and various phytochemicals beneficial to human health [14–18].

Due to the widespread use of pesticides in vegetable cultivation, there is a growing need for strict control measures and limitations on excessive pesticide use [4]. Vegetables, such as potatoes, carrots, radishes, and celery, are often treated with herbicides, insecticides, and fungicides throughout the growing season to ensure yield quality and minimize post-harvest losses [2,4,6,19,20]. Because of their underground growth, roots, tubers, and bulb vegetables are particularly prone to pesticide accumulation, raising concerns about residue levels and the potential health risks associated with their consumption [12,21,22].

Maximum residue limits (MRLs), representing the highest concentrations of pesticide residues legally permitted in food or feed, are established by the European Commission under specific regulations [23]. These thresholds are derived from data obtained through Good Agricultural Practices (GAPs) and are essential criteria in the pesticide approval and registration process. Monitoring pesticide residues in vegetables is of critical importance, given the potential health risks associated with dietary exposure to these compounds [4,24]. Deterministic and probabilistic risk assessment approaches are commonly applied to assess the potential health impacts of pesticide residues. The deterministic method uses point estimates of residue levels, consumption rates, and toxicological reference values to calculate hazard quotients. In contrast, the probabilistic approach incorporates variability and uncertainty by using statistical distributions and Monte Carlo simulations, offering a more comprehensive evaluation of exposure scenarios and population risk [25–27].

This study introduces a novel, integrative framework for pesticide residue assessment that combines deterministic and probabilistic models, supported by sensitivity analysis and compound-specific risk ranking. Although numerous studies have addressed pesticide residues and related health risks in various vegetable types, investigations that simultaneously apply both deterministic and probabilistic approaches remain relatively scarce. Such methodologies have occasionally been applied to citrus fruits [27], apples [28], cucumbers, cantaloupes, and melons [29]; however, their application to root and tuber crops is notably limited [30]. Comprehensive risk assessments involving multiple modeling approaches are rarely conducted for these crop types. Roots, tubers, and bulb vegetables, due to their growth below ground, have unique exposure pathways and residue retention characteristics that are under-represented in risk assessments. This study addresses this gap by providing a comprehensive risk assessment for various types of root and tuber vegetables, as well as onions.

The novelty of this research lies in the simultaneous evaluation of both acute and chronic health risks for adults and children. Additionally, the quantitative pesticide risk ranking matrix, tailored to residue occurrence, toxicity, and exposure parameters, offers a refined prioritization tool that may inform national and international food safety policies. To the best of our knowledge, this is among the first studies to jointly apply these advanced methodologies to a broad spectrum of underground vegetables within a single framework, aiming to enhance consumer protection strategies in both local and broader regulatory contexts.

The specific aims of this study were to investigate the occurrence of pesticide residues in commonly consumed root vegetables and onions and to assess the potential acute and chronic health risks arising from their dietary intake. Exposure scenarios for both adults and children were quantified using deterministic calculations and Monte Carlo simulations, offering a more refined estimate of the consumer risk. The risk ranking of individual pesticides further supports targeted risk mitigation strategies and informed regulatory decisions.

2. Materials and Methods

2.1. Samples and Reagents

In this study, 580 vegetable samples were collected from the local markets in Serbia. The samples included vegetables like celery (n = 34), ginger (n = 60), potatoes (n = 273), radishes (n = 36), horseradish (n = 7), carrots (n = 20), onions (n = 44), and leeks (n = 106). All the samples were stored under refrigerated conditions at 4 °C until analysis. Market availability and distribution of vegetables during the sampling period were the main factors contributing to the unequal distribution of sample sizes among the vegetables. A larger number of potato samples were intentionally collected, as potatoes are the most widely consumed root vegetable, making them a key focus for assessing potential dietary exposure to pesticide residues. On average, individual vegetable units weighed between 8 and 900 g, depending on the type. For example, potato samples weighed 100–200 g, carrots 70–150 g, celery 500–900 g, onions 90–150 g, leeks 150–280 g, ginger 50–100 g, radishes 8–15 g, and horseradish 300–400 g. Each sample was analyzed as a whole or homogenized composite, depending on size and analytical requirements.

Certified reference standards were obtained from Restek (Bellefonte, PA, USA), Dr. Ehrenstorfer (Augsburg, Germany), and Lab Instruments (Castellana Grotte, Italy) (Table S1). HPLC-grade acetonitrile was purchased from Lachner (Neratovice, Czech Republic), whereas methanol, formic acid, and ammonium formate (with purity exceeding 99%) were obtained from Carlo Erba (Milan, Italy). Other reagents, including sodium chloride, anhydrous magnesium sulfate, disodium hydrogen citrate sesquihydrate ($C_6H_8Na_2O_8$), trisodium citrate dihydrate ($C_6H_5O_7Na_3{\cdot}2H_2O$), primary and secondary amine sorbents (PSAs), C18 material, and graphitized carbon black (GCB), were procured from Sigma–Aldrich (St Louis, MO, USA).

2.2. Pesticide Extraction and Analysis

Vegetable samples were homogenized using a high-capacity homogenizer, and 10 g of the homogenate was transferred into a 50 mL centrifuge tube. Pesticide residues were then extracted, and the extract was cleaned in accordance with the EN 15,662 method [31].

To ensure precise quantification and account for matrix-specific variations, pesticide residues were analyzed separately for each type of vegetable. Prior to the analysis, all vegetable samples were rinsed with tap water; peeling was not performed, except for onions, where the outer dry skin was removed. Approximately 1 kg of each vegetable sample was homogenized (Grindomix GM 300, Retsch, Haan, Germany—4.5 L, 1.1 kW) to ensure sample representativity, following the European Commission guidance on analytical quality control for pesticide residue analysis (SANTE/11312/2021) [32]. Briefly, 10 mL of acetonitrile and a mixture of buffering salts, comprising 0.5 g of disodium hydrogen citrate, 1 g of sodium chloride, 1 g of trisodium citrate dihydrate, and 4 g of magnesium sulfate, were added. The mixture was vigorously vortexed for 1.5 min and centrifuged at 3500 rpm for 3 min. The resulting supernatant was transferred to a clean tube for clean-up. For all analyzed samples (potato, onion, celery, carrot, ginger, radish, and horseradish), the clean-up sorbents included 0.90 g of magnesium sulfate and 0.15 g of PSA. Additionally, for carrot, ginger, radish, and horseradish, the clean-up mixture also contained 0.15 g of C18 and 0.054 g of GCB to effectively remove pigments, lipophilic compounds, and essential oils. After a second vortexing and centrifugation, the final extract was filtered using a 0.22 µm PTFE syringe filter. The prepared extracts were analyzed using HPLC-MS/MS and GC-MS/MS [33]. Each sample was processed and analyzed in triplicate.

Pesticide analysis using HPLC-MS/MS was performed to identify and quantify compounds with medium to high polarity and thermal instability (pyrimethanil, boscalid, tebuconazole, azoxystrobin, fluopyram, difenoconazole, tebufenpyrad, tebufenozide, iprovalicarb, clothianidin, fluopicolide, fosthiazate, fluazifop, epoxiconazole, dimethomorph, thiamethoxam, imidacloprid, cyromazine, fenhexamid, imazalil, metamitron, propamocarb, carbaryl, and isoprocarb). A Thermo Scientific Accela HPLC system coupled with a TSQ Quantum Access MAX triple quadrupole mass spectrometer (San Jose, CA, USA) was utilized, operating in multiple reaction monitoring (MRM) mode to ensure high sensitivity and selectivity. Chromatographic separation was achieved on an Accucore aQ C18 column (100 mm × 2.1 mm, 2.6 µm particle size, Thermo Scientific, Waltham, MA, USA) using gradient elution with water and methanol, both containing 0.1% formic acid and 5 mM ammonium formate as modifiers. The system was equipped with a heated electrospray ionization (HESI) interface, which operated in both positive and negative ionization modes depending on the physicochemical characteristics of the analytes. A gradient program was applied, starting with 100% A, transitioning to 70% B over 7 min, reaching 100% B at 9 min, and holding for 3 min, before returning to 100% A at 12.5 min and maintaining it for an additional 4 min. The column was maintained at 40 °C, with a flow rate of 0.3 mL/min and an injection volume of 10 µL. This method provided low detection limits and reliable quantification across a wide range of pesticide residues in complex food matrices such as root, tuber, and bulb samples.

For the determination of volatile and thermally stable pesticide residues (metolachlor, prosulfocarb, linuron, pirimiphos-methyl, chlorpropham, ethoxyquin, chlorpyrifos, resmethrin, and piperonyl butoxide), GC-MS/MS analysis was carried out using a Thermo Trace 1310 GC system linked to a TSQ 8000 Evo mass spectrometer and equipped with a TriPlus AS autosampler. The instrument was operated in electron ionization (EI) mode with detection in multiple reaction monitoring (MRM) to enhance specificity and minimize matrix interferences. Analytes were separated on a Trace TR Pesticide II capillary column (30 m × 0.25 mm ID × 0.25 µm film thickness, Thermo Scientific) optimized for pesticide determination. The injector operated in splitless mode to achieve maximum sensitivity, and a multi-step oven temperature program was applied to ensure optimal elution of compounds with varying volatilities. The injector temperature was programmed to increase from 75 °C (1 min hold) to 330 °C over 2 min. The oven temperature was initially held at 60 °C for 2.3 min, ramped to 90 °C at 25 °C/min with a 1.5 min hold, further increased to 180 °C at the same rate, then gradually increased to 280 °C at 5 °C/min, and finally increased to 300 °C at 10 °C/min with a 5 min hold. Helium (purity ≥ 99.999%) was used as the carrier gas at a constant flow rate of 1.20 mL/min. The temperatures of the transfer line and the ion source were maintained at 250 °C and 300 °C, respectively. The injection volume was 1 μL. This technique enabled the accurate identification and quantification of non-polar pesticides, effectively complementing the HPLC-MS/MS method for comprehensive multi-residue analysis.

2.3. Quality Assurance/Quality Control

The analytical method for pesticide residue determination in vegetable samples was validated following the SANTE/11312/2021 guidelines to ensure reliability and performance. The key validation parameters included linearity, matrix effects, selectivity, sensitivity, accuracy, and precision [32]. Calibration curves were constructed using pesticide standards dissolved in both pure solvent and matrix-matched extracts, covering a concentration range of 5–200 μg/L for both HPLC-MS/MS and GC-MS/MS analyses. A correlation coefficient (r^2) greater than 0.99 was considered indicative of acceptable linearity. Blank matrices were confirmed by screening the samples in the absence of target pesticides before spiking. Matrix effects (MEs) were calculated as the ratio of the analytical signal obtained in the matrix to that in the solvent and expressed as a percentage to assess signal enhancement or suppression. Selectivity was confirmed by comparing the chromatograms of the blank and control samples to those containing spiked pesticides, ensuring no interference at the target retention times. Accuracy and precision were evaluated through recovery experiments by fortifying blank samples at two concentration levels (10 and 100 μg/kg), with six replicates (n = 6) at each level. Recovery rates were required to fall within the range of 70–120%, and precision was expressed as the relative standard deviation (RSD), which had to remain below 20%. The sensitivity of the method was determined by the limits of detection (LOD) and quantification (LOQ). The LOD and LOQ were calculated using an S/N of 3 and S/N of 10 in the lowest spiked recovery experiment.

2.4. Health Risk Assessment

2.4.1. Prioritization of Pesticides Using Risk Ranking

To estimate the potential health risks posed by each detected pesticide, a risk ranking matrix developed by the UK Veterinary Residues Committee was employed [34]. The total residual risk score (S) was calculated using Equation (1).

$$S = (A + B) \times (C + D + E) \times F \quad (1)$$

In this model, parameters A and B reflect toxicity, while C, D, E, and F capture various dimensions of human exposure. Specifically, A represents the acute toxicity score based on the oral LD_{50} values, classified into toxicity categories and sourced from the World Health Organization [35], US Environmental Protection Agency [36–38], or scientific opinion on hexachlorocyclohexanes [39]. B denotes chronic toxic potency, determined from the Acceptable Daily Intake (ADI) values available in the EU Pesticide Database [40], Pesticide Properties DataBase (PPDB) [41], or Joint FAO/WHO Meeting on Pesticide Residues (JMPR) database [42]. Parameter C quantifies dietary exposure by assessing the proportion of specific food items (e.g., root, tuber vegetable, or onion) in the total diet, using national consumption data from the Serbian Household Budget Survey [43]. D reflects the pesticide application frequency, calculated using the formula FOD = (N/T) × 100, where N is the number of pesticide treatments and T is the crop growth period in days. E captures the potential exposure of sensitive populations and was conservatively assigned a constant score (E = 3) due to limited demographic-specific exposure data. The final parameter,

F, addresses the actual pesticide residue levels in relation to the established Maximum Residue Limits (MRLs) [40], using a weighted scoring system:

$$F = \frac{F_0 \times 1 + F_1 \times 2 + F_2 \times 3 + F_3 \times 4}{n} \quad (2)$$

Here, F_0 represents the number of samples with no detectable residues, F_1 those with residues < 1 × MRL, F_2 for residues between 1 and 10 × MRL, and F_3 those exceeding 10 × MRL. The average F score was used to account for contamination severity across all analyzed samples.

Each pesticide was evaluated and ranked based on its total score, with higher scores indicating greater potential risk. Complete definitions, scoring criteria, and supporting values (e.g., LD_{50}, ADI, and MRL) are provided in Supplementary Tables S2–S6.

2.4.2. Chronic and Acute Risk Assessment

Acute dietary risk (short-term risk) was estimated using the International Estimated Short-Term Intake (IESTI) model [44], which calculates the intake of pesticide residues over a single day and compares it with the established Acute Reference Dose (ARfD, mg/kg body weight) [40,45–47]. The percentage of the ARfD (%ARfD) was determined using parameters such as the highest residue concentration (HR), large portion consumption (LP; 97.5th percentile of eaters), unit edible portion (Ue), and a variability factor (v = 3), adjusted for different age groups (bw: body weight). Two formulas were used depending on whether the unit portion was higher or lower than LP. For adult consumers, the acute dietary risk assessment was conducted using case 2a (Equation (3)) of the IESTI model for ginger, potato, leek, and carrot, while case 2b (Equation (4)) was applied for celery, onion, horseradish, and radish. In the case of children, case 2b was consistently used for all assessed vegetable types. If the %ARfD exceeded 100% (or hazard quotient acute, HQa, exceeded 1), short-term exposure was considered unacceptable.

Case 2a: 25 g $\leq$ Ue < LP

$$IESTI = \frac{((U_e \times HR \times v) + (LP - U_e) \times HR)}{bw} \quad (3)$$

Case 2b: 25 g $\leq$ Ue > LP

$$IESTI = \frac{LP \times HR \times v}{bw} \quad (4)$$

$$HQa = \frac{IESTI}{ARfD} \quad (5)$$

$$\%ARfD = HQa \times 100 \quad (6)$$

Chronic risk was evaluated by estimating the Estimated Daily Intake (EDI, mg/kg bw) of each pesticide, based on the supervised trials median residue (STMR) and average food daily intake (Fi, kg/day), normalized by body weight (bw). The average daily intake was determined using national consumption data from the Serbian Household Budget Survey [44]; a portion for children was calculated as 40% of the adult portion [48]. Chronic risk was then expressed as a percentage of the Acceptable Daily Intake (%ADI), calculated by comparing the EDI to the respective ADI values (mg/kg bw). A %ADI below 100% (or hazard quotient chronic, HQc < 1) indicated that long-term exposure did not pose a health concern. Both the %ARfD and %ADI were used to assess overall dietary risk. Values exceeding 100% suggested an unacceptable risk, while lower values indicated an acceptable safety margin for consumers. These calculations provided a comprehensive evaluation of dietary risks, accounting for both high-consumption scenarios and long-term exposure patterns across different population groups.

$$\mathrm{EDI} = \frac{\mathrm{STMR} \times \mathrm{Fi}}{\mathrm{bw}} \tag{7}$$

$$\mathrm{HQc} = \frac{\mathrm{EDI}}{\mathrm{ADI}} \tag{8}$$

$$\%\mathrm{ADI} = \mathrm{HQc} \times 100 \tag{9}$$

The cumulative chronic hazard index (HIc) and cumulative acute hazard index (HIa) were determined using Equation (10):

$$\mathrm{HI} = \sum \mathrm{HQ} \tag{10}$$

These cumulative risk values represent the total sum of the chronic (HQc) or acute (HQa) hazard quotients for all pesticides detected within a single vegetable sample. If the HI exceeds 1 (meaning that the %HI is greater than 100%), it indicates an increased likelihood of potential adverse health effects from pesticide exposure. In such circumstances, food products may be considered unsuitable for consumption.

2.5. Monte Carlo Risk Simulation

Uncertainty and sensitivity analyses related to chronic dietary risk assessments were conducted using a Monte Carlo simulation approach [25]. The simulations were performed using Oracle Crystal Ball software (version 11.1.3.0.0, Oracle Inc., Redwood Shores, CA, USA), employing 10,000 iterations to ensure the robustness and reliability of the results. In the simulation process, different probability distributions were assigned to the input variables: residue concentrations and body weight were modeled using a lognormal distribution to account for their right-skewed nature, while ingestion rates were modeled using a triangular distribution. The input variables without assigned probability distributions were treated as fixed-point values throughout the Monte Carlo simulations. This probabilistic method allows for a comprehensive evaluation of the variability and uncertainty inherent in chronic risk assessment.

3. Results and Discussion

3.1. Method Validation

Excellent linearity was observed for all the analyzed pesticides, with correlation coefficients (R^2) ranging from 0.9935 to 0.9992 (Table 1). Limits of detection (LODs) were between 0.8 and 3.4 μg/kg, while limits of quantification (LOQs) did not exceed 10 μg/kg. Accuracy and precision were confirmed by recovery experiments conducted at two spiking levels. The recovery rates ranged from 84.7% (boscalid) to 115% (tebuconazole) at lower fortifications, with relative standard deviations (RSDs) between 1.8% (azoxystrobin) and 14.3% (tebufenpyrad). At higher spiked concentrations, the recoveries varied between 92.5% (pyrimethanil) and 114% (chlorpropham), while the RSD values ranged from 3.2% (azoxystrobin) to 11.9% (cyromazine). These validation results satisfied the acceptance criteria for pesticide residue analysis (mean recovery within 70–120% and RSD below 20%). Based on the calculated results, the investigated pesticides exhibited predominantly low to moderate matrix effects, ranging from −44.1 to +55.8 %. Most compounds displayed either medium or low matrix effects. Thus, matrix-matched calibration was employed to enhance the quantification accuracy.

Table 1. Validation of GC-MS/MS and LC/MS/MS methods for analysis of pesticide residues in vegetable samples.

Pesticide	ME %	R^2	LOD µg/kg	R # %	RSD # %	R ## %	RSD ## %
Celery							
Pyrimethanil	−27.4	0.9936	1.2	106	2.64	92.5	9.31
Boscalid	9.30	0.9955	0.9	94.7	5.75	110	3.95
Tebuconazole	15.5	0.9962	1.5	115	9.51	93.3	5.78
Metolachlor	−10.3	0.9992	2.7	111	7.34	105	11.1
Azoxystrobin	−13.2	0.9940	1.4	101	1.80	97.8	8.46
Fluopyram	11.3	0.9988	1.2	87.7	2.45	93.1	9.74
Difenoconazole	16.8	0.9939	2.9	89.5	11.2	113	10.1
Prosulfocarb	−33.5	0.9943	1.8	99.7	3.12	105	11.5
Linuron	−10.5	0.9937	1.5	104.3	7.78	95.1	7.35
Carbaryl	12.4	0.9953	1.4	114	8.20	94.2	3.65
Resmethrin	−28.4	0.9982	1.9	93.2	11.3	102	7.12
Potato							
Pyrimethanil	11.5	0.9991	1.6	102	2.43	94.3	4.95
Metolachlor	−27.1	0.9963	2.5	109	4.01	103	6.01
Azoxystrobin	27.3	0.9987	1.3	98.3	3.98	95.4	4.75
Fluopyram	−15.2	0.9974	2.3	90.1	3.40	97.2	5.20
Chlorpropham	35.0	0.9991	2.5	88.0	3.33	114	6.62
Clothianidin	45.2	0.9989	1.0	94.1	12.4	93.9	10.0
Fluopicolide	−44.1	0.9948	3.1	113	4.98	113	3.94
Fosthiazate	17.6	0.9935	2.2	95.2	10.2	101	5.51
Fluazifop	38.3	0.9953	1.4	97.9	13.7	97.2	3.20
Epoxiconazole	16.2	0.9972	1.2	103	6.31	92.9	7.70
Dimethomorph	31.4	0.9990	2.8	111	4.05	101	3.90
Thiamethoxam	32.2	0.9937	3.0	85.9	6.72	93.3	5.19
Piperonyl butoxide	48.3	9.9986	1.4	90.2	2.90	94.0	4.20
Imidacloprid	55.8	0.9935	0.8	91.8	11.0	110	7.32
Cyromazine	−17.3	0.9974	1.9	86.5	7.83	105	11.9
Horseradish							
Metamitron	−18.4	0.9990	1.1	89.8	3.38	92.8	8.15
Ginger							
Metolachlor	−24.5	0.9981	2.7	112	5.15	110	12.4
Tebufenpyrad	−55.6	0.9935	1.1	95.1	14.3	107	5.52
Tebufenozide	−23.1	0.9985	1.5	105	3.50	111	7.17
Iprovalicarb	−29.9	0.9949	2.7	110	7.91	97.3	3.85
Pirimiphos-methyl	22.8	0.9951	1.3	93.7	10.5	93.0	5.19
Onion							
Fluopicolide	−7.5	0.9958	2.6	91.2	5.91	104	4.91
Dimethomorph	−8.8	0.9977	1.3	85.3	3.89	94.9	6.30
Piperonyl butoxide	13.5	0.9964	3.1	99.1	1.99	98.3	3.73
Imidacloprid	−13.3	0.9986	0.9	103	6.01	105	3.79
Cyromazine	−15.6	0.9974	3.2	94.7	7.20	104	11.9
Fenhexamid	−11.2	0.9943	2.3	112	3.34	111	9.91
Ethoxyquin	12.2	0.9983	0.9	87.4	5.07	103	5.54
Imazalil	25.1	0.9949	3.4	114	9.63	93.5	4.11
Chlorpyrifos	21.1	0.9986	2.5	95.7	13.4	113	11.0
Propamocarb	−7.0	0.9940	3.0	92.1	2.95	98.1	9.78
Radish							
Boscalid	−12.2	0.9973	1.2	84.7	3.96	97.2	12.8
Azoxystrobin	−14.5	0.9951	1.7	98.3	7.05	110	3.20

Table 1. *Cont.*

Pesticide	ME %	R^2	LOD µg/kg	R # %	RSD # %	R ## %	RSD ## %
Thiamethoxam	−11.1	0.9959	2.9	104	8.70	112	6.86
Chlorpyrifos	14.8	0.9968	2.3	93.5	4.12	94.2	7.33
Propamocarb	−8.7	0.9973	1.6	107	5.95	112	11.5
Carrot							
Boscalid	−15.9	0.9984	1.1	93.9	8.14	98.0	5.60
Tebuconazole	−33.5	0.9975	1.7	95.1	4.83	107	4.77
Azoxystrobin	−6.4	0.9961	1.2	105	6.24	113	5.95
Prosulfocarb	18.1	0.9992	2.0	111	5.13	105	8.48
Isoprocarb	11.0	0.9957	2.8	89.0	11.6	94.9	2.85

#—Spiking level 10 µg/kg; ##—spiking level 100 µg/kg; ME—matrix effect; LOD—limit of detection; RSD—relative standard deviation; R—recovery; R^2—coefficient of determination.

3.2. Pesticide Distribution

A total of 580 vegetable samples were examined for pesticide residues, confirming the presence of 33 different active substances in the 86 samples. Positive samples contained at least one residue, with concentrations ranging from 0.01 mg/kg (boscalid) to 9.8 mg/kg (metolachlor). The detected pesticides included 13 fungicides (pyrimethanil, boscalid, tebuconazole, azoxystrobin, fluopyram, difenoconazole, iprovalicarb, fluopicolide, epoxiconazole, dimethomorph, fenhexamid, imazalil, and propamocarb), 12 insecticides (carbaryl, resmethrin, tebufenpyrad, tebufenozide, pirimiphos-methyl, clothianidin, fosthiazate, thiamethoxam, imidacloprid, cyromazine, chlorpyrifos, and isoprocarb), 6 herbicides (metolachlor, prosulfocarb, linuron, chlorpropham, fluazifop, and metamitron), 1 synergist (piperonyl butoxide), and 1 non-classical pesticide (ethoxyquin). Piperonyl butoxide is primarily used as a synergist to boost the effectiveness of insecticides, especially pyrethrins and pyrethroids [49]. Ethoxyquin is mainly applied as an antioxidant preservative in animal feed and, to a lesser extent, as a post-harvest treatment to reduce spoilage in certain fruits [50]. Figure 1 illustrates the number of analyzed samples across different vegetable groups (inner circle) and the percentages of positive samples within each group of vegetables (outer circle). The outer circle shows the extent of pesticide contamination in each vegetable group. Potatoes exhibited the highest number of pesticides (16 compounds), followed by onions (11 compounds), celery (11 compounds), and radishes (6 compounds). Fewer residues were detected in ginger, carrot (five compounds), leek (four compounds), and horseradish (one compound). Details regarding the specific pesticides detected in each vegetable type are provided in Table 2.

Table 2. Pesticides detected in the analyzed vegetables.

Potatoes	Onion	Celery	Radish	Ginger	Carrot	Leek	Horseradish
PYM	PYM	PYM	BOS	MTL	BOS	BOS	MTT
MTL	BOS	BOS	AZX	TBP	TEB	TEB	
AZX	AZX	TEB	TMX	TBZ	AZX	AZX	
FPM	FPM	MTL	CPF	IPV	PSC	PBO	
CHP	FPC	AZX	MTT	PPM	ISP		
CLT	DMM	FPM	PPC				
FPC	IMI	DFZ					
FST	FHX	PSC					

Table 2. *Cont.*

Potatoes	Onion	Celery	Radish	Ginger	Carrot	Leek	Horseradish
FLP	ETQ	LIN					
EPX	IMZ	CRB					
DMM	CPF	RMT					
TMX							
PBO							
IMI							
CYR							
PPC							

Metolachlor (MTL), tebufenpyrad (TBP), imazalil (IMZ), propamocarb (PPC), chlorpropham (CPH), boscalid (BOS), azoxystrobin (AZX), prosulfocarb (PSC), tebuconazole (TEB), dimethomorph (DMM), difenoconazole (DFZ), fluazifop (FLP), piperonyl butoxide (PBO), chlorpyrifos (CPF), fluopyram (FPM), linuron (LIN), epoxiconazole (EPX), pyrimethanil (PYM), isoprocarb (ISP), carbaryl (CRB), thiamethoxam (TMX), tebufenozide (TBZ), pirimiphos-methyl (PPM), metamitron (MTT), ethoxyquin (ETQ), fluopicolide (FPC), resmethrin (RMT), iprovalicarb (IPV), cyromazine (CYR), fenhexamid (FHX), fosthiazate (FST), clothianidin (CLT), and imidacloprid (IMI).

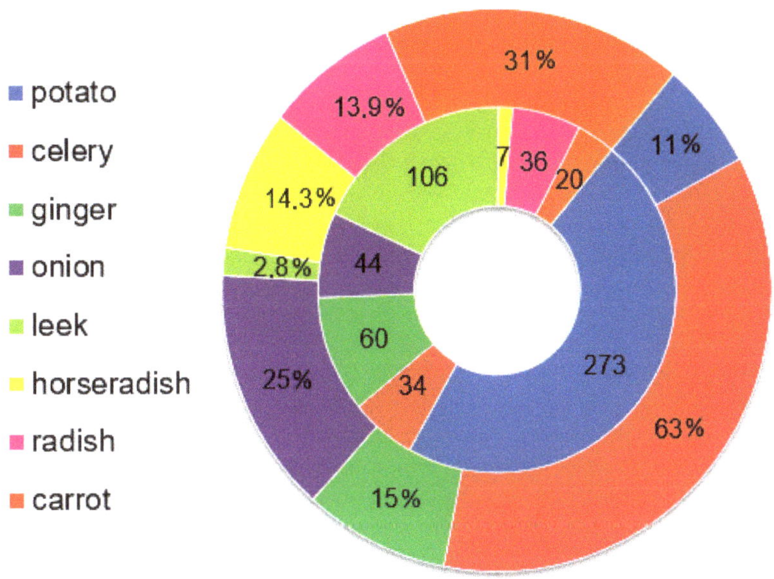

Figure 1. Number of analyzed root, tuber, and onion vegetable samples (inner circle) and percentage of positive samples per vegetable group (outer circle).

In terms of contamination, celery exhibited the highest percentage of contaminated samples (63%), as well as the greatest variety of pesticides relative to the number of analyzed samples. Carrots and bulb onions followed, with 31% and 25% of the samples testing positive, respectively. For the remaining vegetables, the proportion of positive samples ranged from 2.8 to 15%. In comparison with previous investigations, our study revealed a slightly higher proportion of pesticide-positive celery samples than that reported by Fang et al. (2015) [51], who detected residues in 58% of celery samples collected in China. In contrast, a much lower level of contamination was observed for carrots in the study by Kazimierczak et al. (2022) [52] conducted in Poland, where only 5.0% of the samples tested positive. Cui et al. (2024) [21] reported a considerably higher contamination rate for ginger in China, with 66.5% of the samples containing pesticide residues. Regarding onion, the proportion of positive samples in our study was comparable to findings from Ethiopia reported by Jirata et al. (2024) [53], where 22–28% were contaminated. Chu et al. (2023) observed a slightly higher contamination rate in China [54]. Potato, the most consumed vegetable among those analyzed, showed a lower pesticide residue rate in our study than the 26.2% positive samples reported in the 2020 EU report [55]. Root and tuber vegetables such as carrots and potatoes are generally regarded as less problematic in terms of pesticide residues compared to leafy and fruiting vegetables. Potatoes are root vegetables regularly monitored by EFSA because of their high consumption levels [56], as well as their specific characteristics that involve pesticide applications at various stages of production. Pesticide residues may remain on the peel of the tubers and sometimes in their inner tissues due to their physiological characteristics, cultivation practices, post-harvest treatments with fungicides and sprout inhibitors, as well as due to storage in facilities with a history of pesticide usage [56–58]. However, according to recent EFSA reports, only 0.8% of potato samples in the EU exceeded the maximum residue level (MRL) in 2020, meaning that the majority of samples adhere to regulatory limits. Carrots and onions are also among the commodities routinely monitored under the EU-coordinated multiannual control program (EU MACP) [59]. However, monitoring data suggest that these vegetables generally present a low pesticide-related risk, with MRL exceedances observed in only 1.2% of the carrot samples and 0.2% of the onion samples in 2020. Therefore, they pose minimal risks in terms of pesticide residue exposure [59].

Among the positive samples, nineteen contained multiple pesticide residues: one sample contained five pesticides (bulb onion), one contained four (celery), four contained three, and thirteen contained two. This frequent co-occurrence highlights the widespread use of pesticide mixtures and the need for a cumulative dietary risk assessment. The most frequently detected pesticide was azoxystrobin (14 samples), followed by boscalid (10 samples). Tebufenpyrad had the highest number of MRL (Supplementary Materials, Table S4) exceedances among all detected pesticides, exceeding the legal limit in seven ginger samples (MRL = 0.01 mg/kg). MRL exceedances were recorded for 29 samples, including celery (n = 5), ginger (n = 8), potato (n = 9), bulb onion (n = 3), leek (n = 1), horseradish (n = 1), radish (n = 1), and carrot (n = 1). In previous studies, Danek et al. (2021) [6] observed multi-residue presence in 8 of 15 potato samples, while Cui et al. (2024) [21] reported that 38.8% of ginger samples contained multiple residues. Elevated pesticide concentrations above the MRLs have also been reported in other studies, such as P et al. (2022) [24] and Kazimierczak et al. (2022) [52], the latter noting chlorpyrifos exceedance in organically grown carrots.

Summary statistics (Supplementary Table S7) revealed that mean pesticide concentrations ranged from 0.013 mg/kg (clothianidin) to 2.45 mg/kg (metolachlor). The distribution profiles of the individual pesticides are shown in Figure 2 (box plots, logarithmic scale), illustrating substantial variability. To improve clarity and focus on the more frequently occurring residues, pesticides detected in only a single sample, specifically imazalil, fluazifop, isoprocarb, carbaryl, pirimiphos-methyl, metamitron, ethoxyquin, resmethrin, iprovalicarb, cyromazine, and fenhexamid, were excluded from the plot. Metolachlor exhibited the highest mean (2.45 mg/kg) and maximum (9.8 mg/kg) concentrations, followed by notable residues of tebufenpyrad (1.10 mg/kg), imazalil (0.97 mg/kg), propamocarb (0.80 mg/kg), chlorpropham (0.67 mg/kg), boscalid (0.57 mg/kg), and azoxystrobin (0.55 mg/kg). Moderate maximum concentrations were observed for prosulfocarb (0.41 mg/kg), tebuconazole (0.32 mg/kg), dimethomorph (0.30 mg/kg), and fluazifop (0.13 mg/kg). These results emphasize the necessity for regular monitoring programs and cumulative risk assessment strategies for pesticide residues in vegetables.

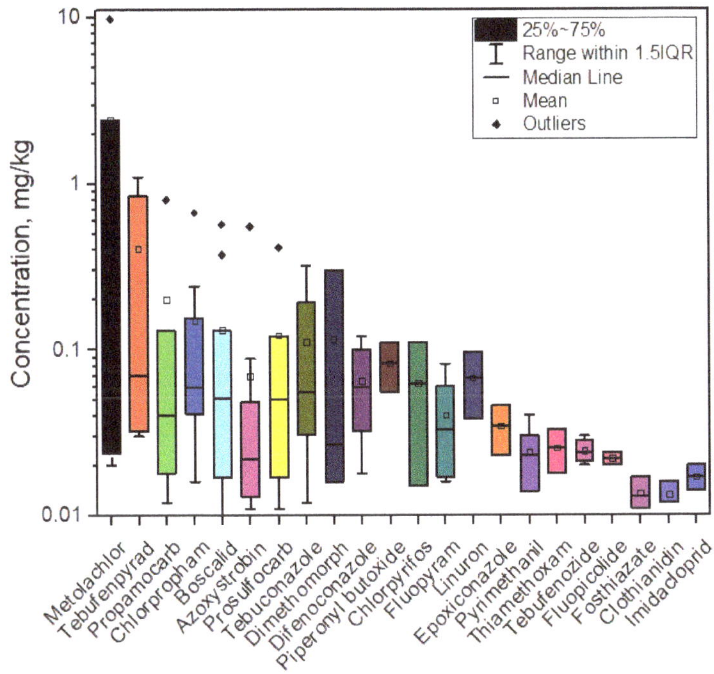

Figure 2. Box plots showing pesticide concentration distributions in root vegetables and onions (logarithmic scale).

3.3. Cluster Analysis of Pesticide Residues

Hierarchical cluster analysis (HCA) was performed to explore the similarity in pesticide residue profiles among vegetable samples. The analysis was performed using Ward's linkage method and Pearson's distance as dissimilarity measures [60]. The resulting dendrogram (Figure 3) grouped the samples into five clusters (A–E), based on compositional similarities in the detected pesticide concentrations (Supplementary Materials, Table S8). Cluster A predominantly consisted of potato samples, indicating a high degree of internal homogeneity, likely attributable to uniform agricultural practices, similar cultivars, or comparable environmental growth conditions [61]. Cluster B was mainly composed of celery samples, which also displayed consistent characteristics within this group. In contrast, Cluster C exhibited a more heterogeneous composition, comprising a mixture of potatoes, ginger, onions, celery, isolated carrot samples, and horseradish. This diversity suggests potential overlap in chemical properties or pesticide residue patterns across these vegetables, possibly resulting from shared environmental exposures or cross-contamination during cultivation or processing [58]. Cluster D was relatively small, consisting of two celery samples and one onion, indicating localized similarity among these samples. Cluster E was the most compositionally diverse, encompassing various vegetable types, including potatoes, celery, onions, ginger, carrots, and radishes. Several sub-clusters were observed within this cluster, with the leftmost sub-cluster primarily containing potato samples. The grouping observed in this cluster may reflect broader or overlapping residue profiles, potentially due to mixed production sources or greater variability in agricultural inputs. These mixed clusters indicate that certain pesticide profiles are common across multiple vegetable types, likely reflecting the use of similar pesticide treatments or environmental factors that affect diverse crops within the same agricultural system.

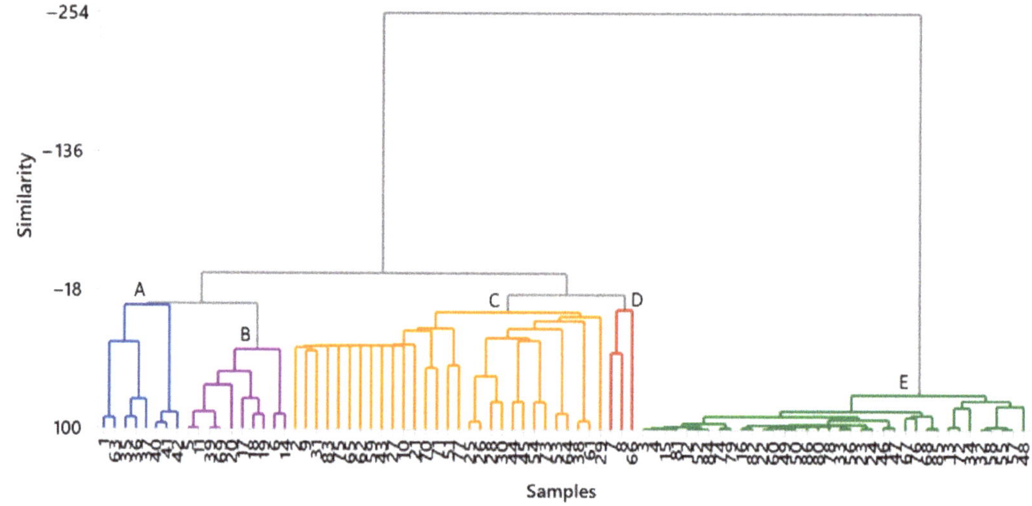

Figure 3. Cluster analysis of underground vegetables based on pesticide residues.

Overall, cluster analysis revealed crop-specific residue patterns for potato and celery, whereas mixed clusters indicated potential cross-contamination or shared sources of pesticide exposure. These findings emphasize the need for crop-specific monitoring strategies and underscore the relevance of cluster-based classification for assessing pesticide distribution in complex food matrices.

3.4. Risk Ranking

Figure 4 and Supplementary Table S5 present the risk ranking of the pesticide residues based on their total scores. The pesticides were categorized into three groups according to their risk levels. Group C, which included chlorpyrifos, isoprocarb, metolachlor, fosthiazate, and ethoxyquin, represents compounds with the highest total scores, indicating a high potential health risk. The pesticides in this group had overall risk scores above 20, accounting for 15.1% of the detected pesticides. These pesticides also exceeded MRL values or had low LD_{50} or ADI values. Group B consisted of pesticides with moderate risk scores, with the total risk score ranging from 15.0 to 19.9. This group included tebufenpyrad, prosulfocarb, difenoconazole, pirimiphos-methyl, carbaryl, clothianidin, epoxiconazole, and fluazifop (24.2% of the pesticides detected). Group A included the majority of the analyzed pesticides (60.6% of the detected pesticides), characterized by lower total scores and a relatively lower health risk. The low-risk group comprised pesticides with total scores below 15.0. This classification highlights the need for closer monitoring of Group C compounds because of their greater contribution to potential dietary exposure risks. The pesticides detected in this study included 15 unapproved substances. Among these, four were classified as high risk (chlorpyrifos, isoprocarb, metolachlor, and ethoxyquin), four as medium risk (carbaryl, clothianidin, epoxiconazole, and fluazifop), and seven as low risk (imidacloprid, thiamethoxam, linuron, chlorpropham, cyromazine, dimethomorph, and resmethrin).

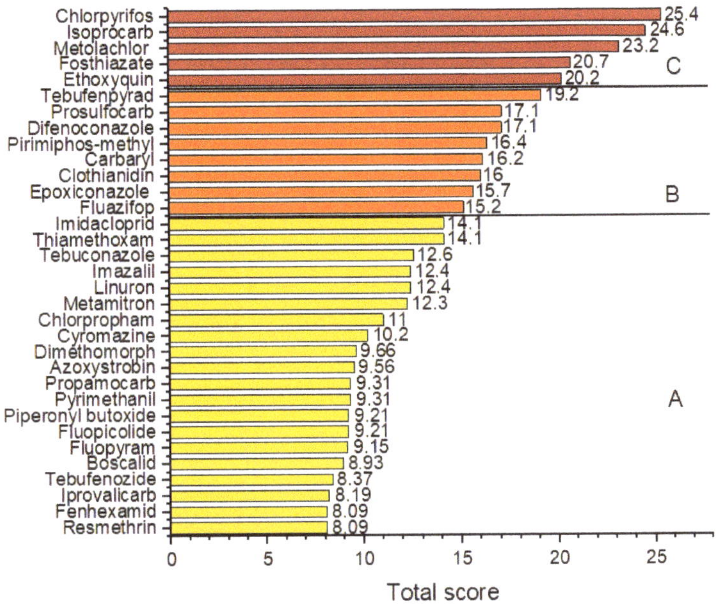

Figure 4. Risk ranking of pesticide residues detected in root vegetables and onions based on total risk scores of vegetables purchased in Serbia. A—low-risk group, B—medium-risk group, and C—high-risk group.

Figure 5 and Supplementary Table S6 present the classification of pesticides detected in root vegetables and onions based on data from the RASFF database for the period 2020–2025 [62]. Since the samples included in this dataset mostly involved cases where MRLs were exceeded, a significantly higher number of pesticides were classified as high risk. Additionally, the database includes pesticides that are banned from use in the European Union. During the observed period, 181 samples were reported in the RASFF system, and 46 different pesticides were identified. Seventeen pesticides were classified as high risk (37% of reported pesticides), including ethylene oxide, omethoate, oxamyl, chlorpyrifos, dieldrin, ethoprophos, HCH, beta-HCH, gamma-HCH, griseofulvin, fenamiphos, diafenthiuron, monocrotophos, prochloraz, lambda-cyhalothrin, bromide, and carbofuran. Among these, eight pesticides—ethylene oxide, chlorpyrifos, dieldrin, HCH, beta-HCH, gamma-HCH (lindane), monocrotophos, and carbofuran—are explicitly banned in the EU [40]. Additionally, the remaining high-risk pesticides (with total risk scores ≥ 20) were not approved for use in the EU, except for lambda-cyhalothrin. The root vegetable samples listed in the RASFF database were primarily imported products from non-EU countries, predominantly from Asia and Africa. The medium-risk group included 13 pesticides, comprising banned/non-approved (diazinon, fipronil, indoxacarb, imidacloprid, and fluazifop) and approved substances (fosthiazate, imazalil, pirimiphos-methyl, acetamiprid, emamectin, cypermethrin, chlormequat, and flonicamid) for use in the European Union [40]. The low-risk pesticide group included both non-approved (linuron, thiamethoxam, bifenthrin, chlorfenapyr, quintozene, propiconazole, carbendazim, thiophanate-methyl, lufenuron, and iprodione) and approved pesticides (metalaxyl, tebuconazole, thiabendazole, malathion, flutolanil, and chlorantraniliprole).

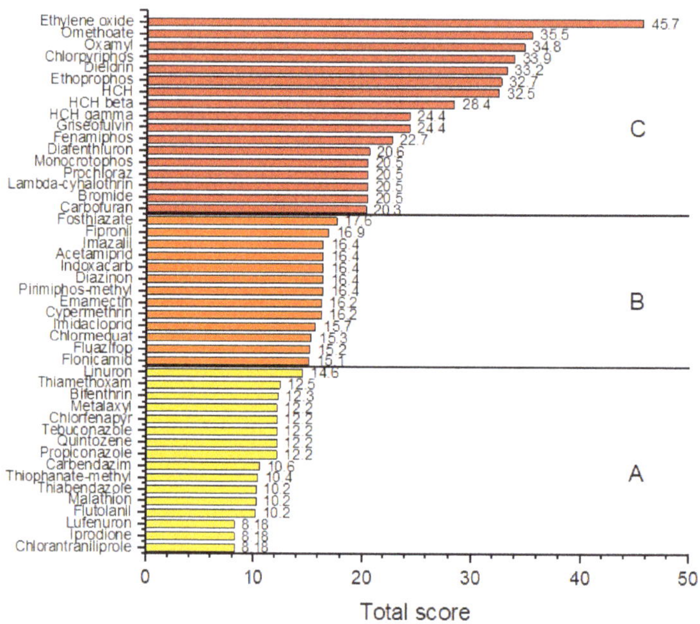

Figure 5. Risk ranking of pesticide residues in root vegetables and onions based on total risk scores. Data are derived from RASFF notifications (2020–2025). Pesticides were categorized into three risk groups: A—low-risk group, B—medium-risk group, and C—high-risk group.

3.5. Acute and Chronic Risks of Pesticide Residues in Root Vegetables and Onions

The average values for acute (short-term intake) risk (HQa) ranged from 1.13×10^{-6} for metamitron to 1.53×10^{-2} for tebufenpyrad. The corresponding chronic (long-term intake) risk values (HQc) ranged from 6.03×10^{-7} for fenhexamid to 2.02×10^{-3} for tebufenpyrad (Supplementary Materials, Table S9). These findings suggest that the detected pesticide residues are unlikely to pose either short- or long-term health risks through dietary exposure. However, this applies to the presence of individual pesticides, while a realistic risk assessment requires considering the combined presence of all pesticides in a single sample, that is, evaluating the total cumulative risk. The mean, maximum, and minimum values of total acute and chronic health risks for both adults and children are presented in Tables 3 and 4, respectively. For adults, the acute hazard index (HIa) ranged from 1.66×10^{-4} to 5.11×10^{-1}, corresponding to %ARfD values between 0.0166% and 51.1%. These results indicated that all analyzed samples were within the safe limits for consumption (HIa < 1 or %ARfD < 100%). However, the maximum acute risk values indicate potential concern associated with higher consumption of ginger. An acute risk assessment for adults revealed that three ginger samples, each containing pesticide residues exceeding their respective MRLs, had acute risk values of 51.1, 39.3, and 33.3%. For children, HIa values ranged from 6.23×10^{-5} to 6.31×10^{-1} and for %ARfD from 0.00623 to 63.1%. The highest acute risk in children (63.1%) was observed in a celery sample that also exceeded the MRL limits.

The observed differences in acute risk (Table 3) between adults and children are attributed to variations in dietary intake patterns, particularly the lower consumption of ginger among children than among adults. Regarding chronic exposure (Table 4), the chronic hazard index (HIc) for adults ranged from 1.80×10^{-5} to 5.90×10^{-2} (corresponding to %ADI values between 0.00180% and 5.90%) and for children from 3.36×10^{-5} to 1.10×10^{-1} (0.00336% to 11.0% of the ADI). These findings indicate that cumulative chronic dietary exposure to pesticide residues from the consumption of root vegetables does not pose a health risk, as the %ADI values remained well below 100% for both adults and children in Serbia, based on average intake levels. In the present study, although MRL exceedances were observed in 29 samples, both the acute and chronic health risks remained low in most cases. This can be attributed to the relatively low dietary intake of the most affected vegetables, namely celery and ginger.

Table 3. Hazard index values of acute risks for adults and children.

Acute Risk for Adults			
Vegetable	**HIa Min**	**HIa Max**	**HIa Mean**
Celery	2.96×10^{-4}	2.19×10^{-1}	2.87×10^{-2}
Ginger	4.25×10^{-4}	5.11×10^{-1}	2.86×10^{-1}
Potato	1.93×10^{-4}	1.23×10^{-1}	1.30×10^{-2}
Onion and leek	1.82×10^{-4}	1.21×10^{-1}	1.84×10^{-2}
Radish	1.66×10^{-4}	9.11×10^{-2}	2.32×10^{-2}
Carrot	2.32×10^{-3}	5.46×10^{-3}	3.89×10^{-3}
All root vegetables	1.66×10^{-4}	5.11×10^{-1}	5.04×10^{-2}
Acute Risk for Children			
Vegetable	**HIa Min**	**HIa Max**	**HIa Mean**
Celery	8.52×10^{-4}	6.31×10^{-1}	8.26×10^{-2}
Ginger	6.23×10^{-5}	7.47×10^{-2}	2.41×10^{-2}
Potato	4.59×10^{-4}	2.92×10^{-1}	3.09×10^{-2}
Onion and leek	3.26×10^{-4}	2.16×10^{-1}	3.50×10^{-2}
Radish	9.72×10^{-5}	2.83×10^{-1}	7.14×10^{-2}
Carrot	7.98×10^{-3}	1.88×10^{-2}	1.34×10^{-2}
All root vegetables	6.23×10^{-5}	6.31×10^{-1}	4.49×10^{-2}

Table 4. Hazard index values of chronic risks for adults and children.

Chronic Risk for Adults			
Vegetable	**HIc Min**	**HIc Max**	**HIc Mean**
Celery	1.96×10^{-5}	5.90×10^{-2}	5.73×10^{-3}
Ginger	7.52×10^{-5}	3.89×10^{-2}	1.08×10^{-2}
Potato	4.68×10^{-5}	1.51×10^{-2}	2.47×10^{-3}
Onion and leek	5.19×10^{-5}	2.40×10^{-2}	3.31×10^{-3}
Radish	4.51×10^{-5}	3.63×10^{-3}	7.42×10^{-4}
Carrot	1.80×10^{-5}	6.38×10^{-3}	1.38×10^{-3}
All root vegetables	1.80×10^{-5}	5.90×10^{-2}	4.08×10^{-3}
Chronic Risk for Children			
Vegetable	**HIc Min**	**HIc Max**	**HIc Mean**
Celery	3.66×10^{-5}	1.10×10^{-1}	1.07×10^{-2}
Ginger	1.40×10^{-4}	7.27×10^{-2}	2.02×10^{-2}
Potato	8.73×10^{-5}	2.83×10^{-2}	4.62×10^{-3}
Onion and leek	9.68×10^{-5}	4.48×10^{-2}	6.18×10^{-3}
Radish	8.42×10^{-5}	6.77×10^{-3}	1.39×10^{-3}
Carrot	3.36×10^{-5}	1.19×10^{-2}	2.57×10^{-3}
All root vegetables	3.36×10^{-5}	1.10×10^{-1}	7.62×10^{-3}

Our findings are consistent with those of previous studies. For example, a health risk assessment of pesticide residues in celery conducted in China reported chronic and acute risks for adults within acceptable limits (%ADI ranged from 0.01 to 1.80% and %ARfD from 0.05 to 28%) [51]. Similar conclusions were drawn in studies from Greece [63] and Turkey [64,65], which also found no significant health risks associated with dietary exposure to pesticide residues in vegetables. For example, in a study from Greece, the acute risk (HQa) for celery and potatoes was 0.0130 and 0.0113, respectively, while the chronic risk (HQc) for the same types of vegetables was 0.117 and 0.0446. These values are well below the threshold of 1, indicating a low health risk [63]. On the other hand, a risk assessment study on pesticide residues in onions conducted in China [54] concluded that although the majority of samples met safety standards, acute health risks were identified for cyhalothrin (%ARfD was 297.95% in children and 140.17% in adults) and carbofuran (%ARfD was 107.94% for children). Similarly, a study on radishes reported that while the HQ and cumulative risk index were well below the threshold, the %ARfD for triazophos exceeded 100 for all age groups (%ARfDs for adults, children, and toddlers were 203.38, 241.0, and 533.97%, respectively), indicating a potential acute risk [22]. Additionally, the acute risk for toddlers was exceeded for carbofuran, aldicarb, monocrotophos, and parathion, with %ARfD values of 152.23, 179.16, 178.57, and 135.83, respectively.

Pesticide exposure from root and tuber vegetables can vary significantly depending on individual dietary habits, with frequent or high-volume consumption potentially leading to greater intake. However, several studies have demonstrated that common household processing methods—such as washing, peeling, and thermal treatment (e.g., boiling or frying)—can substantially reduce pesticide residue levels in these vegetables. For instance, Terfe et al. (2023) reported that peeling and boiling potatoes led to near-complete removal of organochlorine pesticides [66]. Phopin et al. (2022) showed that boiling, blanching, and stir-frying reduced pesticide residues by 18–100% in vegetables, with thermal treatments being especially effective [67]. Similarly, Chu et al. (2024) found that boiling is the most effective method for reducing boscalid levels in Welsh onions [68]. According to Jensen et al. (2022), even high vegetable consumers remain within safe exposure limits, particularly when the food is properly processed [69].

Figure 6 illustrates the contribution of individual pesticides to the overall acute and chronic health risk. The most influential compounds for acute risk were chlorpropham, tebufenpyrad, difenoconazole, fluopyram, and propamocarb. In terms of chronic risk, the highest contributions were associated with azoxystrobin, chlorpropham, boscalid, tebufenpyrad, and fluopyram. An acute risk assessment was performed for 23 detected pesticides; the remaining compounds ($n = 10$) were excluded (boscalid, azoxystrobin, linuron, resmethrin, tebufenozide, iprovalicarb, piperonyl butoxide, fenhexamid, ethoxyquin, and isoprocarb) from the acute risk evaluation due to the absence of established ARfD values.

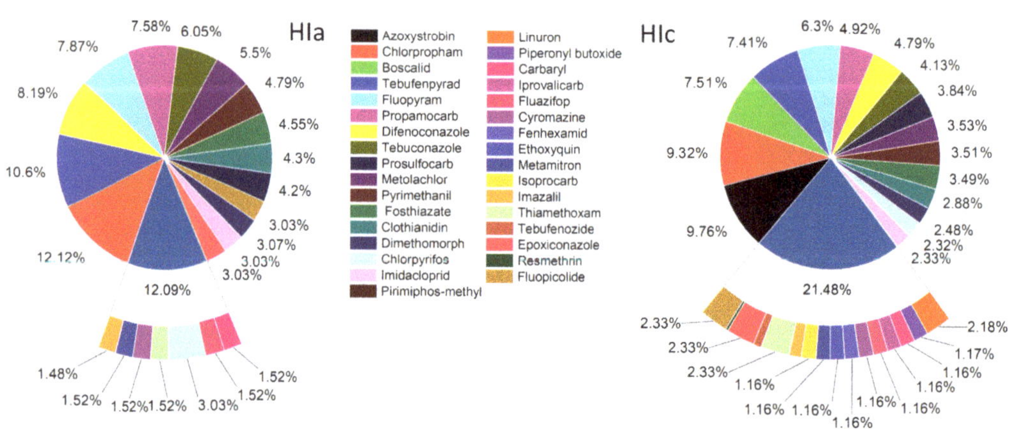

Figure 6. Contribution of pesticides to estimated acute and chronic risks.

3.6. Monte Carlo Simulation

Figure 7 illustrates the cumulative probability distributions of the chronic hazard index (HIc) for various vegetables and population groups (adults and children), based on the Monte Carlo simulation results. Children exhibit higher HIc values than adults for all vegetables, indicating a greater risk of chronic dietary exposure [48]. Ginger and celery had the highest HIc values among adults and children. On the other hand, carrots and radishes generally had the lowest HIc values. For adults, the HIc ranged from approximately 6.48×10^{-4} (10th percentile) to 1.25×10^{-2} (90th percentile), indicating low variability and a low overall risk (Supplementary Table S10). In contrast, children exhibited higher risk values, with the HIc ranging from 1.21×10^{-3} to 2.33×10^{-2} (Table S10). Although all values remained well below the safety threshold (HIc < 1), the results confirmed that children are more vulnerable to chronic pesticide exposure, primarily due to their lower body weight and higher intake relative to body mass. This finding highlights the importance of including sensitive population groups in dietary risk assessments.

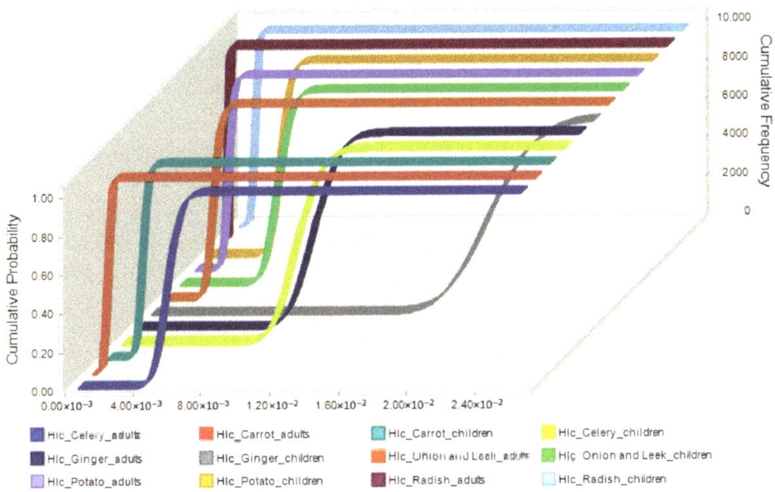

Figure 7. Monte Carlo simulation for chronic health risk for adults and children.

Figure 8 presents the results of a sensitivity analysis for the total chronic health risk (HIc) associated with pesticide residues in root vegetables, conducted separately for adults (A) and children (B). The analysis identified the most influential input parameters that affect the variability in the risk estimates. For both population groups, body weight (BW) was identified as the most influential factor, followed by ingestion of celery, ginger, potato, and onion + leek. Body weight was the most sensitive parameter in the model, and even small variations could significantly affect the final HIc value. This is expected, as body weight is in the denominator of the risk equation (EDI = STMR × Fi/BW), meaning that a lower body weight results in a higher estimated risk. Based on the length of the left bar (orange) in the tornado chart for body weight, it can be observed that the HIc is slightly more sensitive to an increase in body weight than to a decrease. The higher influence of celery and ginger ingestion rates on chronic health risk was due to the higher contamination of these vegetables compared to other types of vegetables. The influence of these parameters is followed by the ingestion of potatoes, the root vegetables most consumed by the average population.

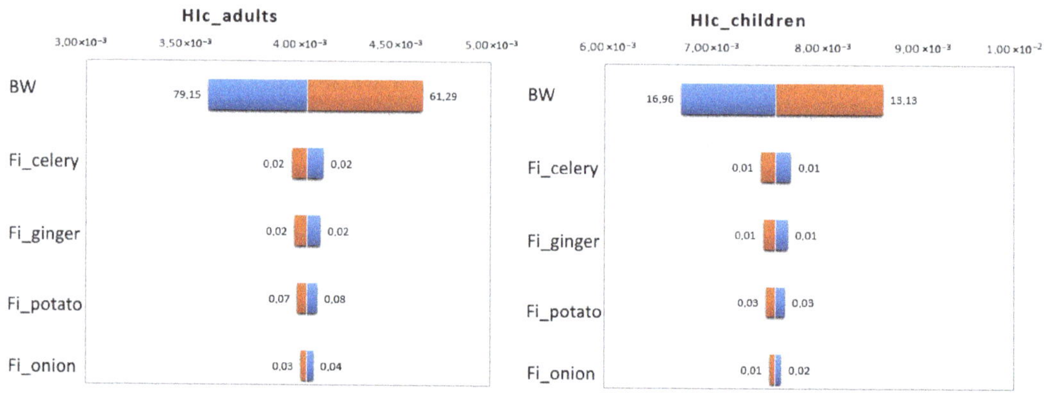

Figure 8. Sensitivity analysis by tornado chart for chronic health risk from pesticides in root vegetables.

4. Conclusions

This study provided a comprehensive assessment of pesticide residues in commonly consumed root vegetables and onions by combining advanced analytical techniques with a probabilistic health risk assessment model. A total of 33 pesticide residues were identified across 580 vegetable samples, with multiple residues and MRL exceedances observed in 19 and 29 samples, respectively. Although individual pesticides generally did not pose significant health risks, cumulative risk analysis revealed that certain samples, particularly ginger and celery, may contribute more substantially to dietary exposure. Monte Carlo simulations confirmed that chronic exposure levels for both adults and children remained below acceptable thresholds, with children identified as the more vulnerable group. Sensitivity analysis further emphasized the critical influence of body weight and vegetable consumption rates on risk estimates. The findings underscore the importance of continuous monitoring, stricter enforcement of pesticide regulations, and implementation of cumulative exposure assessments in routine food safety evaluations. Although dietary patterns were not evaluated in this study, the results may bolster public health initiatives that focus on mitigating dietary risks. Strategies such as promoting dietary diversification, educating consumers on portion sizes and consumption frequency, and addressing the requirements of vulnerable populations, particularly children, could be considered.

Supplementary Materials: The following supporting information can be downloaded at: https://www.mdpi.com/article/10.3390/jox15040125/s1, Table S1: Pesticide standards used for analysis of root and tuber vegetables; Table S2: Definition and individual scores of indices for the pesticide residual risk ranking; Table S3: Summary table for LD50, ADI, and ARfD values and assigned scores for indices A and B. The LD50 values are adopted from the WHO database [35], the EPA Fact Sheet [36–38], and scientific opinion on hexachlorocyclohexanes [39]. Most of the ADI values are sourced from the EU Pesticide database [40], and the values that did not exist in this database were taken from the JMPR database [42]. ARfD values were sourced from from the EU Pesticide database [40], (EFSA, 2024) [45], (WHO & FAO, 2004) [46], and (EFSA, 2019) [47]; Table S4: The MRL values (mg/kg) were sourced from the EU pesticide database [40]. Empty cells: those pesticides were not detected in those vegetables; Table S5: Assigned scores for indices A-F for the calculation of the total risk score of pesticides in our study; Table S6: Assigned scores for indices A-F for the calculation of the total risk score of pesticides in vegetables from RASFF; Table S7: Descriptive statistics of detected pesticides in vegetables (mg/kg); Table S8: Vegetables positive for pesticide residues, hierarchical cluster analysis; Table S9: HQa (acute hazard quotient) and HQc (chronic hazard quotient) for adults and children for each evaluated pesticide: Table S10: Monte Carlo simulation: 10th and 90th percentiles of HIc.

Author Contributions: Conceptualization, A.O.; methodology, M.L.; software, M.L.; validation, M.L.; formal analysis, M.L.; investigation, M.L.; resources, A.O.; data curation, M.L.; writing—original draft preparation, M.L.; writing—review and editing, A.O.; visualization, M.L.; supervision, A.O.; project administration, A.O.; funding acquisition, A.O. All authors have read and agreed to the published version of the manuscript.

Funding: This research was funded by the Ministry of Science, Technological Development and Innovation of the Republic of Serbia (Contracts 451-03-136/2025-03/200287 and 451-03-136/2025-03/200135).

Institutional Review Board Statement: Not applicable.

Informed Consent Statement: Not applicable.

Data Availability Statement: The original contributions presented in this study are included in the article/Supplementary Materials. Further inquiries can be directed to the corresponding author(s).

Conflicts of Interest: The authors declare that they have no competing interests.

References

1. Kim, K.H.; Kabir, E.; Jahan, S.A. Exposure to Pesticides and the Associated Human Health Effects. *Sci. Total Environ.* **2017**, *575*, 525–535. [CrossRef]
2. Ma, C.; Wei, D.; Liu, P.; Fan, K.; Nie, L.; Song, Y.; Wang, M.; Wang, L.; Xu, Q.; Wang, J.; et al. Pesticide Residues in Commonly Consumed Vegetables in Henan Province of China in 2020. *Front. Public Health* **2022**, *10*, 901485. [CrossRef]
3. Chandra, R.; Sharpanabharathi, N.; Prusty, B.A.K.; Azeez, P.A.; Kurakalva, R.M. Organochlorine Pesticide Residues in Plants and Their Possible Ecotoxicological and Agri Food Impacts. *Sci. Rep.* **2021**, *11*, 17841. [CrossRef] [PubMed]
4. Horská, T.; Kocourek, F.; Stará, J.; Holý, K.; Mráz, P.; Krátký, F.; Kocourek, V.; Hajšlová, J. Evaluation of Pesticide Residue Dynamics in Lettuce, Onion, Leek, Carrot and Parsley. *Foods* **2020**, *9*, 680. [CrossRef]
5. Sookhtanlou, M.; Allahyari, M.S.; Surujlal, J. Health Risk of Potato Farmers Exposed to Overuse of Chemical Pesticides in Iran. *Saf. Health Work* **2022**, *13*, 23–31. [CrossRef] [PubMed]
6. Danek, M.; Fang, X.; Tang, J.; Plonka, J.; Barchanska, H. Simultaneous Determination of Pesticides and Their Degradation Products in Potatoes by MSPD-LC-MS/MS. *J. Food Compos. Anal.* **2021**, *104*, 104129. [CrossRef]
7. Kumar, V.; Yadav, C.S.; Banerjee, B.D. Xeno-Estrogenic Pesticides and the Risk of Related Human Cancers. *J. Xenobiotics* **2022**, *12*, 344–355. [CrossRef]
8. Sandoval-Insausti, H.; Chiu, Y.H.; Wang, Y.X.; Hart, J.E.; Bhupathiraju, S.N.; Mínguez-Alarcón, L.; Ding, M.; Willett, W.C.; Laden, F.; Chavarro, J.E. Intake of Fruits and Vegetables According to Pesticide Residue Status in Relation to All-Cause and Disease-Specific Mortality: Results from Three Prospective Cohort Studies. *Environ. Int.* **2022**, *159*, 107024. [CrossRef]
9. Fandiño-Del-Rio, M.; Tore, G.; Peng, R.D.; Meeker, J.D.; Matsui, E.C.; Quirós-Alcalá, L. Characterization of Pesticide Exposures and Their Associations with Asthma Morbidity in a Predominantly Low-Income Urban Pediatric Cohort in Baltimore City. *Environ. Res.* **2024**, *263*, 120096. [CrossRef]
10. de Rodrigues, M.B.; de Carvalho, D.S.; Chong-Silva, D.C.; Urrutia-Pereira, M.; de Albuquerque, G.S.C.; Cieslak, F.; Chong-Neto, H.J. Association between Exposure to Pesticides and Allergic Diseases in Children and Adolescents: A Systematic Review with Meta-Analysis. *J. Pediatr. (Rio. J.)* **2022**, *98*, 551–564. [CrossRef]
11. Albadrani, M.S.; Aljassim, M.T.; El-Tokhy, A.I. Pesticide Exposure and Spontaneous Abortion Risk: A Comprehensive Systematic Review and Meta-Analysis. *Ecotoxicol. Environ. Saf.* **2024**, *284*, 117000. [CrossRef]
12. Knez, E.; Kadac-Czapska, K.; Dmochowska-Ślęzak, K.; Grembecka, M. Root Vegetables—Composition, Health Effects, and Contaminants. *Int. J. Environ. Res. Public Health* **2022**, *19*, 15531. [CrossRef]
13. Hossain, M.Z.; Ferdous, F.; Rayhan, M.I. Pesticide Knowledge and Attitude among the Potato Growing Farmers of Bangladesh and Determinant Factors. *Front. Public Health* **2024**, *12*, 1408096. [CrossRef]
14. Yang, Z.; Guo, Z.; Yan, J.; Xie, J. Nutritional Components, Phytochemical Compositions, Biological Properties, and Potential Food Applications of Ginger (Zingiber Officinale): A Comprehensive Review. *J. Food Compos. Anal.* **2024**, *128*, 106057. [CrossRef]
15. Nićetin, M.; Pezo, L.; Pergal, M.; Lončar, B.; Filipović, V.; Knežević, V.; Demir, H.; Jelena, F.; Manojlović, D. Celery Root Phenols Content, Antioxidant Capacities and Their Correlations after Osmotic Dehydration in Molasses. *Foods* **2022**, *11*, 1945. [CrossRef] [PubMed]
16. Górska-Warsewicz, H.; Rejman, K.; Kaczorowska, J.; Laskowski, W. Vegetables, Potatoes and Their Products as Sources of Energy and Nutrients to the Average Diet in Poland. *Int. J. Environ. Res. Public Health* **2021**, *18*, 3217. [CrossRef] [PubMed]

17. Ikram, A.; Rasheed, A.; Ahmad Khan, A.; Khan, R.; Ahmad, M.; Bashir, R.; Hassan Mohamed, M. Exploring the Health Benefits and Utility of Carrots and Carrot Pomace: A Systematic Review. *Int. J. Food Prop.* **2024**, *27*, 180–193. [CrossRef]
18. Dong, C.; Hu, J. Residue Levels of Four Pesticides from Two Commercial Formulations on Gingers and Their Dietary Intake Risk Assessment under Open Field Conditions. *J. Food Compos. Anal.* **2024**, *128*, 106016. [CrossRef]
19. Kang, L.; Liu, H.; Zhao, D.; Pan, C.; Wang, C. Pesticide Residue Behavior and Risk Assessment in Celery after Se Nanoparticles Application. *Foods* **2021**, *10*, 1987. [CrossRef]
20. Rybacki, P.; Przygodziński, P.; Blecharczyk, A.; Kowalik, I.; Osuch, A.; Osuch, E. Strip Spraying Technology for Precise Herbicide Application in Carrot Field. *Open Chem.* **2022**, *20*, 287–296. [CrossRef]
21. Cui, K.; Wang, J.; Ma, G.; Guan, S.; Liang, J.; Fang, L.; Ding, R.; Li, T.; Dong, Z.; Wu, X.; et al. Residue Levels, Processing Factors and Risk Assessment of Pesticides in Ginger from Market to Table. *J. Hazard. Mater.* **2024**, *470*, 134268. [CrossRef] [PubMed]
22. Jiang, Y.; Zhuang, M.; Xiao, P.; Wang, K.; Song, J.; Liu, H.; Zhao, J.; Chu, Z. Pesticide Residues and Dietary Risk Assessment in Radishes in Shandong. *J. Food Sci.* **2022**, *87*, 4751–4760. [CrossRef] [PubMed]
23. European Commission Pesticide Residues in Food—Maximum Residue Levels (MRLs). Regulation (EC) No 396/2005 of the European Parliament and of the Council of 23 February 2005 on Maximum Residue Levels of Pesticides in or on Food and Feed of Plant and Animal Origin. *Off. J. Eur. Union* **2025**, *10*, 1–21.
24. P, S.; Thasale, R.; Kumar, D.; Mehta, T.G.; Limbachiya, R. Human Health Risk Assessment of Pesticide Residues in Vegetable and Fruit Samples in Gujarat State, India. *Heliyon* **2022**, *8*, e10876. [CrossRef]
25. Lučić, M.; Momčilović, M.; Marković, J.; Jović, M.; Smičiklas, I.; Onjia, A. Monte Carlo Simulation of Health Risk from Cadmium, Lead, and Nickel in Cigarettes. *Toxicol. Environ. Chem.* **2023**, *105*, 92–110. [CrossRef]
26. Miletić, A.; Lučić, M.; Onjia, A. Exposure Factors in Health Risk Assessment of Heavy Metal(Loid)s in Soil and Sediment. *Metals* **2023**, *13*, 1266. [CrossRef]
27. Radulović, J.; Lučić, M.; Nešić, A.; Onjia, A. Multivariate Assessment and Risk Ranking of Pesticide Residues in Citrus Fruits. *Foods* **2023**, *12*, 2454. [CrossRef]
28. Maleki, N.S.; Shakerkhatibi, M.; Dolatkhah, M.; Safari, G.H. Cumulative Health Risk Assessment of Pesticide Residues in Apple Products in the Northwest of Iran Using Monte Carlo Simulation. *Food Addit. Contam.-Part A* **2023**, *40*, 992–1010. [CrossRef]
29. Mahdavi, V.; Eslami, Z.; Gordan, H.; Ramezani, S.; Peivasteh-roudsari, L.; Ma'mani, L.; Mousavi Khaneghah, A. Pesticide Residues in Green-House Cucumber, Cantaloupe, and Melon Samples from Iran: A Risk Assessment by Monte Carlo Simulation. *Environ. Res.* **2023**, *206*, 112563. [CrossRef]
30. Wang, P.; Li, Y.; Sun, J.; Zhang, G. Pesticide Residues in Vegetables from Gansu Province, China and Risk Assessment by Monte Carlo Simulation. *Food Addit. Contam. Part B Surveill.* **2024**, *17*, 251–260. [CrossRef]
31. *EN 15662:2008*; Foods of Plant Origin—Determination of Pesticide Residues Using GC-MS and/or LC-MS/MS Following Acetonitrile Extraction/Partitioning and Clean-up by Dispersive SPE—QuEChERS-Method. European Committee for Standardization (CEN): Brussels, Belgium, 2008.
32. European Commission Analytical Quality Control and Method Validation Procedures for Pesticide Residues Analysis in Food and Feed. SANTE/11312/2021 Rev. 2. Implemented by 01/01/2024. Available online: https://food.ec.europa.eu/system/files/2023-11/pesticides_mrl_guidelines_wrkdoc_2021-11312.pdf (accessed on 1 June 2025).
33. Radulović, J.; Lučić, M.; Onjia, A. GC-MS/MS and LC-MS/MS Analysis Followed by Risk Ranking of Mepiquat and Pyrethroids in Coffee. *J. Food Compos. Anal.* **2024**, *129*, 106100. [CrossRef]
34. Veterinary Residues Committee. *Annual Report on Surveillance for Veterinary Residues in Food in the UK 2007*; Veterinary Residues Committee: Addlestone, UK, 2007.
35. WHO. *The WHO Recommended Classification of Pesticides by Hazard and Guidelines to Classification*, 2019th ed.; WHO: Geneva, Switzerland, 2020; ISBN 978-92-4-000566-2.
36. EPA. *Pesticide Fact Sheet: Epoxiconazole*; EPA: Washington, DC, USA, 2003; p. 19.
37. EPA. *Fact Sheet for Ethoxyquin*; EPA: Washington, DC, USA, 2004.
38. EPA. *Pesticide Fact Sheet Fosthiazate*; EPA: Washington, DC, USA, 2004.
39. Alexander, J.; Autrup, H.; Bard, D.; Carere, A.; Guido Costa, L.; Cravedi, J.-P.; Di Domenico, A.; Fanelli, R.; Fink-Gremmels, J.; Gilbert, J.; et al. Opinion of the Scientific Panel on Contaminants in the Food Chain on a Request from the Commission Related to Gamma-HCH and Other Hexachlorcyclohexanes as Undesirable Substances in Animal Feed. *EFSA J.* **2005**, *4*, 365. [CrossRef]
40. European Commission (EC) EU Pesticides Database. Available online: https://food.ec.europa.eu/plants/pesticides/eu-pesticides-database_en (accessed on 26 May 2025).
41. AERU Agriculture and Environment Research Unit at the University of Hertfordshire: PPDB—Pesticide Properties Database. Available online: http://sitem.herts.ac.uk/aeru/ppdb/en/index.htm (accessed on 3 March 2025).

42. JMPR; WHO. Inventory of Evaluations Performed by the Joint Meeting on Pesticide Residues (JMPR). Available online: https://apps.who.int/pesticide-residues-jmpr-database/Home/Search (accessed on 3 March 2025).
43. Statistical Office of the Republic of Serbia. *Household Budget Survey*; Statistical Office of the Republic of Serbia: Belgrade, Serbia, 2023.
44. WHO; FAO. International Estimated Short-Term Intake (IESTI). Available online: https://cdn.who.int/media/docs/default-source/food-safety/gems-food/guidance-iesti-2014.pdf?sfvrsn=9b24629a_2 (accessed on 10 August 2023).
45. EFSA. Peer Review of the Pesticide Risk Assessment of the Active Substance Pyrimethanil. *EFSA J.* **2024**, *22*, e8998. [CrossRef] [PubMed]
46. WHO; FAO. *Pesticide Residues in Food 2003*; FAO: Rome, Italy, 2004.
47. EFSA. Statement on the Available Outcomes of the Human Health Assessment in the Context of the Pesticides Peer Review of the Active Substance Chlorpyrifos. *EFSA J.* **2019**, *17*, 5809. [CrossRef] [PubMed]
48. Lučić, M.; Miletić, A.; Savić, A.; Lević, S.; Sredović Ignjatović, I.; Onjia, A. Dietary Intake and Health Risk Assessment of Essential and Toxic Elements in Pepper (Capsicum Annuum). *J. Food Compos. Anal.* **2022**, *111*, 104598. [CrossRef]
49. Yu, J.J.; Bong, L.J.; Panthawong, A.; Chareonviriyaphap, T.; Liu, W.T.; Neoh, K.B. Effects of Piperonyl Butoxide Synergism and Cuticular Thickening on the Contact Irritancy Response of Field Aedes Aegypti (Diptera: Culicidae) to Deltamethrin. *Pest Manag. Sci.* **2021**, *77*, 5557–5565. [CrossRef]
50. EFSA. Panel Safety and Efficacy of Ethoxyquin 6-ethoxy-1 2-dihydro-2 2 4-trimethylquinoline for All Animal.Pdf. *EFSA J.* **2015**, *13*, 4272. [CrossRef]
51. Fang, L.; Zhang, S.; Chen, Z.; Du, H.; Zhu, Q.; Dong, Z.; Li, H. Risk Assessment of Pesticide Residues in Dietary Intake of Celery in China. *Regul. Toxicol. Pharmacol.* **2015**, *73*, 578–586. [CrossRef]
52. Kazimierczak, R.; Średnicka-Tober, D.; Golba, J.; Nowacka, A.; Hołodyńska-Kulas, A.; Kopczyńska, K.; Góralska-Walczak, R.; Gnusowski, B. Evaluation of Pesticide Residues Occurrence in Random Samples of Organic Fruits and Vegetables Marketed in Poland. *Foods* **2022**, *11*, 1963. [CrossRef]
53. Jirata, U.; Asere, T.G.; Balcha, Y.B.; Gure, A. Levels of Organochlorine Pesticides in Onion and Tomato Samples from Selected Towns of Jimma Zone, Ethiopia. *Heliyon* **2024**, *10*, e35033. [CrossRef]
54. Chu, N.; Shu, X.; Meng, X.; Zhang, X.; Yang, J.; Li, B. Determination and Dietary Exposure Assessment of 79 Pesticide Residues in Chinese Onion (*Allium fistulosum* L.). *CYTA-J. Food* **2023**, *21*, 41–48. [CrossRef]
55. EFSA. The 2020 European Union Report on Pesticide Residues in Food. Available online: https://multimedia.efsa.europa.eu/pesticides-report-2020/commodity/potatoes/ (accessed on 21 June 2025).
56. Carrasco Cabrera, L.; Di Piazza, G.; Dujardin, B.; Medina Pastor, P. The 2021 European Union Report on Pesticide Residues in Food. *EFSA J.* **2023**, *21*, 7939. [CrossRef] [PubMed]
57. Anastassiadou, M.; Bernasconi, G.; Brancato, A.; Carrasco Cabrera, L.; Greco, L.; Jarrah, S.; Kazocina, A.; Leuschner, R.; Magrans, J.O.; Miron, I.; et al. Reasoned Opinion on the Setting of Temporary Maximum Residue Levels for Chlorpropham in Potatoes. *EFSA J.* **2020**, *18*, e06061. [CrossRef] [PubMed]
58. Carrasco Cabrera, L.; Di Piazza, G.; Dujardin, B.; Marchese, E.; Medina Pastor, P. The 2022 European Union Report on Pesticide Residues in Food. *EFSA J.* **2024**, *22*, e8753. [CrossRef]
59. Carrasco Cabrera, L.; Medina Pastor, P. The 2020 European Union Report on Pesticide Residues in Food. *EFSA J.* **2022**, *20*, e07215. [CrossRef]
60. Onjia, A.; Huang, X.; Trujillo González, J.M.; Egbueri, J.C. Chemometric Approach to Distribution, Source Apportionment, Ecological and Health Risk of Trace Pollutants. *Front. Environ. Sci.* **2022**, *10*, 1107465. [CrossRef]
61. Sayed, A.; Chys, M.; De Rop, J.; Goeteyn, L.; Spanoghe, P.; Sampers, I. Pesticide Residues in (Treated) Wastewater and Products of Belgian Vegetable- and Potato Processing Companies. *Chemosphere* **2021**, *280*, 130619. [CrossRef]
62. European Commission (EC) RASFF—Rapid Alert System for Food and Feed. Available online: https://food.ec.europa.eu/food-safety/rasff_en (accessed on 21 June 2025).
63. Gkountouras, D.; Boti, V.; Albanis, T. Pesticides and Transformation Products Footprint in Greek Market Basket Vegetables: Comprehensive Screening by HRMS and Health Risk Assessment. *Sci. Total Environ.* **2024**, *953*, 176085. [CrossRef]
64. Golge, O.; Kabak, B. Evaluation of QuEChERS Sample Preparation and Liquid Chromatography-Triple-Quadrupole Mass Spectrometry Method for the Determination of 109 Pesticide Residues in Tomatoes. *Food Chem.* **2015**, *176*, 319–332. [CrossRef]
65. Golge, O.; Hepsag, F.; Kabak, B. Health Risk Assessment of Selected Pesticide Residues in Green Pepper and Cucumber. *Food Chem. Toxicol.* **2018**, *121*, 51–64. [CrossRef]
66. Terfe, A.; Mekonen, S.; Jemal, T. Pesticide Residues and Effect of Household Processing in Commonly Consumed Vegetables in Jimma Zone, Southwest Ethiopia. *J. Environ. Public Health* **2023**, *2023*, 7503426. [CrossRef]
67. Phopin, K.; Wanwimolruk, S.; Norkaew, C.; Buddhaprom, J.; Isarankura-Na-ayudhya, C. Boiling, Blanching, and Stir-Frying Markedly Reduce Pesticide Residues in Vegetables. *Foods* **2022**, *11*, 1463. [CrossRef]

68. Cho, M.; Kim, M.; Im, J.; Seo, C.; Moon, Y.S.; Im, M.H. Variation of Boscalid Residues in Welsh Onion during Cooking Processes. *Food Sci. Biotechnol.* **2025**, *34*, 1351–1358. [CrossRef]
69. Jensen, B.H.; Petersen, A.; Petersen, P.B.; Christensen, T.; Fagt, S.; Trolle, E.; Poulsen, M.E.; Hinge Andersen, J. Cumulative Dietary Risk Assessment of Pesticides in Food for the Danish Population for the Period 2012–2017. *Food Chem. Toxicol.* **2022**, *168*, 113359. [CrossRef]

Journal of *Xenobiotics*

MDPI

Article

Do Isopropylammonium Glyphosate and LiCl Impact the Spore Diversity and Functions of Aquatic Fungi Involved in Plant Litter Decomposition in Streams?

Jorge Rodrigues [1,2], Hernâni Gerós [1,2], Manuela Côrte-Real [1,2] and Fernanda Cássio [1,2,*]

1 CBMA—Centre of Molecular and Environmental Biology, University of Minho, Campus of Gualtar, 4710-057 Braga, Portugal; jorgerodrigues@bio.uminho.pt (J.R.); geros@bio.uminho.pt (H.G.); mcortereal@bio.uminho.pt (M.C.-R.)

2 IB-S—Institute of Science and Innovation for Bio-Sustainability, University of Minho, Campus of Gualtar, 4710-057 Braga, Portugal

* Correspondence: fcassio@bio.uminho.pt

Abstract: Glyphosate based-herbicides are stressors of great concern because they can impact aquatic ecosystems. Similarly, lithium, a metal, is currently of concern because of its increasing use worldwide. Because glyphosate-based herbicides and lithium might co-occur in aquatic environments, there is a need to assess their impacts on aquatic organisms, such as aquatic fungi, as they play a key role in plant litter decomposition in streams. Microcosm assays were used to examine the effects of lithium and the herbicide isopropylammonium glyphosate (IPAG), alone or in mixtures, on microbial leaf mass loss, total fungal sporulation and biomass production. IPAG (alone and combined with LiCl) neither affected plant litter decomposition nor fungal biomass production, but boosted total fungal sporulation. *Dimorphospora foliicola*, the most tolerant species among the twelfth leaf inhabitant fungal species, is the major contributor to total fungal sporulation. IPAG interacts with LiCl in the total fungal sporulation and sporulation of *D. foliicola*, *A. tetracladia*, and *F. curvula*, indicating a species dependent-effect. IPAG alone or combined with LiCl greatly decreased the diversity of spores, as did as LiCl alone, but to a lesser extent. Finally, aquatic fungal communities reveal redundancy and resiliency to IPAG and LiCL, maintaining the health of aquatic ecosystems.

Keywords: aquatic fungi; glyphosate-based herbicide; LiCl; fungal activity; fungal spore diversity; plant litter decomposition

1. Introduction

Chemical pollution by pesticides is one of the main threats to freshwater ecosystems [1]. This is of special concern if it surpasses the frontiers of natural integrity because we may lose the ecological functions and services provided by natural ecosystems [2]. Limits for pesticide contamination, above which ecosystem integrity is compromised and the ecological services are likely lost forever, have not been determined yet [3]. Aquatic biota, including microorganisms, drives crucial ecological processes, such as organic matter turnover [4]. Plant litter decomposition is a process mediated by microorganisms, namely by bacteria and fungi, in which aquatic fungi play a major role in providing palatable resources for macroinvertebrate detritivores. This process drives the flux of carbon and energy across the entire food chain [5]. Aquatic fungal communities and the processes they drive have been used for studying the effects of stressors (e.g., microplastics, [6]. Indeed, due to the rapid reproduction rate and rapid adaptation to stressors, those microbial communities may forecast and allow extrapolating the consequences of pesticide contamination for stream ecological functions [3,7].

Pesticides comprising insecticides, herbicides, and fungicides are a group of potentially toxic substances that are capable of disrupting the microbial community structure and function in aquatic habitats (for a revision, see [8]). They may reach freshwater habitats through several pathways, including spray drift and runoff [9,10]. Indeed, field studies in streams near agricultural sites or vineyards have shown that the actual concentrations of a variety of pesticides affect aquatic fungi and lead to the reduction of plant litter decomposition [11]. Agriculture in large scale mostly depends on the use of pesticides and the herbicide glyphosate (N-(phosphonomethyl)glycine) stands out as one of the most used by farmers, as it is highly effective and cost-efficient [12]. The 5-enolpyruvylshikimate-3-phosphate synthase (EPSPS) of the shikimate pathway is the target enzyme of glyphosate, thereby interrupting the biosynthesis of the aromatic amino acids, phenylalanine, tryptophan, and tyrosine, and inhibiting plant secondary metabolism [13]. Glyphosate also damages cellular structures and disrupts chloroplasts, membranes, and cell walls, reduces chlorophyl content, and compromises nucleic acid synthesis, photosynthesis, and respiration [4]. The massive application of glyphosate-based herbicides results in the contamination of streams by glyphosate. Glyphosate occurrence near cultivated areas was reported to be between 0.42 to 2.93 mgL^{-1} [14,15]. In the worst-case scenarios, predicting an accidental spill or direct overspray in water, the glyphosate concentration can reach 7.12 to 24.9 mgL^{-1} [16–19]. As commercial formulations of glyphosate are mixtures of the active ingredient, the concentrations applied to the environment, and, in particular, in the case of isopropylammonium glyphosate (IPAG) used in the present study, would have been 1.43-fold higher than that of the active ingredient.

Lithium is abundant in the Earth's crust, reaching soils and water bodies by weathering processes, contaminating them and affecting living organisms [20]. In addition, discharges of industrial effluents from smelting and mining operations [21] and inadequate disposal of products that use lithium in their manufacture (e.g., batteries, detectors, or atomic reactors) [22] contribute to environmental contamination. As such, worldwide, lithium contamination is increasing and imposing serious threats to the environment and humans [23]. In particular, certain regions of Portugal possess abundant lithium, and prospecting activities are set to begin, raising concerns about the anticipated increase in environmental contamination with lithium in the near future. The lithium concentration in water bodies depends on geology, topography, and hydrogeology, along with other variables [24]. Lithium concentrations in surface waters, namely rivers and ponds, can range from 1 to a maximum threshold of 10 μgL^{-1}. Indeed, lithium has been found at concentrations ranging from 20 to 91 μgL^{-1} in surface waters and between 0.02 and 97 μgL^{-1} in ground waters of the South East of Ireland [25], though it can reach concentrations up to 500 μgL^{-1} ([26] reviewed in [27]). As far as we know, there is no established regulatory maximum admissible concentration for lithium in surface waters. In particular, different countries around the world do not have maximum permissible limits or guideline recommendations for lithium concentrations in drinking water ([28] reviewed in [29]). One of the exceptions is in the United States, where the EPA has set 40 μgL^{-1} of lithium as a regional screening level for residential tap water (reviewed in [29]).

The rising demand for renewable energy systems and more efficient technologies has significantly increased the demand for lithium-ion batteries. To meet this need, battery production has scaled up, resulting in a growing volume of spent batteries. As this waste stream expands, effective management becomes increasingly important. Unfortunately, there are still no universal or standardized disposal strategies in place. Improper handling and disposal will lead, in the near future, to environmental pollution, especially the contamination of water sources, which can disrupt entire ecosystems and harm multiple trophic levels [27]. Although the co-occurrence of glyphosate and lithium has not been documented in streams, their simultaneous presence is plausible, particularly because of their increased release into the environment.

As many aquatic habitats are embedded within agricultural landscapes, they may unintentionally be exposed to agricultural pesticides, which may affect organisms, including aquatic fungal decomposers [10]. However, despite the fundamental role of microorganisms in aquatic food webs, our understanding of how their activity and diversity are affected by the exposure to pesticides and lithium is quite limited [22,29]. Here, we aimed to clarify if IPAG and lithium, alone or in mixtures, might affect fungal fitness and plant litter decomposition under the hypotheses that (i) the exposure to IPAG and lithium can affect microbial activity, namely leaf litter decomposition, fungal sporulation, and fungal biomass production; and (ii) IPAG and lithium may interact with each other in mixtures. We used microcosm assays composed by oak leaves colonized by stream-dwelling aquatic microbes and monitored plant litter decomposition, fungal sporulation, and fungal biomass production after exposure to IPAG and LiCl alone and in combination at environmentally realistic concentrations. The novelty of our study is to assess the effects of glyphosate and lithium alone and in mixtures in a relevant ecological process, the plant litter decomposition, and, in particular, on spore diversity and functions of aquatic fungi driving the process. This can be particularly useful for managers of ecosystems to maintain biodiversity and ecological functions.

2. Methods

2.1. Leaf Processing, Study Area, and Field Procedures

Senescent oak leaves were collected (in October/November 2016), dried at room temperature, and stored in a dry atmosphere. About 2 years later (January 2019), the leaves were immersed in deionized water to soften for 24 h, cut into discs of 12 mm diameter, and then placed inside fine mesh bags (0.5 mm, 16 × 20 cm size) to prevent macroinvertebrate colonization. The bags were immersed (1 February 2019) for 12 days in the Oliveira stream, Northwest Portugal, for microbial colonization. The stream water physicochemical parameters from where microbial communities were retrieved were measured in situ using field probes (Multiline F/set 3 no. 400327 (WTW) British-American multinational company, Washington, DC, USA). After 12 days, colonized leaf discs from each bag were rinsed with deionized water to remove sediments and adhering invertebrates before carrying out the microcosm assays.

2.2. Microcosm Assays

Leaf discs were placed in plastic Erlenmeyer flasks containing 100 mL of a sterilized commercial water (Fastio, Chamoim, Portugal) with IPAG and/or LiCl under shaking at 17 °C for 49 days. Mineral water was used instead of stream water to ensure a controlled medium chemical composition. To determine ergosterol and leaf mass loss, 6 and 70 leaf discs were used per replicate, respectively, with 3 replicates for each condition. We tested five concentrations of IPAG (0.1, 2.5, 50, 500, 1500 μgL^{-1}; Genyen, Loural, Portugal), one of the commercial formulation of the herbicide-based Sereno, and three concentrations of LiCl (0.1, 5, 100 μgL^{-1}; Sigma Aldrich, St. Louis and Burlington, MA, USA), alone and in all possible combinations, with 3 replicates per treatment in a total of 72 microcosm assays. Control microcosm assays contained microbial colonized leaf discs without chemicals. IPAG and LiCl concentrations chosen were within the range reported in the literature in environmental samples. As glyphosate can be degraded, the medium (mineral water with IPAG and/or LiCl) used in the microcosm assays was renewed every week by decantation and subsequently addition of new water and toxicants at the desired concentrations according to other studies [6] to ensure continued exposure to the selected toxicant concentrations. Every week, 40 mL of the decanted medium from each treatment was stored in glass bottles for spore identification and counting. The spore suspensions were preserved by adding 800 μL of formaldehyde and 40 μL of Triton X (Sigma-Aldrich) 15% (*w*/*v*) to prevent their germination and adherence to the bottles. Leaf discs for assessment of mass leaf loss and fungal biomass were frozen at −20 °C. The leaf discs were lyophilized (Bioblock Scientific Christ Alpha 2-4 LD Plus) and kept in a desiccator until further use.

2.3. Leaf Mass Loss

Leaf mass loss was estimated by weighing, with an accuracy of 0.01 g (Mettler Toledo AG245), a subset of lyophilized leaf discs before and after the microcosm assay.

2.4. Total Fungal Sporulation and Fungal Diversity

A suspension of spores from each treatment was used to identify and count fungal spores. Spore suspensions were filtered through 0.45 μm filters and stained with 0.1% cotton blue in lactic acid. Then, the filters were scanned under a light microscope for spore identification and counting, at a magnification of 400× (light microscope; Leica Biomed, Heerbrugg, Switzerland).

2.5. Fungal Biomass

Fungal biomass was based on ergosterol content on leaves according to [30]. Five lyophilized leaf discs were refluxed in 10 mL of 0.8% KOH-methanol for 30 min at 80 °C. Then, the resulting lipid extract was purified by solid-phase extraction followed by high-performance liquid chromatography (HPLC; Beckman Golden System, Beckman Coulter, Inc., Brea, CA, USA), using a Lichrospher RP-18 column (25 by 0.40 cm; Merck, Darmstadt, Germany). The system was run isocratically with HPLC-grade methanol as the mobile phase (33 °C; 1.4 mL min^{-1}). Ergosterol was detected at 282 nm and quantified based on a standard curve of ergosterol (Sigma, Kanagawa, Japan) in isopropanol.

2.6. Data Treatment

The effects of increasing concentrations of IPAG and LiCl on fungal biomass, sporulation rates, and leaf mass loss were tested by ANOVAs. Two-way ANOVAs were used to compare the main effects of the independent variables, IPAG and LiCl, in comparison with the control. Bonferroni post hoc correction was only performed when the two independent variables interacted, and was used to assess differences among groups (IPAG and LiCl). Data were previously checked for normal distribution (Shapiro test) and homoscedasticity (Bartlett test). When the assumptions of ANOVA tests were not met, data were transformed to natural logarithms. When this transformation was not possible (values below 1), the Kruskal–Wallis non-parametric test was performed instead of ANOVA. In this case, if significant, multiple comparisons were performed with an equivalent test to the Tukey test.

All statistics were performed in R, and all graphics were performed in GraphPad Prism Software version 8.0.1.

3. Results

To study the effects of isopropylammonium glyphosate (IPAG) and LiCl, alone and in combination, on microbial decomposition of plant litter and on the aquatic fungal parameters, namely total sporulation and spore diversity as well ergosterol content, microcosm assays were carried out with oak leaves previously colonized by aquatic microbes in the Oliveira stream. The physicochemical parameters of the stream indicated a moderate acidity (pH 6.86), temperature of 9.9 °C, high dissolved oxygen concentration (86.3%), and low conductivity (32 μScm^{-1}). The results obtained are described in the following sections.

3.1. Effects of Isopropylammonium Glyphosate and LiCl Alone and in Combination on Microbial Decomposition of Plant Litter and Total Sporulation and Spore Diversity of Aquatic Fungi

As noted in the introduction, the concentrations of glyphosate most often found in the environment (0.42 to 2.93 mgL^{-1}, [15]) correspond to concentrations of isopropylammonium glyphosate (IPAG) 1.43-fold higher (0.57 to 3.96 mgL^{-1}). Also, as noted in the introduction, the environmental lithium concentrations reported in the literature range from 0.02 to 97 μgL^{-1} [25–27]. In the present study, the effects of IPAG and LiCl, alone or in combination, on plant litter decomposition by aquatic microbes was assessed by the loss of leaf mass in dry weight. For this purpose, 0.1, 2.5, 50, 500, and 1500 μgL^{-1} of IPAG and 0.1, 5, and 100 μgL^{-1} of LiCl were used, alone or in combination. All the concentrations of IPAG and LiCl tested, as well of their combinations, did not affect the leaf mass loss ($p > 0.05$, Figure 1).

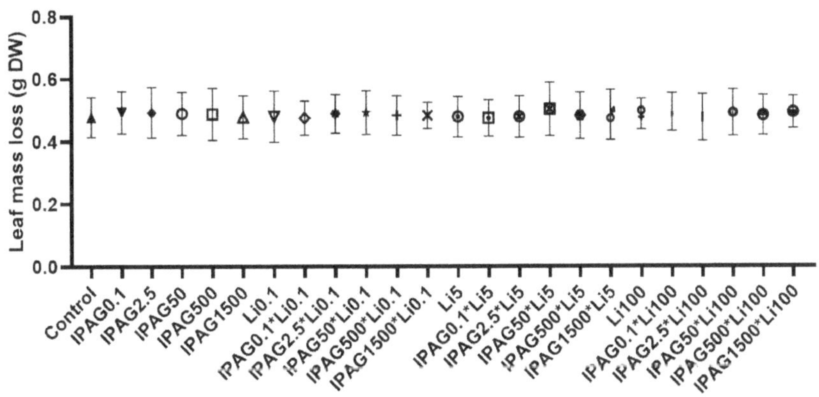

Figure 1. Effects of isopropylammonium glyphosate (IPAG) 1500 μgL^{-1} or LiCl (Li) alone (0.1 or 100 μgL^{-1}) and in combination on plant litter decomposition by colonized aquatic microbes, assessed by loss of leaf mass (g of leaf mass in dry weight, DW) 49 days after non-exposure or exposure to both chemicals in microcosms assays. Control refers to loss of leaf mass in the absence of IPAG and Li. In the XX axis, * means combination of IPAG and LiCl. Data presented are the mean values of three replicates and the respective standard deviations.

Total fungal sporulation rate, assessed by estimation of the number of spores per day and per gram of leaf mass, increased significantly 49 days after exposure to IPAG (two-way ANOVA, F = 29.570, $p = 0.000151$) and LiCl (two-way ANOVA, f = 6.086, $p = 0.014971$). There was no interaction between IPAG and LiCl (two-way ANOVA, F = 0.650, $p = 0.539619$). However, Bonferroni post hoc tests revealed that the adjusted p values were significantly different when comparing the concentrations of IPAG (0 and 1500 μgL^{-1}) ($p = 0.00054$), while there were no significant differences between LiCl (0 and 0.1 and 100 μgL^{-1}) ($p > 0.05$) and between IPAG alone and in combination with LiCl (0.1 and 100 μgL^{-1}) ($p > 0.05$) (Figure 2).

Based on fungal spore morphology, the relative sporulation rates (%) of 12 fungal species (*Dimorphospora foliicola*, *Infundibura* spp., *Triscelophorus* cf. *acuminatus*, *Flagellospora curvula*, *Articulospora tetracladia*, *Clavatospora longibrachiata*, *Flagellospora* spp., *Anguillospora filiformis*, *Alatospora pulchella*, *Goniopila monticula*, *Heliscella stellata*, and *Heliscus submersus*) identified on decomposing leaves 49 days after non-exposure and exposure to IPAG and Li, alone or in combination, were estimated. The species *D. foliicola*, *Infundibura* spp., *T.* cf. *acuminatus*, *F. curvula*, *A. tetracladia*, and *C. longibrachiata* were the six most abundant, as they showed the highest relative sporulation rates. The abundance of *D. foliicola* was significantly higher than that of other species ($p < 0.05$, Figure 3A). Infundibura spp. was the second most predominant species, with a sporulation similar to that of *T.* cf. *acuminatus* (Bonferroni post hoc test, adjusted $p = 1$; Figure 3A). Both species had significantly lower sporulation than *D. foliicola* (Bonferroni post hoc test, adjusted $p = 3.0 \times 10^{-7}$ and adjusted $p = 1.9 \times 10^{-8}$, respectively (Figure 3A)).

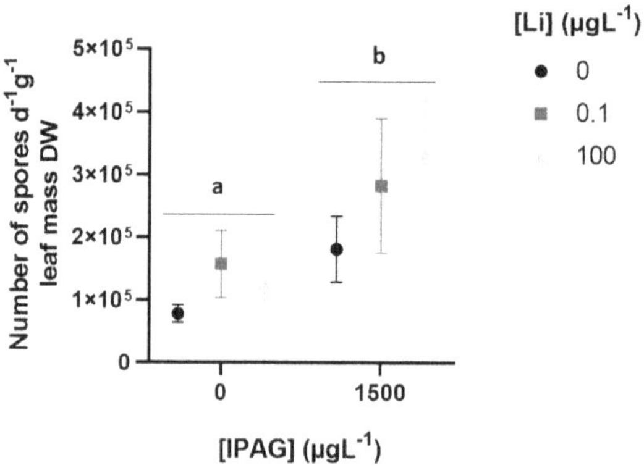

Figure 2. Effects of isopropylammonium glyphosate (IPAG) and LiCl (Li) alone and in combination on the total sporulation of aquatic fungi, expressed as the total number of spores per day and per gram of leaf mass in dry weight (DW). Samples treated with IPAG 1500 μgL^{-1}, Li 0.1 μgL^{-1} or Li 100 μgL^{-1}, or both with IPAG and Li (in grey colors), were compared with non-treated samples used as controls (in black). Data presented are the mean values of three replicates and the respective standard deviations. Significant statistical differences are indicated with small caps (a,b) ($p < 0.05$).

IPAG (1500 μgL^{-1}) significantly increased the relative contribution of *D. foliicola* to the total sporulation (Kruskal–Wallis chi-squared = 33.154, df = 11, $p = 0.0004968$; Figure 3B). All the other species had similar sporulation rates (Bonferroni post hoc test, adjusted $p > 0.05$), with IPAG (1500 μgL^{-1}) significantly decreasing the relative contribution of *A. tetracladia*, *F. curvula*, *T.* cf. *acuminatus*, *C. longibrachiata*, and *Infundibura* spp. (Figure 3B). In the treatment with LiCl 0.1 μgL^{-1}, *D. foliicola* was dominant (Bonferroni post hoc test, adjusted $p = 2 \times 10^{-16}$). *T.* cf. *acuminatus* followed *D. foliicola* as the second most abundant species, similarly to *Infundibura* spp. (Bonferroni post hoc test, adjusted $p = 1$) and *Flagellospora* spp. (Bonferroni post hoc test, adjusted $p = 1$), but had a higher sporulation rate than all the other species (Bonferroni post hoc test, adjusted $p > 0.05$; Figure 3C). In the treatment with LiCl 100 μgL^{-1}, the contributions of more species increased and equaled the contribution of *T.* cf. *acuminatus* and *Infundibura* spp. (Figure 3E). *D. foliicola* remained as the higher contributor to total sporulation (Bonferroni post hoc test, adjusted $p < 0.05$), followed by *T.* cf. *acuminatus* with a contribution similar to that of *Infundibura* spp., *A. tetracladia*, and *Flagellospora* spp. (Bonferroni post hoc test, adjusted $p > 0.05$). The relative sporulation of *Infundibura* spp., *A. tetracladia*, and *Flaggellospora* spp. was not different from that of *C. longibrachiata* and *F. curvula* (Bonferroni post hoc test, adjusted $p > 0.05$). LiCl 100 μgL^{-1} was the treatment that led to a higher number of species contributing to the total sporulation (Figure 3E).

When IPAG 1500 μgL^{-1} was combined with LiCl 0.1 or 100 μgL^{-1}, the contribution of *D. foliicola* to the total sporulation was similar to that with IPAG 1500 μgL^{-1} (Bonferroni post hoc test, adjusted $p = 1$), and so, higher than the control (Bonferroni post hoc test, adjusted $p = 9.1 \times 10^{-7}$; Figure 3A,B,D,F). Under these conditions IPAG impacted the effect of LiCl 0.1 and 100 μgL^{-1}, reducing the number of species contributing to sporulation, which indicates an interaction between IPAG and LiCl (two-way ANOVA, F = 10.420, $p = 0.00238$). In the treatments with IPAG 1500 μgL^{-1} without and with LiCl 0.1 and 100 μgL^{-1} (Figure 3B,D,F), there was mainly no clear difference between *Infundibura* spp. and *T.* cf. *acuminatus* and the other species, while in the treatments without IPAG (control and LiCl 0.1 μgL^{-1} and 100 μgL^{-1}) (Figure 3A,C,E), the contribution of these two species to the total sporulation rate was more evident compared to *D. foliicola*.

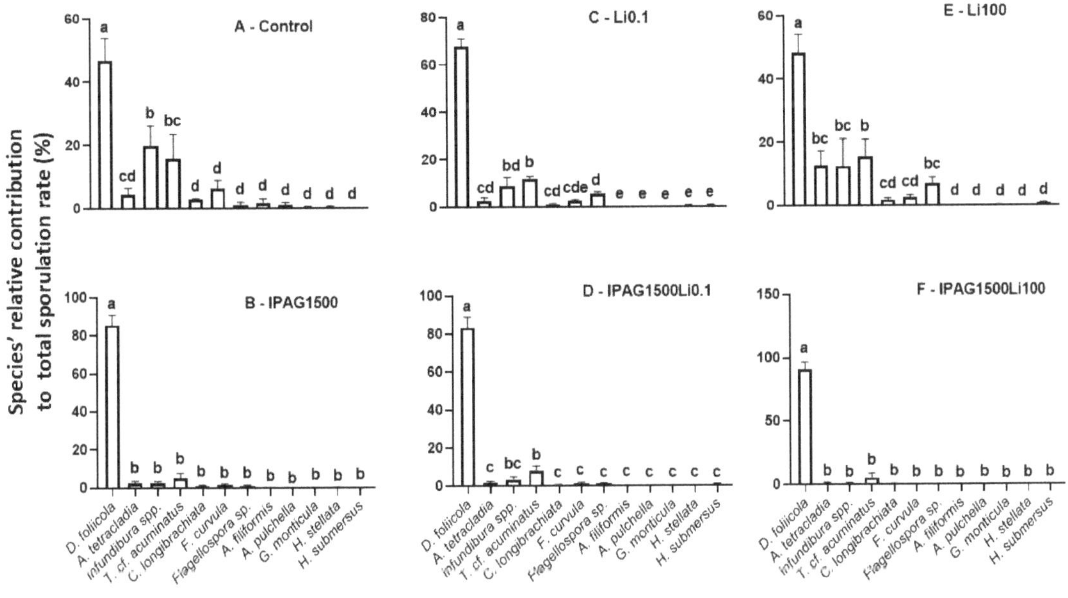

Figure 3. Effects of isopropylammonium glyphosate (IPAG) and LiCl (Li) alone and in combination on aquatic fungal spore diversity through the assessment of the relative contribution of each species to the total sporulation rate. Relative contribution (%) of each aquatic fungal species to total sporulation rate in samples: (**A**), untreated (Control); (**B**), exposed to IPAG 1500 μgL^{-1}; (**C**), Li 0.1 μgL^{-1}; (**D**), IPAG 1500 μgL^{-1} and Li 0.1 μgL^{-1}; (**E**), exposed to Li 100 μgL^{-1}; (**F**), exposed to IPAG 1500 μgL^{-1} and Li 100 μgL^{-1}. Data presented are the mean values of three replicates and the respective standard deviations. Significant differences of the relative contribution of each species to the total sporulation are indicated by small caps ($p < 0.05$).

Although IPAG and Li had an interactive effect on the total fungal sporulation rate, whether it depended on the fungal species remained unknown. To address this question, the impact of the combined treatments with IPAG and LiCl on the relative contribution of each of the six most relevant fungal species was assessed and compared with the treatments with each toxicant alone (Figure 4).

The relative sporulation rates of *D. foliicola* for the different conditions tested were higher (above 50%) than those of *T.* cf. *acuminatus*, *Infundibura* spp., *A. tetracladia* (below 20%), *F. curvula*, and *C. longibrachiata* (below 6%). Moreover, the sporulation rates of *T.* cf. *acuminatus*, *Infundibura* spp., *F. curvula*, *A. tetracladia*, and *C. longibrachiata* were lower in the treatments with IPAG 1500 μgL^{-1} (with or without LiCl) than in the control and treatments with Li alone (Figure 4B–F).

D. foliicola data showed there was an interaction of IPAG and LiCl (two-way ANOVA, F = 10.420, p = 0.00238). The treatments with IPAG (with and without LiCl 0.1 and 100 μgL^{-1}) were higher than the control (Bonferroni post hoc test, adjusted $p = 9.1 \times 10^{-7}$). The treatments with LiCl 0.1 and 100 μgL^{-1} (without IPAG) were equal to the control (Bonferroni post hoc test, adjusted p = 1), and LiCl 0.1 was equal to LiCl 100 μgL^{-1} (Bonferroni post hoc test, adjusted p = 1) (Figure 4A).

T. cf. *acuminatus* data showed there was no interaction of IPAG and LiCl (two-way ANOVA, F = 1.074, p = 0.37243). The treatments with IPAG (with and without LiCl 0.1 and 100 μgL^{-1}) were lower than control (Bonferroni post hoc test, adjusted p = 0.00095). The treatments with LiCl 0.1 and 100 μgL^{-1} (without IPAG) were equal to control (Bonferroni post hoc test, adjusted p = 1; Figure 4B).

Infundibura spp. showed no interaction between IPAG and LiCl (two-way ANOVA, F = 2.869, p = 0.0959). The contribution of this species to total sporulation was low in treatments with IPAG (with or without LiCl) (Bonferroni post hoc test, adjusted $p = 1.2 \times 10^{-5}$) compared with control, but also, there were no differences between control and the treatments with LiCl (0.1 and 100 μgL^{-1}, without IPAG), (Bonferroni post hoc test, adjusted $p > 0.05$; Figure 4C).

F. curvula data showed there was interaction of IPAG and LiCl (two-way ANOVA, F = 4.158, p = 0.04247). The treatments with IPAG 1500 μgL^{-1} (with or without LiCl 0.1 and 100 μgL^{-1}) presented lower contribution to total sporulation than the treatments without IPAG (control, LiCl 0.1 and 100 μgL^{-1}) (Bonferroni post hoc test, adjusted p = 0.0011). The treatments with LiCl 0.1 and 100 μgL^{-1}, without IPAG, had equal effects compared to control (Bonferroni post hoc test, adjusted $p > 0.05$; Figure 4D).

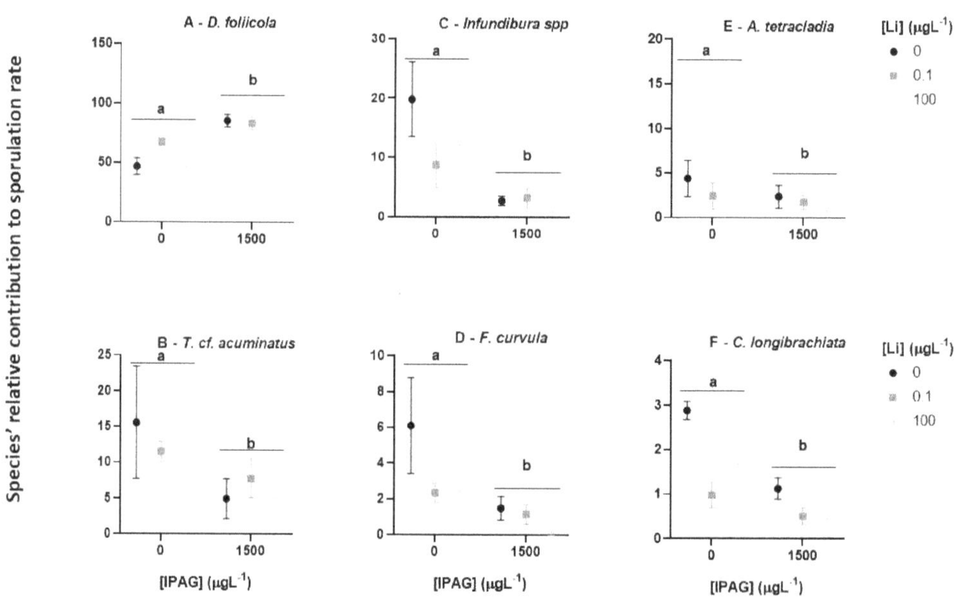

Figure 4. Effects of isopropylammonium glyphosate (IPAG) and LiCl (Li) alone and in combination on the relative sporulation rate (%) of each fungal species (**A–F**). Control, IPAG 1500 μgL^{-1}, Li 0.1 μgL^{-1}, Li 0.1 μgL^{-1} with IPAG 1500 μgL^{-1}, Li 100 μgL^{-1}, and Li 100 μgL^{-1} with IPAG 1500 μgL^{-1}. Data presented are the mean values of three replicates and the respective standard deviations. Significant differences are indicated with small caps (a,b). ($p < 0.05$).

A. tetracladia data showed that there was an interaction between IPAG and LiCl (two-way ANOVA, F = 8.821, $p = 0.004401$). IPAG 1500 μgL^{-1} decreased the sporulation of this species compared to the control (Bonferroni post hoc test, adjusted $p = 0.0042$) and in control and treatments with LiCl 0.1 and 100 μgL^{-1} without IPAG, the sporulation rates were similar (Bonferroni post-Hoc test, adjusted $p > 0.05$; Figure 4E).

C. longibrachiata showed no interaction of IPAG and LiCl (two-way ANOVA, F = 1.361, $p = 0.293358$). IPAG 1500 μgL^{-1} (with or without LiCl 0.1 and 100 μgL^{-1}) decreased fungal sporulation (Bonferroni post hoc test, adjusted $p = 0.0038$), but the treatments of LiCl 0.1 and 100 μgL^{-1} (without IPAG) had no effects (Bonferroni post hoc test, adjusted $p > 0.05$; Figure 4F).

3.2. Effects of Isopropylammonium Glyphosate and LiCl Alone or in Combination on Fungal Biomass

The effects of IPAG and LiCl, alone or in combination, on fungal biomass were assessed by ergosterol content of fungi associated with oak leaves. Effects of IPAG on ergosterol content were marginally significant (two-way ANOVA, F = 2.202, $p = 0.0694$), and the effect of LiCl was significant (two way-ANOVA, F = 3.012, $p = 0.0391$), and there was an interaction between the two factors (two Way-ANOVA, F = 2.260, $p = 0.0167$). However, no differences were observed regarding fungal biomass between the IPAG groups (Bonferroni post hoc test, adjusted $p > 0.05$) and the LiCl groups (Bonferroni post hoc test, adjusted $p > 0.05$). The only group that was marginally different from control was the LiCl 5 μgL^{-1} group (Bonferroni post hoc test, adjusted $p = 0.089$) (Figure 5).

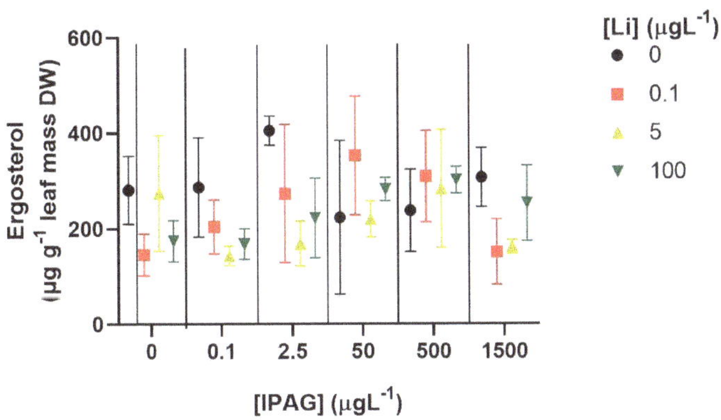

Figure 5. Effects of isopropylammonium glyphosate (IPAG) 0–1500 μgL^{-1} and LiCl (Li) 0.1–100 μgL^{-1} alone or in combination on the fungal biomass, assessed by the concentration of ergosterol on decomposing leaves, 49 days after incubation. Data presented are the mean values of three replicates and the respective standard deviations.

4. Discussion

Though pesticides and metals are toxic to aquatic organisms that are relevant for nutrient cycling and therefore essential to life, the effects of the most commonly used pesticides have not been completely unveiled. We used a natural microbial community composed by bacteria and fungi to test the effects of isopropylammonium glyphosate (IPAG) and LiCl. We want to point out that aquatic fungi have the main role on plant litter decomposition in streams, making them an ideal microorganism to study the effects of these pesticides. Earlier studies reported that exposure to pesticides did not affect leaf microbial decomposition rates [31,32]. Accordingly, isopropylammonium glyphosate (IPAG) and LiCl had no effects on leaf litter decomposition by aquatic microbes. In contrast, treatments with IPAG alone or in combination with LiCl did change the total fungal sporulation significantly. The failure in sporulation does not necessarily translate in the absence of fungal taxa and non-contribution to plant litter decomposition [33,34]. This suggests that these parameters may not be related, and may explain why the treatment with IPAG alone or in combination with LiCl affects the total fungal sporulation, but not leaf microbial breakdown. As opposed to IPAG, treatments with LiCl alone did not change the total fungal sporulation significantly, but in the presence of IPAG, they did, which suggests an interaction between LiCl and IPAG. To ascertain the weight of each species in the total sporulation changes, and eventually in the number species contributing to it, the toxic effects of both IPAG and LiCl on the relative contribution of each of the six species most relevant to the total sporulation were assessed. The results revealed that the significant increase in total fungal sporulation ($p < 0.05$) upon treatment with IPAG alone and in combination LiCl, was mainly due to *D. foliicola*, the most abundant species found in both untreated and treated samples. Accordingly, *D. foliicola* was also found to be a predominant species in a polluted site of the Este River [35]. Altogether, these data support the notion that this species is tolerant to contaminants. However, in contrast to *D. foliicola*, *T.* cf. *acuminatus*, *Infundibura* spp., *F. curvula*, *A. tetracladia*, and *C. longibrachiata* decreased their relative contribution to the total sporulation. Moreover, the analysis of the effects of each treatment on the relative contribution of each species to the total sporulation showed a change in the number of the natural inhabitant fungal species, thus impacting the diversity of spores and function of the fungal community. Indeed, in samples treated with IPAG 1500 μgL^{-1}, alone and in combination with LiCl (0.1 and 100 μgL^{-1}), there was a decrease from six predominant species to two or one species (*D. foliicola* in treatments with IPAG 1500 combined with LiCl 0.1 μgL^{-1}, and *D. foliicola* and *T.* cf. *acuminatus* in the combined treatment with IPAG 1500 μgL^{-1} and LiCl 100 μgL^{-1}, respectively), while for the treatments with LiCl 0.1 or 100 μgL^{-1} alone, there was a decrease to four and five predominant species, respectively. However, in treatments with LiCl 0.1 and 100 μgL^{-1}, *Flagellospora* spp. became predominant, replacing *A. tetracladia* and *F. curvula* in LiCl 0.1 μgL^{-1} and *F. curvula* in LiCl 100 μgL^{-1}. Altogether, the combined treatments also showed that the effects of IPAG mainly overlapped the effects of LiCl 0.1 and 100 μgL^{-1}, as IPAG combined with LiCl displayed the same reduction as IPAG alone in the number of predominant species. Since *D. foliicola* followed by *Infundibura* spp., *T.* cf. *acuminatus* and *A. tetracladia* were the most predominant species found in both untreated and treated samples, they are likely to be more tolerant to IPAG and LiCl alone and in combination than the other inhabitant species, indicating that the fungal response to pesticides is species-dependent. Some earlier evidence indicates that species with different levels of tolerance to contaminants may co-occur in streams, serving as a buffer against environmental changes [36]. Analysis of the effects of each treatment on the relative contribution of each species to the total sporulation revealed that with *D. foliicola*, *A. tetracladia*, and *F. curvula*, there was an interaction between IPAG and Li, while with *T.* cf. *accuminatus*, *Infundibura* spp., and *C. longibrachiata*, there was not. In accordance with leaf mass loss, fungal biomass is not affected by IPAG or LiCl alone or in combination. Glyphosate is co-metabolized as a carbon, nitrogen, and phosphorus source in microbiological cultures including bacteria, actinomycetes, fungi, and other non-identified microorganisms (for a revision, see [14]). However, to our knowledge, there are no data on degradation by aquatic fungi. This factor could explain the lack of effects on leaf mass loss and fungal biomass.

IPAG and LiCl alone and in combination did not affect fungal biomass. Other studies revealed that the use of structural and functional metrics might give uncoupled outputs, as the response in a metric may not be equal in another metric [37]. In our study, the effects of IPAG and LiCl alone and in combination on the structural metric (fungal sporulation) diverged from the functional metric (fungal biomass production), as the treatments that increased total fungal sporulation did not cause an increase in fungal biomass. Nonetheless, the observed negative effects on fungal sporulation might compromise the spread of fungi in natural ecosystems in the long term, but the absence of negative effects on fungal biomass should not lead to lower leaf mass decomposition and less suitable resources for higher trophic levels. These differential effects may be partially explained by the complexity of the ecosystem process studied.

5. Conclusions

The effects of pesticides on aquatic microbial diversity can lead to an impoverished fungal community, compromising its overall performance, unless redundant species can ensure functions. In the present study, changes in aquatic fungal community composition upon exposure to IPAG and LiCl were found. Unexpectedly, this may contribute to maintaining fungal functions under stressed conditions imposed by pesticides, such as IPAG and lithium alone or in combination, as to some extent, sensitive species are substituted by others more tolerant to stressors. Because different fungal species and the total fungal sporulation varied with treatments, we can conclude that shifts in community structure might have ensured the community function (leaf litter decomposition), probably because redundant species in the community could maintain the functions of those species that disappeared. Accordingly, several studies showed considerable redundancy between aquatic fungi during leaf litter decomposition. Both the aforementioned fungal structural and functional community responses may be a safeguard for the functional stability of fungal-mediated processes, namely plant litter decomposition and nutrient cycling in streams chronically affected by fungicides.

In summary, we conclude that aquatic fungal responses varied with the IPAG and LiCl concentrations as well as with their co-occurrence, which reinforces the need to assess the effect of mixtures in the environment. LiCl did not affect the studied fungal parameters. IPAG, alone and in combination with LiCl, had a high impact on fungal spore diversity as well as on the relative contribution of each fungal species to the process of plant litter decomposition, either increasing or decreasing their contribution to total sporulation, though it did not affect either fungal biomass or plant litter decomposition by aquatic microbes. Fungal aquatic communities, involved in the nutrient cycling important to higher trophic levels and ultimately to the quality of our lives, are at risk due to their exposure to environmental contaminants. The present study, together with the assessment of leaf mass loss and biomass formation, explored the fact that spores of aquatic fungi have specific morphological characteristics, which allowed us to identify the sporulating species through simple light microscopic observation. Nonetheless, the spores identified reflected the species that had colonized the plant litter earlier, but not the species that were actually present at the time of sampling. To this end, the use of molecular tools could be an enhanced approach, allowing one to record the history of the contaminants in litter samples. The present study offers novel insights into the resilience of aquatic fungal communities to glyphosate-based herbicides and lithium pollution, two increasingly prevalent environmental contaminants whose combined effects on freshwater ecosystems have been scarcely investigated. Finally, this study is important for ecotoxicologists, freshwater ecologists, and environmental policymakers seeking to understand the related ecological risks.

Author Contributions: J.R.: methodology, investigation, formal analysis, and writing—original draft. H.G.: supervision, writing—original draft, and writing—review and editing. M.C.-R.: supervision, writing—original draft, and writing—review and editing. F.C.: conceptualization, funding acquisition, methodology, formal analysis, supervision, writing—original draft, and writing—review and editing. All authors have read and agreed to the published version of the manuscript.

Funding: This study was supported by the strategic program UIDB/04050/2020 and a PhD fellowship to J.R. (SFRH/BD/131562/2017).

Institutional Review Board Statement: Not applicable.

Informed Consent Statement: Not applicable.

Data Availability Statement: The data presented in this study are openly available in Repositorium at https://repositorium.sdum.uminho.pt.

Conflicts of Interest: The authors declare no conflict of interest.

References

1. Dudgeon, D. Multiple threats imperil freshwater biodiversity in the Anthropocene. *Curr. Biol.* **2019**, *29*, R942–R995. [CrossRef] [PubMed]
2. Deeksha; Shukla, A.K. Ecosystem Services: A Systematic Literature Review and Future Dimension in Freshwater Ecosystems. *Appl. Sci.* **2022**, *12*, 8518. [CrossRef]
3. Schreiner, V.C.; Feckler, A.; Fernández, D.; Frisch, K.; Muñoz, K.; Szöcs, E.; Zubrod, J.P.; Bundschuh, M.; Rasmussen, J.; Kefford, B.; et al. Similar recovery time of microbial functions from fungicide stress across biogeographical regions. *Sci. Rep.* **2018**, *8*, 17021. [CrossRef] [PubMed]
4. Wang, C.; Lin, X.; Li, L.; Lin, S. Differential growth responses of marine phytoplankton to herbicide glyphosate. *PLoS ONE* **2016**, *11*, e0151633. [CrossRef]
5. Pascoal, C.; Cássio, F.; Marcotegui, A.; Sanz, B.; Gomes, P. Role of fungi, bacteria and invertebrates in leaf litter breakdown in a polluted river. *J. N. Am. Benthol. Soc.* **2005**, *24*, 784–797. [CrossRef]
6. Trabulo, J.; Pradhan, A.; Pascoal, C.; Cássio, F. Can microplastics from personal care products affect stream microbial decomposers in the presence of silver nanoparticles? *Sci. Total Environ.* **2022**, *832*, 155038. [CrossRef]
7. Graham, E.B.; Knelman, J.E.; Schindlbacher, A.; Sicilliano, S.; Breulmann, M.; Yannarell, A.; Beman, J.M.; Abell, G.; Philippot, L.; Prosser, J.; et al. Microbes as engines of ecosystem function: When does community structure enhance predictions of ecosystem processes? *Front. Microbiol.* **2016**, *7*, 214. [CrossRef]
8. Pimentão, A.R.; Pascoal, C.; Castro, B.B.; Cássio, F. Fungistatic effect of agrochemical and pharmaceutical fungicides on non-target aquatic decomposers does not translate into decreased fungi- or invertebrate-mediated decomposition. *Sci. Total Environ.* **2020**, *712*, 135676. [CrossRef]
9. Huber, A.; Bach, M.; Frede, H.G. Pollution of surface waters with pesticides in Germany: Modeling non-point source inputs. Agriculture. *Ecosyst. Environ.* **2000**, *80*, 191–204. [CrossRef]
10. Schulz, R. Field studies on exposure, effects, and risk mitigation of aquatic nonpoint-source insecticide pollution—A review. *J. Environ. Qual.* **2004**, *33*, 419–448. [CrossRef]
11. Rasmussen, J.J.; Wiberg-Larsen, P.; Baattrup-Pedersen, A.; Monverg, R.J.; Kronvang, B. Impacts of pesticides and natural stressors on leaf litter decomposition in agricultural streams. *Sci. Total Environ.* **2012**, *416*, 148–155. [CrossRef] [PubMed]
12. Helander, M.; Pauna, A.; Saikkonen, K.; Saloniemi, I. Glyphosate residues in soil affect crop plant germination and growth. *Sci. Rep.* **2019**, *9*, 19653. [CrossRef] [PubMed]
13. Leino, L.; Tall, T.; Helander, M.; Saloniemi, I.; Saikkonen, K.; Ruuskanen, S.; Puigbò, P. Classification of the glyphosate target enzyme (5-enolpyruvylshikimate-3-phosphate synthase) for assessing sensitivity of organisms to the herbicide. *J. Hazard. Mater.* **2021**, *408*, 124556. [CrossRef] [PubMed]
14. Peruzzo, P.J.; Porta, A.A.; Ronco, A.E. Levels of glyphosate in surface waters, sediments and soils associated with direct sowing soybean cultivation in north pampasic region of Argentina. *Environ. Pollut.* **2008**, *156*, 61–66. [CrossRef]
15. Giesy, J.P.; Dobson, S.; Solomon, K.R. Ecotoxicological risk assessment for roundup herbicide. In *Reviews of Environmental Contamination and Toxicology*; Springer: New York, NY, USA, 2000; Volume 167, pp. 35–120.
16. Guilherme, S.; Santos, G.M.A.; Pacheco, M. European eel (*Angilla anguilla*) genotoxic and pro-oxidant responses following short-term exposure to Roundup—A glyphosate-based herbicide. *Mutagenesis* **2010**, *25*, 523–530. [CrossRef]

17. Glozier, N.E.; Struger, J.; Cessna, A.J.; Gledhill, M.; Rondeau, M.; Ernst, W.R.; Sekela, M.A.; Cagampan, S.J.; Sverko, E.; Murphy, C.; et al. Occurrence of glyphosate and acidic herbicides in select urban rivers and streams in Canada. *Environ. Sci. Pollut. Res. Int.* **2012**, *19*, 821–834. [CrossRef]
18. Harayashiki, C.A.Y.; Varela, A.S., Jr.; de Souza Machado, A.A.; da Costa Cabrera, L.; Primel, E.G.; Bianchini, A.; Corcini, C.D. Toxic effects of the herbicide Roundup in the guppy *Poecilia vivipara* acclimated to fresh water. *Aquat. Toxicol.* **2013**, *142–143*, 176–184. [CrossRef]
19. Schrauzer, G.N. Lithium: Occurrence, Dietary Intakes, Nutritional Essentiality. *J. Am. Coll. Nutr.* **2002**, *21*, 14–21. [CrossRef]
20. Tanveer, M.; Hasanuzzama, M.; Wang, L. Lithium in environment and potential targets to reduce lithium toxicity in plants. *J. Plant Growth Regul.* **2019**, *38*, 1574–1586. [CrossRef]
21. Leal, V.M.; Ribeiro, J.S.; Coelho, E.L.D.; Freitas, M.B.J.G. Recycling of spent lithium-ion batteries as a sustainable solution to obtain raw materials for different applications. *J. Energy Chem.* **2023**, *79*, 118–134. [CrossRef]
22. Shahzad, B.; Ughal, M.N.; Tanveer, M.; Gupta, D.; Abbas, G. Is lithium biologically an important or toxic element to living organisms? An overview. *Environ. Sci. Pollut. Res. Int.* **2017**, *24*, 103–115. [CrossRef]
23. Brown, E.E.; Gerretsen, P.; Pollock, B.; Graff-Guerrero, A. Psychiatry benefits of lithium in water supplies may be due to protection from the neurotoxicity of lead exposure. *Med. Hypotheses* **2018**, *115*, 94–102. [CrossRef]
24. Kavanagh, L.; Keohane, J.; Cleary, J.; Cabellos, G.G.; Lloyd, A. Lithium in the natural waters of the South East of Ireland. *Int. J. Environ. Res. Public Health* **2017**, *14*, 561. [CrossRef] [PubMed]
25. Liu, D.; Gao, L.; Zhang, Z.; Tao, S.; Pang, Q.; Li, A.; Deng, H.; Yu, H. Lithium promotes the production of reactive oxygen species via GSK-3b/TSC2/TOR signaling in the gill of zebrafish (*Danio rerio*). *Chemosphere* **2018**, *195*, 854–863. [CrossRef]
26. Ewuzie, U.; Nnorom, I.C.; Eze, S.O. Lithium in drinking water sources in rural and urban communities in Southeastern Nigeria. *Chemosphere* **2020**, *245*, 125593. [CrossRef] [PubMed]
27. Sobolev, O.I.; Gutyj, B.V.; Darmohray, L.M.; Sobolieva, S.V.; Ivanina, V.V.; Kuzmenko, O.A.; Karkach, P.M.; Fesenko, V.F.; Bilkevych, V.V.; Mashkin, Y.O.; et al. Lithium in the natural environment and its migration in the trophic. *Ukr. J. Ecol.* **2019**, *9*, 195–203.
28. Barbosa, H.; Soares, A.M.V.M.; Pereira, E.; Freitas, R. Lithium: A review on concentrations and impacts in marine and coastal systems. *Sci. Total Environ.* **2023**, *857*, 159374. [CrossRef]
29. Muturi, E.J.; Donthu, R.K.; Fields, C.J.; Moise, I.K.; Kim, C.-H. Effects of pesticides on microbial communities in container aquatic habitats. *Sci. Rep.* **2017**, *7*, 44565. [CrossRef]
30. Gessner, M.O. Ergosterol as a measure of fungal biomass. In *Methods in Study Litter Decomposition: A Practical Guide*; Graça, M.A.S., Barlocher, F., Gessner, M.O., Eds.; Springer: Berlin/Heidelberg, Germany, 2005; pp. 189–195.
31. Feckler, A.; Bundschuh, M. Decoupled structure and function of leaf-associated microorganisms under anthropogenic pressure: Potential hurdles for environmental monitoring. *Freshw. Sci.* **2020**, *39*, 652–664. [CrossRef]
32. Fernandes, I.; Pascoal, C.; Cássio, F. Intraspecific traits change effects on ecosystem functioning under metal stress. *Oecologia* **2011**, *166*, 1019–1028. [CrossRef]
33. Duarte, S.; Cássio, F.; Pascoal, C.; Barlocher, F. Taxa-area relationship of aquatic fungi on deciduous leaves. *PLoS ONE* **2017**, *12*, e0181545. [CrossRef] [PubMed]
34. Obojska, A.; Lejczak, B.; Kubrak, M. Degradation of phosphonates by Streptomycetes isolates. *Appl. Microbiol. Biotechnol.* **1999**, *9*, 872e876. [CrossRef] [PubMed]
35. Ternan, N.G.; Mc Grath, J.W.; McMullan, G.; Quin, J.P. Review: Organophosphonates: Occurrence, synthesis and biodegradation by microorganisms. *World J. Microbiol. Biotechnol.* **1998**, *14*, 635–647. [CrossRef]
36. Bier, R.L.; Bernhardt, E.S.; Boot, C.M.; Graham, E.B.; Hall, E.K.; Lennon, J.T.; Nemergut, D.R.; Osborne, B.B.; Ruíz-González, C.; Schimel, J.P.; et al. Linking microbial community structure ad microbial processes: An empirical and conceptual overview. *FEMS Microbiol. Ecol.* **2015**, *91*, fiv113. [CrossRef]
37. Pascoal, C.; Cássio, F.; Marvanová, L. Anthropogenic stress may affect aquatic hyphomycetes diversity more than leaf decomposition in a low-order stream. *Arch. Hydrobiol.* **2005**, *162*, 481–496. [CrossRef]

Journal of *Xenobiotics*

MDPI

Article

Multisampling Strategies for Determining Contaminants of Emerging Concern (CECs) in the Marine Environment

Enrique J. Díaz-Montaña [1,*] and Sofía Domínguez-Gil [2]

[1] Department of Analytical Chemistry, Faculty of Pharmacy, University of Seville, 41012 Seville, Spain
[2] School of Chemical Science, Dulin City University, D09 YT18 Dublin, Ireland; sofiadominguezgil@gmail.com
* Correspondence: edmontana@us.es

Abstract: The determination of contaminants of emerging concern (CECs) in the marine environment is performed through many different sampling approaches. Therefore, the main objective of this study was to compare different existing sampling strategies: biofilm mesocosms, considering micro- and macrofouling; passive sampling; and grab marine water. The sampling of grab water was performed considering spatial and time-line variations. The spatial analysis of CECs showed that three sun agents and caffeine represent the biggest proportion of CECs in the Malaga Mediterranean coastal area, ranging from 0.391 to 0.495 ng/L. The time-line analysis did not show any upward or downward trend in CEC concentration. The mesocosm study comprised a separate evaluation of micro- and macrofouling that showed similar profiles, in which the sun agents presented the highest concentrations. While certain compounds were detected at comparable levels in both fouling types, such as clotrimazole around 0.001 ng/L, others exhibited significant differences in concentration, like caffeine. The passive sampling was also performed, obtaining similar results to those observed in the biofilm mesocosm macrofouling. Finally, all the obtained results from the different samplings were statistically compared, showing that passive sampling presented greater similarities with macrofouling and that there are significant differences between the sampling approach employed.

Keywords: sampling evaluation; grab water; passive sampling; biofilm mesocosm; contaminants of emerging concern; marine environment

1. Introduction

Coastal environments are characterized by the dynamic interface between land and sea and harbour rich biodiversity and essential ecosystem services vital for human well-being. However, these ecosystems face pressure from anthropogenic activities, including urbanization, industrialization, and pollution, which threaten ecological integrity and resilience [1]. Among all the pollutants reaching the coastal and marine environments, new chemicals known as contaminants of emerging concern (CECs) stand out as a threat to human life and wildlife. These potential risks to aquatic ecosystems and human health are due to their persistence, bioaccumulation, and adverse effects. These compounds originate from various sources including pharmaceuticals, personal care products, and chemistry industries, and have garnered increasing attention for their ubiquity and potential ecological impacts [2]. The presence of CECs in coastal waters emphasizes the need for comprehensive monitoring programs to assess their occurrence, fate, and ecological risks [3].

The monitoring of CECs strongly depends on the matrix to be analysed, i.e., water (surface water, ground water, sea, open sea, wastewater, etc.), soil, or air. The common practice to monitor CECs in water is grab sampling, which is bottling water from the location of interest, or passive sampling [4]. Passive sampling (PS) techniques have gained interest in recent years because they allow the quantification of compounds in water, offering more efficient storage and deployment without compromising sensitivity [5,6]. There are many commercially available receiving phases, and modifications of the existing ones are increasing their utility, enhancing the use of PS for water sampling purposes [4]. However, PS also has disadvantages, such as the need of calibration prior to its placement, limited information on the capture efficiency of many sorbent materials, and the inability to collect biological analytes [7]. Furthermore, biota has emerged as a new monitoring strategy for complex environments, like sea/coastal waters, due to the inconvenience presented by PS. A biofilm mesocosm is a combination of multiple organisms (biota) and microorganisms, and it offers advantages as a sample medium for CEC monitoring. A biofilm is a complex assemblage of microorganisms embedded within a matrix of extracellular polymeric substances, forming resilient structures known for their diverse ecological roles [8]. These structures, pervasive in aquatic environments, act as dynamic reservoirs of microbial diversity and metabolic activity, influencing nutrient cycling, pollutant degradation, and biofilm formation processes [9].

Therefore, the objective of this study was to compare different sampling approaches: grab water, biofilm mesocosms (micro- and macrofouling organisms), and passive sampling. Due to the mobilization of contaminants in transitional and coastal areas, the grab sample was taken considering two perspectives: (i) to conduct a thorough examination of multiple sites along the coast of Malaga and (ii) to evaluate pollutant levels over an extended duration of five weeks. By leveraging the biofilm mesocosm's capacity to retain and concentrate contaminants, this study aims to enhance the sensitivity and specificity of environmental monitoring efforts, providing data for assessing the occurrence of CECs and the impact for health and resilience of Malaga's coastal ecosystem.

2. Materials and Methods

2.1. Chemicals

The chemical compounds identified by high-performance liquid chromatography (HPLC; Agilent 1260 series Infinity binary pump (Santa Clara, CA, USA)) coupled to a triple quadrupole mass spectrometer (MS/MS; Agilent 6470 (Santa Clara, CA, USA)) were confirmed using analytical standards (MS degree) purchased from Merck (Darmstadt, Germany) (Table 1). The quantification was performed employing a calibration curve with eighteen points, from 0.01 to 100 ng/L, for each compound. All the analysis, as well as the calibration curves, were performed nine times (in nonuplicate).

Table 1. Description of the measured CECs: chemical group, functional category, empirical formula, CAS number, highest *m/z* fragment, MS transition (quantifier and qualifier), and internal standard (with the deuterated (d) position) employed for each compound.

Chemical Compound	General Group	Functional Category	Empirical Formula	CAS No.	Highest *m/z* Fragment	MS Transition (Quantifier and Qualifier)	Internal Standard
Caffeine	Pharmaceuticals	Stimulant	$C_8H_{10}N_4O_2$	558-08-2	195.0	195.0–138.0 195.0–110.0	Caffeine-d10
Carbamazepine		Anticonvulsant	$C_{15}H_{12}N_2O$	298-46-4	194.0	237.1–194.0 237.1–193.2	Carbamazepine-d10
Clarithromycin		Antibiotic	$C_{38}H_{69}NO_{13}$	81103-11-9	748.5	748.5–590.4 748.5–158.1	Clarithromycin-d3

Table 1. *Cont.*

Chemical Compound	General Group	Functional Category	Empirical Formula	CAS No.	Highest *m/z* Fragment	MS Transition (Quantifier and Qualifier)	Internal Standard
Erythromycin		Antibiotic	$C_{37}H_{67}NO_{13}$	114-07-8	734.5	734.5–576.4 734.5–158.1	Erythromycin-d3
Gabapentin		Anticonvulsant	$C_9H_{17}NO_2$	60142-96-3	172.1	172.1–154.1 172.1–137.0	Gabapentin-d4
O-Desvenlafaxine		Antidepressant	$C_{16}H_{25}NO_2$	93413-62-8	163.8	264.2–107.0 264.2–77.0	O-Desvenlafaxine-d6
Ofloxacin		Antibiotic	$C_{18}H_{20}FN_3O_4$	82419-36-1	362.2	362.2–318.3 362.2–261.3	Ofloxacin-d3
Sulfamethoxazole		Antibiotic	$C_{10}H_{11}N_3O_3S$	723-46-6	156.0	254.0–156.0 254.0–108.4	Sulfamethoxazole-d13
Trimethoprim		Antibiotic	$C_{14}H_{18}N_4O_3$	738-70-5	291.2	291.2–258.0 291.2–230.1	Trimethoprim-d9
Venlafaxine		Antidepressant	$C_{17}H_{27}NO_2$	99300-78-4	215.0	278.2–215.0 278.2–121.0	Venlafaxine-d6
Avobenzone	Sun agents	UV filter	$C_{20}H_{22}O_3$	70356-09-1	161.0	311.0–161.0 311.0–135.0	Avobenzone-d3
Octocrylene		UV filter	$C_{24}H_{27}NO_2$	6197-30-4	250.1	362.2–250.1 362.2–231.0	Octocrylene-d10
Oxybenzone		UV filter	$C_{14}H_{12}O_3$	131-57-7	151.2	229.2–151.2 229.2–128.0	Oxybenzone-d5
Azoxystrobin	Pesticides	Fungicide	$C_{22}H_{17}N_3O_5$	131860-33-8	372.1	404.1–372.1 404.1–344.1	Azoxystrobin-d4
Clotrimazole		Fungicide	$C_{22}H_{17}ClN_2$	23593-75-1	277.0	277.0–241.1 277.0–199.0	Clotrimazole-d10
Bifenthrin		Insecticide	$C_{23}H_{22}ClF_3O_2$	99267-18-2	181.1	442.1–181.1 440.1–198.2	Bifenthrin-d5
Fluconazole		Fungicide	$C_{13}H_{12}F_2N_6O$	86386-73-4	238.1	307.1–238.1 307.1–220.1	Fluconazole-d4
Miconazole		Fungicide	$C_{18}H_{14}C_{14}N_2O$	22916-47-8	159.0	415.0–159.0 415.0–69.0	Miconazole-d5
Penconazole		Fungicide	$C_{13}H_{15}Cl_2N_3$	66246-88-6	159.0	284.1–159.0 284.1–70.0	Penconazole-d7
Prochloraz		Fungicide	$C_{15}H_{16}Cl_3N_3O$	67749-09-5	376.0	376.0–308.0 376.0–266.0	Prochloraz(ethylene)-d4

2.2. Monitoring Sites

The spatial analysis of Malaga (Spain) coast was performed, as mentioned, in order to obtain information from the Mediterranean Sea about the pollutants ending up in it. Table 2 lists the selected sampling locations with the assigned code, the exact coordinates (longitude and latitude), and the main activity performed there. Figure 1 (included in Table 1) shows the sampling points, and the spatial analysis was conducted using Google Earth Pro software v. 7.3.6.10201.

Table 2. Sampling points for the spatial analysis of Malaga coastal area and their main activities.

Sampling Code	Longitude	Latitude	Main Activity of the Area
L1	36.673905 N	4.449118 W	Beach/Swimming
L2	36.682097 N	4.444226 W	Beach/Swimming
L3	36.690012 N	4.440707 W	Beach/Swimming
L4	36.696688 N	4.436844 W	Beach/Swimming
L5	36.702469 N	4.426030 W	Swimming/Navigation
L6	36.701368 N	4.418906 W	Harbour/Navigation
L7	36.708868 N	4.418048 W	Harbour/Navigation
L8	36.717607	4.416588	Harbour/Navigation

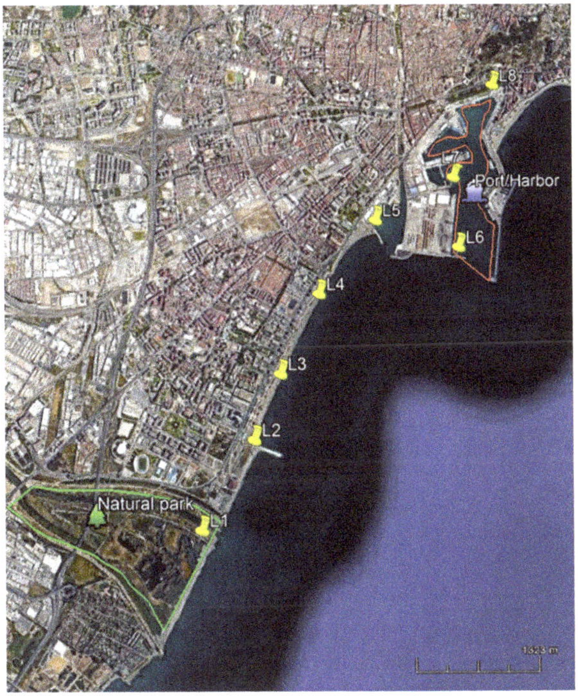

Figure 1. Sampling locations. Green line delimitates the green areas, whereas the red line delimitates the harbour/port area.

The concentration variability along different weeks (time-line analysis) was performed in L5 (36°70′24.7″ N 4°42′60.3″ W), at the end of Guadalmedina River (transitional water), where swimming and navigation activities are performed close to the Malaga Port. This site was chosen because it is a tidal, highly dynamic ecosystem, with anthropogenic pressure caused by shipping activities or tourism, as well as the influence of the surface water from the river.

2.3. Sampling

Three different types of sampling have been performed, i.e., grab water sampling, biofilm mesocosm sampling, and passive sampling (Figure S1). The grab water sampling has been performed considering time-line analysis and spatial analysis, to improve the understanding of the location of interest. Figure S1 schematizes the different sampling types and deployment and retrieval periods.

2.3.1. Grab Water Sampling (Spatial and Time-Line Analyses)

The time-line analysis samples were taken during five consecutive weeks starting in March 2025. However, the spatial analysis samples were taken in April 2025, coinciding with the last week of the time-line analysis.

Both spatial and time-line analysis sampling were performed using the "grab water approach", using glass bottles, which were previously rinsed with methanol and double-distilled water, taking 1 L of seawater each time. Water samples were taken at 1 m from the surface and 10 m from the beach side. Samples were stored at −18 °C until extraction.

2.3.2. Biofilm Mesocosm Sampling

The biofilm mesocosm cage holder was a modified lobster cage. It was soaked overnight (10 h) in DCN 90 surfactant detergent (Merck, Germany). Afterwards, it was washed thoroughly with ultrapure water, rinsed with methanol in an ultrasonic bath for 10 min, and, finally, air dried. Once the biofilm cage was cleaned, it was deployed in April 2024 at L5 using a rope at a depth of 1 m from the sea surface and was collected after twelve months, allowing the development of biofilm mesocosms, microorganisms, and macroorganisms.

2.3.3. Passive Sampling

The passive sampling deployment was also conducted in March 2025 and at the L5 location site at 10 m from the biofilm holder to avoid possible bias; thus, the comparison between the different sampling procedures could be made. It was submerged for one month, leaving 1 m between the sampler and the water surface. As the target compound were pharmaceuticals, sun agents, and pesticides, hydrophilic–lipophilic balance (HLB) disks (Affinisep, France) were used. Furthermore, the lobster cages, i.e., the chemcatchers (PTFE bodies; Affinisep, France), were soaked overnight (10 h) in DCN 90 surfactant detergent (Merck, Germany). Afterwards, they were washed thoroughly in ultrapure water, rinsed with methanol in an ultrasonic bath for 10 min, and, finally, air dried. The HLB disks were conditioned as follows: firstly, soaking them overnight (10 h) in methanol; secondly, washing with 50 mL of methanol under gentle vacuum; thirdly, washing with 100 mL of distilled water under gentle vacuum; and finally, storing them in water until the deployment. Once the disks were prepared, they were covered by a polyethersulfone (PES) membrane (Affinisep, France), which was previously rinsed with methanol and water for 30 min each and stored in water until deployment, and then placed in a clean and dry chemcatcher (Affinisep, Le Houlme, France).

2.4. Sample Treatment

Each sample (biofilm mesocosm, PS, and grab water) underwent different initial preparatory procedures for the extraction of the target analytes. The microorganism biofilms, as well as the macroorganisms, were removed from the glass slides using a cell scraper to transfer them into large glass jars and were separated out based on the type of organism collected. Firstly, biofouling organisms were divided into two categories for collection: microfouling organisms (i.e., microalgae and slime) and macrofouling organisms (i.e., barnacles, tubeworms, red and green algae). Secondly, both types of fouling were dried in an oven (Moder Orbital Incubator S1500 Stuart, VWR, Avantor, Barcelona, Spain) at 30 °C until the weight remains constant. A 2 g quantity of (micro or macro) mesocosm was weighed in a glass vial, and 50 mL of mass spectrum (MS)-grade methanol (Honeywell, NC, USA) was added to each sample. Afterwards, the samples were sonicated for 20 min, filtered through a 0.45 μm nylon filter disk (Merck, Germany) into a falcon tube, and centrifuged (Model 5804, Eppendorf, Hamburg, Germany) for 20 min at 4000 rpm. Finally, the supernatant was transferred to a 100 mL brown bottles and spiked with 25 μL of 100 ng/L internal standard mix.

In the case of PS, the sampler surfaces were wiped with paper tissues and rinsed with seawater. The wrapped devices were transported in aluminium foil and placed in cool boxes. Then, the device (chemcatcher) was dissembled, and the PES membranes were discarded. The HLB disks were dried at room temperature on solvent rinsed foil. If the HLB disks need to be stored, once dried, they were kept at −4 °C. Before the extraction, the disks were left at room temperature for 2 h. The extraction of the HLB disks was carried out by placing them in a 100 mL clean beaker with 40 mL of LC-MS/MS-grade methanol and shaking them for 1 h. Then, the membranes were placed in a filtration system, and the methanol employed before was passed through them. The extract was collected in a 50 mL falcon tube and evaporated with nitrogen (N2) to dryness. Finally, it was reconstituted in 1 mL of methanol/water (20:80) and spiked with 25 μL of 100 ng/L internal standard mix.

The extracts from the mesocosm and the water samples were subjected to the same preconcentration step, employing OASIS HLB 500 mg solid-phase extraction (SPE) cartridges (Affisinep, Le Houlme, France). The cartridges were conditioned with 4 mL of methanol, followed by 4 mL of ultrapure water (adjusted with glacial acetic acid to pH 3). The samples (mesocosm supernatant and water) were loaded into the cartridge at a flow rate of 2 mL/min. The cartridges were rinsed with 4 mL of ultrapure water and were dried with vacuum for 30 min. Analytes were eluted with 5 mL of acetonitrile (ACN) (Honeywell, NC, USA) into amber vials (22 mL). Eluates were placed under a N_2 current until dryness. Samples were then reconstituted in 1 mL methanol/water (20:80), vortexed, and left to stand in 22 mL vials for 5 min. The reconstituted sample was syringe-filtered with a 0.2 μm nylon filter (Merck, Germany) and transferred to an LC-MS/MS 2 mL vial for storage at 4 °C until analysis.

2.5. Instrumental Analysis (LC-MS/MS)

All extracts were analysed by high-performance liquid chromatography (HPLC; Agilent 1260 series Infinity binary pump) coupled to a triple quadrupole mass spectrometer (MS/MS; Agilent 6470), equipped with an electrospray ionization source. A 10 μL volume of the extracts was injected into an Agilent Zorbax Eclipse Plus C18 column (3.0 × 50 mm; 1.8 mm), with a guard column of the same materials. Mobile phases for pharmaceutical (Table 1) analysis were as follows: (A): 0.1% formic acid/5 mM ammonium formate (pH 3.2) in water and (B): 0.1% formic acid in acetonitrile; for pesticides and sun agents (Table 2), the mobile phases employed were as follows: (A) 0.1% formic acid/5 mM ammonium formate (pH 3.2) in water and (B): 0.1% of formic acid in methanol; in both cases, the flow rate (0.4 mL/min) and gradient were the same. The gradient elution was as follows: 0–1.5 min, 90–45% A; 1.5–7.0 min, 45–10% A; 7.0–8.5 min, 10–0% A; 8.5–10 min, 0% A; 10–11 min, returns to initial conditions; 11–12 min, equilibration of the column. Column temperature was set at 35 °C. Compound-dependent mass spectrometric parameters (fragmentor, collision energy) as well as compound-selected reaction monitoring transitions were optimized by the direct infusion of individual standard solution of each analyte at 10 ng/L using the Agilent Compound Optimizer software v. 12.1. The ionization mode employed for pharmaceuticals (Table 1) was positive (ESI+) and for pesticides and sun agents (Table 1) negative (ESI-). Source-dependent parameters were determined by flow injection analysis: gas temperature: 230 °C; gas flow: 4 L/min; nebulizer: 15 psi; source temperature, 375 °C; ion spray voltage, 2500 V; and sheath gas flow: 12 L/min. Instrument control data acquisition and data analysis were carried out using the Agilent software (MassHunter Workstation).

The analysis of each sample was performed in nonuplicate. The chemical identification performed by liquid chromatography coupled with a mass spectrum triple quadrupole (LC-MS/MS) was confirmed by the analysis of analytical standards (MS-grade) (Table 1) (Merck, 185 Darmstadt, Germany) and the monitored transitions.

2.6. Quality Assurance and Quality Control

All materials were washed with ultrapure water and rinsed with methanol three times; additionally, the glass materials were heated until dryness and rinsed with methanol again.

Initially, three water blanks were analysed in nonuplicate to assure the absence of analytes of interest. Additionally, for each batch of ten samples, two water blanks were analysed in nonuplicate. Furthermore, all standards were randomly measured in a concentration range between 0.01 and 100 ng/L during the analytical run assuring the correct response of the instrument. All the blanks showed concentration below the method detection limit (MLD) for the different compounds.

Additionally, the analytical method was validated to ensure reliability and performance. The key validation parameters evaluated were linearity, matrix effects, selectivity, sensitivity, accuracy, and precision. As mentioned, the calibration curves were prepared between 0.01 and 100 ng/L. The linearity was considered acceptable when the correlation coefficient (r^2) was greater than 0.99. The sensitivity was determined by the limit of detection (LOD) and limit of quantification (LOQ). The blank signal was multiplied by 3 to determine the limit of detection (LOD = $S_{blank} \times 3$), and by 10 to determine the limit of quantification (LOQ = $S_{blank} \times 10$). Accuracy and precision were assessed considering two concentrations (1 and 10 ng/L) in nonuplicate, being evaluated as recovery experiment-fortifying blanks. Acceptable values of recovery are considered between 70 and 120%, whereas precision (expressed as relative standard deviation (RSD)) has to remain below 20%. Matrix effects (MEs), expressed as percentage, were calculated as the ratio of the analytical signal obtained in a blank matrix to the signal obtained in the solvent. Lastly, selectivity was performed to ensure no interference at the target retention times, by comparing chromatograms of the blank and spiked samples.

2.7. Statistical Treatment

The data obtained from the Agilent software were exported to Excel 365 (Microsoft Corp., Redmond, WA, USA) to obtain graphs and basic statistics (mean, standard deviation, minimum and maximum concentrations). Once the basic statistics were performed and the homogeneity of variance was confirmed, the obtained information was used to perform an ANOVA for selecting the more important variables (compounds). Once the variables were selected, a principal component analysis (PCA) was performed. This multivariate analysis was performed using the Statistica software V8 (Statsoft; Tulsa, OK, USA).

3. Results and Discussion

Three different sampling approaches were considered (grab, mesocosm biofilms, and passive sampling). The grab water sampling was performed considering spatial and time-line distribution. Eight sampling locations along the Malaga Mediterranean coast, including beaches and harbours, were analysed to determine the occurrence of contaminants in the coastal marine area. Time-line analysis results were obtained to study the CEC variation over five consecutive weeks in the context of a busy dynamic coastal area. The twelve months deploying marine mesocosms provides a unique method to address potential differences between micro- and macrofouling uptake. Passive sampling was also employed to determine the occurrence of CECs in the Mediterranean coast. Lastly, the results obtained for the different sampling approaches were statistically compared to detect significant differences between them.

3.1. Method Validation

Before the sample analysis, the analytical method employed was validated (Table 3). The linearity was excellent for all the determined compounds (Table 1) with correlation coefficients ranging from 0.9912 (erythromycin) to 0.9994 (caffeine). The limit of detection (LOD) range was between 0.1 and 1.1×10^{-3} ng/L, whereas the limit of quantification

(LOQ) range was between 0.2 and 3.3 × 10^{-3} ng/L. As mentioned, accuracy and precision were determined by recovery experiments at the fortification levels. The recovery rates were between 82.7% (carbamazepine) and 113.4% (prochloraz) at the lower spiking levels, while they ranged between 85.1% (gabapentin) and 111.9% (trimethoprim) at the higher spiking levels. The relative standard deviations (RSDs) were between 0.9% (o-desvenlafaxine) and 8.6% (avobenzone) at the lower spiking level and between 1.1% (ofloxacin) and 8.5% (oxybenzene) at the higher spiking level. Therefore, the acceptable criteria were satisfied for both accuracy (mean recovery within 70–120%) and precision (RSD below 20%). All the determined contaminants (Table 1) showed low-to-moderate matrix effects (MEs) ranging between −43.5% (carbamazepine) and 51.2% (bifenthrin).

Table 3. Validation of LC-MS/MS method for analysis of CECs.

Chemical Compound	r^2	LOD ($\times10^{-3}$ ng/L)	LOQ ($\times10^{-3}$ ng/L)	R * (%)	RSD * (%)	R ** (%)	RSD ** (%)	ME (%)
Caffeine	0.9994	0.2	0.7	82.7	2.1	94.1	6.0	−35.5
Carbamazepine	0.9924	1.0	3.2	85.6	1.7	88.9	7.8	−43.5
Clarithromycin	0.9956	0.4	1.3	82.9	2.8	104.5	4.9	42.8
Erythromycin	0.9912	0.8	2.9	94.1	4.6	91.6	3.6	−39.3
Gabapentin	0.9969	1.1	3.3	109.8	5.4	85.1	8.5	30.1
O-Desvenlafaxine	0.9967	0.8	2.9	96.2	0.9	86.1	3.1	10.8
Ofloxacin	0.9914	0.1	0.6	91.4	7.9	97.9	1.1	22.4
Sulfamethoxazole	0.9966	0.6	2.0	112.5	5.1	103.7	6.3	−15.1
Trimethoprim	0.9923	0.2	0.7	84.9	3.2	111.9	3.9	−44.5
Venlafaxine	0.9986	0.3	1.0	110.8	2.5	88.5	6.4	−32.2
Avobenzone	0.9943	0.9	3.0	96.4	8.6	100.6	3.5	26.5
Octocrylene	0.9988	0.8	2.9	108.9	6.4	108.4	8.1	−18.7
Oxybenzone	0.9916	0.9	2.9	89.3	1.3	95.2	8.5	45.4
Azoxystrobin	0.9924	0.1	0.4	106.7	7.8	91.3	1.4	−30.5
Clotrimazole	0.9953	1.0	3.1	97.7	3.1	94.7	5.2	−31.3
Bifenthrin	0.9927	0.1	0.3	104.1	5.6	109.5	4.3	51.2
Fluconazole	0.9931	0.4	1.1	97.5	4.7	93.3	7.2	38.1
Miconazole	0.9945	0.6	2.0	111.6	5.0	107.8	4.9	26.1
Penconazole	0.9978	0.3	1.0	98.2	1.9	90.6	2.8	−16.1
Prochloraz	0.9980	0.1	0.2	113.4	2.8	103.4	1.6	27.5

Note: r^2—correlation coefficient; LOD—limit of detection; LOQ—limit of quantification; R—recovery; RSD—relative standard deviation; *—spiking level 1 ng/L; **—spiking level 10 ng/L; ME—matrix effect.

3.2. Spatial Occurrence of Contaminants of Emerging Concern

The twenty pollutants under study were determined in the eight sampling seawater sites, including L1 that is a protected natural park. The compounds that presented the highest concentration were the sun agents, followed by the pharmaceuticals and lastly the pesticides. The three sun agents quantified were avobenzone, octocrylene, and oxybenzone, with octocrylene being the most concentrated one in all the sampling sites. This compound is highly used due to its capacity of filtering UV radiation (UVF) and its capability of preventing erythema and, subsequently, the risk of melanoma due to sun exposure; but it presents an environmental problem due to its lipophilic character (log Kow = 6.88) that makes it resistant to degradation and stable in marine ecosystem [10]. The sun agents were more distributed in beach areas (L1–L4) than in harbours (L6–L8), which is logical given that harbours, unlike beaches, are primarily used for boating rather than swimming activities. The locations that presented the highest level of sun agents were the beach areas (Figure 2), especially L2 and L3 beaches presenting 0.495 ± 0.008 and 0.419 ± 0.009 ng/L of octocrylene, respectively. These results are in concordance with those of Fenni et al. [11] that concluded the necessity of UVF environmental monitoring due to the high usage and the ecotoxicity shown. However, as shown in Figure 2, the distribution was not homogeneous, having significant differences ($p < 0.05$) between the different sampling sites. As mentioned, the high concentrations of avobenzone and octocrylene in the protected natural park (L1) are astonishing. This demonstrates that the remobilization of these compounds into the marine environment threatens wildlife.

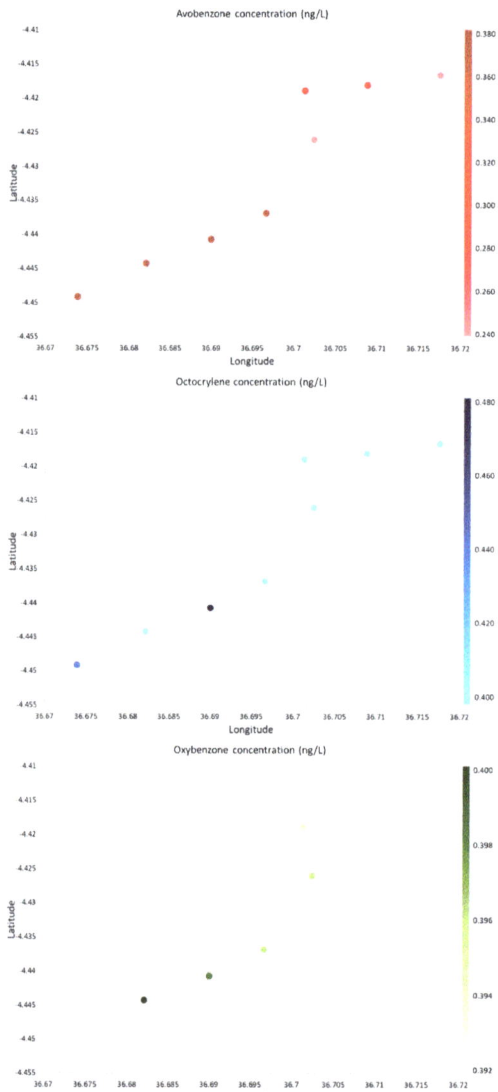

Figure 2. Concentrations (ng/L) of the sun agents in the different sampling points.

As the concentrations of sun agents were higher in the beach areas, the places that presented the highest concentrations of pharmaceuticals were those near the port (L6, L7, and L8), possibly due to the large number of cruises that directly dump their wastewater into the sea. As it can be seen in Table 4, the pharmaceuticals that showed the highest concentrations were caffeine and gabapentin. Additionally, the site L5 also presented a high level of caffeine, reaching a concentration of 0.391 ± 0.007 ng/L, and of gabapentin, i.e., 0.314 ± 0.006 ng/L, possibly due to the contamination coming from and near the Guadalmedina River, where sewage pipe systems directly dump waste into the sea. However, these concentrations are still far from the high caffeine concentration found on the Black Sea coast (13,575 ng/L) [12]. This difference is due to the sampling site, the sampling period (not tourist season—April—or tourist season—August), and the proximity to more or less densely populated areas. In contrast, the compounds that presented the lowest concentration were erythromycin and trimethoprim and reached 0.028 ± 0.004 and 0.081 ± 0.006 ng/L, respectively. These two drugs are antibiotics used for treating the infection of respiratory and urinary tracts, and their low concentration can be due to their fast deterioration rate or their relatively low consumption, compared to other antibiotics such as amoxicillin, in Spain. Additionally, the relative low concentration of these compounds can be due to the degradation of these chemicals in marine water as well as the inherent difficulty of marine water matrix, in concordance with Boti et al. [13].

Table 4. Pharmaceutical concentration ± standard deviation (SD) (ng/L) in the different sampling points.

Locations	L1	L2	L3	L4	L5	L6	L7	L8
Caffeine	0.313 ± 0.010	0.367 ± 0.012	0.376 ± 0.027	0.363 ± 0.016	0.391 ± 0.007	0.322 ± 0.014	0.358 ± 0.025	0.361 ± 0.019
Carbamazepine	0.185 ± 0.009	0.339 ± 0.031	0.191 ± 0.012	0.261 ± 0.010	0.134 ± 0.009	0.189 ± 0.014	0.246 ± 0.022	0.179 ± 0.008
Clarythromycin	0.309 ± 0.007	0.271 ± 0.022	0.270 ± 0.014	0.188 ± 0.016	0.237 ± 0.015	0.270 ± 0.020	0.342 ± 0.028	0.350 ± 0.017
Erythromycin	0.044 ± 0.008	0.050 ± 0.006	0.049 ± 0.007	0.050 ± 0.005	0.028 ± 0.004	0.052 ± 0.006	0.055 ± 0.002	0.032 ± 0.004
Gabapentin	0.298 ± 0.021	0.265 ± 0.014	0.283 ± 0.011	0.251 ± 0.015	0.314 ± 0.006	0.271 ± 0.012	0.292 ± 0.024	0.277 ± 0.026
O-Desmovenlafaxine	0.376 ± 0.019	0.198 ± 0.015	0.374 ± 0.016	0.364 ± 0.028	0.262 ± 0.018	0.334 ± 0.023	0.267 ± 0.029	0.235 ± 0.021
Ofloxacin	0.156 ± 0.011	0.201 ± 0.005	0.286 ± 0.017	0.131 ± 0.021	0.348 ± 0.022	0.314 ± 0.016	0.285 ± 0.009	0.222 ± 0.030
Sulfamethoxazole	0.246 ± 0.013	0.361 ± 0.026	0.293 ± 0.012	0.360 ± 0.034	0.243 ± 0.017	0.211 ± 0.019	0.362 ± 0.020	0.284 ± 0.016
Trimethoprim	0.073 ± 0.008	0.099 ± 0.006	0.096 ± 0.009	0.084 ± 0.004	0.081 ± 0.006	0.112 ± 0.005	0.062 ± 0.04	0.120 ± 0.009
Venlafaxine	0.128 ± 0.004	0.200 ± 0.032	0.321 ± 0.023	0.365 ± 0.026	0.340 ± 0.029	0.142 ± 0.019	0.270 ± 0.030	0.174 ± 0.011

The detection of the seven pesticides (Table 5) across the different sampling sites highlights a clear One Health concern. This framework recognizes that human health, animal health, and environmental integrity are interconnected. Some of these compounds, such as clotrimazole, fluconazole, and miconazole, are antifungals used in human medicine, while others, like bifenthrin or penconazole, are applied in agriculture or for pest control in grains and vegetables. Their simultaneous presence in coastal waters suggests multiple sources of contamination—both medical and agricultural—which converge in shared environments. The detection of bifenthrin at the highest concentrations, especially in L5 and consistently across all sampled beaches, together with the probable remobilization of contaminants via the Guadalmedina River as it passes through farming areas, indicates a transport pathway that integrates urban, rural, and aquatic ecosystems. Supporting this observation, studies on urban coastal and wetland sediments have shown that bifenthrin is frequently the most abundant pesticide, with concentrations capable of exceeding lethal thresholds for sensitive benthic organisms [14]. The environmental persistence of these substances, along with their endocrine-disrupting potential and capacity to act through diverse biological mechanisms, underscores the need for integrated management strategies that protect ecosystem health, safeguard wildlife, and prevent potential risks to human populations.

Therefore, these results underscore the necessity for a control of port areas and the direct dumping of waste, as well as the monitoring of surface water crossing highly populated and polluted areas.

Table 5. Pesticide concentration ± SD (ng/L) in the different sampling points.

Locations	Azoxystrobin	Bifenthrin	Clotrimazole	Fluconazole	Miconazole	Penconazole	Prochloraz
L1	<LOQ	0.008 ± 0.001	<LOQ	<LOQ	<LOQ	<LOQ	0.002 ± 0.001
L2	<LOQ	0.023 ± 0.004	0.003 ± 0.001	<LOQ	<LOQ	<LOQ	0.002 ± 0.001
L3	<LOQ	0.007 ± 0.001	<LOQ	<LOQ	<LOQ	<LOQ	0.003 ± 0.001
L4	<LOQ	0.008 ± 0.001	0.003 ± 0.001	<LOQ	0.002 ± 0.001	0.003 ± 0.001	0.003 ± 0.001
L5	<LOQ	0.027 ± 0.003	<LOQ	0.002 ± 0.001	0.002 ± 0.001	<LOQ	0.004 ± 0.001
L6	<LOQ	0.010 ± 0.001	<LOQ	0.002 ± 0.001	0.004 ± 0.001	<LOQ	0.002 ± 0.001
L7	<LOQ	0.015 ± 0.002	<LOQ	<LOQ	<LOQ	<LOQ	0.003 ± 0.001
L8	<LOQ	0.014 ± 0.001	<LOQ	0.003 ± 0.001	<LOQ	0.004 ± 0.001	0.002 ± 0.001

Note: <LOQ means below the limit of quantification (LOQ); thus, those compounds were not quantified.

3.3. Time-Line Analysis of CECs

The five-week analysis was performed at the L5 sampling site. As seen in the spatial analysis, it is a significant point due to its proximity to the Guadalmedina River, which has remobilized contaminants from land sources, and port and swimming areas, making this point a highly dynamic environmental area crucial for CEC monitoring.

The sun agents were also compounds with the highest concentrations, but in this case, the compound that presented the highest concentration was avobenzone (Figure 3). This might be due to the difference in the sampling, while the previous analysis considered different locations in a short time, and this sampling was performed in five different weeks.

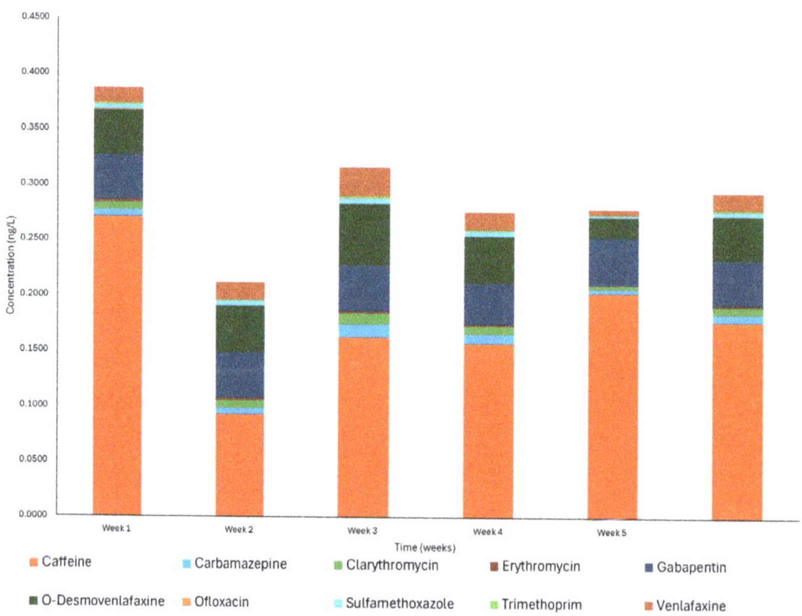

Figure 3. Predominance of individual and total pharmaceuticals (ng/L) in L5 sampling site during five consecutive weeks.

The pharmaceuticals showed the same trend as in the spatial analysis (Table S1), with caffeine the highest concentration, with a mean concentration of 0.178 ± 0.021 ng/L for the five weeks. The second most concentrated pharmaceutical was gabapentin (in the spatial analysis) and O-desvenlafaxine. These results highlight the limitation of grab sampling, not considering time variations. The ones with the lowest concentrations were erythromycin and trimethoprim, with 0.002 ng/L for both. Figure 3 shows a common behaviour among all the pharmaceutical compounds. Additionally, avobenzone may originate from water treatment processes, as studies have indicated that the use of showers, such as those found on beaches, can serve as a source of these contaminants [15].

Figure 3 also shows the cumulative sum across the various sampling weeks. As previously mentioned, the concentrations of these compounds ranged from nondetectable levels to 0.5 ng/L. No discernible trend was observed throughout the sampling period, which is a notable finding. This lack of trend suggests a consistent presence of marine pollutants of diverse origins contaminating the marine environment.

In the case of pesticides (Table S1), bifenthrin was detected at concentrations ranging from 0.030 $\pm$ 0.004 to 0.079 $\pm$ 0.007 ng/L, consistent with the spatial analysis for the same area. Sampling in different weeks enabled the detection of short-time-line variations and the remobilization within the marine environment. This was the case for pesticides, with four new compounds detected, i.e., diflufenican, famoxadone, mecoprop, and permethrin. As mentioned, the high diversity and abundance of pesticides near Malaga can be a consequence of the large number of green areas and of the river that starts in a mountainous/forest area, and in its 30 km length, it goes through several agricultural and farming areas, contributing to the presence of pesticides in Mediterranean Sea. In contrast, other authors have quantified up to thirty-seven pesticides in the Mediterranean French Coast [16]. Postigo et al. [17], in concordance with these results, concluded that the urban and industrial activities (like farming) are responsible for the occurrence of pesticides in the marine environment. On the other hand, the partial elimination of these pollutants in the wastewater treatment plants also contributed to the occurrence of these pollutants in the coastal and marine waters [18].

Additionally, a risk assessment was performed to evaluate the risk of the contaminants reaching the Malaga Mediterranean coastal area. The risk assessment (Table 6) was calculated by the risk quotient (RQ), which is the ratio of the measured environmental concentrations (MECs) between the predicted no-effect concentrations (PNECs): RQ = MEC/PNEC. Four risk levels were considered, i.e., RQ < 0.01—no risk (N.R.); 0.01 < RQ < 0.1—low risk (L.W.); 0.1 < RQ < 1—moderate risk (M.R.); and RQ > 1—high risk (H.R.). Table 4 shows the risk assessment of the compounds that have published and validated PNECs [19]; thus, not all the compounds are plotted. Four compounds showed low risk, i.e., caffeine, o-desvenlafaxine, venlafaxine, and azoxystrobin, even though caffeine was the most concentrated compound, but due to its high PNEC, it did not show high risk. However, the high concentration of caffeine it is a concern due to its relationship with oxidative stress, lipid peroxidation, and neurotoxicity [20,21]. Azoxystrobin was shown to be the most influent one depending on the measured weeks, presenting low risk in three of the five weeks. This needs to be considered when monitoring or with a longer time-line analysis, as certain compounds might show risk in certain periods, such as sun agents in the summer season. On the other hand, avobenzone and bifenthrin showed high risk due to their high concentrations and low PNECs (3.2 and 9.5 μg/L, respectively). The risk associated with avobenzone includes reproductive toxicity, developmental toxicity, neurotoxicity, and endocrine disrupting. Additionally, it has also been related to the development of nervous and retinal systems in zebrafish [22–25]. However, bifenthrin is associated with inducing hyperactivity and neurodevelopment issues (due to alterations in calcium oscillations) as toxic effects [26]. Therefore, these two compounds should be deeply studied to search for the sources of their production and considered for monitoring by the regulation institutions.

Table 6. Risk assessment of CECs during five consecutive weeks.

Compound	Week 1	Week 2	Week 3	Week 4	Week 5
Caffeine	L.R.	L.R.	L.R.	L.R.	L.R.
Carbamazepine	N.R.	N.R.	N.R.	N.R.	N.R.
Erythromycin	N.R.	N.R.	N.R.	N.R.	N.R.
Gabapentin	N.R.	N.R.	N.R.	N.R.	N.R.
O-desvenlafaxine	L.R.	L.R.	L.R.	L.R.	L.R.
Sulfamethoxazole	N.R.	N.R.	N.R.	N.R.	N.R.
Trimethoprim	N.R.	N.R.	N.R.	N.R.	N.R.
Venlafaxine	L.R.	L.R.	L.R.	L.R.	N.R.
Avobenzone	H.R.	H.R.	H.R.	H.R.	H.R.
Octocrylene	N.R.	N.R.	N.R.	N.R.	N.R.
Oxybenzone	N.R.	N.R.	N.R.	N.R.	N.R.
Azoxystrobin	N.R.	L.R.	L.R.	L.R.	N.R.
Bifenthrin	H.R.	H.R.	H.R.	H.R.	H.R.
Clotrimazole	N.R.	N.R.	N.R.	N.R.	N.R.
Fluconazole	N.R.	N.R.	N.R.	N.R.	N.R.
Miconazole	N.R.	N.R.	N.R.	N.R.	N.R.
Penconazole	N.R.	N.R.	N.R.	N.R.	N.R.
Prochloraz	N.R.	N.R.	N.R.	N.R.	N.R.

Note: four risk levels were considered: RQ < 0.01—no risk (N.R.; green color); 0.01 < RQ < 0.1—low risk (L.W.; orange color); 0.1 < RQ < 1—moderate risk (M.R.); RQ > 1—high risk (H.R.; red color).

3.4. *Biofilm Mesocosms as a New Contaminant Monitoring Tool*

Biofilms are usually employed to study the microbial biofilms generated on marine plastics (the platisphere); however, macroorganisms can also develop on the surface of these devices or plastics [1]. Additionally, biofilms have been employed to remove dissolved organic compounds and the associated toxicity from produced water [9]. In this study, the mesocosm biofilms were studied considering the micro- and macrofouling uptake of thirteen target contaminants (seven were below the limit of detection (LOD)). Microfouling is the initial step in the growth of biofouling on hard substrata submerged in marine waters. Thus, for this fouling, bacteria and microalgae were analysed. However, macrofouling is the fouling caused by large organisms, such as macroalgae, oysters, mussels, clams, and barnacles. The thirteen CECs quantified included two sun agents (avobenzone and octocrylene), six pharmaceutical compounds (caffeine, carbamazepine, clarithromycin, gabapentin, O-desvenlafaxine, and trimethoprim), and five pesticides (azoxystrobin, bifenthrin, clotrimazole, penconazole, and prochloraz). Thus, not all the compounds detected in the spatial and time-line analyses were able to be detected in the mesocosm.

In general, micro- and macrofouling showed the same profile (Figure 4), with sun agents being the compounds that presented the highest concentrations. Avobenzone, in the time-line analysis, was the most concentrated compound, with a significantly high concentration, having an average concentration for both fouling of 1.030 $\pm$ 0.030 ng/L. This highlights the differences that can be observed employing one type of sampling as well as the bioaccumulation of certain chemicals by organisms. Avobenzone and other UV filters, as mentioned, show toxicity and potential bioaccumulation in aquatic environments. Avobenzone has been quantified in significant concentrations in different aquatic environments due to the discharge from wastewater effluents, landfills, or recreational activities [27,28]. In contrast, octocrylene did present significant differences ($p < 0.05$) between both fouling ($p < 0.05$). For the mesocosm microfouling, the concentration of octocrylene was 0.381 ± 0.020 ng/L, whereas for macrofouling (mussels, barnacles, tunicates, and macroalgae), it was 0.150 ± 0.021 ng/L. These results are in agreement with the spatial and time-line analyses and have an advantage because they allow the quantification of the sun agents and give an output after a year of exposure ("year average"), whereas by other methodologies, octocrylene's concentration varies depending on the day as observed for O-desvenlafaxine in the time-line analysis. The aquatic toxicity testing is complicated to perform because the octocrylene effect threshold is often above the water solubility limit [29]. In contrast to Pawloski et al. [30], certain bioaccumulation has been detected in the macroorganisms, but at it was mentioned in the study, there might be significant differences between organism bioaccumulation and the concentration quantified in water, possibly due to the season sampling period. However, biofilms consider one-year exposure, and water samples only consider the occurrence at the moment of sampling.

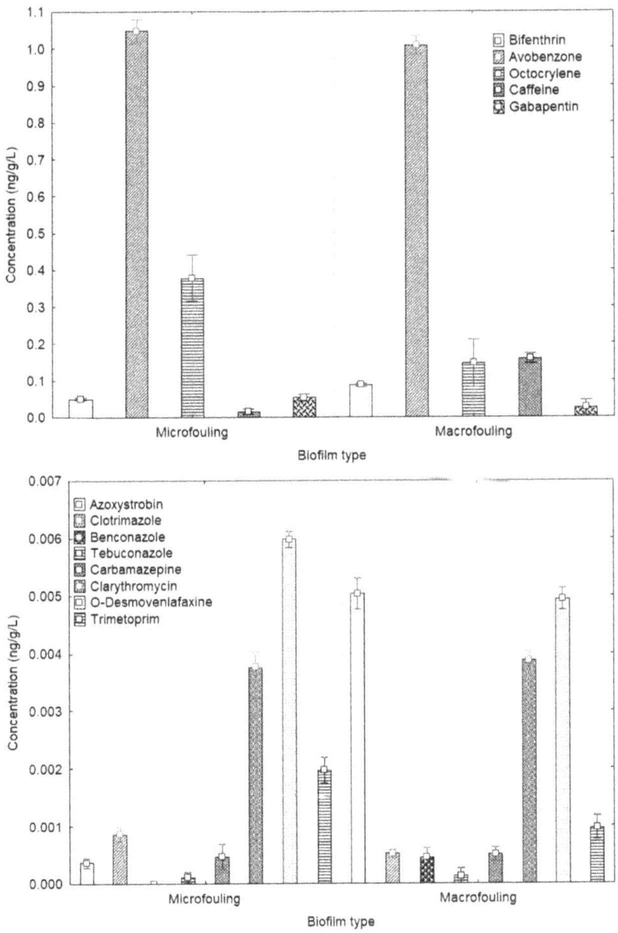

Figure 4. Box plot of high (**up**)- and low (**down**)-concentration compounds (ng/L) in both fouling of the mesocosm (separated by a red dotted line). Note: mean box: 25–75%.

The pharmaceutical compounds exhibited consistent concentration levels across both fouling, except for caffeine. Caffeine showed significant ($p < 0.05$) disparity between microfouling and macrofouling. It was determined at 0.021 ± 0.010 ng/L in microfouling, whereas in macrofouling, the concentration surged tenfold to 0.164 ± 0.042 ng/L. This disparity aligns with the known bioaccumulative nature of caffeine in marine organisms, as reviewed in coastal systems where caffeine has been found in tissues of microalgae, bivalves, and fishes, producing different toxic effects [21]. As for the spatial and timeline analyses, the second most concentrated compound was gabapentin, confirming the previous data. In the case of erythromycin and trimethoprim, the first one was not detected, highlighting the influence on the sampling type and period, whereas trimethoprim was the least concentrated pharmaceutical compound. Trimethoprim was the least concentrated pharmaceutical, yet environmental risk assessments in European waters highlight its widespread presence, arguing for integrative sampling approaches that better reflect long-term exposure [31]. These results shed light on the importance of having a sampling method that could provide information for a long period, not only of the moment of sampling.

Conversely, the detection of pesticides yielded a limited number, with only five compounds identified. The pesticides, in the other sampling approaches, were the compounds that presented the lowest concentration. Among them, bifenthrin was the predominant one, exhibiting a concentration level that was a striking 100 times higher than that of the other pesticides detected, with an average between fouling of 0.074 ± 0.022 ng/L. The remarkably high concentration of bifenthrin in comparison to other pesticides can be due to its remobilization by the Guadalmedina River, sewage pipes, and it relation with residential and industrial areas [32].

When all the compounds are considered collectively, some of them, such as clarithromycin or prochloraz, display similar concentrations in both micro- and macrofouling. Conversely, compounds like caffeine, octocrylene, and azoxystrobin exhibit variations between fouling types. Figure 5 facilitates a comparative analysis of the detected compounds. Therefore, while the mesocosm proves useful as monitoring tool, the scale needs to be considered, whether micro or macro.

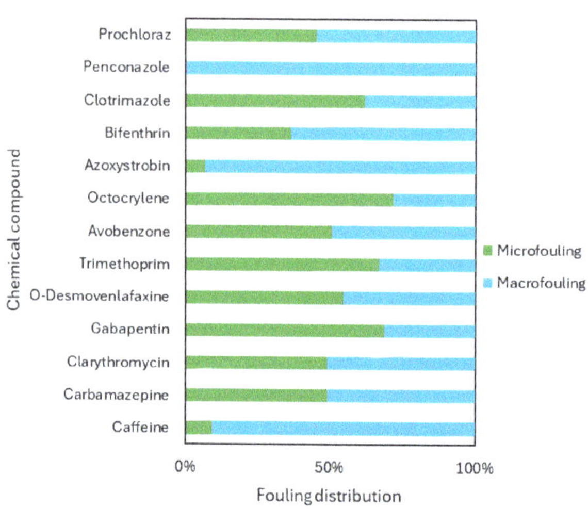

Figure 5. Profile of the relative distribution of individual CECs in the mesocosm macro- (blue) and microfouling (green).

3.5. Passive Sampling and Sampling Comparison

Finally, passive sampling was performed to study the occurrence of CECs in the Malaga Mediterranean coastal area using HLB disks, as this is a more established and widely studied sampling method [5,6]. Passive sampling showed similar concentrations (Table S2) for certain compounds in relation to the biofilm mesocosm. The two sun agents presented the highest concentrations: 1.031 ± 0.111 ng/L for avobenzone and 0.27 ± 0.03 ng/L for octocrylene. These compounds were followed by caffeine, 0.090 ± 0.019 ng/L, other pharmaceuticals, and the pesticides, with the lowest concentration.

Since only thirteen compounds were detected in both the mesocosm and passive samplers (PS), these were selected for the statistical analysis. For the time-line study, the mean concentration over the five sampling weeks was calculated. This mean value, together with the grab samples collected at each location, was considered as the water sampling data. Consequently, the analysis was limited to the following: time-line–water sampling (mean concentration across five consecutive weeks), location–water sampling (one per site), mesocosm–macrofouling, mesocosm–microfouling, and passive sampling. The thirteen compounds identified in the mesocosm biofilm were selected by an ANOVA, and a principal component analysis (PCA; Figure 6) was subsequently performed using these compounds as variables and the different sampling procedures as cases. The PCA scores are presented in Figure S2.

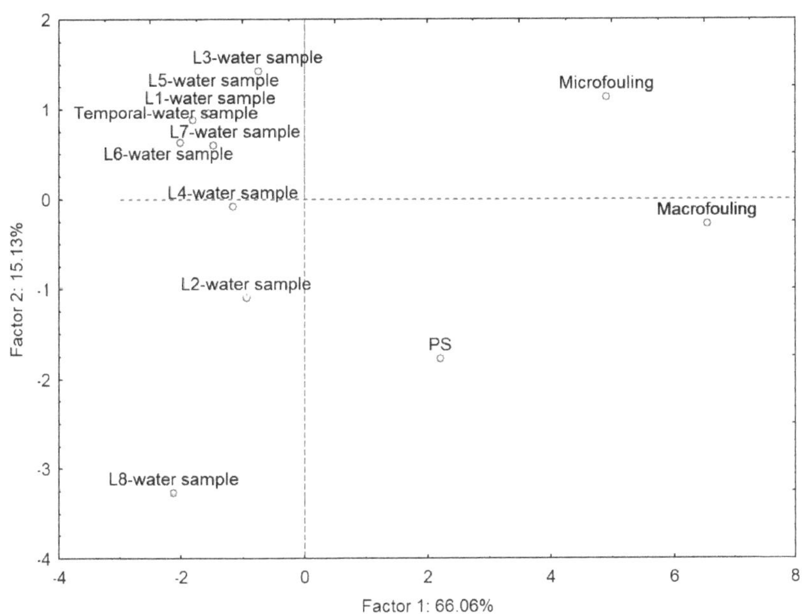

Figure 6. PCA of the sampling procedures (time-line grab sample, different location grab samples, mesocosm microfouling, mesocosm macrofouling and passive sampling (PS)) using thirteen CECs as variables.

The two principal components together explain 81.19% of the variance in the dataset. Factor 1 clearly separates the water samples from the fouling samples, emphasizing the substantial differences between instantaneous water sampling and integrative approaches such as fouling. This axis is primarily driven by compounds like caffeine, octocrylene, or trimethoprim, reflecting their strong influence on water-phase profiles. Factor 2 further distinguishes mesocosm microfouling from both passive sampling and mesocosm macrofouling, indicating compositional divergence between early-stage biofilm communities and more complex fouling assemblages. Notably, macrofouling and passive sampling cluster closely, supporting the notion that passive samplers mimic bioaccumulation patterns in higher-trophic-level organisms, in line with Smedes et al. [33]. The separation of microfouling and macrofouling along Factor 2 also highlights distinct accumulation dynamics: early-colonizing biofilms versus mature assemblages such as mussels, barnacles, tunicates, and macroalgae, consistent with observed differences in compounds including caffeine and octocrylene. Furthermore, the intermediate position of the passive sampler-closer to macrofouling than to water samples demonstrates its effectiveness as an integrative monitoring tool that parallels organismal accumulation. This interpretation is reinforced by Martinaiou et al. [34], who showed that passive samplers, when used alongside spot sampling in marine aquaculture systems, successfully detected pharmaceuticals such as trimethoprim over a 10-month period, even when they remained undetected by water sampling.

In summary, this analysis underscores the superiority of fouling-based and passive sampling approaches for monitoring environmental exposure to contaminants. They provide a broader temporal integration and more ecologically relevant signatures of pollutant accumulation compared to point-in-time water sampling.

4. Conclusions

In conclusion, UV filters, particularly avobenzone and octocrylene, were detected at their highest concentrations, with elevated levels in beach areas compared to harbours. Pharmaceuticals followed in abundance, with caffeine, gabapentin, and o-desvenlafaxine being most prominent, likely reflecting remobilization processes and direct inputs from sewage and cruise activities. The detection of pesticides, especially bifenthrin, raises additional concerns regarding their extensive use and environmental persistence. Time-line analyses revealed no clear temporal trends for either pharmaceuticals or pesticides, suggesting a continuous influx of CECs into the marine environment, while also underlining the influence of sampling time. The mesocosm experiment, which assessed both micro- and macrofouling, showed overall similar contaminant profiles but significant differences for specific compounds, pointing to divergent accumulation dynamics among fouling communities. Passive sampling proved to be a valuable complementary tool for monitoring CECs in marine and transitional waters, exhibiting patterns comparable to mesocosm macrofouling. However, it should not be considered equivalent, as differences were observed between mesocosm microfouling and passive samplers, and deployment duration provides only a partial snapshot of contaminant occurrence.

Therefore, the use of biofilm mesocosms as sampling devices holds significant promise for advancing marine environmental monitoring, providing a more robust and integrative assessment of pollution impacts on coastal and marine ecosystems across diverse spatial and temporal scales.

Supplementary Materials: The following supporting information can be downloaded at: https://www.mdpi.com/article/10.3390/jox15050149/s1. Figure S1: Schematic overview of the experimental setup. Figure S2: Score of the PCA using 13 of the measured CECs. Table S1: Concentration ± SD (ng/L) of the pharmaceuticals and pesticides in the L5 sampling site during five consecutive weeks. Table S2: Concentration ± SD (ng/L) of the compounds detected by passive sampling.

Author Contributions: Conceptualization, E.J.D.-M. and S.D.-G.; methodology, E.J.D.-M.; validation, E.J.D.-M. and S.D.-G.; formal analysis, S.D.-G.; investigation, E.J.D.-M.; data curation, E.J.D.-M.; writing—original draft preparation, E.J.D.-M. and S.D.-G.; writing—review and editing, E.J.D.-M.

and S.D.-G.; visualization, S.D.-G.; supervision, E.J.D.-M. All authors have read and agreed to the published version of the manuscript.

Funding: This research received no external funding.

Institutional Review Board Statement: Not applicable.

Informed Consent Statement: Not applicable.

Data Availability Statement: The original contributions presented in this study are included in the article/Supplementary Material. Further inquiries can be directed to the corresponding author(s).

Conflicts of Interest: The authors declare no conflicts of interest.

Abbreviations

The following abbreviations are used in this manuscript:

CECs	Contaminants of emerging concern
PS	Passive sampling
SPE	Solid-phase extraction
HPLC-MS	High-performance liquid chromatography couple with triple quadrupole mass spectrum
UVF	Ultraviolet filter
LOD	Limit of detection
LOQ	Limit of quantification
ME	Matrix effect
RSD	Relative standard deviation
r^2	Correlation coefficient
N.R.	No risk
L.R.	Low risk
M.R.	Moderate risk
H.R.	High risk

References

1. Oberbeckmann, S.; Loeder, M.G.J.; Gerdts, G.; Osborn, A.M. Spatial and seasonal variation in diversity and structure of microbial biofilms on marine plastics in Northern European waters. *FEMS Microbiol. Ecol.* **2014**, *90*, 478–492. [CrossRef]
2. Carvalho Ferreira, H.; Lôbo-Hajdu, G. Microplastics in coastal and oceanic surface waters and their role as carriers of pollutants of emerging concern in marine organisms. *Mar. Environ. Res.* **2023**, *188*, 106021. [CrossRef]
3. Han, Q.F.; Zhao, S.; Zhang, X.R.; Wang, X.L.; Song, C.; Wang, S.G. Distribution, combined pollution and risk assessment of antibiotics in typical marine aquaculture farms surrounding the Yellow Sea, North China. *Environ. Int.* **2020**, *138*, 105551. [CrossRef]
4. Godlewska, K.; Stepnowski, P.; Paszkiewicz, M. Carbon nanotube-passive samplers as novel tools for sampling and determining micropollutants in the aquatic environment. *Sci. Total Environ.* **2022**, *836*, 155551. [CrossRef]
5. Charriau, A.; Lissalde, S.; Poulier, G.; Mazzella, N.; Buzier, R.; Guibaud, G. Overview of the Chemcatcher® for the passive sampling of various pollutants in aquatic environments Part A: Principles, calibration, preparation and analysis of the sampler. *Talanta* **2016**, *148*, 556–571. [CrossRef]
6. Cole, R.F.; Mills, G.A.; Parker, R.; Bolam, T.; Birchenough, A.; Kröger, S.; Fones, G.R. Trends in the analysis and monitoring of organotins in the aquatic environment. *Trends Environ. Anal. Chem.* **2015**, *8*, 1–11. [CrossRef]
7. Bivins, A.; Kaya, D.; Ahmed, W.; Brown, J.; Butler, C.; Greaves, J.; Leal, R.; Maas, K.; Rao, G.; Sherchan, S.; et al. Passive sampling to scale wastewater surveillance of infectious disease: Lessons learned from COVID-19. *Sci. Total Environ.* **2022**, *835*, 155347. [CrossRef]
8. González-Gaya, B.; Cherta, L.; Nozal, L.; Rico, A. An optimized sample treatment method for the determination of antibiotics in seawater, marine sediments and biological samples using LC-TOF/MS. *Sci. Total Environ.* **2018**, *643*, 994–1004. [CrossRef] [PubMed]

9. Ferreira, A.R.; Skjolding, L.M.; Sanchez, D.F.; Bernar Ntynez, A.G.; Ivanova, Y.D.; Feilberg, K.L.; Chhetri, R.K.; Andersen, H.R. Offshore produced water treatment by a biofilm reactor on the seabed: The effect of temperature and matrix characteristics. *J. Environ. Manag.* **2024**, *365*, 121391. [CrossRef] [PubMed]
10. Gago-Ferrero, P.; Alonso, M.B.; Bertozzi, C.P.; Marigo, J.; Barbosa, L.; Cremer, M.; Secchi, E.R.; Azevedo, A.; Lailson-Brito, J., Jr.; Torres, J.P.M.; et al. First Determination of UV Filters in Marine Mammals. Octocrylene Levels in Franciscana Dolphins. *Environ. Sci. Technol.* **2013**, *47*, 5619–5625. [CrossRef] [PubMed]
11. Fenni, F.; Sunyer-Caldú, A.; Ben Mansour, H.; Diaz-Cruz, M.S. Contaminants of emerging concern in marine areas: First evidence of UV filters and paraben preservatives in seawater and sediment on the eastern coast of Tunisia. *Environ. Pollut.* **2022**, *309*, 119749. [CrossRef]
12. Iancu, V.-I.; Chiriac, L.F.; Paun, I.; Dinu, C.; Pirvu, F.; Cojocary, V.; Tenea, A.G.; Cimpean, I.A. Pharmaceutical Contaminants Occurrence and Ecological Risk Assessment Along the Romanian Black Sea Coast. *Toxics* **2025**, *13*, 498. [CrossRef]
13. Boti, V.; Toli, V.; Efthymiou, C.; Albanis, T. Screening of Commonly Used Antibiotics in Fresh and Saltwater Samples Impacted by Aquacultures: Analytical Methodology, Occurrence and Environmental Risk Assessment. *Sustainability* **2023**, *15*, 9199. [CrossRef]
14. Jeppe, K.; Kellar, C.R.; Marshall, S.; Colombo, V.; Sinclair, G.M.; Pettigrove, V. Bifenthrin Causes Toxicity in Urban Stormwater Wetlands: Field and Laboratory Assessment Using *Austrochiltonia* (Amphipoda). *Environ. Sci. Technol.* **2017**, *51*, 7254–7262. [CrossRef]
15. Downs, C.A.; Diaz-Cruz, M.S.; White, W.T.; Rice, M.; Jim, L.; Punihaole, C.; Dant, M.; Gautam, K.; Woodley, C.M.; Walsh, K.O.; et al. Beach showers as sources of contamination for sunscreen pollution in marine protected areas and areas of intensive beach tourism in Hawaii, USA. *J. Hazard. Mater.* **2022**, *438*, 129546. [CrossRef] [PubMed]
16. Munaron, D.; Mérigot, B.; Derolez, V.; Tapie, N.; Budzinski, H.; Fiandrino, A. Evaluating pesticide mixture risks in French Mediterranean coastal lagoons waters. *Sci. Total Environ.* **2023**, *867*, 161303. [CrossRef] [PubMed]
17. Postigo, C.; Ginebreda, A.; Barbieri, M.V.; Barceló, D.; Martín-Alonso, J.; de la Cal, A.; Boleda, M.R.; Otero, N.; Carrey, R.; Solá, V.; et al. Investigative monitoring of pesticide and nitrogen pollution sources in a complex multi-stressed catchment: The lower Llobregat River basin case study (Barcelona, Spain). *Sci Total Environ.* **2021**, *755*, 142377. [CrossRef]
18. Campo, J.; Masiá, A.; Blasco, C.; Picó, Y. Occurrence and removal efficiency of pesticides in sewage treatment plants of four Mediterranean River Basins. *J. Hazard. Mater.* **2013**, *263*, 146–157. [CrossRef]
19. NORMAN Network. NORMAN Database System (NDS). Available online: https://www.norman-network.com/nds/ (accessed on 15 June 2025).
20. Aguirre-Martínez, G.; DelValss, T.A.; Martín-Díaz, M.L. General stress, detoxification pathways, neurotoxicity and genotoxicity evaluated in *Ruditapes philippinarum* exposed to human pharmaceuticals. *Ecotoxicol. Environ. Saf.* **2016**, *124*, 18–31. [CrossRef]
21. Vieira, L.R.; Soares, A.M.V.M.; Freitas, R. Caffeine as a contaminant of concern: A review on concentrations and impacts in marine coastal systems. *Chemosphere* **2022**, *286*, 131675. [CrossRef]
22. de Paula, V.C.S.; Gomes, M.F.; Martins, L.R.R.; Yamamoto, F.Y.; de Freitas, A.M. Acute toxicity characterization of organic UV-filters and chronic exposure revealing multigenerational effects in *Daphnia magna*. *Ecotoxicology* **2022**, *31*, 1413–1425. [CrossRef]
23. Ka, Y.; Ji, K. Waterborne exposure to avobenzone and octinoxate induces thyroid endocrine disruption in wild-type and thra$\alpha^{-/-}$ zebrafish larvae. *Ecotoxicology* **2022**, *31*, 948–955. [CrossRef]
24. Klopcic, I.; Dolenc, M.S. Endocrine activity of AVB, 2MR, BHA, and their mixtures. *Toxicol. Sci.* **2017**, 240–251. [CrossRef]
25. Liu, Y.; Wang, Y.; Li, N.; Jiang, S. Avobenzone and nanoplastics affect the development of zebrafish nervous system and retinal system and inhibit their locomotor behavior. *Sci. Total Environ.* **2022**, *806*, 150681. [CrossRef] [PubMed]
26. Mundy, P.C.; Carte, M.F.; Brander, S.M.; Hung, T.-C.; Fangue, N.; Connon, R.E. Bifenthrin exposure causes hyperactivity in early larval stages of an endangered fish species at concentrations that occur during their hatching season. *Aquat. Toxicol.* **2020**, *228*, 105611. [CrossRef] [PubMed]
27. Azevedo, T.; Gonçalves, M.; Silva-Reis, R.; Medeiros-Fonseca, B.; Roboredo, M.; Sousa, J.R.; Oliveira, P.A.; Pinto, M.D.L.; Peixoto, F.; Gaivão, I.; et al. Do endocrine disrupting compounds impact earthworms? A comprehensive evidence review. *Rev. Environ. Sci. Biotechnol.* **2024**, *23*, 633–677. [CrossRef]
28. Boyd, A.; Stewart, C.B.; Philibert, D.A.; How, Z.T.; El-Din, M.G.; Tierney, K.B.; Blewett, T.A. A burning issue: The effect of organic ultraviolet filter exposure on the behaviour and physiology of *Daphnia magna*. *Sci. Total Environ.* **2021**, *750*, 141707. [CrossRef]
29. Böhm, L.; Schlechtriem, C.; Düring, R.-A. Sorption of highly hydrophobic organic chemicals to organic matter relevant for fish bioconcentration studies. *Environ. Sci. Technol.* **2016**, *50*, 8316–8323. [CrossRef] [PubMed]
30. Pawlowski, S.; Lanzinger, A.C.; Dolich, T.; Fübl, S.; Salinas, E.R.; Zok, S.; Weiss, B.; Hefner, N.; Van Sloun, P.; Hombeck, H.; et al. Evaluation of the bioaccumulation of octocrylene after dietary and aqueous exposure. *Sci. Total Environ.* **2019**, *672*, 669–679. [CrossRef]

31. Straub, J.O. An Environmental Risk Assessment for Human-Use Trimethoprim in European Surface Waters. *Antibiotics* **2013**, *2*, 115–162. [CrossRef]
32. Sardiña, P.; Leahy, P.; Metzeling, L.; Stevenson, G.; Hinwood, A. Emerging and legacy contaminants across land-use gradients and the risk to aquatic ecosystems. *Sci. Total Environ.* **2019**, *695*, 133842. [CrossRef] [PubMed]
33. Smedes, F.; Sobotka, J.; Rusina, T.P.; Fialová, P.; Carlsson, P.; Kopp, R.; Vrana, B. Unraveling the Relationship between the Concentrations of Hydrophobic Organic Contaminants in Freshwater Fish of Different Trophic Levels and Water Using Passive Sampling. *Environ. Sci. Technol.* **2020**, *54*, 7942–7951. [CrossRef] [PubMed]
34. Martinaiou, P.; Manoli, P.; Boti, V.; Hela, D.; Makou, E.; Albanis, T.; Konstantinou, I. Quality Control of Emerging Contaminants in Marine Aquaculture Systems by Spot Sampling-Optimized Solid Phase Extraction and Passive Sampling. *Sustainability* **2022**, *14*, 3452. [CrossRef]

Journal of *Xenobiotics*

MDPI

Article

Bacterial Cytochrome P450 Involvement in the Biodegradation of Fluorinated Pyrethroids

Mohd Faheem Khan [1,*], Jun Liao [1,2], Zhenyang Liu [1,2] and Gaurav Chugh [3]

1 School of Agriculture and Food Science, University College Dublin, Belfield, Dublin 4, D04 V1W8 Dublin, Ireland
2 China Guangzhou Dublin International College of Life Sciences & Technology, South China Agricultural University, 483 Wushan Road, Tianhe District, Guangzhou 510642, China
3 School of Biomolecular and Biomedical Science, University College Dublin, Belfield, Dublin 4, D04 V1W8 Dublin, Ireland
* Correspondence: mohd.khan@ucd.ie

Abstract: Fluorinated pyrethroids, such as cyfluthrin and cyhalothrin, are more effective insecticides due to their enhanced stability and lipophilicity. However, they pose greater risks to non-target organisms. Their persistence in the environment and accumulation in tissues can lead to increased toxicity and ecological concerns. This study investigates the biodegradation of the fluorinated pyrethroids β-cyfluthrin (BCF) and λ-cyhalothrin (LCH) using a newly isolated *Bacillus* sp. MFK14 from a garden soil microbial consortium. Initial screening using ^{19}F NMR analysis showed that the microbial consortium degraded both pyrethroids, leading to the isolation of *Bacillus* sp. MFK14. Subsequent GC-MS analysis revealed various degradation intermediates in both pyrethroids after incubation with *Bacillus* sp. MFK14. Notably, *Bacillus* sp. MFK14 completely degraded β-cyfluthrin and λ-cyhalothrin within 48 h at 30 °C. Fluoride ions from β-cyfluthrin and trifluoroacetic acid (TFA) from λ-cyhalothrin were detected as the end-products by ^{19}F NMR analysis of the aqueous fraction. The pathway of the degradation was proposed for both the pyrethroids indicating shared biodegradation pathways despite different fluorinations. Inhibition studies with 1-ABT suggested the involvement of bacterial cytochrome P450 (CYP) enzymes in their biodegradation. The CYPome of *Bacillus* sp. MFK14 includes 23 CYP variants that showed significant sequence similarity to known bacterial CYPs, suggesting potential roles in pyrethroid biodegradation and environmental persistence. These findings highlight the potential for bioremediation of fluorinated pesticides, offering an environmentally sustainable approach to mitigate their ecological impact.

Keywords: pesticides; cyfluthrin; cyhalothrin; soil bacteria; biodegradation; cytochrome P450

1. Introduction

Excessive pesticide use in agriculture adversely impacts insects, wildlife, aquatic life, plants, and human health. Among these pesticides, pyrethroids are a potent class of synthetic insecticides that effectively target a wide array of agricultural pests, mosquitoes, and flies by disrupting their nervous systems, leading to paralysis and death [1]. Compared to other insecticides, pyrethroids exhibit relatively mild toxicity to humans and mammals. However, fluorination of pyrethroids enhances their properties by increasing lipophilicity, stability, and binding affinity to biological targets, which improves their insecticidal efficacy but also heightens toxicity to non-target organisms, including humans and animals [2]. Notably, fluorinated pyrethroids—such as transfluthrin, tefluthrin, flumethrin, cyhalothrin, cyfluthrin, bifenthrin, and fluvalinate penetrate membranes more effectively, persist longer in the environment, and accumulate in biological tissues, leading to increased neurotoxicity, metabolic resistance, and potential endocrine disruption [2]. For instance, Martínez et al. investigated cyfluthrin-induced neurotoxicity in humans, revealing that it led to significant increases in ROS, lipid peroxides, and nitric oxide, along with altered gene expression related to apoptosis and inflammation due to induced oxidative stress [3]. Numerous other studies have reported their high toxicity to aquatic organisms and the adverse effects of long-term exposure on human semen quality [4,5]. Due to their environmental persistence, pyrethroids are widely detected in soil, water, and air, impacting various ecosystems and raising significant health and environmental concerns.

Microbial biodegradation has emerged as a potential strategy for mitigating pyrethroid contamination, particularly given the resistance of these compounds to metabolic detoxification pathways. Various studies have demonstrated that bacteria such as *Pseudomonas aeruginosa, Bacillus licheniformis*, and *Bacillus subtilis*, as well as fungi like *Aspergillus niger, Cunninghamella elegans*, and *Fusarium proliferatum*, can effectively degrade pyrethroid residues in water and soil [6]. For instance, Li et al. reported that a novel soil bacterial consortium effectively degraded β-cyfluthrin, with different bacterial genera like *Enterobacter, Microbacterium, Ochrobactrum*, and *Pseudomonas* playing key roles at successive stages of the degradation process [7]. In bacteria such as *Bacillus subtilis* BSF01, *Bacillus licheniformis* B-1, and *Aspergillus niger* YAT, pyrethroids are initially hydrolysed by carboxylesterase or pyrethroid hydrolase, which cleaves the main ester bond linking the cyclopropane and aromatic moieties [8,9]. Similarly, in fungi like *Cunninghamella elegans*, cytochrome P450 enzymes (CYP5208A3) mediate cleavage of the central ester bonds [10,11].

Cytochrome P450 enzymes (CYP) play a crucial role in the biodegradation of various environmental pollutants by catalysing oxidation reactions that enhance the solubility and breakdown of complex organic compounds, including pesticides, hydrocarbons, and pharmaceuticals [12]. While the role of fungal cytochrome P450 enzymes in pyrethroid degradation is well established, there are limited reports on bacterial cytochrome P450 enzymes in this process. Li et al. highlighted the significance of these enzymes in the late-stage metabolism of β-cyfluthrin, particularly in the breakdown of benzoate intermediates. However, the involvement of bacterial cytochrome P450 enzymes in the initial cleavage of pyrethroid ester bonds remains largely unreported [7]. Further degradation generally involves oxidation and cleavage of pyrethroid aromatic rings facilitated by enzymes such as phenol hydrolases, monooxygenases, and dioxygenases [13].

In addition, fluorinated xenobiotic compounds, including pesticides, pose a significant challenge to biodegradation due to the stability of the carbon–fluorine (C-F) bond—the strongest in organic chemistry. Microbial enzymes have not naturally evolved to efficiently cleave C-F bonds, making these compounds persist in the environment and complicating their microbial degradation [12].

In this study, we explore the biodegradation of the fluorinated pyrethroids β-cyfluthrin and λ-cyhalothrin using microbial isolates from garden soil. We identify bacterial strains capable of metabolising these compounds and examine the degradation pathways involved. The identification of degradation intermediates, including fluoride ions and trifluoroacetic acid, was achieved using GC-MS and ^{19}F NMR analysis. Furthermore, inhibition studies confirmed the involvement of cytochrome P450 enzymes in multiple steps of the biodegradation process. These findings highlight the potential for an environmentally sustainable approach to mitigate the ecological impact of fluorinated pesticides. By elucidating the microbial degradation pathways and identifying key enzymatic players, this study paves the way for developing bioremediation strategies that harness naturally occurring bacteria to detoxify contaminated environments. Such approaches could reduce pesticide accumulation in soil and water, minimising harmful effects on ecosystems while promoting sustainable agricultural and environmental management practices.

2. Materials and Methods

2.1. Chemicals and Reagents

The pyrethroid insecticides β-cyfluthrin (PESTANAL®) and λ-cyhalothrin (PESTANAL®) were purchased from Sigma-Aldrich (Wicklow, Ireland). Other reagents and chemicals, including N-methyl-N-(trimethylsilyl)trifluoroacetamide (MSTFA), 1-aminobenzotriazole (1-ABT), phenylmethylsulfonyl fluoride (PMSF), p-nitrophenyl acetate (p-NPA), deuterium oxide (D_2O), and methanol-d_4, were also acquired from Sigma-Aldrich (Ireland).

2.2. Soil Microbe Isolation and Cultivation

Details on soil sample collection and cultivation of the soil microbial consortium were previously described by Khan et al. [14]. The microbial consortium isolated from the garden soil sample was cultured in 20 mL of tryptone soya broth (TSB) in a 100 mL Erlenmeyer flask at 30 °C with shaking at 200 rpm for 48 h. The inoculum was diluted up to 10^{-6}, spread on tryptone soya agar (TSA) plates, and incubated overnight at 30 °C. Different colonies were grown overnight on M9 minimal media supplemented with pyrethroids as the sole carbon source (2 mg/mL), and a potential bacterial strain was selected based on growth, which was estimated by measuring optical density (OD_{600}) using a Jenway 7315 spectrophotometer [15].

2.3. Whole Genome Sequencing and Phylogenetic Analysis

The isolated bacterial strain, capable of growing on pyrethroid as the sole carbon source, was named "MFK14" and streaked on a fresh TSA plate. A single colony was selected and sent for whole genome Illumina sequencing to MicrobesNG (Birmingham, UK). The genome sequence was then submitted to NCBI GenBank (Accession number is JBLMDV000000000). To find the 16S rRNA gene sequence, the genome sequence (in fasta format) was uploaded in NCBI BLASTn and chosen "16S ribosomal RNA sequences (Bacteria and Archaea)" from the Database menu. The homologous 16S rRNA sequences of the known bacteria were obtained using BLASTn for phylogenetic analysis. The Clustal Omega multiple sequence alignment program (https://www.ebi.ac.uk/jdispatcher/msa/clustalo; accessed on 3 January 2025) was used to align all the 16S rRNA gene sequences for the generation of a phylogenetic tree using iTOL v6 (https://itol.embl.de/).

2.4. Biodegradation of Pyrethroids and Metabolite Extraction

The isolated bacterial strain, now identified as *Bacillus* sp. MFK14, was grown in 20 mL TSB for 48 h at 30 °C with shaking at 200 rpm. The cultures were transferred into 50 mL Falcon tubes, and the cells were collected by centrifugation at 4000 rpm for 5 min. The cells were then washed and resuspended in 20 mL phosphate-buffered saline (PBS) solution (pH 7.4). The pesticides (2 mg) were incubated with the bacterial suspension at 30 °C with shaking at 200 rpm for 48 h. The control experiment was carried out without the addition of pesticides.

The supernatants were transferred to fresh Falcon tubes, and the cells were removed by centrifugation at 4000 rpm for 5 min. Organic-soluble metabolites were extracted from the supernatants by adding 20 mL of ethyl acetate, shaking the tubes for 5 min, and centrifuging at 4000 rpm for 2 min to separate the organic and aqueous layers. The top organic layer was collected into fresh Falcon tubes, and the solvent evaporated under a stream of N_2 gas until approximately 1 mL of the sample remained. The samples were then transferred to 2 mL glass vials and further concentrated under N_2 gas.

2.5. ^{19}F NMR and GC-MS Analysis

The ^{19}F nuclear magnetic resonance spectroscopy (^{19}F NMR) spectra were acquired using a Varian 400 MHz spectrometer, and the sample preparation method for organic-soluble and aqueous metabolites was described previously by Khan and Murphy [15]. For the organic metabolites, the residue left after evaporation of extract in glass vials was redissolved in 800 µL of methanol-d_4 and transferred to NMR tubes. For the aqueous metabolites, culture supernatants were freeze-dried, and the resulting residue was redissolved in 800 µL of D_2O and transferred into NMR tubes for ^{19}F NMR analysis.

The gas chromatography-mass spectrometry (GC–MS) was performed with a 7890B Agilent GC equipped with an HP-5MS column, and a 5977A mass-selective detector was used to identify organic biodegradation intermediates. The dried organic extracts were silylated by adding 20 µL MSTFA and heating at 100 °C for 30–60 min; the sample volume was adjusted to 0.5 mL with ethyl acetate prior to injection onto the HP-5MS capillary column (30 m × 0.25 mm × 0.33 µm). The sample injector was set to splitless mode and injected 1 µL onto the column, with the oven temperature held at 80 °C for 3 min and then increased to 300 °C at 10 °C/min, while the mass spectrometer operated in scan mode [16].

2.6. Inhibition Assays and CYPome Prediction

Bacillus sp. MFK14 cultures were pre-grown for 48 h, the growth media was replaced with PBS, and 1 mM 1-ABT was added. The cultures were then incubated at 30 °C, 200 rpm shaking for 1 h before the addition of fluorinated pyrethroids, and the incubation continued for 48 h. The supernatants were extracted, silylated, and analysed by GC–MS as previously described.

To study the primary ester bond hydrolysis of pyrethroids, the role of esterases or CYPs in the cell-free lysate of *Bacillus* sp. MFK14 (1 mL) was assessed using a colourimetric assay with p-NPA as the substrate. The cleavage of the ester bond released p-nitrophenol (p-NP), resulting in the development of a yellow colour. To evaluate the involvement of these enzymes, 1 mM 1-ABT was used to inhibit CYP activity, while 1 mM PMSF was used to inhibit esterase activity. If both enzymes are involved, their activity would be inhibited by the respective inhibitors when tested with the cell-free lysate (sonicated at 35% amplitude, 5 s on/5 s off for 10 min).

The sequences of bacterial cytochrome P450 (CYP) enzymes were collected from UniProtKB (https://www.uniprot.org/uniprotkb) and the Cytochrome P450 Homepage available on the University of Tennessee Health Science Center website (https://drnelson.uthsc.edu/bacteria/). Each sequence of the collected CYP was searched for CYP orthologs in the uncharacterised genome of *Bacillus* sp. MFK14 using NCBI tBLASTn. The percentage protein identity of CYP orthologs in *Bacillus* sp. MFK14 was then determined.

3. Results

3.1. Pyrethroid Biodegradation by Microbial Consortium Isolated from Garden Soil

A consortium of garden soil microbes was screened for degradation of the fluorinated pyrethroids β-cyfluthrin and λ-cyhalothrin by enrichment. For β-cyfluthrin, the ^{19}F NMR spectra showed the disappearance of the characteristic resonance at $\delta = -130.5$ ppm, corresponding to the C-F bond on the aromatic moiety in the 0 h sample, and the appearance of two new resonances at δ = −130.2 and −131.0 ppm after 24 h of incubation with the soil microbial consortium. After 48 h, the resonance at δ = −131.0 ppm persisted, while the resonance at δ = −130.2 ppm significantly decreased, with the emergence of a major resonance at δ = −129.9 ppm and a minor resonance at δ = −129.6 ppm, indicating the formation of different fluorinated compounds due to microbial degradation (Figure 1A).

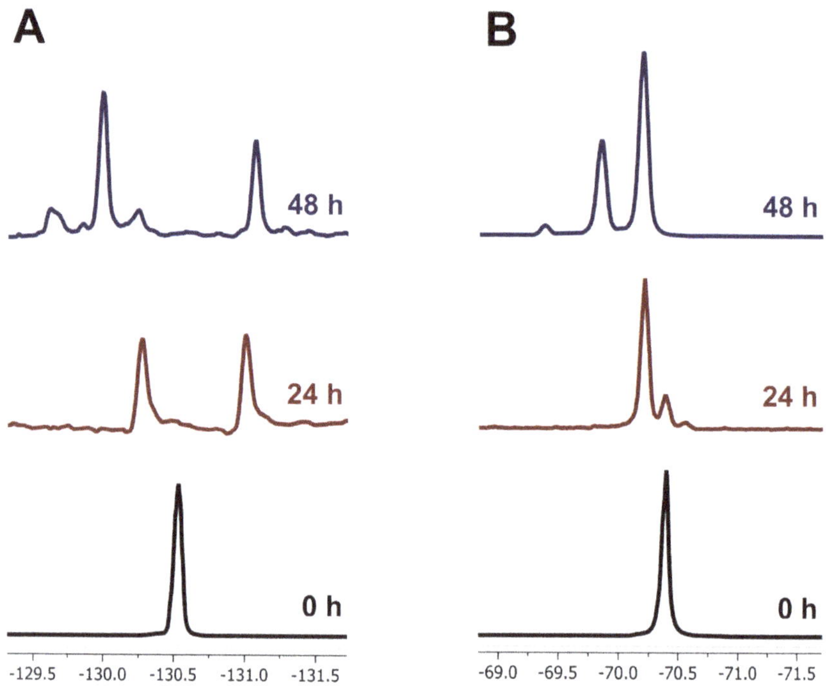

Figure 1. ^{19}F NMR analysis of biodegradation of fluorinated pyrethroid (**A**) β-cyfluthrin, and (**B**) λ-cyhalothrin by garden soil microbial consortium at different time incubations.

Similarly, for λ-cyhalothrin, the ^{19}F NMR spectra showed the disappearance of the characteristic resonance at δ = −70.4 ppm, corresponding to the -CF_3 group in the 0 h sample, and the emergence of a new resonance at δ = −70.1 ppm after 24 h, followed by the appearance of two additional resonances at δ = −69.8 and −69.4 ppm after 48 h, further indicating the microbial degradation of the substrate (Figure 1B).

3.2. Isolation and Identification of Bacterium Utilising Fluorinated Pyrethroids as a Carbon Source

The bacterium responsible for the degradation of fluorinated pyrethroids was isolated using a 10-fold serial dilution method. The soil microbe consortium was diluted up to 10^{-6} times and then cultured on TSA plates. After overnight incubation, a total of eighteen colonies were observed. Each colony was inoculated into M9 minimal media supplemented with pyrethroid as the sole carbon source. Among the colonies, the MFK14 colony exhibited growth after overnight incubation, suggesting that this bacterium could utilise the fluorinated pyrethroids β-cyfluthrin and λ-cyhalothrin as carbon sources for growth, as indicated by the OD_{600} measurements shown in Figure 2.

Whole genome sequencing confirmed that the isolated bacterium belongs to the genus *Bacillus* and was designated as *Bacillus* sp. MFK14. Phylogenetic analysis of 16S rRNA gene sequences revealed that *Bacillus* sp. MFK14 clusters closely with several species, including *B. nematocida*, *B. amyloliquefaciens*, *B. halotolerans*, and *B. subtilis*, suggesting a close evolutionary relationship within the genus (Figure 3). Notably, *B. nematocida* was identified as the closest phylogenetic neighbour, with a 16S rRNA sequence similarity of 97.2% to *Bacillus* sp. MFK14. Although this high similarity indicated a close genetic relationship, it falls within the range where a novel species could still be distinguished. Further taxonomic investigations, including phenotypic, chemotaxonomic, and genome-based analyses, are warranted to confirm its classification.

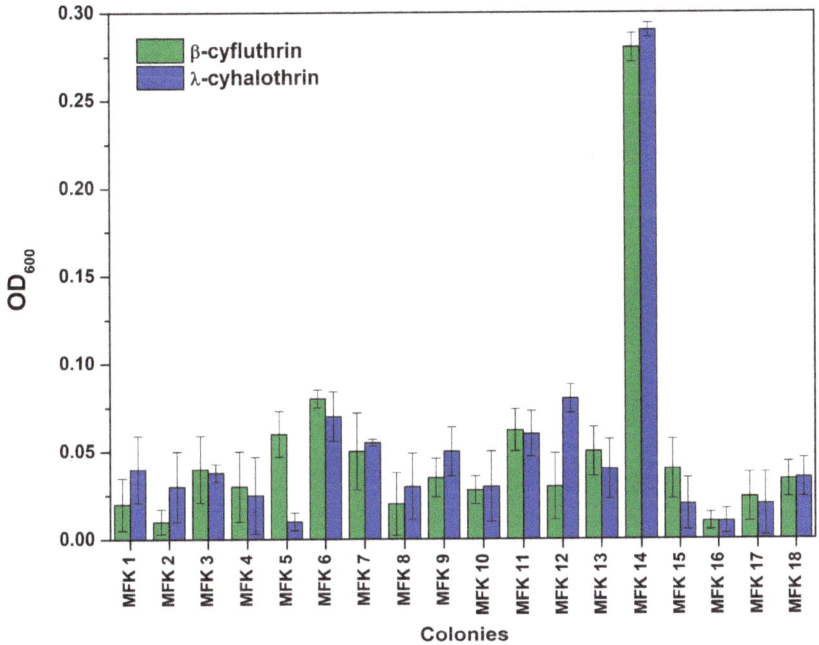

Figure 2. Screening of colonies grown on M9 minimal media supplemented with a pyrethroid (β-cyfluthrin and λ-cyhalothrin) as the sole carbon source.

3.3. Metabolite Identification in Pyrethroid Degradation by Bacillus sp. MFK14

To investigate the biodegradation of fluorinated pyrethroids, organic extracts from *Bacillus* sp. MFK14 cultures were analysed by GC–MS following MSTFA silylation. After 48 h of incubation with β-cyfluthrin and λ-cyhalothrin with *Bacillus* sp. MFK14, several degradation metabolites were identified which were absent in control samples (without any pyrethroids). Thirteen metabolites from β-cyfluthrin were labelled M1–M13 based on their retention times in the gas chromatogram, while thirteen metabolites from λ-cyhalothrin were labelled N1–N13 (Figure 4). The complete disappearance of the substrate peaks for β-cyfluthrin and λ-cyhalothrin after 48 h of incubation with *Bacillus* sp. MFK14 indicates that both compounds were fully degraded within this timeframe. This was confirmed by comparing the retention times of the authentic standards, with β-cyfluthrin observed at 23.45 min and λ-cyhalothrin at 22.54 min (Figure S1).

Metabolite identification was confirmed either through comparison with authentic standards or by matching the mass spectra with reference data from the NIST library. The mass spectra corresponding to each metabolite are provided in Supplementary Figures S2–S23. Among the identified metabolites, four were common to both pyrethroids: M1/N3 (hydroquinone) at 10.16 min, M5/N6 (hydroxyquinol) at 13.27 min, M6/N8 (4-hydroxy-

6-oxohexa-2,4-dienoic acid) at 14.46 min, and M9/N9 (2-formyl-3-(hydroxymethyl)-2-methylcyclopropane-1-carboxylic acid) at 15.59 min. Both β-cyfluthrin and λ-cyhalothrin were extensively degraded by *Bacillus* sp. MFK14 over the incubation period, with distinct metabolic profiles for each compound (M-series and N-series metabolites). The chromatograms confirmed the bacterium's capacity to metabolise fluorinated pyrethroids effectively, with no interference observed in the control sample.

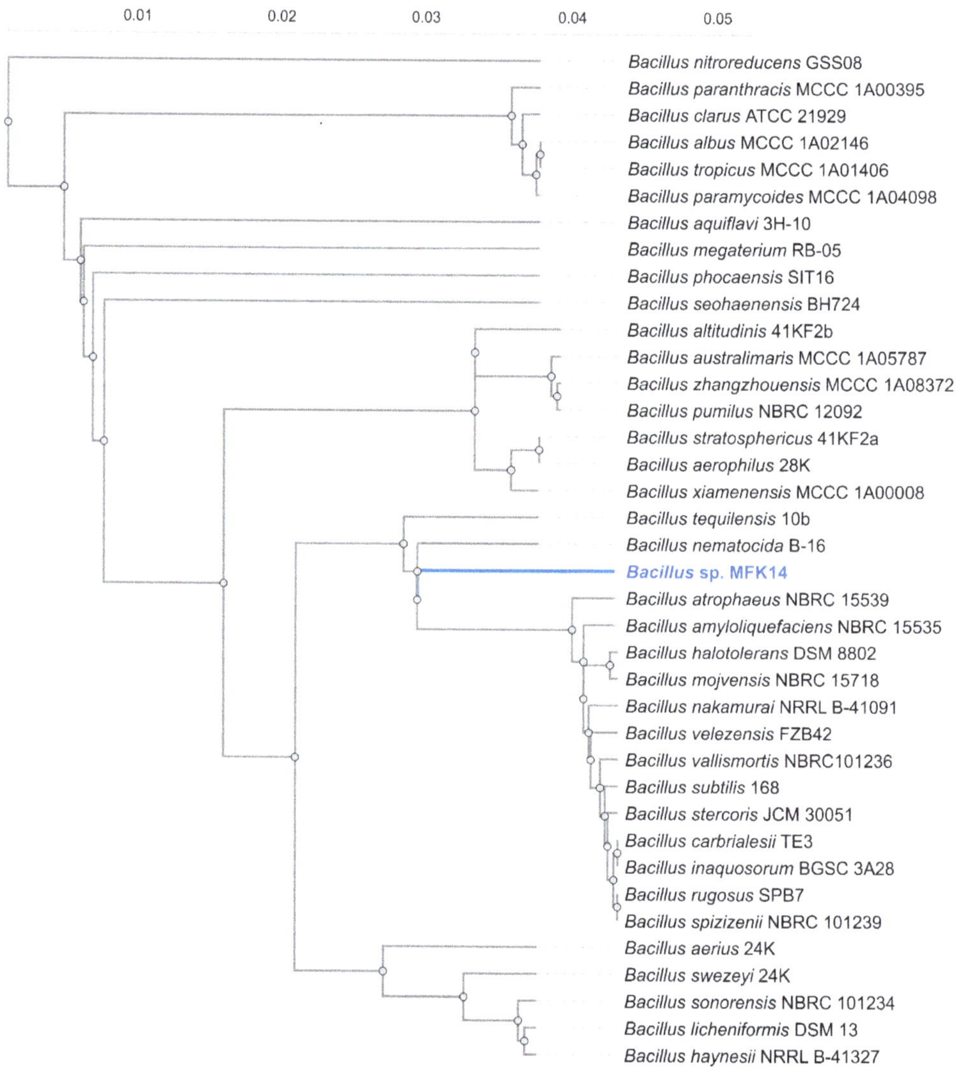

Figure 3. Phylogenetic analysis of isolated *Bacillus* sp. MFK14 using homologous 16S rRNA sequences from another *Bacillus* spp.

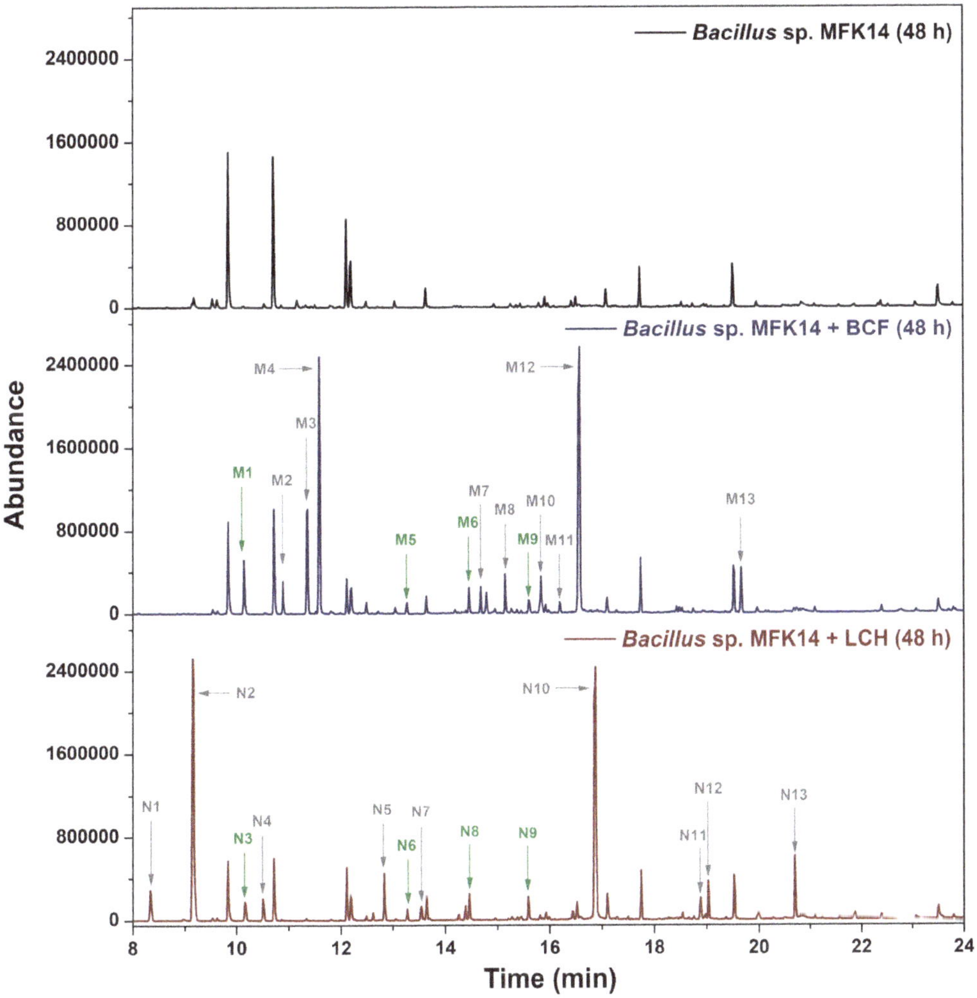

Figure 4. Total-ion chromatogram of *Bacillus* sp. MFK14 incubation with β-cyfluthrin (BCF) and λ-cyhalothrin (LCH) for 48 h. The control experiment was kept with bacteria without pyrethroids. The metabolites produced from β-cyfluthrin are numbered with "M": M1–M13, and λ-cyhalothrin are numbered with "N": N1–N13. Common metabolites are represented by green, whereas the specific metabolites of BCF and LCH are represented by grey.

Figure 5 illustrates the proposed common pathway for the biodegradation of β-cyfluthrin and λ-cyhalothrin by *Bacillus* sp. MFK14. This figure summarises the complexity of the study, detailing metabolite structures, labels, names, and fluorinated end products, along with the types of reactions such as hydroxylation, oxidation, reduction, or decarboxylation.

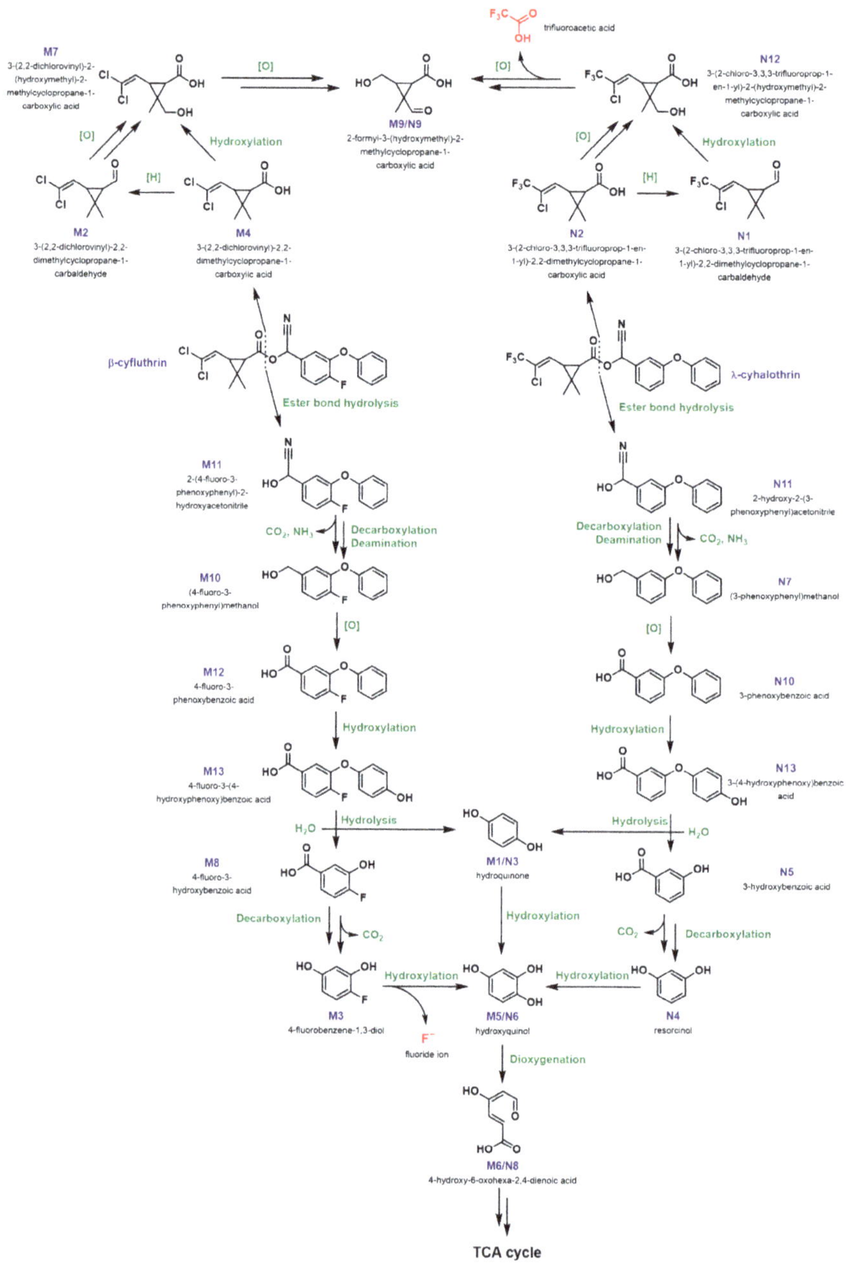

Figure 5. A common proposed pathway for biodegradation of fluorinated pyrethroid β-cyfluthrin and λ-cyhalothrin. The metabolites produced from β-cyfluthrin are numbered with "M": M1–M13, and λ-cyhalothrin are numbered with "N": N1–N13.

3.3.1. Biodegradation Pathway for β-cyfluthrin

In the biodegradation of β-cyfluthrin, key metabolites identified included M4 (3-(2,2-dichlorovinyl)-2,2-dimethylcyclopropane-1-carboxylic acid) at 11.59 min and M12 (4-fluoro-

3-phenoxybenzoic acid) at 16.59 min, likely formed by cleavage of the central ester bond followed by oxidoreductive removal of the cyano (-CN) group. M4 was then converted into M7 (3-(2,2-dichlorovinyl)-2-(hydroxymethyl)-2-methylcyclopropane-1-carboxylic acid) via M2 (3-(2,2-dichlorovinyl)-2,2-dimethylcyclopropane-1-carbaldehyde) through reduction of the carboxyl group to an aldehyde, followed by oxidation and hydroxylation steps, or directly into M7 by a hydroxylation step. One of the end products, M9 (2-formyl-3-(hydroxymethyl)-2-methylcyclopropane-1-carboxylic acid), was formed from M7 through multiple steps of oxidative dechlorination. On the other hand, M12 was transformed into M13 (4-fluoro-3-(4-hydroxyphenoxy)benzoic acid) by hydroxylation, likely at the aromatic ring other than the fluorinated one. This was followed by hydrolysis of the diphenyl ether produced M1 (hydroquinone) and M8 (4-fluoro-3-hydroxybenzoic acid). M8 was subsequently converted to M5 (hydroxyquinol) via M3 (4-fluorobenzene-1,3-diol) through oxidative decarboxylation and defluorination, while M1 was directly transformed into M5 by hydroxylation to form hydroxyquinol. M5 underwent ring cleavage by a dioxygenation step to form M6 (4-hydroxy-6-oxohexa-2,4-dienoic acid), which was eventually catabolised through the TCA cycle.

3.3.2. Biodegradation Pathway for λ-cyhalothrin

In the degradation of λ-cyhalothrin by *Bacillus* sp. MFK14, the major metabolites N2 (3-(2-chloro-3,3,3-trifluoroprop-1-en-1-yl)-2,2-dimethylcyclopropane-1-carboxylic acid) at 9.15 min and N10 (3-phenoxybenzoic acid) at 16.88 min were produced through a similar cleavage of the central ester bond and subsequent oxidoreductive removal of the -CN group. N2 was then converted into N12 (3-(2-chloro-3,3,3-trifluoroprop-1-en-1-yl)-2-(hydroxymethyl)-2-methylcyclopropane-1-carboxylic acid) via N1 (3-(2-chloro-3,3,3-trifluoroprop-1-en-1-yl)-2,2-dimethylcyclopropane-1-carbaldehyde) by reduction of the carboxyl group to an aldehyde, followed by oxidation and hydroxylation steps, or directly to N12 through hydroxylation. One of the end products, N9 (2-formyl-3-(hydroxymethyl)-2-methylcyclopropane-1-carboxylic acid), was formed from N12 through multiple steps of oxidative dechlorination. On the other hand, N10 was transformed into N13 (3-(4-hydroxyphenoxy)benzoic acid) by hydroxylation at one of the aromatic rings. Subsequent hydrolysis of the diphenyl ether produced N3 (hydroquinone) and N5 (3-hydroxybenzoic acid). N5 was then converted to N6 (hydroxyquinol) via N4 (resorcinol) through oxidative decarboxylation and hydroxylation, while N3 was directly transformed into N6 by hydroxylation to form hydroxyquinol. N6 underwent ring cleavage by a dioxygenation step to form N8 (4-hydroxy-6-oxohexa-2,4-dienoic acid), which was ultimately catabolised through the TCA cycle, like the pathway for β-cyfluthrin.

Furthermore, the phenolic metabolites M1/N3 (hydroquinone) at 10.16 min, N4 (resorcinol) at 10.50 min, and M5/N6 (hydroxyquinol) at 13.27 min were confirmed by comparing their retention times and mass spectra with those of authentic standards (Figure S24). These findings indicated the presence of key metabolites in the culture extracts, suggesting that *Bacillus* sp. MFK14 had the potential to biodegrade the fluorinated pyrethroids β-cyfluthrin and λ-cyhalothrin. The observed pathways appeared to involve the formation and subsequent transformation of phenolic compounds.

Water-soluble metabolites were identified by freeze-drying the culture supernatants of *Bacillus* sp. MFK14 was incubated with β-cyfluthrin and λ-cyhalothrin for 48 h. The dried residues were redissolved in D_2O for ^{19}F NMR analysis (Figure S25). Notably, fluoride ions (resonance at $\delta = -122.4$ ppm) were detected in the β-cyfluthrin biodegradation sample, likely formed during the transformation of metabolite M3 to M5 through oxidative defluorination. However, trifluoroacetic acid (resonance at $\delta = -75.8$ ppm) was detected in the λ-cyhalothrin biodegradation sample, possibly formed during the conversion of metabolite N12 to N9 via multiple steps of oxidative dechlorination and deacetylation. Both aqueous metabolites were confirmed against authentic standards, with NaF (resonance at $\delta = -122.4$ ppm) and TFA (resonance at $\delta = -75.8$ ppm) serving as references in the ^{19}F NMR analysis. The detection of fluoride ions in the β-cyfluthrin sample indicated its mineralisation, while the presence of trifluoroacetic acid, although resistant to microbial degradation, suggested that λ-cyhalothrin has been effectively catabolised into a smaller fluorinated end-product. The comprehensive degradation pathway indicates the potential of *Bacillus* sp. MFK14 in breaking down persistent fluorinated pesticides, indicating its relevance for environmental bioremediation efforts.

3.4. CYP Involvement in the Biodegradation of β-cyfluthrin and λ-cyhalothrin

The role of cytochrome P450 enzymes (CYPs) in the biodegradation of β-cyfluthrin and λ-cyhalothrin by *Bacillus* sp. MFK14 was evaluated using the CYP inhibitor 1-aminobenzotriazole (1-ABT). To assess this, CYPs were inhibited by pre-incubating *Bacillus* sp. MFK14 with 1-ABT for one hour before adding the pesticides. These cultures were then used for biodegradation experiments.

GC–MS analysis revealed that most metabolites were undetected following CYP inhibition, except for early-stage degradation products. In the β-cyfluthrin biodegradation pathway, metabolites such as M2 (3-(2,2-dichlorovinyl)-2,2-dimethylcyclopropane-1-carbaldehyde), M4 (3-(2,2-dichlorovinyl)-2,2-dimethylcyclopropane-1-carboxylic acid), M10 (4-fluoro-3-phenoxyphenyl)methanol), and M11 (2-(4-fluoro-3-phenoxyphenyl)-2-hydroxyacetonitrile) were still detected (Figure 6A). The presence of M4 and M11, formed immediately after ester bond hydrolysis, indicated that esterases facilitated this step, even in the presence of the CYP inhibitor. However, the substantial peak of the parent compound β-cyfluthrin (at 23.45 min) after 48 h, compared to normal biodegradation without 1-ABT (Figure 6), suggested that CYP inhibition slowed the rate of ester bond hydrolysis. Further transformations of M4 to M2 and M11 to M10, which do not require CYPs, were observed. However, the degradation of M2 and M10—steps solely dependent on CYPs—was inhibited, preventing the formation of downstream metabolites such as M1, M3, M5, M6, M7, M8, M9, M12, and M13.

Similarly, in the biodegradation of λ-cyhalothrin, early-stage metabolites such as N1 (3-(2-chloro-3,3,3-trifluoroprop-1-en-1-yl)-2,2-dimethylcyclopropane-1-carbaldehyde), N2 (3-(2-chloro-3,3,3-trifluoroprop-1-en-1-yl)-2,2-dimethylcyclopropane-1-carboxylic acid), N7 ((3-phenoxyphenyl)methanol), and N11 (2-hydroxy-2-(3-phenoxyphenyl)acetonitrile) were detected (Figure 6A). This suggested that esterases were responsible for the initial ester bond hydrolysis, even when CYPs were inhibited. However, the significant persistence of λ-cyhalothrin (at 22.54 min) after 48 h, along with the absence of further degradation of N1 and N7, indicated that CYP inhibition slowed the biodegradation process, preventing the formation of downstream metabolites such as N3, N4, N5, N6, N8, N9, N10, N12, and N13.

To further confirm the involvement of cytochrome P450 enzymes (CYPs) and esterases in ester bond hydrolysis, a separate enzyme inhibition assay was performed using *Bacillus* sp. MFK14 cell-free lysate. Three reactions were set up. In the first reaction, the lysate was incubated with 1-ABT to inhibit CYPs; in the second reaction, the lysate was treated with PMSF to inhibit esterases; and in the third reaction, the lysate was treated with both 1-ABT and PMSF to achieve dual inhibition (Figure 6B). After 10 min of incubation, ester bond hydrolysis was assessed using p-NPA as a substrate, which releases p-nitrophenol (p-NP) upon ester bond hydrolysis, producing a yellow colour. The reaction with 1-ABT and p-NPA developed a yellow colour, indicating that esterases were active in cleaving the ester bond. Similarly, the reaction with PMSF and p-NPA also produced a yellow colour, suggesting that CYPs were involved in ester bond hydrolysis. However, the reaction with both 1-ABT and PMSF did not produce any yellow colour, confirming that both CYPs and esterases were required for ester bond hydrolysis. These results confirm that both CYPs and esterases are essential for ester bond hydrolysis of p-NPA. Similarly, in the biodegradation of β-cyfluthrin and λ-cyhalothrin, both enzyme systems (CYPs and esterases) were involved in the initial cleavage of ester bonds, with CYPs playing a key role in the subsequent degradation steps.

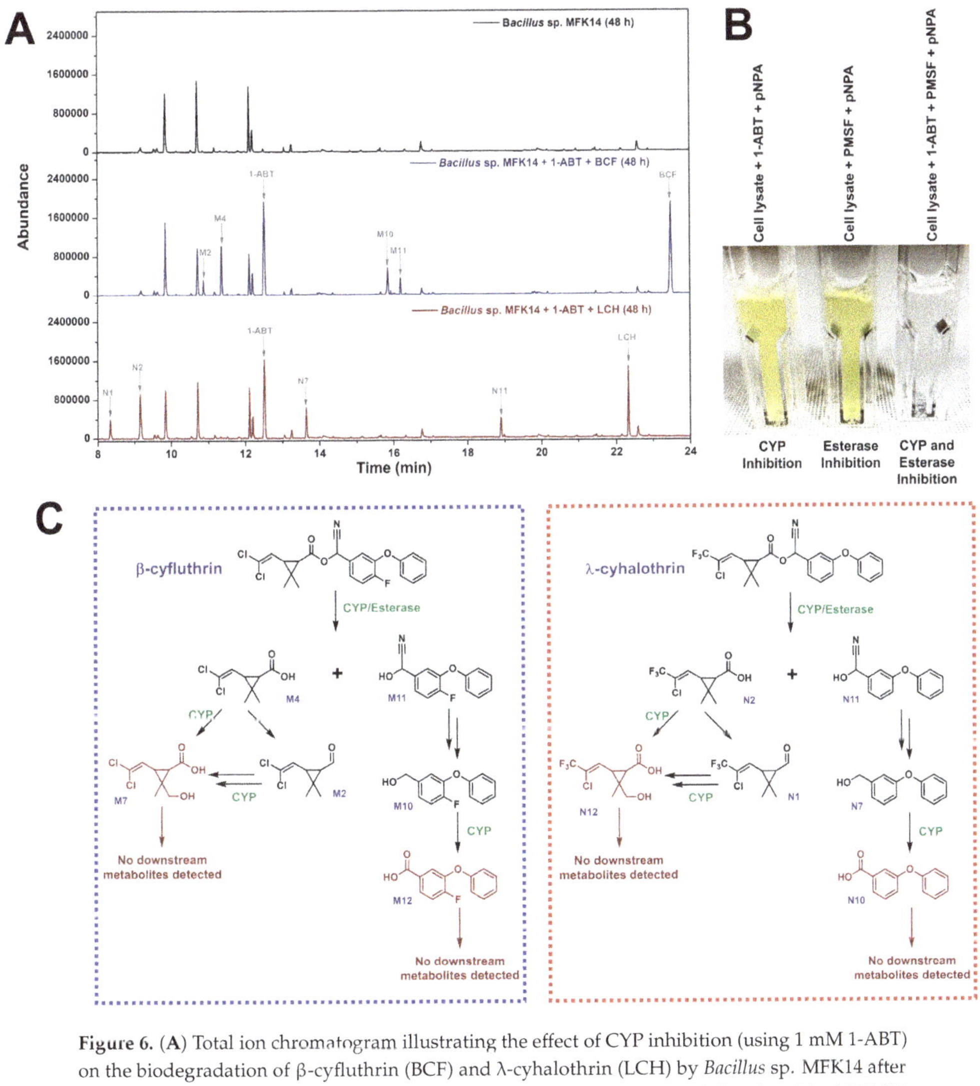

Figure 6. (**A**) Total ion chromatogram illustrating the effect of CYP inhibition (using 1 mM 1-ABT) on the biodegradation of β-cyfluthrin (BCF) and λ-cyhalothrin (LCH) by *Bacillus* sp. MFK14 after 48 h of incubation. (**B**) Enzyme inhibition assay using 1 mM 1-ABT (CYP inhibitor) and 1 mM PMSF (esterase inhibitor) to demonstrate ester bond hydrolysis in the cell lysate of *Bacillus* sp. MFK14, with p-NPA as the substrate. This experiment demonstrates that both CYP and esterase independently exhibit ester bond hydrolysis. (**C**) Proposed role of CYPs in various steps of the biodegradation pathways of β-cyfluthrin (in the blue box) and λ-cyhalothrin (in the red box).

Figure 6C illustrates the proposed role of *Bacillus* sp. MFK14 CYPs in various steps of the biodegradation pathways of β-cyfluthrin and λ-cyhalothrin. For β-cyfluthrin, CYPs facilitated transformations such as β-cyfluthrin to M4 and M11, M4 to M7, M2 to M7, and M10 to M12. In λ-cyhalothrin biodegradation, CYPs mediated the conversion of λ-cyhalothrin to N11 and N2, N2 to N12, N1 to N12, and N7 to N10. Fluorinated metabolites resulting from the biodegradation of β-cyfluthrin included fluoride ions, while trifluoroacetic acid was produced from λ-cyhalothrin. This production was likely mediated by multiple enzymes, with CYPs being a potential contributor to the process. Esterases play a crucial role in the initial breakdown of β-cyfluthrin into M4 and M11 and λ-cyhalothrin into N11 and N2 by cleaving the ester bonds. While esterase activity is independent, it worked in conjunction with CYPs, which also contributed to ester bond hydrolysis, complementing the overall biodegradation process.

3.5. Predicted CYPome of *Bacillus* sp. MFK14

To explore the uncharacterised genome of *Bacillus* sp. MFK14, a search for CYP gene sequences was conducted to support the inhibition experiments. Using tBLASTn, CYP orthologs were identified to predict the complete CYP enzyme set, or "CYPome," as listed in Table 1. In silico analysis revealed a high degree of conservation among CYP genes in closely related species, including *B. subtilis*, *B. cereus*, *B. anthracis*, *B. weihenstephanensis*, *B. megaterium*, and *B. licheniformis*. Notably, the identified CYPs in *Bacillus* sp. MFK14 included variants of CYP102, CYP106, CYP107, CYP109, CYP134, CYP152, and CYP197. The putative gene sequences for these CYPs, extracted from the newly sequenced genome of *Bacillus* sp. MFK14 are provided in Supplementary Table S1. Among these, the CYPs of *B. subtilis* exhibited high sequence identity (above 90%) with the identified CYPs in *Bacillus* sp. MFK14, except for CYP109A1 (only 42.1% identity). The genomic evidence confirmed the presence of CYP genes in *Bacillus* sp. MFK14, indicating their crucial role in the biodegradation of β-cyfluthrin and λ-cyhalothrin. These findings align with the inhibition experiments, which showed CYPs are essential for degrading these pesticides, highlighting their importance in environmental cleanup.

Table 1. CYPs in *Bacillus* sp. MFK14 and their sequence similarity to orthologs in other *Bacillus* spp.

CYP Name	Protein ID *	Source Organism	Protein Identity of CYP Orthologs (%) in *Bacillus* sp. MFK14
CYP102A1	P14779	*Bacillus megaterium*	65.0
CYP102A2	O08394	*Bacillus subtilis*	98.5
CYP102A3	O08336	*Bacillus subtilis*	99.3
CYP102A4	A0A6L7HJT4	*Bacillus anthracis*	75.2
CYP102A5	Q81BF4	*Bacillus cereus*	75.5
CYP102A7	NA	*Bacillus licheniformis*	61.8
CYP102A8	NA	*Bacillus thuringiensis*	75.8
CYP102A9	NA	*Bacillus weihenstephanensis*	75.0
CYP102A15	NA	*Bacillus pumilus*	60.7
CYP106A1	P14762	*Bacillus megaterium*	40.8
CYP106A2	Q06069	*Bacillus megaterium*	42.0
CYP106B1	A0A6L7HJD6	*Bacillus anthracis*	47.1
CYP107DY1	D5E3H2	*Bacillus subtilis*	42.8
CYP107H1	P53554	*Bacillus subtilis*	98.9
CYP107J1	O08469	*Bacillus subtilis*	94.6
CYP107J2	A0A6L7H515	*Bacillus anthracis*	57.6
CYP107J3	Q81CY0	*Bacillus cereus*	57.6
CYP107K1	O31785	*Bacillus subtilis*	98.9
CYP109A1	P27632	*Bacillus subtilis*	42.1
CYP109B1	O34374	*Bacillus subtilis*	93.0
CYP134A1	O34926	*Bacillus subtilis*	96.8
CYP152A1	O31440	*Bacillus subtilis*	95.2
CYP197A1	Q9KFA6	*Bacillus halodurans*	27.7

* NA—protein ID is not available in the UniProt, but its sequence is available in the Cytochrome P450 database (https://drnelson.uthsc.edu/bacteria/; accessed on 10 January 2025).

4. Discussion

The use of pyrethroids in agriculture practices for pest control leads to environmental pollution, negatively impacting human health and reducing soil microbial populations, which diminishes soil fertility [17]. Some soil bacteria can tolerate these harmful metabolites, evolving mechanisms to degrade pyrethroids [18]. In this study, we utilised a bacterial consortium isolated from garden soil to degrade the fluorinated pyrethroids β-cyfluthrin and λ-cyhalothrin. Similar studies, such as Birolli et al., Liu et al., and Zhang et al., have demonstrated the degradation of various pyrethroids using bacterial consortia from different environments [19–21]. Our study found that newly isolated *Bacillus* sp. MFK14 from garden soil completely degraded the β-cyfluthrin and λ-cyhalothrin within 48 h, producing 13 metabolites in each pesticide. In comparison, previous research reported that *Bacillus subtilis* BSF01 degraded 89.4% of 50 mg/L β-cypermethrin in 7 days [9], *Brevibacterium aureum* DG-12 achieved 88.6% degradation of 50 mg/L β-cyfluthrin within 5 days [22], and *Bacillus thuringiensis* ZS-19 completely degraded 100 mg/L of λ-cyhalothrin in 72 h [23]. The degradation efficiency of λ-cyhalothrin and β-cyfluthrin in bacteria and fungi is summarised in Table 2, comparing the findings of this study with previously reported microbial strains.

Table 2. Comparison of β-cyfluthrin and λ-cyhalothrin degradation by different microorganisms.

Microorganism	Type	Concentration (mg/L)	Degradation (%)	Time (days)	Temperature (°C)	Reference
β-Cyfluthrin						
***Bacillus* sp. MFK14**	**Bacteria**	**100**	**100**	**2**	**30**	**Current study**
Photobacterium ganghwense 6046	Bacteria	100	82.8	3	30	[24]
Brevibacterium aureum DG-12	Bacteria	50	88.6	5	27	[22]
Lysinibacillus sphaericus FLQ-11-1	Bacteria	50	80.4	5	35	[25]
Cunninghamella spp.	Fungus	100	NA [1]	5	28	[10]
λ-cyhalothrin						
***Bacillus* sp. MFK14**	**Bacteria**	**100**	**100**	**2**	**30**	**Current study**
Bacillus thuringiensis ZS-19	Bacteria	100	100	3	30	[23]
Paracoccus acridae SCU-M53	Bacteria	75	79.84	2	28	[26]
Cunninghamella elegans DSM1908	Fungus	100	NA [1]	5	28	[27]
Aspergillus sp. CBMAI 1829	Fungus	100	44.8	14	32	[28]

[1] NA = not available.

Esterases play a vital role in pyrethroid degradation by cleaving the primary ester bonds in these compounds [29]. A carboxylesterase from *Bacillus cereus* BCC01, which functioned at pH 8 and 30 °C, was characterised for its pyrethroid-degrading activity [30]. Similarly, a cypermethrin-degrading esterase from *Bacillus subtilis* was most effective at pH 7.0 and 32 °C [31]. However, beyond esterase activity, our study also demonstrated that the degradation of β-cyfluthrin and λ-cyhalothrin involved cytochrome P450 (CYP)-mediated cleavage of ester bonds in *Bacillus* sp. MFK14. This was confirmed using 1-ABT, a CYP inhibitor, which showed that β-cyfluthrin was transformed into M4 and M11, while λ-cyhalothrin was converted to N11 and N2 through CYP-mediated biotransformation. Khan and Murphy previously identified similar ester bond hydrolysis in transfluthrin and β-cyfluthrin by *Cunninghamella* spp. (*C. elegans*, *C. blakesleeana*, and *C. echinulata*), attributing this activity to cytochrome P450s (CYPs) [10]. Further research revealed that CYP5208A3 and CYP5313D1, along with their redox partner CYP reductase (CPR_C) from

C. elegans, were involved in the oxidative ester cleavage of the transfluthrin pyrethroid [10]. Additionally, Hedges et al. demonstrated significant variability in the activity of recombinant human CYPs across different pyrethroids (including λ-cyhalothrin, bifenthrin, and β-cyfluthrin), with CYP2C19 showing the highest effectiveness [32]. Our study further revealed that CYPs in *Bacillus* sp. MFK14 not only cleaved ester bonds but also participated in the transformation of various metabolites in β-cyfluthrin and λ-cyhalothrin biodegradation. Specifically, β-cyfluthrin was transformed through M4 to M7, M2 to M7, and M10 to M12, while λ-cyhalothrin was converted from N2 to N12, N1 to N12, and N7 to N10. These conversions were supported by CYP inhibition experiments (Figure 6C). In silico analysis of the CYPome in *Bacillus* sp. MFK14 genome revealed a significant number of CYP genes (23 variants), indicating a high potential for CYPs to participate in ester hydrolysis, as well as monooxygenation, defluorination, and other reactions during pyrethroid biodegradation.

The presence of fluoride ions in β-cyfluthrin and trifluoroacetic acid in λ-cyhalothrin following 48 h of incubation with *Bacillus* sp. MFK14 was confirmed by ^{19}F-NMR analysis of the aqueous fraction. The release of fluoride ions likely occurred during the conversion of 4-fluorobenzene-1,3-diol (M3) to hydroxyquinol (M5) in β-cyfluthrin biodegradation, potentially catalysed by cytochrome P450s (CYPs), as previously described by Harkey et al. in the CYP-mediated oxidative defluorination of 4-fluorophenol to hydroquinone [33]. Similarly, Li et al. observed defluorinated metabolites in the bacterial degradation of β-cyfluthrin [7].

Dechlorination of the 2,2-dichlorovinyl group in β-cyfluthrin (conversion from M7 to M9) and the 2-chloro-3,3,3-trifluoroprop-1-en-1-yl group in λ-cyhalothrin (conversion from N12 to N9) was rare, with limited reports documenting these dechlorinated metabolites. For example, Birolli et al. demonstrated that a *Bacillus* sp. CSA-1 consortium of 10 bacterial strains achieved rapid CYP-mediated biodegradation of cypermethrin, identifying dechlorinated metabolites via LC-MS/MS analysis [19]. Additionally, Li et al. identified a dechlorinated metabolite, (2,2,3,3-tetramethyl-cyclopropyl)-methanol, resulting from the removal of terminal chlorines during CYP-mediated β-cyfluthrin degradation by a bacterial consortium [7]. Zhang et al. reviewed the dechlorination of 2,2-dichlorovinyl groups in dichlorvos by various microorganisms, including *Pseudomonas*, *Bacillus*, *Fusarium*, *Aspergillus*, and *Trichoderma*. Consequently, the dechlorination of the 2-chloro-3,3,3-trifluoroprop-1-en-1-yl group in λ-cyhalothrin may have resulted in the release of trifluoroacetic acid, as identified in our ^{19}F-NMR analysis [34].

In addition to CYPs, the degradation processes for both pyrethroids involved a possible range of enzymes, including hydrolases, dioxygenases, etc. Earlier work by Schmidt et al. described the nonspecific attack of dioxygenase on halodiphenyl ethers, leading to the formation of halophenols during biodegradation by *Sphingomonas* sp. strain SS3 [35]. This mechanism is commonly observed in both bacterial and fungal degradation [36,37]. In our study, *Bacillus* sp. MFK14 exhibited a similar diphenyl ether moiety degradation pattern in both β-cyfluthrin and λ-cyhalothrin. Following diphenyl ether cleavage, the phenolic metabolites hydroquinone (M1/N3) at 10.16 min, resorcinol (N4) at 10.50 min, and hydroxyquinol (M5/N6) at 13.27 min were identified by comparing their retention times and mass spectra with those of authentic standards. These findings provide strong evidence for the production of these phenolic metabolites during the biodegradation of β-cyfluthrin and λ-cyhalothrin by *Bacillus* sp. MFK14.

The genome analysis of *Bacillus* sp. MFK14 also revealed the presence of catechol-2,3-dioxygenase with high sequence identity (99%) to *Bacillus subtilis* strain 168 catechol-2,3-dioxygenase (Protein ID: P54721). This supported the observed ring cleavage from hydroxyquinol (M5/N6) to 4-hydroxy-6-oxohexa-2,4-dienoic acid (M6/N8) in both pesticides, which was subsequently catabolised through the TCA cycle. In summary, the proposed biodegradation pathways for β-cyfluthrin and λ-cyhalothrin by *Bacillus* sp. MFK14 illustrated the complex interplay of various enzymes in breaking down these fluorinated pyrethroids. Understanding these pathways enhanced our knowledge of microbial degradation mechanisms and could have informed strategies for environmental remediation.

5. Conclusions

This study highlighted the crucial role of cytochrome P450 enzymes in the biodegradation of fluorinated pyrethroids by *Bacillus* sp. MFK14, a newly isolated strain from garden soil. The strain effectively degraded β-cyfluthrin and λ-cyhalothrin, releasing fluoride ions and trifluoroacetic acid (TFA), demonstrating the potential of microbial bioremediation to mitigate the environmental impact of these persistent insecticides. Notably, *Bacillus* sp. MFK14 completely degraded both β-cyfluthrin and λ-cyhalothrin within 48 h at 30 °C, showcasing its efficiency in degrading these hazardous compounds. The findings also revealed that, despite differing fluorinations, β-cyfluthrin and λ-cyhalothrin shared degradation pathways, showcasing the adaptability and specificity of bacterial CYP enzymes in breaking down complex compounds. Inhibition studies with 1-ABT further confirm the involvement of CYP enzymes in this process, while genomic analysis of *Bacillus* sp. MFK14's CYPome showed significant sequence similarity with known bacterial CYPs, highlighting the evolutionary conservation and functional importance of these enzymes.

6. Future Perspectives

Microbial degradation offers an eco-friendly, cost-effective solution for pesticide, herbicide, and insecticide remediation, with advantages over traditional methods by breaking down pollutants into less toxic forms. It can be applied in situ, reducing environmental persistence and preserving soil and water quality. However, its efficiency can vary with environmental conditions, and challenges such as toxic intermediates and slow microbial adaptation to synthetic pollutants exist. Genetic engineering and optimising microbial consortia could improve the process and enhance its sustainability. In line with these considerations, the environmental challenges posed by fluorinated pyrethroids make this research a promising foundation for developing biotechnological applications aimed at reducing their persistence and toxicity. Future studies should focus on the genetic engineering of *Bacillus* sp. MFK14 and heterologous expression of predicted CYPs to enhance degradation efficiency. Moreover, investigating the interaction of these bacterial CYPs with other classes of persistent organic pollutants could broaden the scope of bioremediation strategies, ultimately contributing to more sustainable agricultural practices.

Supplementary Materials: The following supporting information can be downloaded at https://www.mdpi.com/article/10.3390/jox15020058/s1: Figure S1: GC-MS analysis of standards: (A) β-cyfluthrin, and (B) λ-cyhalothrin; Figure S2: Mass spectrum of metabolite M1 or N3 (t_R = 10.16 min; M^+ = 254); Figure S3: Mass spectrum of metabolite M2 (t_R = 10.91 min; M^+ = 192); Figure S4: Mass spectrum of metabolite M3 (t_R = 11.36 min; M^+ = 272 (257+15)); Figure S5: Mass spectrum of metabolite M4 (t_R = 11.59 min; M^+ = 280 (265+15)); Figure S6: Mass spectrum of metabolite M5 or N6 (t_R = 13.27 min; M^+ = 342); Figure S7: Mass spectrum of metabolite M6 or N8 (t_R = 14.46 min; M^+ = 286); Figure S8: Mass spectrum of metabolite M7 (t_R = 14.69 min; M^+ = 368 (353+15)); Figure S9: Mass spectrum of metabolite M8 (t_R = 15.14 min; M^+ = 300); Figure S10: Mass spectrum of metabolite M9 or N9 (t_R = 15.59 min; M^+ = 302); Figure S11: Mass spectrum of metabolite M10 (t_R = 15.85 min; M^+ = 290); Figure S12: Mass spectrum of metabolite M11 (t_R = 16.19 min; M^+ = 315); Figure S13: Mass spectrum of metabolite M12 (t_R = 16.59 min; M^+ = 304); Figure S14: Mass spectrum of metabolite M13

(t_R = 19.66 min; M^+ = 392); Figure S15: Mass spectrum of metabolite N1 (t_R = 8.33 min; M^+ = 226); Figure S16: Mass spectrum of metabolite N2 (t_R = 9.15 min; M^+ = 314 (299+15)); Figure S17: Mass spectrum of metabolite N4 (t_R = 10.50 min; M^+ = 254); Figure S18: Mass spectrum of metabolite N5 (t_R = 12.83 min; M^+ = 282); Figure S19: Mass spectrum of metabolite N7 (t_R = 13.53 min; M^+ = 272); Figure S20: Mass spectrum of metabolite N10 (t_R = 16.88 min; M^+ = 286); Figure S21: Mass spectrum of metabolite N11 (t_R = 18.88 min; M^+ = 297); Figure S22: Mass spectrum of metabolite N12 (t_R = 19.03 min; M^+ = 402 (387+15)); Figure S23: Mass spectrum of metabolite N13 (t_R = 20.70 min; M^+ = 374); Figure S24: GC-MS analysis of standards: (A) resorcinol, (B) hydroquinone, and (C) hydroxyquinol; Figure S25: ^{19}F-NMR analysis confirmed the presence of (A) fluoride ion in β-cyfluthrin, and (B) trifluoroacetic acid in λ-cyhalothrin after 48 h incubation with *Bacillus* sp. MFK14; Table S1: Putative CYP sequences were extracted from the newly sequenced genome of *Bacillus* sp. MFK14.

Author Contributions: Conceptualisation, supervision, resources, data curation, project administration, writing—original draft preparation, M.F.K.; methodology, investigation, validation, formal analysis, writing—review and editing, M.F.K., J.L., Z.L. and G.C. All authors have read and agreed to the published version of the manuscript.

Funding: The UCD Internal supported this research financially (Award number: 82930-NP).

Institutional Review Board Statement: Not applicable.

Informed Consent Statement: Not applicable.

Data Availability Statement: Data will be made available on request.

Acknowledgments: Mohd Faheem Khan expresses gratitude to Cormac D. Murphy for his valuable discussions and for providing essential laboratory facilities and access to instrumentation, which were crucial for the success of this research.

Conflicts of Interest: The authors declare no conflicts of interest.

Abbreviations

The following abbreviations are used in this manuscript:

1-ABT	1-aminobenzotriazole
^{19}F NMR	^{19}F nuclear magnetic resonance spectroscopy
BCF	β-cyfluthrin
LCH	λ-cyhalothrin
CYP	Cytochrome P450
GC-MS	Gas chromatography-mass spectrometry
PBS	Phosphate-buffered saline
p-NPA	p-Nitrophenyl acetate
PMSF	Phenylmethylsulfonyl fluoride
TFA	Trifluoroacetic acid
TSB	Tryptone soya broth
TSA	Tryptone soya agar

References

1. Asem, D.; Kumari, M.; Singh, L.R.; Bhushan, M. Pesticides: Pollution, risks, and abatement measures. In *Emerging Aquatic Contaminants*, 2nd ed.; Elsevier: Amsterdam, The Netherlands, 2023; pp. 307–326.
2. Galadima, M.; Singh, S.; Pawar, A.; Khasnabis, S.; Dhanjal, D.S.; Anil, A.G.; Rai, P.; Ramamurthy, P.C.; Singh, J. Toxicity, microbial degradation and analytical detection of pyrethroids: A review. *Environ. Adv.* **2021**, *5*, 100105. [CrossRef]
3. Martínez, M.A.; Rodríguez, J.L.; Lopez-Torres, B.; Martínez, M.; Martínez-Larrañaga, M.R.; Anadón, A.; Ares, I. Oxidative stress and related gene expression effects of cyfluthrin in human neuroblastoma SH-SY5Y cells: Protective effect of melatonin. *Environ. Res.* **2019**, *177*, 108579. [CrossRef] [PubMed]

4. Ham, J.; You, S.; Lim, W.; Song, G. Bifenthrin impairs the functions of Leydig and Sertoli cells in mice via mitochondrion-endoplasmic reticulum dysregulation. *Environ. Pollut.* **2020**, *266*, 115174. [CrossRef] [PubMed]
5. Wang, R.; Wang, F.; Lu, Y.; Zhang, S.; Cai, M.; Guo, D.; Zheng, H. Spatial distribution and risk assessment of pyrethroid insecticides in surface waters of East China Sea estuaries. *Environ. Pollut.* **2024**, *344*, 123302. [CrossRef]
6. Huang, Y.; Chen, S. Pyrethroid-degrading microorganisms and their potential application for the bioremediation of contaminated environments. In *Enzymes for Pollutant Degradation*; Springer Nature: Singapore, 2022; pp. 119–137.
7. Li, H.; Ma, Y.; Yao, T.; Ma, L.; Zhang, J.; Li, C. Biodegradation pathway and detoxification of β-cyfluthrin by the bacterial consortium and its bacterial community structure. *J. Agric. Food Chem.* **2022**, *70*, 7626–7635. [CrossRef]
8. Deng, W.; Lin, D.; Yao, K.; Yuan, H.; Wang, Z.; Li, J.; Zou, L.; Han, X.; Zhou, K.; He, L.; et al. Characterization of a novel β-cypermethrin-degrading *Aspergillus niger* YAT strain and the biochemical degradation pathway of β-cypermethrin. *Appl. Microbiol. Biotechnol.* **2015**, *99*, 8187–8198. [CrossRef]
9. Xiao, Y.; Chen, S.; Gao, Y.; Hu, W.; Hu, M.; Zhong, G. Isolation of a novel beta-cypermethrin degrading strain *Bacillus subtilis* BSF01 and its biodegradation pathway. *Appl. Microbiol. Biotechnol.* **2015**, *99*, 2849–2859. [CrossRef]
10. Khan, M.F.; Murphy, C.D. *Cunninghamella* spp. produce mammalian-equivalent metabolites from fluorinated pyrethroid pesticides. *AMB Express* **2021**, *11*, 101. [CrossRef]
11. Khan, M.F.; Murphy, C.D. Cytochrome P450 5208A3 is a promiscuous xenobiotic biotransforming enzyme in *Cunninghamella elegans*. *Enzyme Microb. Technol.* **2022**, *161*, 110102. [CrossRef]
12. Khan, M.F.; Hof, C.; Niemcova, P.; Murphy, C.D. Biotransformation of fluorinated drugs and xenobiotics by the model fungus *Cunninghamella elegans*. *Methods Enzymol.* **2024**, *696*, 251–285.
13. Tang, J.; Liu, B.; Chen, T.T.; Yao, K.; Zeng, L.; Zeng, C.Y.; Zhang, Q. Screening of a beta-cypermethrin-degrading bacterial strain *Brevibacillus parabrevis* BCP-09 and its biochemical degradation pathway. *Biodegradation* **2018**, *29*, 525–541. [CrossRef] [PubMed]
14. Khan, M.F.; Chowdhary, S.; Koksch, B.; Murphy, C.D. Biodegradation of amphipathic fluorinated peptides reveals a new bacterial defluorinating activity and a new source of natural organofluorine compounds. *Environ. Sci. Technol.* **2023**, *57*, 9762–9772. [CrossRef] [PubMed]
15. Khan, M.F.; Murphy, C.D. Bacterial degradation of the anti-depressant drug fluoxetine produces trifluoroacetic acid and fluoride ion. *Appl. Microbiol. Biotechnol.* **2021**, *105*, 9359–9369. [CrossRef] [PubMed]
16. Hof, C.; Khan, M.F.; Murphy, C.D. Endogenous production of 2-phenylethanol by *Cunninghamella echinulata* inhibits biofilm growth of the fungus. *Fungal Biol.* **2023**, *127*, 1384–1388. [CrossRef]
17. Singh, S.; Mukherjee, A.; Jaiswal, D.K.; de Araujo Pereira, A.P.; Prasad, R.; Sharma, M.; Kuhad, R.C.; Shukla, A.C.; Verma, J.P. Advances and future prospects of pyrethroids: Toxicity and microbial degradation. *Sci. Total Environ.* **2022**, *829*, 154561. [CrossRef]
18. Cycoń, M.; Piotrowska-Seget, Z. Pyrethroid-degrading microorganisms and their potential for the bioremediation of contaminated soils: A review. *Front. Microbiol.* **2016**, *7*, 1463. [CrossRef]
19. Birolli, W.G.; da Silva, B.F.; Rodrigues Filho, E. Biodegradation of the pyrethroid cypermethrin by bacterial consortia collected from orange crops. *Environ. Res.* **2022**, *215*, 114388. [CrossRef]
20. Liu, J.; Ding, Y.; Ma, L.; Gao, G.; Wang, Y. Combination of biochar and immobilized bacteria in cypermethrin-contaminated soil remediation. *Int. Biodeter. Biodegr.* **2017**, *120*, 15–20. [CrossRef]
21. Zhang, H.; Zhang, Y.; Hou, Z.; Wang, X.; Wang, J.; Lu, Z.; Zhao, X.; Sun, F.; Pan, H. Biodegradation potential of deltamethrin by the *Bacillus cereus* strain Y1 in both culture and contaminated soil. *Int. Biodeter. Biodegr.* **2016**, *106*, 53–59. [CrossRef]
22. Chen, S.; Dong, Y.H.; Chang, C.; Deng, Y.; Zhang, X.F.; Zhong, G.; Song, H.; Hu, M.; Zhang, L.H. Characterization of a novel cyfluthrin-degrading bacterial strain *Brevibacterium aureum* and its biochemical degradation pathway. *Bioresour. Technol.* **2013**, *132*, 16–23. [CrossRef]
23. Chen, S.; Deng, Y.; Chang, C.; Lee, J.; Cheng, Y.; Cui, Z.; Zhou, J.; He, F.; Hu, M.; Zhang, L.H. Pathway and kinetics of cyhalothrin biodegradation by *Bacillus thuringiensis* strain ZS-19. *Sci. Rep.* **2015**, *5*, 8784. [CrossRef] [PubMed]
24. Wang, T.; Hu, C.; Zhang, R.; Sun, A.; Li, D.; Shi, X. Mechanism study of cyfluthrin biodegradation by *Photobacterium ganghwense* with comparative metabolomics. *Appl. Microbiol. Biotechnol.* **2019**, *103*, 473–488. [CrossRef] [PubMed]
25. Hu, G.P.; Zhao, Y.; Song, F.Q.; Liu, B.; Vasseur, L.; Douglas, C.; You, M.S. Isolation, identification and cyfluthrin-degrading potential of a novel *Lysinibacillus sphaericus* strain, FLQ-11-1. *Res. Microbiol.* **2014**, *165*, 110–118. [CrossRef]
26. Tian, J.W.; Long, X.F.; Zhang, S.; Qin, Q.M.; Gan, L.Z.; Tian, Y.Q. Screening cyhalothrin degradation strains from locust epiphytic bacteria and studying *Paracoccus acridae* SCU-M53 cyhalothrin degradation process. *Environ. Sci. Pollut. Res.* **2018**, *25*, 11505–11515. [CrossRef]
27. Palmer-Brown, W.; de Melo Souza, P.L.; Murphy, C.D. Cyhalothrin biodegradation in *Cunninghamella elegans*. *Environ. Sci. Pollut. Res. Int.* **2019**, *26*, 1414–1421. [CrossRef]

28. Birolli, W.G.; Vacondio, B.; Alvarenga, N.; Seleghim, M.H.R.; Porto, A.L.M. Enantioselective biodegradation of the pyrethroid (+/)-lambda-cyhalothrin by marine-derived fungi. *Chemosphere* **2018**, *197*, 651–660. [CrossRef]
29. Bhatt, P.; Bhatt, K.; Huang, Y.; Lin, Z.; Chen, S. Esterase is a powerful tool for the biodegradation of pyrethroid insecticides. *Chemosphere* **2020**, *244*, 125507. [CrossRef]
30. Hu, W.; Lu, Q.; Zhong, G.; Hu, M.; Yi, X. Biodegradation of pyrethroids by a hydrolyzing carboxylesterase EstA from *Bacillus cereus* BCC01. *Appl. Sci.* **2019**, *9*, 477. [CrossRef]
31. Gangola, S.; Sharma, A.; Bhatt, P.; Khati, P.; Chaudhary, P. Presence of esterase and laccase in *Bacillus subtilis* facilitates biodegradation and detoxification of cypermethrin. *Sci. Rep.* **2018**, *8*, 12755. [CrossRef]
32. Hedges, L.; Brown, S.; MacLeod, A.K.; Moreau, M.; Yoon, M.; Creek, M.R.; Osimitz, T.G.; Lake, B.G. Metabolism of bifenthrin, β-cyfluthrin, λ-cyhalothrin, cyphenothrin and esfenvalerate by rat and human cytochrome P450 and carboxylesterase enzymes. *Xenobiotica* **2020**, *50*, 1434–1442. [CrossRef]
33. Harkey, A.; Kim, H.J.; Kandagatla, S.; Raner, G.M. Defluorination of 4-fluorophenol by cytochrome P450BM3-F87G: Activation by long chain fatty aldehydes. *Biotechnol. Lett.* **2012**, *34*, 1725–1731. [CrossRef] [PubMed]
34. Zhang, Y.; Zhang, W.; Li, J.; Pang, S.; Mishra, S.; Bhatt, P.; Zeng, D.; Chen, S. Emerging technologies for degradation of dichlorvos: A review. *Int. J. Environ. Res. Public Health* **2021**, *18*, 5789. [CrossRef] [PubMed]
35. Schmidt, S.; Wittich, R.M.; Erdmann, D.; Wilkes, H.; Francke, W.; Fortnagel, P. Biodegradation of diphenyl ether and its monohalogenated derivatives by *Sphingomonas* sp. strain SS3. *Appl. Environ. Microbiol.* **1992**, *58*, 2744–2750. [CrossRef]
36. Gu, C.; Jin, Z.; Fan, X.; Ti, Q.; Yang, X.; Sun, C.; Jiang, X. Comparative evaluation and prioritization of key influences on biodegradation of 2, 2′, 4, 4′-tetrabrominated diphenyl ether by bacterial isolate B. xenovorans LB400. *J. Environ. Manag.* **2023**, *331*, 117320. [CrossRef]
37. Hundt, K.; Jonas, U.; Hammer, E.; Schauer, F. Transformation of diphenyl ethers by *Trametes versicolor* and characterization of ring cleavage products. *Biodegradation* **1999**, *10*, 279–286. [CrossRef]

MDPI AG
Grosspeteranlage 5
4052 Basel
Switzerland
Tel.: +41 61 683 77 34

Journal of Xenobiotics Editorial Office
E-mail: jox@mdpi.com
www.mdpi.com/journal/jox

MDPI

www.ingramcontent.com/pod-product-compliance
Lightning Source LLC
Chambersburg PA
CBHW061601220926
51850CB00005B/391
* 9 7 8 3 7 2 5 8 7 8 0 9 3 *